S. RAMAMOORTHY
E.G. BADDALOO

VOLUME I
AQUATIC SPECIES

HANDBOOK OF CHEMICAL TOXICITY PROFILES OF BIOLOGICAL SPECIES

Boca Raton New York London Tokyo

Library of Congress Cataloging-in-Publication Data

Handbook of chemical toxicity profiles of biological species, volume I: aquatic/
by Sub Ramamoorthy and Earle G. Baddaloo
p. cm.
Includes bibliographical references.
ISBN 1-56670-013-2 (v. 1)
1. Toxicity testing — Handbooks, manuals, etc.
QH541.15. T68B335 1995
591.2′4—dc20 94-41896
CIP

Lewis Publishers is an imprint of CRC Press

International Standard Book Number 1-56670-013-2
Library of Congress Card Number 94-41896
Printed in the United States of America 1 2 3 4 5 6 7 8 9 0
Printed on acid-free paper

Preface

This is the first of two volumes of handbooks on the chemical toxicity profiles of biological species. The first volume deals with the chemical toxicity data, arranged in descending order, for aquatic species and further subclassification of species.

The volume begins with an introduction explaining the need for a handbook of this format. The introduction is followed by chapters on biological concepts, an index to aquatic species, chemical concepts, and an index of chemicals (name, chemical Abstract Service number, common name, synonyms, uses, molecular formulae, molecular weight, and properties). References, a key to the abbreviations used in the references, and a glossery of terms used follow the toxicity data.

Authors

Sub Ramamoorthy received his Ph.D. degree in 1969 in physical chemistry from the University of Madras, India. He was awarded his D.Sc. degree in 1984 for his research work on "Heavy Metals in the Environment—Interdisciplinary Research Studies". In 1970, Dr. Ramamoorthy joined the Department of Environment–Inland Waters Directorate in Ottawa as a National Research Council Fellow. He was an adjunct professor at the University of Ottawa, while employed at the Division of Biological Sciences, Ecological Kinetics Group, National Research Council of Canada, Ottawa, from 1976 to 1979. In 1979, Dr. Ramamoorthy joined the Alberta Department of Environment as the Head of the Limnology Section, Animal Sciences Division, at the Alberta Environmental Centre in Vegreville. He is currently a Senior Manager at the Environmental Regulatory Services, Alberta Environmental Protection, Edmonton, which he joined in 1986. Dr. Ramamoorthy is a co-author of three books on Environmental Management and has authored more than 70 publications in primary journals in addition to over 30 reports in the field of environmental chemistry and toxicology. Dr. Ramamoorthy is the co-chair of the 15th International Symposium on Chlorinated Dioxins and Related Compounds which will be held in Edmonton, Alberta, in 1995. He is an active member of the International Society of Regulatory Toxicology and Pharmacology, the International Society of Environmental Toxicology and Chemistry, and the International Humic Substances Society. He is listed in the *American Men and Women of Sciences*, 13th ed., and in the CNTC *Directory of Toxicological Expertise in Canada.* Dr. Ramamoorthy has served on federal/provincial environmental committees, QA/QC committees, a panel of referees for the International Association on Water Quality, England (IAWQ), *Journal of Water Science and Technology* (UK), and *Water Research* (UK).

Earle G. Baddaloo obtained his Masters of Science from Laurentian University in Ontario, Canada, in 1975. He commenced his working career with Ontario's Department of the Environment, carrying out air and water quality monitoring programs. Previous to his current position as a Standards Development Coordinator with Alberta Environmental Protection in Edmonton, Mr. Baddaloo worked for the Bechtel Corporation (1975–1978) and for the oil industry in Alberta (1978–1985). Some of the environmental projects he has worked on include: Quebec's James Bay hydro development, Monroe coal handling, Rio Algom's mining development, Interprovincial's Sarnia–Montreal oil pipeline, various legs of the Alaska Highway gas pipeline, and the Beaufort Sea-MacKenzie Delta Region hydrocarbon development. Mr. Baddaloo chaired the 19th Annual Aquatic Toxicity Workshop in Edmonton in 1992, and in 1993 he was one of the two international consultants chosen by the Inter-American Development Bank to carry out technical needs analyses for the development of an environmental management agency for the Republics of Trinidad and Tobago. He is a member of the Society of Toxicology of Canada (STC) and has authored or co-authored more than 50 publications, including reports, EIA sections, presentations, and a book.

Contents

Acknowledgment

We would like to acknowledge the valuable contribution made by Sita Ramamoorthy.

Dedicated to

Kavita, Yamini, Julie, and Marc

1 Introduction

Ever-improving analytical capabilities have changed the priorities in environmental research and regulations. In the early 1970s, we were largely concerned with the gross pollution of air and water. Although most scientists were aware of chemical-related diseases, this area of research did not get the attention it deserved. As a result, episodes such as Minamata disease (arising from consuming methylmercury-contaminated fish in Minamata Bay, Japan, receiving industrial discharges); Love Canal, New York (contamination of homes and schools from buried chemicals); Times Beach, Missouri (contamination of the entire town from chemical residues spread on roads); Seveso, Italy (accidental release of chemicals injuring thousands of residents), and Bhopal, India (accidental chemical release killing more than 2000 residents) led to the awareness of possible adverse effects from exposure to hitherto unknown chemicals. Since then, the focus has shifted to understanding toxic chemicals, possible routes of exposures, and their potential to cause chemical injury on humans and to the ecosystem.

In the past 15 years the detection limits have been improved by more than six orders of magnitude in the analysis of organic and inorganic compounds. With the routine use of ultratrace analytical techniques, it is hoped that there will be very few "chemical surprises". However, the ever-refining analytical capability has produced the "list syndrome". This syndrome has created a dilemma about how environmental assessment is to be conducted. The two scenarios currently operating are, first, the analyst, after having detected a new chemical in an environmental sample, could initiate an extensive monitoring program followed by toxicological studies to assess the impact. Second, the field biologist observes a biological impact in the natural environment and transmits a request to the chemist and toxicologist to search the cause through diagnostic services. The question to be asked is which scenario is more cost-effective in protecting the environment and human health. For example, the discovery of the chemical named mirex hiding beneath a polychlorinated biphenyl (PCB) peak was a brilliant example of analytical sleuthing,[1] but subsequent toxicological testing showed that mirex was not of any toxicological significance to the biology of Lake Ontario. However, this turn of events did divert scarce resources away from the search for the chick edema factor during the late 1970s.[1] To be cost-effective, we have to keep our focus on critical compounds.

Another disadvantage with the list syndrome is the complexity of the resulting database, most of which could not be quality-controlled due to lack of proper analytical standards. This, in turn, weakens the interpretative aspect of the database, leading only to an educated guesswork on the part of the chemist.

Our objective in writing this handbook is to provide a viable alternative to the use of the list syndrome approach in environmental impact assessment. In this handbook, the data are arranged in decreasing order of toxicity for aquatic species such as amphibians, algae, bacteria, crustaceans, fish, insects, molluscs, protozoa, and others. Species are arranged as phylum, class, order, suborder, section. For example, arthropods: phylum (Arthropoda), class (Crustacean),- subclass (Branchiopoda), order (Decapoda), suborder (Natantia), section (Caridea), *Crangon vulgaris* and *Crangon crangon*.

For fish, data have been arranged as follows: phylum (Chlordata), subphylum (Agantha) and subphylum (Gnathostomata), class (Osteichthyes), subclass (Actinoterygii), order (Salmoniformes), family (Salmininae), genus (*Oncorhynchus*), species (*kisutch*).

For surface water impact assessment, one has to decide on the type and number of species to be protected and then select the most toxic chemicals listed under the species of concern for analysis. The number of chemicals chosen for analysis will depend on the resources available as well as on the toxicity concentration ratio of the most toxic chemical to the cut-off chemical for a given biological species. It is important to compare identical toxicity testing methods and duration of exposure (for example, LC_{50} 96 hours) to arrive at a ratio. The ambient concentration and its ratio to the toxic level will assist in deciding on the cut-off chemical.

It is hoped that this approach will maximize the use of resources and generate data that are relevant to the protection of the species and the ecosystem under study.

REFERENCE

1. M. Gilbertson, *J. Fish Aquat. Sci.,* 42, 1681–1692, 1985.

2 Biological Concepts

The aquatic environment is not only complex and diverse, but its main component, water, commands a unique place among our natural resources. Within the aquatic environment, there are two distinct types of waters — fresh and salt waters. On many occasions these meet to form brackish areas known as estuaries. Fresh waters include streams, rivers, ponds, and lakes, while areas such as coasts, bays, oceans, and some seas incorporate salt waters. These different waters may form several distinct types of ecosystems within which may exist many different biotic and abiotic communities. The biotic or living component consists of various species of microorganisms, invertebrates, plants, and animals that inhabit specific ecological niches in each ecosystem. The abiotic or nonliving part incorporates the physical environment, which includes water, suspended material, sediment, and substrate. Within each ecosystem there are continuous interactions between the biotic and the abiotic components.

Aquatic ecosystems also carry out various functions that are required for maintaining harmony and continuance of the environment. They recharge ground and surface water reservoirs, maintain flows in streams and rivers, control flooding, recycle nutrients, purify water, provide habitat for various species of both aquatic and terrestrial flora and fauna, and furnish recreational areas for human beings. Some of these functions have been impaired as a result of various factors such as rapid population growth and increased industrial, commercial, and residential developments. These activities have led to the discharge of nutrient–rich municipal sewage, pesticides, fertilizers, toxic leachates, and the indiscriminate withdrawal of both surface and groundwaters for consumptive purposes.

The physical conditions of the aquatic environment include morphology and substrate types of basins and channels, movement and depth of water, suspended solids, temperature, and the effects of light. There are also naturally occurring chemicals and processes such as dissolved solids and gases, pH, sorption, and volatilization. All of these conditions and processes work in harmony to positively influence biological species and activities.

Within the aquatic environment there are biological processes that are continuously being carried out by various species that inhabit specific ecosystems. Some of these include such life history activities as feeding, growth, reproduction, and migration. There is also ongoing competition for food and habitat, and there are further stresses on the biota as a result of natural prey and predator selections. All species that inhabit the aquatic environment are not designed to leave and, as such, remain unable to escape the aquatic medium for their entire cycles. This should be emphasized because these systems appear to many as ideal hiding places for various chemicals and discarded wastes.

The influence of chemical pollutants and xenobiotics as a result of anthropogenic activities could have profound effects on naturally occurring chemicals, physical and biological components, and processes. Factors that affect the magnitude of such insults include the properties of the pollutant and its products, the quantity of the chemical, the frequency of the discharge, and the properties of the affected ecosystem and its location with respect to the pollutant. Furthermore, two ecosystems that appear to be identical may not be similarly affected by the same quantity of a chemical. Because ecosystems involve complex interactions of various processes, it is difficult to understand the response of the system because very minor differences in both the environment and the species composition could result in contrasting fate and effects of the chemical. Therefore, all conditions of any system should be considered before attempting to evaluate the effects of a chemical.

BIOLOGICAL FACTORS INFLUENCED BY CHEMICALS

There are three biological factors that are influenced by a chemical: behavior, morphology, and physiology. In most cases populations may adapt to minor influences of pollutants, carrying out their various life cycle activities without any noticeable changes. This may be so because each species is uniquely adapted to the environment in which it has evolved and this natural environment is quite variable. When influences to this particular habitat, such as the introduction of a chemical dose, are beyond the adaptive capabilities or tolerances of the species the particular species may fail to adjust and their population may be adversely affected.

Aquatic species display various types of behavioral activities throughout their life cycles, including avoidance, feeding, migratory, reproductive, territorial, and many other instinctive and learned behaviors. These are influenced by factors that are relayed through the central nervous system. Information from the outside environment is received through sensory receptors and fed back to the central nervous system before a behavioral decision is made. These processes may be adversely affected by chemicals in the aquatic system.

Each species has developed both morphological and physiological adaptations to ideally suit its particular niche. For instance, a blackfly larva can remain attached to the top of a rock utilizing a disc of hooks at its posterior end; at the same time it may move along a silk strand and strain its food from the gushing waters with two fan–like structures near its mouth. Similarly, the Pacific salmon travels thousands of kilometers to make its way to its original birthplace to spawn; in some cases it may traverse virtually impassable rapids to arrive at that precise location.

Every animal species has developed mechanisms to process toxic substances both from the natural environment and from its own metabolic pathways. These adaptations were not designed to counteract the large quantities of effluents resulting from human progress and industrial revolution. In many cases, however, they have attempted to a lesser or greater extent to handle manmade influences, and as a result changes have occurred to various natural populations. As the populations persist through generations, subtle biological changes would have occurred affecting both morphological and physiological adaptations, and sooner or later the entire population would have changed completely. When manmade chemical influences persist and concentrations become too high to enable adaptation, species are eliminated, followed by natural populations, which leave the affected areas ecologically unbalanced and, in many cases, void of biological activity.

MODE OF CHEMICAL ACTION

Toxic substances influence changes in morphology and physiology by absorption and redistribution through the body, binding and localizing into different tissues; then possibly the substances inactivate themselves through biotransformation followed by excretion as metabolites. These processes involve the movement of toxicants across various plasma membranes either by pressure diffusion or by active transport movements. Toxic substances that are soluble in lipids dissolve in the lipid membranes, while the water-soluble substances pass through the pores along with the water.

Uptake of chemicals from the aquatic medium may occur directly from the water or indirectly through the consumption of food. Direct absorption takes place through the surfaces of respiration and through the skin or body surface of the animal. Absorption through food occurs through the stomach and the intestinal surfaces. The rate of absorption is found to be dependent upon the concentration of the toxicant in both the water column and the gut cavity, and also upon what complex compounds it may form upon combination with other substances. Other variables such as pH are also important in determining rate of absorption and site of deposit.

Within the body itself, distribution of a toxic substance and its resting place are dependent upon body fluids, transport mechanisms, and differences in permeability of the various membranes. Some toxicants may be lipid soluble, and considerable storage may occur in fat tissue; others may become bonded to plasma proteins, and still others may find rest in particular cell types, tissues, and various organs such as the liver and the kidney. Some aquatic animals are capable of eliminating foreign substances into the open water by passive diffusion through the lipoidal membranes of the gills. Excretion also occurs through the skin and through other passages of the animal.

In aquatic insects, biotransformation occurs in the microsomes of cells where possible alterations to the structure of the toxic substances may take place. In crayfish and lobsters, biotransformation occurs in the hepatopancreas by enzymic actions. Most biotransformation processes result in detoxification, such as the transformation of lipid-soluble substances into water-soluble metabolites, enhancing excretion by reducing binding to plasmaprotein and inhibiting their ability to cross cell membranes. This results in a decrease in the amount of toxic substances reaching target sites within the cells.

Toxic substances may act at various levels in the affected animal and could affect many physiological functions at different levels of integration. Some are believed to combine with enzymes or other cellular functions to produce biochemical changes and physiological effects. In other instances they may alter the movement of substances that are required for energy production or other cellular functions in and out of the cells. For instance, cyanide affects oxidative phosphorylation and ultimately the metabolic processes of energy transfer.

Death as a result of physiological causes is usually not specific, and many different chemicals may react in the same way. Many organic and inorganic chemicals will cause extensive damage to the gills at acute concentrations, but the affected animal may die from the lack of oxygen as a result of damages to the epithelium although the water may be very saturated with dissolved oxygen. The animal could also suffer from an electrolyte imbalance as a result of loss of osmoregulatory ability. On the other hand, if concentrations were lower, say at sublethal levels, and the exposure period was longer, it is very likely that the cause of death might be due to factors other than altered gas exchange or osmoregulation.

MODIFYING FACTORS OF TOXICITY

There are various factors that modify the effects of a toxic substance. These may be classified as biotic (those within the organisms) and abiotic (those outside of the organisms). The biotic factors start with the type of organism such as fish, aquatic invertebrates, or macrophytes. These factors vary among species, with one responding very differently to a chemical than the other. Other biotic factors may include physiological differences, size, nutritional variation, species health, and differences in life history stages. Biotic factors that commence within individuals even of the same species and under the same conditions will rarely respond in the same way to the same concentration of a toxicant. Abiotic factors that can act as modifiers are physicochemical parameters such as temperature, dissolved oxygen, pH, and hardness. Others, such as organic and inorganic materials, suspended solids, dissolved nutrients and gases, and photoperiod, also influence toxicity.

In order to appreciate some of the effects of modifiers, one may look at the responses of some species to certain chemicals. It is well known that sunfish are more tolerant to copper than salmonids and minnows. With organophosphate pesticides, goldfish and fathead minnows are more tolerant than bluegills and guppies. The life and size of the organism also modify or influence the effect that a toxic substance may have on an organism. The moulting of aquatic arthropods influences the results of acute toxicity tests, and for fish the sensitive life stages appear to be the embryo to early larval stage. With reference to size, larger fish are expected to be more tolerant to toxic substances than smaller fish.

In order to minimize modifying effects, fish that are chosen for toxicity testing should be nourished and healthy. Research has shown that stocks that were diseased or infected with parasites suffered greater losses during testing than those that were healthy. It has been shown also that test fish that were nourished were more tolerant to lethal levels of some metals than malnourished specimens.

TOXICITY ASSESSMENT

The effects of chemicals may be assessed by (1) the nature of the chemical, (2) the exposure period of the receptor, and (3) the dose of the chemical. For example, exposure to a high concentration of a contaminant for a relatively short period would result in noticeable effects that are experienced promptly; this type of an exposure is classified as "acute". On the other hand, exposure to low concentrations over a considerable period may only show effects after a long latent period; these types of effects are classified as "chronic".

There is no single approach to evaluating toxicity. There are a number of variables that may have different effects on a chemical, on a combination of chemicals, and on a biological system. However, tests should reflect the level of exposure to the chemical that is expected to be present in the environment, the possible effects and consequences of this chemical, and the concentration that is likely to produce adverse effects. Toxicity testing must be developed in such a way that it could incorporate new advances and changes regarding previous findings. Pretesting is an important step in the design and development of an appropriate battery of tests. Toxicity testing should not be utilized to provide absolute answers or be considered the final step in the prevention of toxic effects; instead, it should lead towards the provision of information that allows for the development of appropriate degrees of safety.

ACUTE TOXICITY

The acute toxicity test is designed to determine the concentration of a chemical that produces adverse effects on test organisms that are exposed for a short period of time in a controlled environment. Basically, acute toxicity tests are carried out by exposing organisms to various concentrations of a chemical that is mixed into treated water for a measured period of time. The acute lethality is determined by measuring the percentage of organisms that die within the time during which the test is run. The lethal concentration required to produce death to 50% of the test organisms over an exposure period of 96 hours is recorded and referred to as the 96-h median lethal concentration (96-h LC_{50}). In some instances, the exposure period may be shorter (24, 48, or 72 hours), expressing the LC_{50} for that specific period of time. In some species it might be difficult to determine death or definite effects on the number of organisms, for example, invertebrates, bacteria, etc. For these assessments, the results are expressed as a median effective concentration (EC_{50}), and the effect used for estimating the EC_{50} for certain invertebrates is immobilization.

Some scientists may choose to express the lethal concentration (LC) from 0% (that is no observed deaths, LC_0) to 100% (all test animals died, LC_{100}). For example, they may record their results for any percentage between 0 and 100, such as LC_{10}, LC_{40}, LC_{90}, etc. These concentrations may have been obtained from actual counts or from dose-response curves. The LC_0 concentration may also be defined as the no observed effect level (NOEL) or the no observed adverse effect level (NOAEL). The NOEL/NOAEL may be expressed as the highest concentration in a toxicity test that results in no statistically significant adverse effects to the exposed test organisms when compared to controls.

Acute toxicity tests are usually carried out in a laboratory where conditions are controlled and various natural situations can be simulated through the use of different combinations of variables. The variables that affect the type of response are the concentration of the chemical that is being used, the duration of the test/exposure, the species of the organism, the variation in exposure apparatus (static, recirculation, renewal, flowthrough), and the test conditions.

SUBCHRONIC TOXICITY

Subchronic toxicity procedures are generally designed to evaluate the adverse effects of chemicals administered to biological organisms during repeated exposures on a daily basis from a period of a few days to about 3 to 4 months. In many instances, subchronic exposures are also classified as prolonged exposures, and the tests are usually designed to incorporate the systemic effects as a result of cumulative exposure. The effective doses of subchronic exposures are also lower than those of acute toxicity studies; these doses are developed for prolonged experiments, and lethal effects are not the expected end results.

In order to develop a battery of subchronic toxicity tests, it is usually necessary to experiment with several short–term dose-finding pilot studies. These preliminary tests provide information regarding body weight, target organs, organ damage, behavioral changes, biochemistry, hematology, and toxicological and physiological responses. Based on the information on target organ, critical concentration, and associated effects, administration of the test chemical to the animal must be designed prior to the commencement of the exposure studies. An autopsy should be performed on all animals at the end of the pilot study, noting all lesions and histological and pathological changes.

The choice of animal for the study is dependent on variables such as the physical and chemical properties of the test substance, length of exposure, laboratory treatment facilities, and similarity of metabolic pathway of the test organism to those of the organisms that might be influenced by the chemical under investigation. The study period may vary between 21 and 30 days, with an increase in dose level at the end of the period and continuation of the treatment for an additional period if the effects are questionable. The dose range may consist of several levels, commencing at a no observed effect level and concluding with a maximum tolerable level. During the entire study, gross observations are continually recorded in order to ensure toxicity signs are not omitted.

CHRONIC TOXICITY

Chronic toxicity usually occurs because of repeated or prolonged exposures to chemicals which might result in deleterious effects to the exposed organism. The observed toxic response of a chemical during a chronic exposure could result from direct effects of the chemical, altered form of the chemical, redistribution of the metabolites in the animal body, and continued aggravation of target organs, enzyme systems, and hormonal systems by the chemical. At the end of a chronic exposure, all animals are evaluated for gross pathology and histological effects. Also, any animal that might have died during the experiment is completely autopsied and all of the respective organs are examined and analyzed physically, pathologically, and histologically.

Chronic studies must be designed in such a way as to include various levels of exposure. Experiments should range from exposures at levels that would be expected to produce no adverse effects, to those that might cause deleterious or harmful effects to the test organisms. This allows for a wide range of adverse effects to be observed and thus should enable a thorough evaluation of the test chemical under investigation.

It is important to carry out clinical evaluation of the test organisms prior to the commencement of the study. This should be continued daily during the initial stages of the study and at least biweekly during the rest of the experiment. Both symptomatic responses and behavior should be continually followed throughout the study period. Other measurements such as weight, food consumption, and appropriate biochemistry should be performed at routine intervals throughout the study. Eventually, all organisms of both the exposed and control groups should be subjected to complete pathological and histological examinations.

Generally, chronic toxicity tests are carried out in order to evaluate the effects of a chemical on an ecosystem and ultimately the subsequent effects on species in that system. As a result, the species selected and tested under controlled conditions should produce information that would indicate absorption rate, metabolic rate, metabolic pathways, time taken to react, target organ effects, etc., similar to that of the organism to which the result will be compared.

ECOTOXICOLOGY

Various hazardous chemicals are released into the environment during transportation, manufacturing, use, accidents, industrial activities, domestic uses, or disposal. These result in degrading the environment, adverse effects to various species of biota, loss of both agricultural products and lands, and esthetically displeasing scenery.

Ecotoxicology deals with the multicausal simultaneous effects of contaminants in the system. First, a chemical is released into the environment; the amount, forms, and sites of such a release must be known if its subsequent environmental fate is to be understood. Second, the chemical is transported geographically and into different biota, and perhaps chemically transformed, giving rise to compounds that have quite different environmental behavioral patterns and toxic properties. The nature of such processes is unknown for the majority of environmental contaminants, and the dangers arising from ignorance of the ultimate fate of certain chemicals have been well documented in recent years. The third part of the process is the exposure of one or more target organisms. To assess this process, one must first identify the nature of the target and the type of the exposure that is to be examined. Finally, one has to assess the response of the individual organism, population, or community exposed to that pollutant over the appropriate time scale.

Ecotoxicological assessment involves a combination of the factors just discussed; therefore, the selection of indicators should provide wide representation and include various biological processes. An attempt to investigate the toxic effects to a multifaceted system is further complicated by changes that might occur as a result of adaptation, the range of differences in the responses that might be observed by components, and the diversity of the components within any one of the ecosystems.

In broad terms, the choice of test methods not only should balance considerations of costs, precision, and accuracy, but also should take into account the fate and transport that are influencing the chemical in the environment. One must know how the chemical is distributed within the ecosystem and how it may affect the population utilizing that specific ecosystem. Also, the original state of the chemical might be modified with other chemicals or biological components as it is transported through any one ecosystem to produce new forms that might have totally different effects, thus developing new and different concerns.

ALGAL INHIBITION

The growing algae are exposed to various concentrations of a test substance over several generations under a defined set of conditions. The growth among the controls and the test solutions is then compared for a fixed period of time. A test solution is considered to be toxic when statistically significant inhibition of algal growth occurs in it. The cell multiplication inhibition test (CMIT) is also referred as the toxicity threshold (TT). The endpoint of the algal growth inhibition test is the median inhibition concentration (IC_{50}), that is, the concentration estimated to cause a 50% reduction in growth compared to the control over a specified period of time. The no observed effect concentration, the lowest observed effect concentration, and the toxicity threshold have also been used as endpoints.

Algae are selected to represent primary producers because they are at the entry level, both for the uptake of nutrients as well as contaminants. Thus any adverse effects that might be detected in the algal community truly represent an early warning of a potential contaminant presence in the ecosystem.

LEMNA INHIBITION

Duckweed are exposed to various concentrations of a test substance either in a static or dynamic test system. The response is measured through root growth, frond number, and/or dry weight. The endpoint of the Lemna inhibition is the median inhibition concentration (IC_{50}), that is, the concentration estimated to cause 50% reduction in growth compared to the control over a specified period of time.

The selection of Lemna is based on the information of its genetics and physiology. Also it is easy to culture and cheap to maintain. The major shortcoming is that it is not completely typical of higher plants and, as such, introduces uncertainty into its use as a test organism.

TESTING CONSIDERATIONS

In summary, the following considerations might be useful in determining tests and ensuring that overall environmental effects are addressed:

- Tests should be cost-effective.
- Tests should be sensitive enough in order to determine the effects.
- Tests should be applicable to a wide range of chemicals.
- Test organisms should be representative of the ecosystem and form an important part of the system/food chain in relation to man.
- The organism should be readily available and previous scientific information should be obtained.

SPECIES COMMONLY USED IN TOXICITY TESTING

Vertebrates

In aquatic toxicology, vertebrates that are used for testing include fish, amphibians, and reptiles. These test animals should be representative of the areas exposed to the impact, sensitive to the chemical, and indigenous or closely related to species found in the exposed areas. They may be collected from wild populations, purchased from suppliers, or cultured in a laboratory.

Fish

Species of fish that are most commonly used in acute toxicity testing are listed in Table 1.

Table 1 Fish Species Most Commonly Used in Acute Aquatic Toxicity Tests

Freshwater		Saltwater	
Rainbow trout	*Oncorhynchus mykiss*	Sheephead minnow	*Cyprinodon variegatus*
Brook trout	*Salvelinus fontinalis*	Threespine stickleback	*Gasterosteus aculeatus*
Channel catfish	*Ictalurus punctatus*	Mummichog	*Fundulus heteroclitus*
Fathead minnow	*Pimephales promelas*	Longnose killifish	*Fundulus similis*
American flagfish	*Jordanella floridae*	Silverside	*Menidia* sp.
Goldfish	*Carassius auratus*	Pinfish	*Lagondon rhomboides*
Bluegill	*Lepomis macrochirus*	Sand dab	*Citharichthys stigmaeus*
		Spot	*Leiostomus xanthurus*

Amphibians and Reptiles

Amphibians and reptiles are being used more frequently in aquatic toxicology. It is very likely that their use will increase because of chronic and long-term effects of various pollutants in these specific groups. Table 2 lists species that are most commonly used.

Invertebrates

Invertebrates are being consistently used for assessing the effects of chemicals in the aquatic environment. These lower animals are common indicator species because they have been found to be more susceptible to the effects of many chemicals than fish or other vertebrates. They cover a wide range of organisms including protozoans, mesozoans, parazoans, hydroids, platyhelminthes, aschelminthes, molluscs, annelids, and arthropods. Table 3 lists aquatic invertebrates that are commonly utilized in aquatic toxicity testing.

Table 2 Amphibians and Reptiles Most Commonly Used in Toxicity Testing

Mexican axolotl	*Ambystoma mexicanum*	Edible frog	*Rana esculenta*
African frog	*Xenopus laevis*	Common toad	*Bufo bufo*
Leopard frog	*Rana pipiens*	American toad	*Bufo americanus*
Common frog	*Rana temporaria*	Fowler's toad	*Bufo fowleri*
Pickerel frog	*Rana palustris*	Tree frog	*Hyla versicolor*
Green frog	*Rana calmitans*	Spring peeper	*Hyla crucifer*
Wood frog	*Rana sylvatica*	Box turtle	*Terrapene* sp.
Bullfrog	*Rana catesbiana*		

Table 3 Invertebrates Most Commonly Used in Aquatic Toxicity Testing

Freshwater		Saltwater		Other	
Daphnids	*Daphnia magna*	Copepods	*Acartia tonsa*	Bacteria	*Pseudomonas putida*
	Daphnia pulex		*Acartia clausi*		
	Daphnia pulicaria	Polychaetes	*Capitella capitata*		*Microcystis aeruginosa*
Amphipods	*Gammarus lacustris*		*Neanthes* sp.		
	Gammarus fasciatus	Shrimp	*Penaeus setiferus*		*Salmonella typhimurium*
	Gammarus		*Penaeus duorarum*		
	pseudolimnaeus		*Penaeus aztecus*		*Photobacterium*
Crayfish	*Orconectes* sp.		*Palaemonetes pugio*		*phosphoreum*
	Cambarus sp.		*Palaemonetes*		
	Procambarus sp.		*vulgaris*		
	Pacifasiacus		*Palaemonetes*		
	leniusculus		*intermedius*		
Midges	*Chironomus* sp.		*Crangon septemspinosa*		
Stoneflies	*Plecoptera*		*Crangon nigricauda*		
Mayflies	*Ephemeroptera*		*Crangon crangon*		
Dragonflies	*Odonata*		*Pandalus jordani*		
Caddis flies	*Trichoptera*		*Pandalus danae*		
Flies	*Diptera*		*Mysidopsis bahia*		
Molluscs	*Gastropoda*	Crab	*Callinectes sapidus*		
Algae	*Selanastrum*		*Hemigrapsus* sp.		
	capricornutum		*Pachygrapsus* sp.		
	Scenedesmus		*Carcinus maenas*		
	quadricauda		*Uca* sp.		
		Oyster	*Crassostrea virginica*		
			Crassostrea gigas		

3 Index to Aquatic Species

CLASSIFICATION

Figure 1 illustrates an abridged classification of species that utilize the aquatic environment. This genealogic tree attempts to outline the probable relationships and relative positions of major groups. The species that have been referenced in this book will fall under different phyla, which have been systematically classified according to standard taxonomic procedures. An attempt has been made to correlate common names with Latin nomenclature for each species that has been referenced.

DATA ARRANGEMENT

The following outlines the position of various phyla according to standard taxonomic procedures. The data in this book, however, have been listed alphabetically under the different groups.

Phylum Protozoa: protozoans
Class Mastigophora — flagellates
Class Sarcodina — rhizopods and amoebas
Class Sporozoa — spore-formers
Class Ciliata — ciliates

Phylum Mesozoa: mesozoans — worm-like

Phylum Porifera: sponges

Phylum Coelenterata: hydroids
Class Hydrozoa — hydras and siphonophores
Class Scyphozoa — jellyfishes
Class Anthozoa — sea anemones and corals

Phylum Platyhelminthes: flatworms — flukes and tapeworms

Phylum Aschelminthes: roundworms
Class Nematoda — roundworms
Class Rotifera — rotifers

Phylum Mollusca: molluscs
Class Gastropoda — gastropods
Class Bivalvia — bivalves
Subclass Lamellibranchia
Class Cephalopoda — squids, cuttlefishes, octopuses

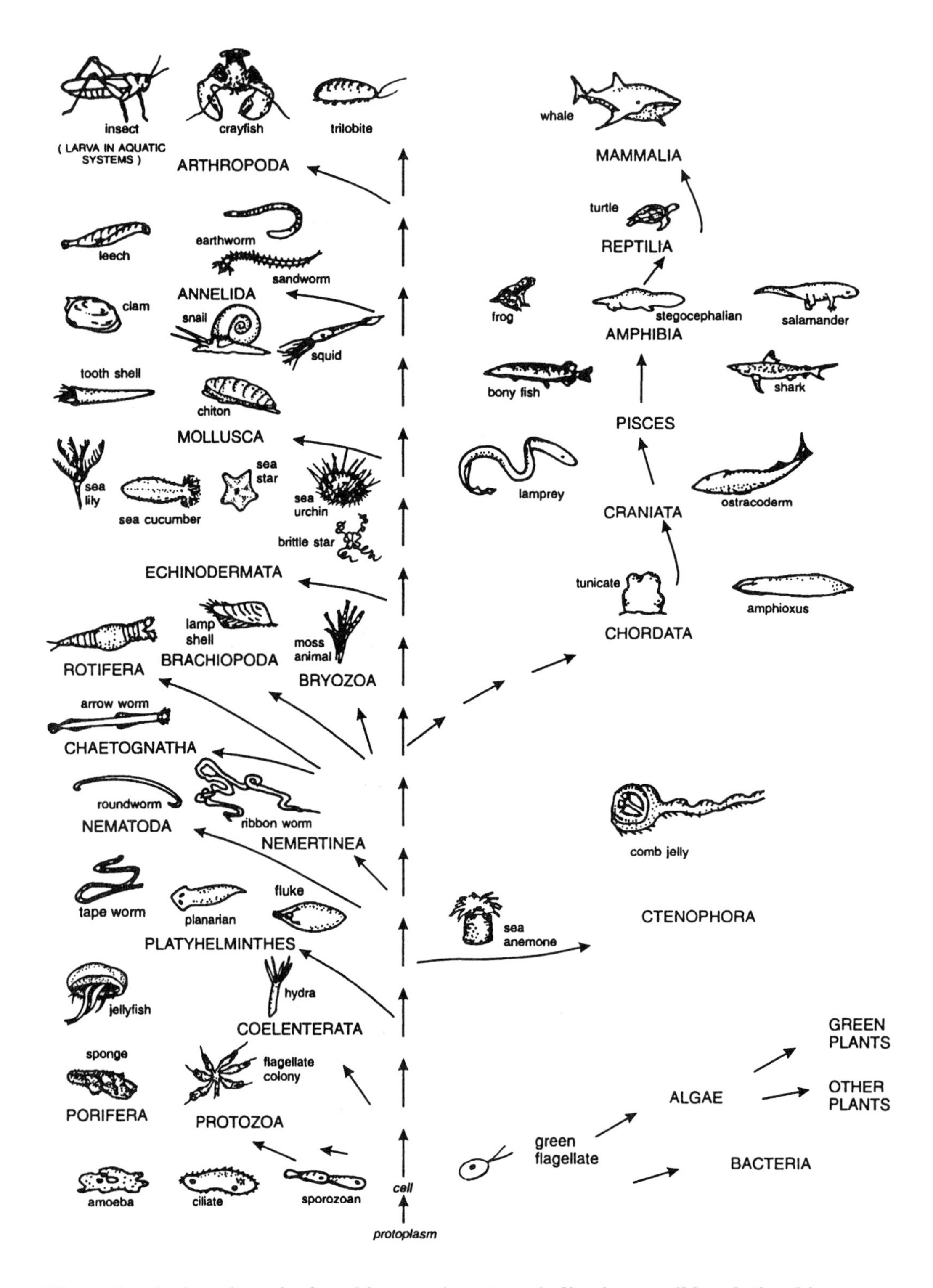

Figure 1. A view of species found in aquatic systems indicating possible relationships among groups.

Phylum Annelida: segmented worms
Class Polychaeta — polychaetes
Family Nereidae — nereids
Class Oligochaeta — oligochaetes including earthworms
Family Tubificidae — Tubifex
Class Hirudinea — leeches

Phylum Arthropoda: arthropods
Class Insecta — insects
— exopterygote insects
Order Ephemeroptera — mayflies
Order Odonata — dragonflies
Order Plecoptera — stoneflies
Order Lepidoptera — pyralidids
Order Diptera — flies
Order Coleoptera — beetles
Class Crustacea — crustaceans
Subclass Branchiopoda
Order Anostraca — brineshrimp
Order Notostraca
Order Conchostraca
Order Cladocera — waterfleas
Subclass Copepoda — *Daphnia* copepods
Order Calanoida — *Calanus*
Order Cyclopoida — *Cyclops*
Subclass Malacostraca — malacostracans or higher crustaceans
Order Isopoda — isopods, sowbugs
Order Amphipoda — amphipods, scuds, and sideswimmers
Suborder Gammaridea — *Gammarus*
Order Decapoda — decapods
Suborder Natantia — Crangon
Suborder Reptantia
Section Astacura — *Homarus*
Section Brachyura — green crab
Class Arachnida — spiders
Order Araneae — spiders

Phylum Echinoderrmata: starfishes, sea urchins
Class Asteroidea — starfishes
Class Echinoidea — sea urchins, sand dollars

Phylum Chordata: chordates
Subphylum Vertebrata — vertebrates
Classes Marsipobranchii, Selachii, Bradyodonti, and
Pisces — fishes
Class Amphibia — amphibians
Order Caudata — newts and salamanders
Order Salientia — frogs and toads
Class Reptilia — reptiles
Order Chelonia – Testudines — tortoises, terrapins, and turtles

Order Loricata – Crocodylia — crocodilians
Order Cetacea — whales
Suborder Pinnipedia — seals, sealions, walruses

The following is a list of organisms in alphabetical order by Latin names of species. These species have been utilized in experimental work with chemicals and have been referenced in environmental surveys. They also include species referenced in this document.

Latin Name	English Name
Acanthurus sp.	sturgeon fish
Adrichetta forsteri	yellow–eye mullet
Aedes sp.	mosquito
Aedes	mosquito
Aequipecten gibbus	calico scallop
Aequipecten (Pecten) irradians	bay scallop
Agonus cataptaractus	armed bullhead
Alburnus alburnus	bleak
Ambassis safga	minnow
Ambystoma mexicanum	Mexican axolotl
Amia calva	bowfin
Amiurus sp.	bullhead
Amiurus melas	black bullhead
Amiurus nebulosus	American catfish
Anchoa (Anchoiella) mitchilly	bay anchovy
Anguilla anguilla	eel
Anguilla japonica	Japanese eel
Anguilla rostrata	American eel
Anguilla vulgaris	eel
Annelida	segmented worm
Anodonta cygnea	freshwater clam
Anolis carolinensis	anole
Anopheles sp.	mosquito
Anthocidaris crassispina	
Anthopleura aureodiata	anemone
Anthopleura elegantissma	sea anemone
Aplocheilus latipes	medaka
Arbacid puntulata	sea urchin
Aschelminthes	roundworms
Asterias forbesi	starfish
Atherinamosa microstoma	atherinids
Balanus sp.	barnacle
Barbus conchonius	rosy barb
Barbus machecola	barb
Brevoortia patronus	gulf menhaden
Brevoortia tyrannus	Atlantic menhaden
Bufo americanus	American toad
Bufo bufo	common toad
Bufo fowleri	Fowler's toad
Callinectes sapidus	blue crab
Campostoma anomalum	stone roller
Cancer magister	Dungeness crab
Caranx sp.	pompano, jack cravally
Carassius auratus	goldfish
Carassius carassius	crucian carp
Carcinides maenas	green crab
Carcinus maenas	shore crab
Carpiodes cyprinus	white carp
Catostomus sp.	sucker
Catostomus catostomus	longnose sucker
Catostomus commersoni	white sucker
Centropomus undecimalis	common snook
Chaarogobius heptacanthus	gobi
Channa gachua	snakehead
Channa punctatus	snakehead
Chanos chanos	milkfish
Chelon labrosus	grey mullet
Chelydra serpentina	snapping turtle
Chingatta	
Chironomus plumosus	midge
Chondrus crispus	sea moss
Cirrhinus mrigala	cyprinid
Clarius batrachus	walking catfish
Clinocardium nuttali	cockle clam
Cloeon sp.	mayfly
Clupea harengus	herring
Clupea pallasii	Pacific herring
Clupea sprattus	sprat
Colisa fasciata	striped gourami
Colpoda	protozoa
Conus sp.	cone shells
Coregonus hoyi	coregonid

Latin Name	English Name
Coregonus lavaretus	coregonid
Coregonus peled	coregonid
Couesius plumbeus	lake chub
Crangon sp.	shrimp
Crangon crangon	brown shrimp
Crangon septemspinosa	sand shrimp
Crassostrea gigas	Pacific oyster
Crassostrea virginica	eastern oyster
Culex sp.	mosquito
Cymatogaster aggregata	shiner perch
Cyprinodon macularius	desert pupfish
Cyprinodon variegatus	sheepshead minnow
Cyprinus carpio	carp
Daphnia sp.	water flea
Echinometra sp.	sea urchin
Enallagma sp.	damsel fly
Engraulis encrasicholus	anchovy
Entosiphon sulcatum	protozoa
Epeorus sp.	mayfly
Ephemerella sp.	mayfly
Ephemerella walkeri	mayfly
Esox lucius	northern pike
Euplotes vannus	ciliate
Fundulus sp.	killifish
Fundulus diaphanus diaphanus	Eastern banded killifish
Fundulus grandis	gulf killifish
Fundulus heteroclitus	mummichog
Fundulus majalis	striped killifish
Fundulus similis	longnose killifish
Gadus merlangus	whiting
Gadus morrhua	cod
Gadus pollachius	pollack
Gadus virens	coalfish
Gambusia affinis	mosquito fish
Gammarus pulex	water shrimp
Gasterosteus aculeatus	three-spine stickle-back
Glassosiphonia complanata	mollusca
Glycera dibranchiata	bloodworm
Gobiidae sp.	gobi
Gobio gobio	gobi
Gobius minutus	gobi
Gracilaria foliifera	red seaweed
Gracilaria virrucosa	red seaweed
Gymnodium breve	dinoflagellate
Haliotis sp.	abalone
Harpacticoid	copepod
Helcioniscus argentatus	saltwater limpet
Helcioniscus exaratus	saltwater limpet
Helix sp.	land snail
Hemigrapsis sp.	shore crab
Heteropneustes fossilis	catfish
Hexagenia sp.	mayfly
Homarus americanus	northern lobster
Homarus gammarus	European lobster
Hydra oligactis	hydra
Hydropsyche sp.	caddis fly
Hyla crucifer	spring peeper
Hyla versicolor	treefrog
Ictalurus sp.	bullheads
Ictalurus ameiurus	catfish
Ictalurus melas	black bullhead
Ictalurus natalis	yellow bullhead
Ictalurus nebulosus	brown bullhead
Ictalurus punctatus	channel catfish
Ictiobus sp.	sucker
Ictiobus cyprinellus	bigmouth buffalo
Idus idus melanotus	orfe
Ischnura sp.	damsel fly
Jordanella floridae	American flagfish
Kuhlia sandvicensis	mountain bass
Lagodon rhomboides	pinfish
Lagodon rhomboides	marine pin perch
Laminaria agardhii	Atlantic kelp
Laminaria digitata	Atlantic kelp
Leander adspersus	shrimp
Lebistes reticulatus	guppy
Leiostomus xantharus	spot
Lepomis cyanellus	green sunfish
Lepomis gibbosus	pumpkinseed sunfish
Lepomis humilis	common sunfish
Lepomis macrochirus	bluegill sunfish
Lepomis megalotis	longear sunfish
Lepomis microlophus	redear sunfish
Leptocottus armatus	staghorn sculpin
Leucaspius delineatus	minnow
Leuciscus cephalas	chub
Leuciscus idus melanotus	golden orfe
Leuciscus leuciscus	dace
Limanda limanda	dab
Limnophilus sp.	caddis fly
Limnoria tripunctata	marine bore

Latin Name	English Name	Latin Name	English Name
Littoridina sp.	gastropod	*Ophiocephalus punctatus*	snakehead
Lucania parva	rainwater killifish	*Ophiogomphus* sp.	dragonfly
Lumbricus terrestris	common earthworm	*Ophiura texturata*	starfish
Lutianus campechanus	red snapper	*Orchestiodea californiana*	amphipod
Lymnaea sp.	gastropod	*Orconectes nais*	crayfish
Lytechnius sp.	sea urchin	*Orconectes rusticus*	crayfish
Macrocystis pyrifera	kelp	*Oryzias latipes*	paddy fish
Menidia beryllina	tidewater silverside	*Ovalipes ocellatus*	calico crab
Menidia menidia	Atlantic silverside	*Pagurus longicarpus*	hermit crab
Menidia peninsulae	silverside	*Palaemonetes kadiakensis*	shrimp
Mercenaria mercenaria	hard clam	*Palaemonetes macrodactylus*	Korean shrimp
Micrometrus minimus	dwarf perch	*Palaemonetes pugio*	grass shrimp
Micropogon undulatus	croaker	*Palaemonetes vulgaris*	grass shrimp
Micropterus dolomieui	smallmouth bass	*Pandalus* sp.	shrimp
Micropterus salmoides	largemouth bass	*Pandalus goniurus*	bay shrimp
Mollienesia latipinna	sail fin molly	*Panulirus japonicus*	Japanese lobster
Morone laborax	bass	*Panulirus pencillatus*	lobster
Morone saxatilis	striped bass	*Paragrapsis quadridentatus*	grapsid crab
Mugil cephalus	striped mullet	*Paralichthys dentatus*	summer flounder
Mugil crinatus	mullet	*Paralichthys lethostigma*	southern flounder
Mugil curema	grey mullet	*Paralithoides camtschatica*	king crab
Mulloidichthys sp.	goatfish	*Paramecium caudatum*	paramecium
Musca domestica	housefly	*Parhyale hawaiensis*	amphipod
Mya arenaria	soft shell clam	*Parophrys vetulus*	English sole
Myocaster coypus	nutria	*Penaeus aztecus*	brown shrimp
Mysidopsis almyra	shrimp	*Penaeus duorarum*	pink shrimp
Mysidopsis bahia	mysid shrimp	*Penaeus merguiennis*	banana prawn
Mytilus californicus	California sea mussel	*Penaeus monodum*	decapod
Mytilus edulis	bay mussel	*Penaeus setiferus*	white shrimp
Neanthes arenaceodentata	polychaete	*Perca flavescens*	yellow perch
Nereis vexillosa	sandworm	*Perca fluviatilis*	perch
Nereis virens	sandworm	*Pertunus sanquinolentes*	crab
Nitrocra spinipes	harpacticoid copepod	*Petromyzon marinus*	sea lamprey
Notemigonus crysoleucas	golden shiner	*Phaeroides maculatus*	northern puffer
Notopterus notopterus	mooneye	*Phoxinus phoxinus*	minnow
Notropis atherinoides	emerald shiner	*Pimephales notatus*	bluntnose minnow
Ocorhynchus tschawytscha	chinook salmon	*Pimephales promelas*	fathead minnow
Oedothorax insecticeps	rice field spider	*Platichthys flesus*	flounder
Oncorhynchus gorbuscha	pink salmon	*Platichthys stellatus*	starry flounder
Oncorhynchus keta	chum salmon	*Platyhelminthes*	flatworms
Oncorhynchus kisutch	coho salmon	*Pleunonectes platessa*	plaice
Oncorhynchus mykiss	rainbow trout		
Oncorhynchus nerka	sockeye salmon		

Latin Name	English Name
Podophthalmus vigil	crab
Poecilia reticulata	poeciliidae (tooth-carp)
Poecilia (Mollienesia) latipinna	sailfin molly
Porphyra sp.	red algae
Portunus sanquinolentus	crab
Procambarus sp.	crayfish
Protothaca staiminea	littleneck clam
Pseudocalanus minutus	copepod
Pseudodiaptimus coronatus	copepod
Pseudopleuronectes americanus	winter flounder
Pteronarcella sp.	stonefly
Pteronarcys sp.	stonefly
Pteronarcys californica	stonefly
Puntius conchonius	cyprinid
Pygosteus pungitius	stickleback (12–spined)
Rana catesbiana	bullfrog
Rana clamitans	greenfrog
Rana esculenta	edible frog
Rana palustris	pickerel frog
Rana pipiens	leopard frog
Rana sylvatica	wood frog
Rana temporaria	common frog
Rangia cuneata	mactrid clam
Ranina serrata	crab
Rasbora heteromorpha	harlequin fish
Rasbora trilineata	rasbora
Rhithropanopeus harrisii	mud crab
Roccus saxatilis	striped bass
Rutilus rutilus	roach
Saccobranchus fossilis	airsac catfish
Salmo aguabonita	golden trout
Salmo clarki	cutthroat trout
Salmo irideus	rainbow trout
Salmo salar	Atlantic salmon
Salmo trutta	brown trout
Salvelinus fontinalis	brook trout
Salvelinus namaycush	lake trout
Sardinops caerula	Pacific sardine
Sarotherodon mossambicus	tilapia
Scylla serrata	lobster
Semotilus atromaculatus	creek chub
Sesarma cinnerum	crab
Siliqua patula	razor clam
Simocephalus serrulatus	waterflea
Simulium sp.	blackfly
Siphonaria normalis	ribbed limpet
Solea solea	sole
Spirula solidissima	surf clam
Squalius cephalus	chub
Squalius leuciscus	dace
Stenotomus chrysops	scup
Stizostedion lucioperca	zander
Stizostedion vitreum	walleye
Stolephorus purpureus	anchovy
Stronglo centrotus purpuratus	purple sea urchin
Sturnis vulgaris	starling
Sus scrofa	miniature swine
Tendipedidae	midge
Tenebrio sp.	mealworm
Terrapene sp.	box turtle
Tetrahymena pyriformis	ciliate
Thalassoma bifasciatum	bluehead
Therapon janbua	crescent perch
Tilapia leucosticta	tilapia
Tilapia mossambica	tilapia
Tilapia rendalli	tilapia
Tilapia sparmanii	tilapia
Tinca tinca	tench
Tisbe furcata	
Tivela stultorum	pismo clam
Trachinotus carolinus	pompano
Trigriopus japonicus	
Uca cramulata	fiddler crab
Uca pugilator	fiddler crab
Ulva sp.	sea lettuce
Umbra pygmaeo	eastern mud minnow
Urunema parduczi	protozoa
Volsella demissa	Atlantic ribbed mussel
Vorticella campanula	protozoa
Xenopus laevis	frog
Xiphophorus helleri	swordtail

4 Chemical Concepts

The persistence and mobility of a chemical in the aquatic environment are governed by a number of physicochemical and biological processes. These processes include sorption-desorption, volatilization, and chemical and biological transformations. Solubility, vapor pressure, and lipid solubility of a chemical determine its exposure concentration to aquatic species. The residence time of the chemical in water will account for the duration of the exposure as well as the transformation processes, both biological and abiological, in the water phase. Thus, any chemical that is released into the environment is likely to be transported through water, sediment/soil, and air. It is also taken up and displaced by biological systems.

The physicochemical processes would include the properties of the aquatic environment such as temperature, salinity, pH, conductivity, dissolved and suspended solids, sediment particle size, and organic content of the sediment and the water column. The mass flow of the water column determines the rate of transport of the chemical discharged into surface water. Processes that remove the chemical from the water column, such as flocculation, volatilization, hydrolysis, and complexation with natural organics, may determine how long the toxicant remains within a compartment of an aquatic ecosystem. The distribution of the chemical within the biological compartment of the ecosystem will depend upon the bioaccumulative capacity of the organism(s).

SORPTION

All chemicals that are released into water are continually being transported and redistributed among various media such as air, sediment, and biota. The term *sorption* used here covers both adsorption and absorption, which are difficult to distinguish in most situations. Many metals and organic chemicals sorb to sediments and suspended solids. This determines the fraction that is available in the water column for other fate processes.

Sorption of chemicals to organic matter depends upon the pH and the type of chemical interaction, either ionic or nonionic. Interaction of nonionic compounds will not be greatly affected by changes in pH, whereas ionic compounds will be repelled by the sorbent surfaces at high pH values due to like charges of the sorbent and the sorbate. Chemicals such as 2,4-dichlorophenoxy acetic acid (2,4-D), 2,4,5-trichlorophenoxy acetic acid (2,4,5-T), dicamba, chloramben, and picloram behave in this manner. Sorption of these compounds increases with a decrease in pH due to the formation of unionized surface and unionized form of the chemical. Sorption coefficients based on the organic content of the sorbent provide a basis for estimating other accumulation parameters such as *n*-octanol–water partition coefficients and bioconcentration factors for the biota. Sorption coefficients also provide a measure of the leachability of chemicals into the water column. Compounds having a K_{oc} value greater than 1000 are quite strongly bound to organic matter of sediment or soil and are considered immobile. Chemicals with a K_{oc} value below 100 are moderately to highly mobile. Thus, K_{oc} values can be useful predictors of the potential leachability of compounds from aqueous sediments or soil. Estimation of values of K_{oc} using correlation equations with other properties such as water solubility, *n*-octanol–water partition coefficient K_{ow}, and bioconcentration factors (BCF) are discussed elsewhere.[1]

Sorption is the process by which a chemical is bound to a surface (biotic or abiotic) by either covalent, electrovalent, or molecular forces. When bound to a biotic surface, the chemical could have significant effects on various physiological processes that take place on the skin or epithelial layers, in addition to contributing to the total body burden. Sorption reduces the concentration of the chemical in the dissolved phase.

Sorption can be expressed in terms of an equation:

$$C_s = K_p \, C_w^{1/n}$$

where C_s and C_w are the concentrations of the chemical in solid and water phases, respectively, K_p is the partition coefficient for sorption, and $1/n$ is the exponential factor.

At environmentally significant concentrations that are low compared with the sorption capacities of the surface components, the term $1/n$ reduces to unity. It is important to allow sufficient time for equilibration between phases to be established in determining K_p. This time could vary from a few minutes to several days, depending on the chemical. For neutral organic compounds, the sorption was shown to correspond with the organic content of the particulates.

VOLATILIZATION

The route of exposure of the chemical can be modified by its volatilization to the vapor phase. The movement of a chemical from the water phase is a kinetic function and is dependent on factors such as diffusion, vapor pressure, dispersion of emulsions, aqueous solubility, and temperature. Volatilizational loss of chemicals from water to air is an important fate process for chemicals with low aqueous solubility, low polarity, and/or high vapor pressures. Many chemicals, despite low vapor pressures, can volatilize rapidly out of the water column due to their high activity coefficients in solution.

The transport of an organic solute across a two-layer system (such as water–air) can be expressed by the Henry's law constant (H), which, in simple terms, is a ratio of its vapor pressure to its water solubility. This ratio is based on the assumption that water is not very soluble in the organic solute, and the vapor pressure used is that of the pure substance. When the solubility of water exceeds a few percent, this assumption may not be valid. In such cases, it is recommended to measure the Henry's law constant by measuring air and water concentrations of the organic chemical at equilibrium close to the interface. The techniques to measure H have been described in the literature. It is important to ensure that the vapor pressures and solubilities used to calculate H are for the same temperature and the same phase.

Henry's law constants range over many orders of magnitude: high for chemicals with high vapor pressures, low aqueous solubilities, and low boiling points, such as alkanes. The value of H is low for chemicals with high aqueous solubility and low vapor pressure, such as alcohols. As stated earlier, chemicals with low vapor pressures but with low aqueous solubilities will have high H values. Such compounds may leave water rapidly for the atmosphere. For example, DDT can evaporate rapidly from lakes and rivers because of its low aqueous solubility, which compensates for its low vapor pressure. 2,4-D, on the other hand, has a very low H value compared to DDT, in spite of their vapor pressures being similar. In general, for highly volatile compounds and for H values greater than 10^{-3} (atm·m^3/g·mol units), the liquid-phase resistance dominates, and for solutes of low H $<10^{-4}$, such as SO_2, the gas-phase resistance dominates.[2] The volatilities of chemicals have been classified according to Henry's law constant, H,[3] which is as follows: chemicals having H >0.01 are readily lost from water surfaces; those with H = 0.01 to 0.00001 are moderately volatile; and those with H <0.00001 are nonvolatile.

BIOTRANSFORMATION

Biotransformation may be defined as the biological alteration or conversion of one chemical from one form to another. It can be distinguished from other chemical conversions because in many cases it requires biological catalysts called enzymes. The transformed forms may have physicochemical properties that are different from the parent compound and therefore may behave differently within the biological system. Methylation of metals, particularly mercury to methylmercury, with totally different lipid solubilities and toxicities is a good example. Parameters that may be affected due to biotransformation include bioaccumulation, biological half-life, tissue distribution, and excretion. It is also possible that the transformed product may have different pharmacokinetic and toxicokinetic properties.

Biodegradation tends to transform lipid-soluble compounds into more water-soluble forms, possibly with reduced toxicities, through oxidation catalyzed by microsomal enzymes. These transformed compounds are more easily excreted by the organism. Some reactions may convert the chemical into more lipid-soluble form, which will increase its biological retention time.[4] The production of metabolites that are more toxic than the parent compound has been experimentally demonstrated.[5] Biodegradation in general, with the exception of bacteria, leads to total degradation of chemicals into carbon dioxide and water. In many instances, biodegraded products that are excreted by organisms are relatively intact molecules that may not be degraded further by other components of an ecosystem.[4]

CHEMICAL TRANSFORMATION

Chemical transformation of a compound in the environment could arise from one or more of the following reactions: (1) redox changes, (2) hydrolysis, (3) halogenation–dehalogenation, or (4) photochemical alteration or degradation. The extent to which the chemical breaks down to simple molecules will determine its persistence and level of exposure. The transformed derivative could be substantially more hazardous and persistent. Examples are photochemical degradation of hydrocarbons and nitrogen oxides to produce a smog that has a more direct effect on the environment and humans. Halogenation of aromatic compounds and aliphatic hydrocarbons is environmentally significant. Chlorinated dioxins and furans and formation of chloroform in the presence of organic matter are examples of chemical transformation processes. Ionization of organic chemicals is likely to alter physicochemical properties such as solubility, sorption, and bioconcentration of the parent compound. For example, the ionized species have longer residence time in water and less ability to migrate into the organic or lipid part of the abiotic (such as sediment) or biotic substrates (such as fish), respectively, than the parent neutral unionized molecule.

REFERENCES

1. S. Ramamoorthy and E. Baddaloo, *Evaluation of Environmental Data for Regulatory and Impact Assessment*, Elsevier Science Publishers, Amsterdam, The Netherlands, 1991.
2. D. Mackay, W.Y. Shiu, and R.J. Sutherland, Estimating volatilization and water column diffusion rate of hydrophobic contaminants, in *Dynamics, Exposure and Hazard Assessment of Toxic Chemicals*, R. Haque, Ed., Ann Arbor Science, Ann Arbor, Michigan, 1980.
3. U.S. E.P.A. Part VI, Environmental Chemistry, *Federal Register*, 40, 26878–26896, 1975.
4. J.J. Lech and M.J. Vodicnik, *Fundamentals of Aquatic Toxicology*, G.M. Rand and F.R. Petrocelli, Eds., Hemisphere, Washington, D.C., 1985, pp. 526-676.
5. W.B. Jacoby, Ed., *Enzymatic Basis for Detoxification*, Vol. 1 and 2, Academic Press, New York, 1980.

5 Index to Chemicals

Chemical name *O,O*-Dimethylphosphorothioate *O,O*-diester with 4,4′-thiophenol
Chemical Abstract # 3383-96-8
Common name Abate
Synonyms Temephos, Abathion, Difenthos, Nimitox
Uses Mosquito larvicide
Molecular formula $C_{16}H_{20}O_6P_2S_3$
Molecular weight 466.46
Properties m.p. 30.0–30.5°C

Chemical name Abietic acid
Chemical Abstract # 514-10-3
Common name Sylvic acid
Synonyms Abietinic acid
Uses
Molecular formula $C_{20}H_{30}O_2$
Molecular weight 302.44

Chemical name Acetaldehyde
Chemical Abstract # 75-07-0
Common name Ethanal
Synonyms Ethylaldehyde
Uses Perfumes, flavors, plastics
Molecular formula C_2H_4O
Molecular weight 44.1
Properties m.p. –123.5°C, b.p. 20.2°C

Chemical name Acetamide
Chemical Abstract # 60-35-5
Common name Ethanamide
Synonyms Aectic acid amine
Uses Wetting agent, lacquers, general solvent
Molecular formula C_2H_5NO
Molecular weight 59.07
Properties m.p. 81.0°C, b.p. 222.0°C

Chemical name Acetanilide
Chemical Abstract # 103-84-4
Common name Antifebrin
Synonyms *N*-Phenylacetamide
Uses Stabilizer for cellulose ester coatings
Molecular formula C_8H_9NO
Molecular weight 135.16
Properties m.p. 114.0°C, b.p. 305.0°C

Chemical name Acetic acid
Chemical Abstract # 64-19-7
Common name Vinegar acid
Synonyms Ethanoic acid
Uses In chemical manufacturing
Molecular formula $C_2H_4O_2$
Molecular weight 60.05
Properties m.p. 16.7°C, b.p. 118.1°C

Chemical name Acetic anhydride
Chemical Abstract # 108-24-7
Common name Acetic oxide
Synonyms Ethanoic anhydride
Uses
Molecular formula $C_4H_6O_3$
Molecular weight 102.09
Properties m.p. –68.0°C, b.p. 139.9°C

Chemical name Acetone
Chemical Abstract # 67-64-1
Common name DMK
Synonyms 2-Propanone, dimethyl ketone
Uses In manufacturing paints, varnishes, organic chemicals, sealants, and adhesives and as solvent
Molecular formula C_3H_6O
Molecular weight 58.08
Properties m.p. –95.0°C, b.p. 56.2°C

Chemical name	Acetone cyanohydrin
Chemical Abstract #	75-86-5
Common name	Isopropylcyanhydrin
Synonyms	2-Hydroxy-2-methylpropanenitrile
Uses	Insecticides, intermediate in organic synthesis
Molecular formula	C_4H_7NO
Molecular weight	85.10
Properties	m.p. –19.0°C, b.p. 95.0°C

Chemical name	Acetonitrile
Chemical Abstract #	75-05-8
Common name	Methyl cyanide
Synonyms	Ethanenitrile
Uses	Solvent, manufacture of synthetic pharmaceuticals
Molecular formula	C_2H_3N
Molecular weight	41.05
Properties	m.p. –45.0°C, b.p. 81.6°C

Chemical name	Acetophenone
Chemical Abstract #	98-86-2
Common name	Hypnone
Synonyms	Methyl phenyl ketone
Uses	Solvent, perfume, and plasticizer manufacturing
Molecular formula	C_8H_8O
Molecular weight	120.15
Properties	m.p. 20.5°C, b.p. 202.0°C

Chemical name	Acetoxime
Chemical Abstract #	127-06-0
Common name	Acetoneoxime
Synonyms	2-Propanoneoxime
Uses	Solvent, intermediate in organic synthesis
Molecular formula	C_3H_7NO
Molecular weight	73.09
Properties	m.p. 60.0°C, b.p. 136.3°C

Chemical name	Acetylene
Chemical Abstract #	74-86-2
Common name	Ethine
Synonyms	Ethyne
Uses	
Molecular formula	C_2H_2
Molecular weight	26.02
Properties	m.p. –81.8°C, b.p. –84.0°C

Chemical name	Acetylene tetrabromide
Chemical Abstract #	79-27-6
Common name	Sym. tetrabromoethane
Synonyms	1,1,2,2-Tetrabromoethane
Uses	
Molecular formula	$C_2H_2Br_4$
Molecular weight	345.7
Properties	m.p. 0.1°C, b.p. 239.0–242.0°C

Chemical name	4-Acetylmorpholine
Chemical Abstract #	1696-20-4
Common name	
Synonyms	*N*-Acetylmorpholine
Uses	
Molecular formula	
Molecular weight	129.16
Properties	m.p. 14.0°C, b.p. 152.0°C (decomposes)

Chemical name	Acridine
Chemical Abstract #	266-94-6
Common name	
Synonyms	
Uses	Dye manufacturing
Molecular formula	$C_{13}H_9N$
Molecular weight	179.21
Properties	m.p. 108.0°C (sublimes), b.p. 346.0°C

Chemical name	Acrolein
Chemical Abstract #	107-02-8
Common name	Aqualin
Synonyms	Acraldehyde
Uses	Manufacture of colloidal forms of metals; making plastics; in organic synthesis
Molecular formula	C_3H_4O
Molecular weight	56.06
Properties	m.p. –87.7°C, b.p. 52.5°C

Chemical name	Acrylamide
Chemical Abstract #	79-06-1
Common name	
Synonyms	Propenamide
Uses	Polymers or copolymers, adhesives, soil conditioning agents, flocculants
Molecular formula	C_3H_5NO
Molecular weight	71.08
Properties	m.p. 84.0°C

Chemical name	Acrylic acid
Chemical Abstract #	79-10-7
Common name	
Synonyms	Propenoic acid, ethylenecarboxylic acid
Uses	Monomer for polyacrylic and other polyacrylic polymers
Molecular formula	$C_3H_4O_2$
Molecular weight	72.06
Properties	m.p. 14.0°C, b.p. 141.0°C

Chemical name	Acrylonitrile
Chemical Abstract #	107-13-1
Common name	Vinylcyanide
Synonyms	2-Propenenitrile, Acrylon, Fumigrain
Uses	In copolymerization, in manufacture of resins, as fumigant
Molecular formula	C_3H_3N
Molecular weight	53.06
Properties	m.p. –83.55°C, b.p. 77.3°C

Chemical name	Actellic acid
Chemical Abstract #	
Common name	PP-511
Synonyms	Pirimiphosmethyl
Uses	Insecticide and acaricide
Molecular formula	$C_9H_{18}PSO_3$
Molecular weight	237.18
Properties	

Chemical name	Adipic acid
Chemical Abstract #	124-04-9
Common name	
Synonyms	Hexanedioic acid, 1,4-butanedicarboxylic acid
Uses	
Molecular formula	$C_6H_{10}O_4$
Molecular weight	146.14
Properties	m.p. 152.0°C, b.p. 265.0°C (100 mm)

Chemical name	Adiponitrile
Chemical Abstract #	111-69-3
Common name	
Synonyms	1,4-Dicyanobutane, hexanedioic acid dinitrile
Uses	
Molecular formula	$C_6H_8N_2$
Molecular weight	108.15
Properties	m.p. 1.0°C, b.p. 295.0–306.0°C

Chemical name	Aldicarb
Chemical Abstract #	116-06-3
Common name	Ambush, Temik
Synonyms	2-Methyl-2-(methylthio)propionaldehyde *O*-(methylcarbamoyl)-oxime
Uses	Systemic insecticide, acaricide, and nematocide
Molecular formula	$C_7H_{14}N_2O_2S$
Molecular weight	190.25
Properties	m.p. 99.0–100.0°C

Chemical name	Aldrin
Chemical Abstract #	309-00-2
Common name	Aldrex, Drinox, Octalene
Synonyms	1,2,3,4,10,10-Hexachloro-1,4,4a,5,8,8a-hexahydro-1,4-endo,exo-5,8-dimethanonaphthalene
Uses	Insecticide and fumigant
Molecular formula	$C_{12}H_8Cl_6$
Molecular weight	364.93
Properties	m.p. 104.0°C

Chemical name	Alkylbenzenesulfonate (ABS), linear
Chemical Abstract #	42615-29-2
Common name	Teepol 715
Synonyms	
Uses	
Molecular formula	
Molecular weight	
Properties	

Chemical name	*d-trans*-Allethrin
Chemical Abstract #	28434-00-6
Common name	Pyrethroid
Synonyms	(*d,l*-2-Allyl-4-hydroxy-3-methyl-2-cyclopenten-1-one ester of *d-trans*-chrysanthemum monocarboxylic acid
Uses	Insecticide
Molecular formula	
Molecular weight	
Properties	b.p. 160.0°C

Chemical name	Allyl acetate
Chemical Abstract #	591-87-7
Common name	
Synonyms	2-Propenyl ethanoate
Uses	
Molecular formula	$C_5H_8O_2$
Molecular weight	100.11
Properties	b.p. 103.0°C

Chemical name	Allyl alcohol
Chemical Abstract #	107-18-6
Common name	Vinyl carbinol
Synonyms	Propenyl alcohol
Uses	Contact pesticide
Molecular formula	C_3H_6O
Molecular weight	58.08
Properties	m.p. –129.0°C, b.p. 96.9°C

Chemical name	Allylamine
Chemical Abstract #	107-11-9
Common name	Aminopropylene
Synonyms	2-Propenylamine
Uses	In the manufacture of mercurial diuretics
Molecular formula	C_3H_7N
Molecular weight	57.09
Properties	b.p. 55.0–58.0°C

Chemical name Allyl chloride
Chemical Abstract # 107-05-1
Common name Chloroallylene
Synonyms 3-Chloro-1-propene, chloropropylene
Uses In the synthesis of allyl compounds
Molecular formula C_3H_5Cl
Molecular weight 76.53
Properties m.p. –136.0°C, b.p. 44.0–45.0°C

Chemical name Allyl glycidyl ether
Chemical Abstract # 106-92-3
Common name AGE
Synonyms Allyl 2,3-epoxypropylether
Uses Component of epoxy resins
Molecular formula $C_6H_{10}O_2$
Molecular weight 114.15
Properties m.p. –100.0°C (forms glass), b.p. 153.9°C

Chemical name Altosid-SR-10
Chemical Abstract # 73309-75-8
Common name Methoprene, ZR-515
Synonyms Isopropyl(2*E*-4*E*)-11-methoxy-3,7,11-trimethyl-dodecyl-2,4-dienoate
Uses Insect growth control, prevents adult emergence of mosquitos, houseflies, stable flies and blackflies
Molecular formula $C_{19}H_{34}O_3$
Molecular weight 310.48
Properties b.p. 100.0°C (at 0.05 mm)

Chemical name Ametryn
Chemical Abstract # 834-12-8
Common name Gesapax, Evik
Synonyms 6-Ethylamino-4-isopropylamino-2-methylthio-1,3,3-triazine
Uses Herbicide
Molecular formula $C_9H_{17}N_5S$
Molecular weight 227.35
Properties m.p. 88.0–89.0°C

Chemical name Metribuzin
Chemical Abstract # 21087-64-9
Common name Sencor, Bay 94337
Synonyms 4-Amino-6-*t*-butyl-3-methylthio-1,2,4-triazin-5(4*H*)-one
Uses Herbicide
Molecular formula $C_8H_{14}N_4OS$
Molecular weight 214.28
Properties m.p. 125.0–126.5°C

Chemical name Aminocarb
Chemical Abstract # 2032-59-9
Common name Metacil
Synonyms 4-Dimethylamino-*m*-tolylmethylcarbamate
Uses Nonsystemic insecticide, molluscicide
Molecular formula $C_{11}H_{16}N_2O_2$
Molecular weight 208.26
Properties m.p. 93.0°C

Chemical name Chloramben
Chemical Abstract # 133-90-4
Common name Vegiben, Amiben
Synonyms 3-Amino-2,5-dichlorobenzoic acid
Uses Selective pre-emergence herbicide
Molecular formula $C_7H_5Cl_2NO_2$
Molecular weight 206.02
Properties m.p. 200.0–201.0°C

Chemical name *p*-Aminodimethylaniline
Chemical Abstract # 121-69-7
Common name Dimethylaminoaniline
Synonyms Dimethyl-*p*-phenylenediamine
Uses Photodeveloper, in making methylene blue, in tests for acetone, uric acid, phthalic salts, lignin, ozone, etc.
Molecular formula $C_8H_{12}N_2$
Molecular weight 136.19
Properties m.p. 53.0°C, b.p. 262.0°C

Chemical name *o*-Aminophenol
Chemical Abstract # 95-55-6
Common name 2-Hydroxyaniline
Synonyms 2-Amino-1-hydroxybenzene
Uses In making azo and sulfur dyes; dyeing furs and hairs
Molecular formula C_6H_7NO
Molecular weight 109.12
Properties m.p. 170.0–174.0°C, b.p. sublimes

Chemical name *m*-Aminophenol
Chemical Abstract # 591-27-5
Common name 3-Hydroxyaniline
Synonyms 3-Amino-1-hydroxybenzene
Uses Dye intermediate
Molecular formula C_6H_7NO
Molecular weight 109.12
Properties m.p. 122.0–123.0°C

Chemical name *p*-Aminophenol
Chemical Abstract # 123-30-8
Common name Rodinal
Synonyms 4-Amino-1-hydroxybenzene
Uses Photographic developer; intermediate in the manufacture of sulfur and azo dyes; in dyeing furs and feathers
Molecular formula C_6H_7NO
Molecular weight 109.12
Properties m.p. 184.0°C (decomposes)

Chemical name 2-Aminopyridine
Chemical Abstract # 504-29-0
Common name
Synonyms α-Pyridylamine
Uses In the manufacture of antihistaminic drugs
Molecular formula $C_5H_6N_2$
Molecular weight 94.11
Properties m.p. 58.1°C, b.p. 210.6°C

Chemical name 4-Aminopyridine
Chemical Abstract # 504-24-5
Common name
Synonyms 4-Pyridineamine
Uses Intermediate in chemical synthesis
Molecular formula $C_5H_7N_2$
Molecular weight 94.12
Properties m.p. 155.0–158.0°C, b.p. 273.5°C

Chemical name	3-Amino-1,2,4-triazole
Chemical Abstract #	61-82-5
Common name	Amitrol, Amerol, Cytrol, Weedazol
Synonyms	Amizol, Herbozole
Uses	Nonselective post-emergence herbicide, plant growth regulator
Molecular formula	$C_2H_4N_4$
Molecular weight	84.08
Properties	m.p. 153.0–156.0°C

Chemical name	4-Amino-3,5-xylenol
Chemical Abstract #	6623-41-2 (isomer)
Common name	
Synonyms	4-Amino-3,5-dimethylphenol
Uses	
Molecular formula	
Molecular weight	
Properties	

Chemical name	Ammonia
Chemical Abstract #	7664-41-7
Common name	
Synonyms	
Uses	In the manufacture of nitric acid, explosives, synthetic fibers, fertilizers; used in refrigeration
Molecular formula	NH_3
Molecular weight	17.03
Properties	m.p. –77.7°C, b.p. –33.4°C

Chemical name	Ammonium acetate
Chemical Abstract #	631-61-8
Common name	
Synonyms	
Uses	Drugs, textile dyeing, foam rubbers
Molecular formula	$C_2H_7NO_2$
Molecular weight	77.08
Properties	m.p. 114.0°C

Chemical name	Ammonium chloride
Chemical Abstract #	12125-02-9
Common name	Amchlor, Daraammon, Salmiac, Salammoniac
Synonyms	Ammonium muriate
Uses	Dry batteries, electroplating, soldering, as a flux for coating sheet iron with zinc; dyeing and freezing mixtures, in safety explosives, lustering cotton, in washing powders, in cement powders
Molecular formula	NH_4Cl
Molecular weight	53.50
Properties	Sublimes at 350.0°C

Chemical name	Ammonium fluoride
Chemical Abstract #	12125-01-8
Common name	
Synonyms	Neutral ammonium fluoride
Uses	As fluorides, antiseptic in brewing, etching and frosting glass; as antiseptic in brewing beer; in printing and dyeing textiles; as mothproofing agent
Molecular formula	NH_4F
Molecular weight	37.04
Properties	Deliquescent

Chemical name	Ammonium picrate
Chemical Abstract #	131-74-8
Common name	Ammonium picronitrate
Synonyms	Ammonium carbazotate
Uses	Explosives, fireworks, rocket propellants, medicine
Molecular formula	$C_6H_6N_4O_7$
Molecular weight	246.14
Properties	

Chemical name	Ammonium sulfate
Chemical Abstract #	7783-20-2
Common name	Mascagnite
Synonyms	Sulfuric acid diammonium salt
Uses	Fertilizers, tanning, fermentation, food additive
Molecular formula	$H_8N_2O_4S$
Molecular weight	132.14
Properties	m.p. 280.0°C (decomposes)

Chemical name	Ammonium sulfite
Chemical Abstract #	10196-04-0
Common name	
Synonyms	
Uses	Chemical intermediate, photography
Molecular formula	$H_8N_2O_3S$
Molecular weight	116.14
Properties	Sublimes at 150.0°C with decomposition

Chemical name	[1-(4-Amino-2-propyl-5-pyrimidinyl)methyl]-2-picoliniumchloride, HCl
Chemical Abstract #	121-25-5
Common name	Amprolium
Synonyms	1-[(4-Amino-2-propyl-5-pyrimidinyl)methyl]-2-methylpyridinium chloride
Uses	As a coccidiostat
Molecular formula	$C_{14}H_{19}ClN_4$
Molecular weight	278.78
Properties	Decomposes at 248.0–249.0°C

Chemical name	*p*-Amyl acetate
Chemical Abstract #	628-63-7
Common name	Amyl acetic ester
Synonyms	*n*-Amyl acetate, 1-pentanol acetate
Uses	
Molecular formula	$C_7H_{14}O_2$
Molecular weight	130.2
Properties	b.p. 148.0°C

Chemical name	*s*-Acetyl amylalcohol
Chemical Abstract #	6032-29-7
Common name	Methylpropylcarbinol
Synonyms	2-Pentanol, 1-methyl-1-butanol
Uses	
Molecular formula	$C_5H_{12}O$
Molecular weight	88.15
Properties	b.p. 119.0°C

Chemical name	*n*-Amylamine
Chemical Abstract #	110-58-7
Common name	Pentylamine
Synonyms	1-Aminopentane
Uses	
Molecular formula	$C_5H_{13}N$
Molecular weight	87.16
Properties	m.p. –55.0°C, b.p. 104.0°C

Chemical name	Amyl chloride
Chemical Abstract #	543-59-9
Common name	Pentylchloride
Synonyms	1-Chloropentane
Uses	
Molecular formula	$C_5H_{11}Cl$
Molecular weight	106.6
Properties	m.p. –99.0°C, b.p. 108.2°C

Chemical name	Amyl xanthate, Na
Chemical Abstract #	
Common name	
Synonyms	
Uses	Flotation agent
Molecular formula	
Molecular weight	
Properties	Solid, decomposes on heating

Chemical name Aminobenzene
Chemical Abstract # 62-53-3
Common name Aniline
Synonyms Phenylamine
Uses
Molecular formula C_6H_7N
Molecular weight 93.1
Properties m.p. –6.0°C, b.p. 184.0°C

Chemical name Aniline hydrochloride
Chemical Abstract #
Common name
Synonyms Aniline chloride
Uses
Molecular formula C_6H_7N,HCl
Molecular weight
Properties m.p. 189.0°C, b.p. 245.0°C

Chemical name Anthracene
Chemical Abstract # 120-12-7
Common name
Synonyms
Uses In making dyes
Molecular formula $C_{14}H_{10}$
Molecular weight 178.23
Properties m.p. 216.0°C, b.p. 340.0°C

Chemical name *o*-Aminobenzoic acid
Chemical Abstract # 118-92-3
Common name Anthranilic acid
Synonyms
Uses In making dyes, perfumes, pharmaceuticals; cadmium salt as an acaricide in swine
Molecular formula $C_7H_7NO_2$
Molecular weight 137.13
Properties m.p. 145.0°C, b.p. sublimes

Chemical name 9,10-Dihydro-9,10-diketoanthracene
Chemical Abstract # 84-65-1
Common name Anthraquinone
Synonyms
Uses Intermediate for dyes and organics, bird repellent for seeds
Molecular formula $C_{14}H_8O_2$
Molecular weight 208.2
Properties m.p. 286.0°C, b.p. sublimes

Chemical name 1,2-Dicarboxy-3,6-endoxocyclohexane
Chemical Abstract # 145-73-3
Common name Aquathol, Endothall, Hydout, Ripenthol, Desicate
Synonyms 3,6-Endoxohexahydrophthalic acid
Uses Herbicide, defoliant, dessicant, growth regulator
Molecular formula $C_8H_{10}O_5$
Molecular weight 186.16
Properties m.p. 144.0°C

Chemical name Aroclor 1221
Chemical Abstract # 11104-28-2
Common name
Synonyms PCB with 21% Cl
Uses Insecticide, heat transfer and hydraulic fluids
Molecular formula
Molecular weight
Properties

Chemical name Aroclor 1232
Chemical Abstract # 11141-16-5
Common name PCB with 32% Cl
Synonyms
Uses Same as Aroclor 1221
Molecular formula
Molecular weight
Properties

Chemical name Aroclor 1242
Chemical Abstract # 53469-21-9
Common name PCB with 42% Cl
Synonyms
Uses Dielectric liquids, thermostatic fluids, in transmission seals, lubricants, oils and greases, as plasticizer
Molecular formula
Molecular weight
Properties

Chemical name Aroclor 1254
Chemical Abstract # 11097-69-1
Common name PCB with 11% tetra, 49% penta, 34% hexa, and 6% heptachloro congeners
Synonyms
Uses Same as Aroclor 1242
Molecular formula
Molecular weight
Properties

Chemical name Aroclor 1260
Chemical Abstract # 11096-82-5
Common name PCB with 12% penta, 38% hexa, 41% hepta, 8% octa, and 1% nonachlorobiphenyls
Synonyms Clophen A60
Uses Same as Aroclor 1242
Molecular formula
Molecular weight
Properties

Chemical name Aroclor 1262
Chemical Abstract # 37324-23-5
Common name
Synonyms
Uses Same as Aroclor 1242
Molecular formula
Molecular weight
Properties

Chemical name Methyl-4-aminobenzenesulfonyl carbamate
Chemical Abstract #
Common name Asulam, Asulox
Synonyms Methyl[(4-aminophenyl)sulfonyl] carbamate
Uses Herbicide
Molecular formula 40% w/v Methyl-4-aminobenzenesulfonyl carbamate
Molecular weight
Properties m.p. 143.0°C

Chemical name 2-Chloro-4-ethylamino-6-isopropylamino-*s*-triazine
Chemical Abstract # 1912-24-9
Common name Atrazine
Synonyms 6-Chloro-*N*-ethyl-*N'*-(1-methylethyl)-1,3,5-triazine-2,4-diamine
Uses As pre-emergence weed control agent for corn, sugarcane, pineapple, etc.
Molecular formula $C_8H_{14}ClN_5$
Molecular weight 215.68
Properties m.p. 171.0–174.0°C

Chemical name	*O,O*-Diethyl-*S*-(4-oxo-3-*H*-1,2,3-benzotriazine-3-yl)methyldithiophosphate
Chemical Abstract #	2642-71-9
Common name	Azinphosethyl, Ethylguthion, Triazition, Gusthion A
Synonyms	S-3,4-Dihydro-4-oxobenzo(*d*)-(1,2,3)-triazin-3-(methyl) diethylphosphorothiolothionate
Uses	Nonsystemic insecticide and acaricide
Molecular formula	$C_{12}H_{16}N_3O_3PS_2$
Molecular weight	335.36
Properties	m.p. 53.0°C, b.p. 111.0°C at 0.001 mm

Chemical name	*O,O*-Dimethyl-*S*-[(4-oxo-1,2,3-benzotrazin-3(4*H*)-yl)methyl]phosphorothioate
Chemical Abstract #	86-50-0
Common name	Azinphosmethyl, Guthion, Gusation M
Synonyms	*S*-3,4-Dihydro-4-oxobenzo(*d*)-(1,2,3)(trazin-3-ylmethyl) dimethylphosphorodithioate
Uses	Nonsystemic insectide and acaricide of long persistence
Molecular formula	$C_{10}H_{12}N_3O_3PS_2$
Molecular weight	317.34
Properties	m.p. approx. 73.0°C

Chemical name	Diphenyldiimide
Chemical Abstract #	103-33-3
Common name	Azobenzene
Synonyms	Benzeneazobenzene
Uses	In the manufacture of dyes, as fumigant and acaricide
Molecular formula	$C_{12}H_{10}N_2$
Molecular weight	182.23
Properties	m.p. 68.3°, b.p. 297.0°C

Chemical name	Dimethylphosphate of 3-hydroxy-*N*-methyl-*cis*-crotonamide-*O,O*-dimethyl-*O*-(2-methylcarbamoyl-1-methylvinyl)phosphate
Chemical Abstract #	6923-22-4
Common name	Azodrin
Synonyms	Monocrotophos
Uses	Systemic insecticide, acaricide
Molecular formula	$C_7H_{14}O_5NP$
Molecular weight	223.16
Properties	

Chemical name	*N*-Butyl-*N*-ethyl-2,6-dinitro-4-trifluoromethylaniline
Chemical Abstract #	1861-40-1
Common name	Balan, Quilan, Benfluralin, Bethrodine
Synonyms	*N*-Butyl-*N*-ethyl-α,α,α-trifluoro-2,6-dinitro-*p*-toluidine
Uses	Herbicide
Molecular formula	$C_{13}H_{16}F_3N_3O_4$
Molecular weight	335.29
Properties	m.p. 66.5°C

Chemical name	2-(1-Methylethoxy)phenol methyl carbamate
Chemical Abstract #	114-26-1
Common name	Baygon, Propoxur, Arprocarb, Suncide, Blattanex
Synonyms	*o*-Isopropylphenyl *N*-methyl carbamate
Uses	Insecticide
Molecular formula	$C_{11}H_{15}NO_3$
Molecular weight	209.24
Properties	m.p. 91.5°C

Chemical name	*O,O*-Dimethyl-*O*-(3-methyl-4(methylthio)phenyl) phosphorothioate
Chemical Abstract #	55-38-9
Common name	Baytex, Fenthion, Mercaptophos, Entex, Baycid, Tiguvon
Synonyms	*O,O*-Dimethyl-*O*-(4-methylmercapto-3-methyl-phenyl) thionophosphate
Uses	Systemic and contact herbicide
Molecular formula	$C_{10}H_{15}O_3PS_2$
Molecular weight	278.34
Properties	b.p. 87.0°C at 0.01 mm

Chemical name	Methyl 1-(butylcarbamoyl)-2-benzinimidazole-carbamate
Chemical Abstract #	17804-35-2
Common name	Benomyl, Benlate, Tersan 1991
Synonyms	1-(Butylcarbamoyl)-2-benzimidazolecarbamic acid, methyl ester
Uses	Systemic fungicide for a broad spectrum of phytopathogenic fungi, insecticide
Molecular formula	$C_{14}H_{18}N_4O_3$
Molecular weight	290.32
Properties	

Chemical name	*N*-(2-Ethylthio)benzene sulfonamide *S,O,O*-diiso-propylphosphorodithioate
Chemical Abstract #	
Common name	Bensulide, Betasan, Prefar, Exporsan
Synonyms	S-(*O,O*-Diisopropylphosphorothioate) of *N*-(2-mercaptoethyl)benzenesulfonamide
Uses	Selective pre-emergence herbicide
Molecular formula	
Molecular weight	
Properties	

Chemical name	Benz(c)acridine
Chemical Abstract #	221-51-4
Common name	Chrysidene
Synonyms	3,4-Benzacridine
Uses	
Molecular formula	
Molecular weight	229.0
Properties	

Chemical name	Benzaldehyde
Chemical Abstract #	100-52-7
Common name	Oil of bitter almonds
Synonyms	Benzene carbonal
Uses	In manufacturing dyes, perfumeries, as solvent, and in flavors
Molecular formula	C_7H_6O
Molecular weight	106.12
Properties	m.p. –26.0°C, b.p. 179.0°C

Chemical name	Benzene
Chemical Abstract #	71-43-2
Common name	Coal naphtha
Synonyms	Cyclohexatriene
Uses	In the manufacture of styrene, phenol, detergents, pesticides, plastics and resins, flavors and perfumes, paints and coatings, pharmaceuticals and photographic chemicals
Molecular formula	C_6H_6
Molecular weight	78.12
Properties	m.p. 5.51°C, b.p. 80.1°C

Chemical name Benzenesulfonic acid
Chemical Abstract # 98-11-3
Common name
Synonyms Phenylsulfonic acid
Uses
Molecular formula $C_6H_6O_3S$
Molecular weight 158.17
Properties m.p. 1.5°C,with H_2O at 43.0–44.0°C, b.p. decomposes

Chemical name Benzethonium chloride
Chemical Abstract # 5929-09-09
Common name Hyamine 1622
Synonyms *N,N*-Dimethyl-*N*-[2-[2-[4-(1,1,3,3-tetramethylbutyl) phenoxy]ethoxy]ethyl] benzene-methanaminium chloride
Uses Cationic detergent, antiseptic
Molecular formula $C_{27}H_{42}ClNO_2$
Molecular weight 448.10
Properties m.p. 164.0–166.0°C

Chemical name Benzidine
Chemical Abstract # 92-87-5
Common name
Synonyms *p,p′*-Bianiline,4,4′-diaminobiphenyl
Uses In manufacturing dyes, in the synthesis of organic chemicals
Molecular formula $C_{12}H_{12}N_2$
Molecular weight 184.23
Properties m.p. 116.0–128.7°C, b.p. 401.7°C

Chemical name Benzoic acid
Chemical Abstract # 65-85-0
Common name Dracylic acid
Synonyms Phenylformic acid
Uses In the manufacture of dyes, pharmaceuticals, cosmetics, alkyl resins, phenol and plasticizer, as food preservative
Molecular formula $C_7H_6O_2$
Molecular weight 122.1
Properties m.p. 121.7°C, b.p. 249.0°C

Chemical name Benzonitrile
Chemical Abstract # 100-47-0
Common name Cyanobenzene
Synonyms Phenylcyanide, Benzoic acid nitrile
Uses
Molecular formula C_7H_5N
Molecular weight 103.13
Properties m.p. –13.0°C, b.p. 190.7°C

Chemical name Benzo(a)pyrene
Chemical Abstract # 50-32-8
Common name B(a)P
Synonyms 3,4-Benzopyrene
Uses
Molecular formula $C_{20}H_{12}$
Molecular weight 252.30
Properties m.p. 179.0°C, b.p. 312.0°C

Chemical name Benzo(e)pyrene
Chemical Abstract # 192-97-2
Common name B(e)P
Synonyms 1,2-Benzopyrene
Uses
Molecular formula $C_{20}H_{12}$
Molecular weight 252.30
Properties m.p. 230.0°C

Chemical name	*p*-Benzoquinone
Chemical Abstract #	106-51-4
Common name	Quinone
Synonyms	2,5-Cyclohexadiene-1,4-dione
Uses	In the manufacture of dyes and hydroquinone
Molecular formula	$C_6H_4O_2$
Molecular weight	108.1
Properties	m.p. 115.7°C, b.p. sublimes

Chemical name	Benzoic trichloride
Chemical Abstract #	98-07-7
Common name	Phenylchloroform
Synonyms	Benzyltrichloride, phenylchloroform
Uses	In dye chemistry, organic syntheses
Molecular formula	$C_7H_5Cl_3$
Molecular weight	195.47
Properties	m.p. –5.0°C, b.p. 221.0°C

Chemical name	Benzoyl chloride
Chemical Abstract #	98-88-4
Common name	Benzenecarbonyl chloride
Synonyms	α-chlorobenzaldehyde
Uses	In organic syntheses
Molecular formula	C_7H_5ClO
Molecular weight	140.57
Properties	m.p. –1.0°C, b.p. 197.2°C

Chemical name	Benzyl alcohol
Chemical Abstract #	100-51-6
Common name	Benzene carbinol
Synonyms	Benzene methanol, hydroxytoluene, phenyl methanol
Uses	In making perfumes and flavors, inks, as solvent and surfactant
Molecular formula	C_7H_8O
Molecular weight	108.15
Properties	m.p. 115.3°C, b.p. 205.7°C

Chemical name	*o*-Aminotoluene
Chemical Abstract #	100-46-9
Common name	Benzylamine, moringine
Synonyms	Phenylmethylamine
Uses	Chemical intermediate for dyes, pharmaceuticals, polymers
Molecular formula	C_7H_9N
Molecular weight	107.2
Properties	b.p. 185.0°C

Chemical name	α-Chlorotoluene
Chemical Abstract #	100-44-7
Common name	Benzyl chloride
Synonyms	α-chlorotoluol, tolyl chloride
Uses	In the manufacture of dyes
Molecular formula	C_7H_7Cl
Molecular weight	126.59
Properties	m.p. –43.0°C, b.p. 179.0°C

Chemical name	2-(1-Methyl-*n*-propyl)-4,6-dinitrophenyl-2-methyl-crotonate
Chemical Abstract #	485-31-4
Common name	Binapacryl, Ambox, Endosan, Dapacril, Morrocid
Synonyms	2(2-Butyl-4,6-dinitrophenyl)-3,3-dimethylacrylate
Uses	As contact miticide, fungicide
Molecular formula	$C_{15}H_{18}N_2O_6$
Molecular weight	322.35
Properties	m.p. 65.0–69.0°C

Chemical name 2-(4-Chloro-6-ethylamino-*s*-triazine-2-yl-amino)-2-methyl propionitrile
Chemical Abstract # 21725-46-2
Common name Bladex, Cyanazine, Fortrol, Payze
Synonyms 2-Chloro-4-(1-cyano-1-methylethylamino)-6-ethylamino-1,3,5-triazine
Uses As herbicide
Molecular formula $C_9H_{13}ClN_6$
Molecular weight 240.73
Properties m.p. 167.0°C

Chemical name 5-Bromo-3-*sec*-butyl-6-methyluracil
Chemical Abstract # 314-40-9
Common name Bromacil, Hyvar, Uragon, Nalkil, Weedkiller
Synonyms 5-Bromo-6-methyl-3-(1-methylpropyl)-2,4-(1*H*,3*H*)-pyrimidine dione
Uses As herbicide for a wide range of grasses and broadleaf weeds
Molecular formula $C_9H_{13}BrN_2O_2$
Molecular weight 261.15
Properties m.p. 158.0–159.0°C

Chemical name Tribromomethane
Chemical Abstract # 75-25-2
Common name Bromoform
Synonyms Methenyl tribromide
Uses In manufacturing pharmaceuticals, fire-resistant preparations, gauge fluids, and as solvent for waxes, greases and oils.
Molecular formula $CHBr_3$
Molecular weight 252.75
Properties m.p. 6–7°C, b.p. 149.5°C

Chemical name *o*-Bromophenol
Chemical Abstract # 95-56-7
Common name
Synonyms 2-Bromophenol
Uses
Molecular formula C_6H_5BrO
Molecular weight 173.02
Properties m.p. 5.6°C, b.p. 195°C

Chemical name *m*-Bromophenol
Chemical Abstract # 591-20-8
Common name
Synonyms 3-Bromophenol
Uses
Molecular formula C_6H_5BrO
Molecular weight 173.02
Properties m.p. 33°C, b.p. 236°C

Chemical name *p*-Bromophenol
Chemical Abstract # 106-41-2
Common name
Synonyms 4-Bromophenol
Uses As disinfectant
Molecular formula C_6H_5BrO
Molecular weight 173.02
Properties m.p. 63.5°C, b.p. 238°C

Chemical name Dimethoxystrychnine
Chemical Abstract # 357-57-3
Common name Brucine
Synonyms 10,11-Dimethoxystrychnine
Uses In medicine, in denaturing alcohol, as additive in lubricants
Molecular formula $C_{23}H_{26}N_2O_4$
Molecular weight 394.51
Properties m.p. 178°C

Chemical name	1,3-Butadiene
Chemical Abstract #	106-99-0
Common name	Divinyl, pyrrolylene
Synonyms	Vinylethylene
Uses	In making styrene-butadiene rubber, in latex paints, resins, and as organic intermediate
Molecular formula	C_4H_6
Molecular weight	54.09
Properties	m.p. –108.9°C, b.p. –4.41°C

Chemical name	*n*-Butyl alcohol
Chemical Abstract #	71-36-3
Common name	*n*-Butanol
Synonyms	Propyl methanol, 1-hydroxybutane
Uses	Solvent for fats, waxes, resins, shellac, etc.; in manufacture of rayons, detergents, and lacquers
Molecular formula	$C_4H_{10}O$
Molecular weight	74.12
Properties	m.p. –89.9°C, b.p. 117.7°C

Chemical name	*sec*-Butyl alcohol
Chemical Abstract #	78-92-2
Common name	*sec*-Butanol, 2-butanol
Synonyms	2-Butyl alcohol
Uses	In the syntheses of flotation agents, flavors, perfumes, and wetting agents; industrial cleaners and paint removers; as solvent
Molecular formula	$C_4H_{10}O$
Molecular weight	74.12
Properties	m.p. –89°C, b.p. 99.5°C

Chemical name	*tert*-Butyl alcohol
Chemical Abstract #	75-65-0
Common name	*t*-Butanol
Synonyms	2-Methyl 2-propanol, trimethyl carbinol
Uses	As blending agent for gasoline to increase octane rating, in making flotation agents, perfumes
Molecular formula	$C_4H_{10}O$
Molecular weight	74.14
Properties	m.p. 25.3°C, b.p. 82.8°C

Chemical name	2-Butanone oxime
Chemical Abstract #	96-29-7
Common name	Ethylmethyl ketoxime
Synonyms	Methylethyl ketoxime
Uses	
Molecular formula	C_4H_9ON
Molecular weight	87.12
Properties	m.p. –29.5°C, b.p. 152.0°C

Chemical name	*n*-Butyl acetate
Chemical Abstract #	123-86-4
Common name	
Synonyms	Butyl ethanoate
Uses	As solvent in production of lacquers, perfumes, gums, and synthetic resins
Molecular formula	$C_6H_{12}O_2$
Molecular weight	116.18
Properties	m.p. –76.8°C, b.p. 126.0°C

Chemical name	*tert*-Butyl acetate
Chemical Abstract #	540-88-5
Common name	Texaco lead appreciator
Synonyms	Acetic acid 1,1-dimethyl ethyl ester
Uses	
Molecular formula	$C_6H_{12}O_2$
Molecular weight	116.18
Properties	b.p. 96.0°C

Chemical name	*n*-Butyl acrylate
Chemical Abstract #	141-32-2
Common name	Butyl acrylate
Synonyms	Butyl-2-propionate
Uses	As a monomer in the manufacture of polymers and resins, in paint formulations
Molecular formula	$C_7H_{12}O_2$
Molecular weight	128.19
Properties	m.p. –64.4°C, b.p. 145.0°C

Chemical name	*n*-Butyl amine
Chemical Abstract #	109-73-9
Common name	Norvalamine
Synonyms	1-Aminobutane
Uses	As intermediate for emulsifying agents, pharmaceuticals, insecticides, dyes, tanning agents
Molecular formula	$C_4H_{11}N$
Molecular weight	73.16
Properties	m.p. –50.0°C, b.p. 77.0°C

Chemical name	*sec*-Butyl amine
Chemical Abstract #	13952-84-6
Common name	Butafume
Synonyms	2-Aminobutane, 1-methylpropylamine, Tutane
Uses	As fungistat
Molecular formula	$C_4H_{11}N$
Molecular weight	73.16
Properties	m.p. –104.0°C, b.p. 63.0°C

Chemical name	*p-tert*-Butylbenzoic acid
Chemical Abstract #	98-73-7
Common name	PTBBA
Synonyms	
Uses	
Molecular formula	$C_{11}H_{14}O_2$
Molecular weight	178.25
Properties	m.p. 166.3°C

Chemical name	Benzylbutyl phthalate
Chemical Abstract #	85-68-7
Common name	Saniticizer 160, Sicol 160, Palatinol BB
Synonyms	1,2-Benzenedicarboxylic acid, butylphenylmethyl ester
Uses	Plasticizer for synthetic resins, mainly PVC
Molecular formula	$C_{19}H_{20}O_4$
Molecular weight	312.39
Properties	m.p. < – 35.0°C, b.p. 370.0°C

Chemical name	2-Butoxyethanol
Chemical Abstract #	111-76-2
Common name	Butylcellosolve, butylglycol
Synonyms	Ethylene glycolmono-*n*-butyl ether
Uses	Solvent for nitrocellulose, resins, gums, oil, albumin; in dry cleaning
Molecular formula	$C_6H_{14}O_2$
Molecular weight	118.17
Properties	m.p. < – 40.0°C, b.p. 170.0°C

Chemical name	*n*-Butyl chloride
Chemical Abstract #	109-69-3
Common name	
Synonyms	1-Chlorobutane
Uses	As butylating agent
Molecular formula	C_4H_9Cl
Molecular weight	92.58
Properties	m.p. –123.1°C, b.p. 78.0°C

Chemical name	2-*tert*-Butyl-4,6-dinitrophenyl acetate
Chemical Abstract #	3204-27-1
Common name	Dinoterb acetate
Synonyms	2-(1,1-Dimethylethyl)-4,6-dinitrophenol acetate
Uses	Herbicide
Molecular formula	$C_{12}H_{14}N_2O_6$
Molecular weight	282.28
Properties	m.p. 133.0–134.5°C

Chemical name	*n*-Butyl ether
Chemical Abstract #	142-96-1
Common name	Di-*n*-butyl ether
Synonyms	1-Butoxybutane
Uses	As solvent for hydrocarbons, fatty materials, as extracting agent
Molecular formula	$C_8H_{18}O$
Molecular weight	130.2
Properties	m.p. –95.0°C, b.p. 141.0°C

Chemical name	*n*-Butyl formate
Chemical Abstract #	592-84-7
Common name	
Synonyms	
Uses	
Molecular formula	$C_5H_{10}O_2$
Molecular weight	102.13
Properties	m.p. –90.0°C, b.p. 106.8°C

Chemical name	*n*-Butyl glycidyl ether
Chemical Abstract #	2426-08-6
Common name	Ageflex BGE,ERL 0810,2,3-epoxypropyl butyl ether
Synonyms	3-Butoxy-1,2-epoxypropane
Uses	Component of epoxy resins
Molecular formula	$C_7H_{14}O_2$
Molecular weight	130.21
Properties	b.p. 164.0–168.0°C

Chemical name	*n*-Butyl mercaptan
Chemical Abstract #	109-79-5
Common name	
Synonyms	Butanethiol
Uses	
Molecular formula	C_4H_9SH
Molecular weight	90.18
Properties	m.p. –116.°C, b.p. 98.0°C

Chemical name	*p*-tert-Butylphenol
Chemical Abstract #	98-54-4
Common name	
Synonyms	4-(α,α-Dimethylethylphenol)
Uses	
Molecular formula	$C_{10}H_{14}O$
Molecular weight	150.24
Properties	m.p. 99.0°C, b.p. 236.0°C

Chemical name	*p*-*tert*-Butyltoluene
Chemical Abstract #	98-51-1
Common name	PTBT
Synonyms	1-Methyl-4-*tert*-butylbenzene
Uses	
Molecular formula	$C_{11}H_{16}$
Molecular weight	148.27
Properties	m.p. –62.53°C, b.p. 192.8°C

Chemical name	Butyraldehyde
Chemical Abstract #	123-72-8
Common name	Butyric aldehyde
Synonyms	1-Butanal
Uses	In the manufacture of rubber accelerators, synthetic resins, solvents and plasticizers
Molecular formula	C_4H_8O
Molecular weight	72.1
Properties	m.p. –97.0 to –99.0°C, b.p. 75.0°C

Chemical name *n*-Butyric acid
Chemical Abstract # 107-92-6
Common name Ethylacetic acid
Synonyms *n*-Butanoic acid
Uses In the manufacture of esters that are used as artificial flavoring agents in liquors, candies, etc.; as decalcifier of hides
Molecular formula $C_4H_8O_2$
Molecular weight 88.1
Properties m.p. –6.0 to –8.0°C, b.p. 163.7°C

Chemical name Butyronitrile
Chemical Abstract # 109-74-0
Common name Cyanopropane
Synonyms *n*-Propyl cyanide
Uses
Molecular formula C_4H_7N
Molecular weight 69.1
Properties m.p. –112.0°C, b.p. 118.0°C

Chemical name Cacodylic acid
Chemical Abstract # 75-60-5
Common name Phytar, Ansar
Synonyms Hydroxydimethylarsine oxide
Uses As contact herbicide, cotton defoliant
Molecular formula $C_2H_7O_2As$
Molecular weight 137.99
Properties m.p. 195.0–196.0°C

Chemical name Methyltheobromine
Chemical Abstract # 58-08-2
Common name Caffeine
Synonyms 1,3,7-Trimethylxanthine
Uses Beverages, medicine
Molecular formula $C_8H_{10}N_4O_2$
Molecular weight 194.22
Properties m.p. 236.8°C

Chemical name 2-Camphanone
Chemical Abstract # 76-22-2
Common name Camphor
Synonyms
Uses As odorant/flavorant, in pharmaceutical industry, as plasticizer for cellulose esters and ethers, insect repellant, incense manufacturing, lacquers and varnishes, explosives and plastics manufacturing
Molecular formula $C_{10}H_{16}O$
Molecular weight 152.26
Properties m.p. 180.0°C, b.p. 204.0°C

Chemical name *n*-Hexanoic acid
Chemical Abstract # 142-62-1
Common name Caproic acid
Synonyms Hexanoic acid
Uses In the manufacture of artificial flavoring agents
Molecular formula $C_6H_{12}O_2$
Molecular weight 116.2
Properties m.p. –6.0°C, b.p. 204.0–208.0°C

Chemical name Cyclohexanone isooxime
Chemical Abstract # 105-60-2
Common name α-Caprolactam
Synonyms 2-Oxohexamethylenimine, 6-aminohexanoic acid cyclic lactam
Uses In manufacturing nylon, plastics, bristles, film, synthetic leather, plasticizers, as curing agent for polyurethanes
Molecular formula $C_6H_{11}NO$
Molecular weight 113.18
Properties m.p. 69.0°C, b.p. 139.0°C

Chemical name *N*-Trichloromethylthioterahydrophthalimide
Chemical Abstract # 133-06-2
Common name Captan, Merpan, Orthocide, Vanicide, Americide
Synonyms 1,2,3,6-Tetrahydro-*N*-(trichloromethylthio)phthalimide
Uses Protectant-eradicant fungicide, bacteriostat in soap
Molecular formula $C_9H_8Cl_3NO_2S$
Molecular weight 300.59
Properties m.p. 175.0°C

Chemical name 1-Naphthyl *N*-methylcarbamate
Chemical Abstract # 63-25-2
Common name Carbaryl, Sevin, Carbatox-60, Crag Sevin
Synonyms Methylcarbamate 1-naphthalenol
Uses Contact insecticide
Molecular formula $C_{12}H_{11}NO_2$
Molecular weight 201.24
Properties m.p. 142.0°C

Chemical name 2,3-Dihydro-2,2-dimethylbenzofuranyl 7-*N*-methyl-carbamate
Chemical Abstract # 1563-66-2
Common name Carbofuran, Furadan, Yaltox, Niagra 10242
Synonyms 2,2-Dimethyl-7-coumaranyl *N*-methylcarbamate
Uses Systemic insecticide to control corn rootworms
Molecular formula $C_{12}H_{15}NO_3$
Molecular weight 221.26
Properties m.p. 150.0–152°C

Chemical name Carbon disulfide
Chemical Abstract # 75-15-0
Common name
Synonyms Dithio carbonic anhydride
Uses In manufacture of rayon, carbon tetrachloride, cellophane, soil disinfectants, grain fumigants, soil conditioners, herbicides, and as solvent for chemicals
Molecular formula CS_2
Molecular weight 76.14
Properties m.p. –108.6 to –117.0°C, b.p. 46.3°C

Chemical name Carbon tetrachloride
Chemical Abstract # 56-23-5
Common name Tetrachloromethane
Synonyms
Uses In dry cleaning operations, in the manufacture of fire extinguishers, refrigerants (chlorofluoromethane), and propellants, and as solvent
Molecular formula CCl_4
Molecular weight 153.82
Properties m.p. –23.0°C, b.p. 76.7°C

Chemical name S-[(*p*-Chlorophenylthio)methyl] *O*, *O*-diethyl phosphorodithioate
Chemical Abstract # 786-19-6
Common name Carbophenothion, trithion
Synonyms *S*-(4-Cholorophenylthiomethyl) diethyl phosphorothiolothionate
Uses Insecticide, acaricide, and miticide
Molecular formula $C_{11}H_{16}ClO_2PS_3$
Molecular weight 342.85
Properties b.p. 82.0°C at 0.01 mm

Chemical name 1,2-Dihydroxybenzene
Chemical Abstract # 120-80-9
Common name Catechol
Synonyms 1,2-Benzenediol, pyrocatechol
Uses In making dyes, pharmaceuticals, anti-oxidants for rubber and lubricating oils, in fur dyeing, and used in specialty inks
Molecular formula $C_6H_6O_2$
Molecular weight 110.12
Properties m.p. 105.0°C, b.p. 246.0°C

Chemical name Trichloroacetaldehyde
Chemical Abstract # 302-17-0
Common name Chloral
Synonyms Trichloroethanal
Uses In the manufacture of DDT and used in organic synthesis
Molecular formula C_2HCl_3O
Molecular weight 147.7
Properties m.p. –57.5°C, b.p. 98.0°C

Chemical name Tetrachloro-*p*-benzoquinone
Chemical Abstract # 118-75-2
Common name Chloranil, spergon
Synonyms 2,3,5,6-Tetrachloro-2,5-cyclohexadiene 1,4-dione
Uses Agricultural fungicide, vulcanizing agent, dye intermediate
Molecular formula $C_6Cl_4O_2$
Molecular weight 245.89
Properties m.p. 290.0°C

Chemical name Chlordane
Chemical Abstract # 57-74-9
Common name Octaklor, Dowklor, Velsicol 1068, Topiclor
Synonyms 1,2,4,5,6,7,8,8-Octachloro-4,7-methano-3a,4,7,7a-tetrahydroindane
Uses Nonsystemic insecticide
Molecular formula $C_{10}H_6Cl_8$
Molecular weight 409.76
Properties b.p. 175.0°C

Chemical name Chlorfenvinphos
Chemical Abstract # 470-90-6
Common name Birlane, Apachlor, Supona, Unitox, Vinylphate
Synonyms 2-Chloro-1(2,4-dichlorophenyl)vinyl diethyl-phosphate
Uses Insecticide and acaricide
Molecular formula $C_{12}H_{14}Cl_3O_4P$
Molecular weight 359.56
Properties m.p. –16.0 to –22.0°C, b.p. 168–170.0°C (at 0.5 mm)

Chemical name *S*-Chloromethyl-*O*,*O*-diethylphosphorothiolo-thionate
Chemical Abstract # 24934-91-6
Common name Chlormephos, Dotan
Synonyms *S*-(Chloromethyl)-*O*,*O*-dimethylphosphorodithioic acid,ester
Uses Insecticide
Molecular formula $C_5H_{12}ClO_2PS_2$
Molecular weight 234.7
Properties b.p. 81.0–85.0°C

Chemical name	*p*-Chloraniline
Chemical Abstract #	106-47-8
Common name	
Synonyms	4-Chlorophenylamine
Uses	In making pharmaceuticals, agricultural chemicals, and dyes
Molecular formula	C_6H_6ClN
Molecular weight	127.58
Properties	m.p. 72.5°C, b.p. 232.0°C

Chemical name	*o*-Chlorobenzaldehyde
Chemical Abstract #	89-98-5
Common name	
Synonyms	
Uses	
Molecular formula	C_7H_5ClO
Molecular weight	140.57
Properties	m.p. 11.0°C, b.p. 209.0–215.0°C

Chemical name	Chlorobenzene
Chemical Abstract #	108-90-7
Common name	
Synonyms	Phenylchloride
Uses	In the manufacture of aniline, phenol, insecticide
Molecular formula	C_6H_5Cl
Molecular weight	112.56
Properties	m.p. –45.0°C, b.p. 132.0°C

Chemical name	*p*-Chlorobenzene sulfonic acid
Chemical Abstract #	5138-90-9
Common name	
Synonyms	
Uses	
Molecular formula	$C_6H_5ClSO_3$
Molecular weight	192.62
Properties	m.p. 68.0°C, b.p. 147.0°C

Chemical name	*o*-Chlorobenzoic acid
Chemical Abstract #	118-91-2
Common name	
Synonyms	
Uses	Preservative for glues and paints
Molecular formula	$C_7H_5ClO_2$
Molecular weight	156.57
Properties	m.p. 142.0°C, b.p. sublimes

Chemical name	*m*-Chlorobenzoic acid
Chemical Abstract #	535-80-8
Common name	
Synonyms	
Uses	
Molecular formula	$C_7H_5ClO_2$
Molecular weight	156.57
Properties	m.p. 158.0°C, b.p. sublimes

Chemical name	*o*-Chlorobenzylidene malonitrile
Chemical Abstract #	2698-41-1
Common name	OCBM
Synonyms	
Uses	
Molecular formula	$C_{10}H_5ClN_2$
Molecular weight	188.62
Properties	m.p. 95.0°C, b.p. 310–315.0°C

Chemical name	2-Chloro-4,6-bis(ethylamino)-*s*-triazine
Chemical Abstract #	122-34-9
Common name	Simazine, Princep
Synonyms	6-Chloro-*N,N'*-diethyl-1,3,5-triazine-2,4-diamine
Uses	As herbicide
Molecular formula	$C_7H_{12}ClN_5$
Molecular weight	201.67
Properties	m.p. 226.0–227.0°C

Chemical name 4-Chloro-*o*-cresol
Chemical Abstract # 1570-64-5
Common name
Synonyms
Uses Microbial metabolite of MCPA
Molecular formula C_7H_7ClO
Molecular weight 142.6
Properties

Chemical name 4-Chloro-*m*-cresol
Chemical Abstract # 59-50-7
Common name PCMC
Synonyms 4-Chloro-3-hydroxytoluene
Uses External germicide, preservative for glues, gums, inks, leather and textile goods
Molecular formula C_7H_7ClO
Molecular weight 142.6
Properties m.p. 66.0°C, b.p. 235.0°C

Chemical name 2-Chloroethanol
Chemical Abstract # 107-07-3
Common name Ethylenechlorhydrin
Synonyms Glycolchlorhydrin
Uses As solvent; in the manufacture of insecticides; for treating sweet potatoes before planting
Molecular formula C_2H_5ClO
Molecular weight 80.52
Properties m.p. 67.5–71°C, b.p. 128.9°C

Chemical name Trichloromethane
Chemical Abstract # 67-66-3
Common name Chloroform
Synonyms
Uses In the manufacture of fluorocarbon refrigerants and propellants, anesthetics and pharmaceuticals, as fumigant, solvent, and as insecticide
Molecular formula $CHCl_3$
Molecular weight 119.38
Properties m.p. –64.0°C, b.p. 62.0°C

Chemical name 3-Chloro-4-methylbenzenamine hydrochloride
Chemical Abstract #
Common name DRC-1339
Synonyms Starlicide
Uses As avian toxicant in cattle and poultry industries
Molecular formula
Molecular weight
Properties

Chemical name Chloromethyl methyl ether
Chemical Abstract # 107-30-2
Common name Chloromethyl ether
Synonyms Chloromethoxymethane
Uses In manufacturing lacrymators (irritant gases)
Molecular formula C_2H_5ClO
Molecular weight 80.52
Properties m.p. –103.5°C, b.p. 59.5°C

Chemical name *o*-Chloronitrobenzene
Chemical Abstract # 25167-93-5
Common name
Synonyms *o*-Nitrochlorobenzene
Uses Used in dye chemistry
Molecular formula $C_6H_4ClNO_2$
Molecular weight 157.56
Properties m.p. 32.5°C, b.p. 245.7°C

Chemical name *m*-Chloronitrobenzene
Chemical Abstract # 121-73-3
Common name
Synonyms *m*-Nitrochlorobenzene
Uses In manufacturing dyes, as intermediate in organic chemicals synthesis
Molecular formula $C_6H_4ClNO_2$
Molecular weight 157.56
Properties m.p. 46.0°C, b.p.236.0°C

Chemical name 1-Chloro-1-nitropropane
Chemical Abstract # 2425-66-3
Common name
Synonyms
Uses
Molecular formula $C_3H_6ClNO_2$
Molecular weight 123.5
Properties b.p. 139.0–143.0°C

Chemical name *o*-Chlorophenol
Chemical Abstract # 95-57-8
Common name
Synonyms 1-Chloro-2-hydroxybenzene
Uses Used in organic synthesis
Molecular formula C_6H_5ClO
Molecular weight 128.56
Properties m.p. 7.0°C, b.p. 175.6°C

Chemical name *m*-Chlorophenol
Chemical Abstract # 108-43-0
Common name
Synonyms 1-Chloro-3-hydroxybenzene
Uses
Molecular formula C_6H_5ClO
Molecular weight 128.56
Properties m.p. 32.8°C, b.p. 214.0°C

Chemical name *p*-Chlorophenol
Chemical Abstract # 106-48-9
Common name
Synonyms 1-Chloro-4-hydroxybenzene
Uses
Molecular formula C_6H_5ClO
Molecular weight 128.56
Properties m.p. 43.0°C, b.p. 217.0°C

Chemical name 4-Chlorophenoxyacetic acid
Chemical Abstract # 122-88-3
Common name 4-CPA
Synonyms
Uses
Molecular formula $C_8H_7ClO_3$
Molecular weight 186.59
Properties m.p. 157.0–159.0°C

Chemical name 3-(*p*-Chlorophenyl)-1,1-dimethylurea
Chemical Abstract # 150-68-5
Common name Monuron, Telvar, Chlorfenidim
Synonyms (*N'*-(4-Chlorophenyl)-*N,N*-dimethylurea
Uses As herbicide, sugarcane flowering suppressant
Molecular formula $C_9H_{11}ClN_2O$
Molecular weight 198.67
Properties m.p. 171.0°C

Chemical name Trichloronitromethane
Chemical Abstract # 76-06-2
Common name Chloropicrin
Synonyms Nitrochloroform
Uses In fumigants, insecticides, fungicides, rat poisons, organic synthesis, in making dyes
Molecular formula CCl_3NO_2
Molecular weight 164.39
Properties m.p. –64.0°C, b.p. 112.0°C

Chemical name	2-Chloropropane
Chemical Abstract #	75-29-6
Common name	
Synonyms	Isopropylchloride
Uses	
Molecular formula	C_3H_7Cl
Molecular weight	78.54
Properties	m.p. –117.0°C, b.p. 36.5°C

Chemical name	Tetrachloroisophthalonitrile
Chemical Abstract #	1897-45-6
Common name	Chlorothalonil, Bravo
Synonyms	2,4,5,6-Tetrachloro-1,3-dicyanobenzene
Uses	As broad-spectrum fungicide
Molecular formula	$C_8Cl_4N_2$
Molecular weight	265.89
Properties	m.p. 250.0°C, b.p. 350.0°C

Chemical name	*O,O*-dimethyl-*O*-(3-chloro-4-nitrophenyl) phosphorothioate
Chemical Abstract #	500-28-7
Common name	Chlorthion
Synonyms	Phosphorothioic acid *o*-(3-chloro-4-nitrophenyl)-*o,o*-dimethyl ester
Uses	As insecticide
Molecular formula	$C_8H_9ClNO_5PS$
Molecular weight	297.68
Properties	m.p. 21.0°C, b.p. 125.0°C (at 0.1 mm)

Chemical name	*o*-Chlorotoluene
Chemical Abstract #	95-49-8
Common name	
Synonyms	2-Chloro-1-methylbenzene
Uses	As solvent, as an intermediate for organic chemicals and dyes
Molecular formula	C_7H_7Cl
Molecular weight	126.59
Properties	m.p. –36.0 to –34.0°C, b.p. 159.0°C

Chemical name	*m*-Chlorotoluene
Chemical Abstract #	108-41-8
Common name	
Synonyms	3-Chloro-1-methylbenzene
Uses	
Molecular formula	C_7H_7Cl
Molecular weight	126.59
Properties	m.p. –48.0°C, b.p. 160.0–162.0°C

Chemical name	*p*-Chlorotoluene
Chemical Abstract #	106-43-4
Common name	
Synonyms	4-Chloro-1-methylbenzene
Uses	As solvent and intermediate for organic chemicals and dyes
Molecular formula	C_7H_7Cl
Molecular weight	126.59
Properties	b.p. 162.0–166.0°C

Chemical Name	3-[*p*-(*p*-Chlorophenoxy)phenyl]-1,1-dimethylurea
Chemical Abstract #	1982-47-4
Common name	Chloroxuron
Synonyms	Tenoran, Norex, Chloroxifenidim, CIBA 1983
Uses	As herbicide
Molecular formula	$C_{15}H_{15}ClN_2O_2$
Molecular weight	290.77
Properties	m.p. 149.0°C

Chemical name *O,O*-Diethyl-*O*-(3,5,6-trichloro-2-pyridyl) phosphorothioate
Chemical Abstract # 2921-88-2
Common name Chlorpyrifos, Dursban
Synonyms Phosphorothioic acid *o,o*-diethyl *o*-(3,5,6-trichloro-2-pyridinyl) ester
Uses As insecticide
Molecular formula $C_9H_{11}Cl_3NO_3PS$
Molecular weight 350.57
Properties m.p. 41.5–43.5°C

Chemical name Chrysene
Chemical Abstract # 218-01-9
Common name
Synonyms 1,2-Benzophenanthrene
Uses
Molecular formula $C_{18}H_{12}$
Molecular weight 228.2
Properties m.p. 254.0°C, b.p. 448.0°C

Chemical name Dimethyl-1-methyl-2-(1-phenylethoxycarbonyl) vinyl phosphate
Chemical Abstract # 7700-17-6
Common name Ciodrin
Synonyms Crotoxyphos
Uses As insecticide
Molecular formula $C_{14}H_{19}O_6P$
Molecular weight 314.3
Properties b.p. 135.0°C (at 0.03 mm)

Chemical name 2,6-Dihydroxyisonicotinic acid
Chemical Abstract # 99-11-6
Common name Citrazinic acid
Synonyms
Uses In color developer solutions
Molecular formula $C_6H_7NO_2$
Molecular weight 155.11
Properties m.p. 300.0°C

Chemical name 2-Hydroxy-1,2,3-propanetricarboxylic acid
Chemical Abstract # 77-92-9
Common name Citric acid
Synonyms β-Hydroxytricarballylic acid
Uses
Molecular formula $C_6H_8O_7$
Molecular weight 192.12
Properties m.p. 153.0°C, b.p. decomposes

Chemical name 3-Chloro-4-methyl-7-coumarinyl diethyl phosphorothionate
Chemical Abstract # 56-72-4
Common name Co-Ral, Coumaphos, Muscatox, Resistox, Asuntol, Meldane, Baymix
Synonyms *O,O*-Diethyl-*O*-[3-chloro-4-methyl-2-oxo(2*H*)-1-benzopyran-7-yl]-phosphorothioate
Uses Livestock insecticide, nematocide
Molecular formula $C_{14}H_{16}ClO_5PS$
Molecular weight 362.78
Properties m.p. 91.0°C

Chemical name	Dimethyltetrachloroterephthalate
Chemical Abstract #	1861-32-1
Common name	DCPA, Fatal, Chlorthalmethyl, Dacthal
Synonyms	2,3,5,6-Tetrachloro-1,4-benzene dicarboxylic acid dimethyl ester
Uses	As selective pre-emergent herbicide
Molecular formula	$C_{10}H_6Cl_4O_4$
Molecular weight	331.99
Properties	m.p. 155.0–156.0°C

Chemical name	2,2-Bis(*p*-chlorophenyl)-1,1-dichloroethane
Chemical Abstract #	6088-51-3
Common name	DDD, Rhothane
Synonyms	Dichlorodiphenyldichloroethane
Uses	As nonsystemic contact and stomach insecticide
Molecular formula	$C_{14}H_{10}Cl_4$
Molecular weight	320.1
Properties	m.p. 112.0°C

Chemical name	1,1,1-Trichloro-2,2-bis(*p*-chlorophenyl)ethane
Chemical Abstract #	50-29-3
Common name	DDT, Dicophane
Synonyms	Dichlorodiphenyltrichloroethane
Uses	As nonsystemic stomach and contact insecticide
Molecular formula	$C_{14}H_9Cl_5$
Molecular weight	354.48
Properties	m.p. 108.5–109.5°C

Chemical name	Decahydronaphthalene
Chemical Abstract #	91-17-8
Common name	Decalin
Synonyms	
Uses	As solvent for oils, fats, waxes, resins, etc.
Molecular formula	$C_{10}H_{18}$
Molecular weight	138.25
Properties	m.p. –43.0°C, b.p. 195.7°C

Chemical name	Decamethrin
Chemical Abstract #	52918-63-5
Common name	Butox, Decis
Synonyms	Deltamethrin
Uses	As insecticide
Molecular formula	$C_{22}H_{19}Br_2NO_3$
Molecular weight	505.24
Properties	m.p. 98.0–101.0°C

Chemical name	13-Isopropylpodocarpa-8,11,13-trien-15-oic acid
Chemical Abstract #	1740-19-8
Common name	Dehydroabietic acid
Synonyms	
Uses	In making thermoplastic resins
Molecular formula	$C_{21}H_{30}O_2$
Molecular weight	314.51
Properties	

Chemical name	Phosphorothioic acid *O,O*-diethyl *O*-[2-(ethylthio)-ethyl] ester mixture with *O,O*-diethyl *S*-[2-(ethylthio) ethyl]phosphate thioate
Chemical Abstract #	8065-48-3
Common name	Demeton, Mercaptofos, Di-septon, Demeton-O
Synonyms	
Uses	As systemic insecticide and acaricide
Molecular formula	$C_8H_{19}O_3PS_2$
Molecular weight	258.36
Properties	b.p. 123.0°C at 1 mm

Chemical name	2-(α-Naphthoxy)-*N,N*,-diethylpropionamide
Chemical Abstract #	15299-99-7
Common name	Devrinol, Napropamide
Synonyms	*n,n*-Diethyl-2-(1-naphthalenyloxy)propanamide
Uses	Herbicide
Molecular formula	$C_{17}H_{21}NO_2$
Molecular weight	271.37
Properties	m.p. 63–64.0°C

Chemical name 4-(Dimethylamino)phenyl diazinesulfonate
Chemical Abstract #
Common name Dexon
Synonyms Sodiumfenaminosulf
Uses
Molecular formula $C_8H_{10}N_3O_3Na$
Molecular weight 227.16
Properties Decomposes > 200.0°C

Chemical name 4-Hydroxy-4-methyl-2-pentanone
Chemical Abstract # 123-42-2
Common name Diacetonealcohol, Pyranton A
Synonyms Dimethylacetonylcarbinol
Uses
Molecular formula $C_6H_{12}O_2$
Molecular weight 116.16
Properties m.p. –57.0 to –43.0°C, b.p. 166.0–169°C

Chemical name Dichloroallyl diisopropylthiocarbamate
Chemical Abstract # 2303-16-4
Common name Diallate, Avadex, Pyradex
Synonyms *S*-2,3-Dichloroallyldiisopropylthiocarbamate
Uses As herbicide
Molecular formula $C_{10}H_{17}Cl_2NOS$
Molecular weight 270.24
Properties m.p. 25.0–30.0°C, b.p. 150.0°C (at 9 mm)

Chemical name Di-2-propenylamine
Chemical Abstract # 124-02-7
Common name Diallylamine
Synonyms *N*-2-Propenyl-2-propen-1-amine
Uses
Molecular formula $C_6H_{11}N$
Molecular weight 97.18
Properties m.p. –88.4°C, b.p. 112.0°C

Chemical name Diallyl phthalate
Chemical Abstract # 131-17-9
Common name Dapon 35, Dapon R,NCI-C50657
Synonyms Di-2-propenyl ester, 1,2-benzenedicarboxylic acid
Uses
Molecular formula $C_{14}H_{14}O_4$
Molecular weight 246.28
Properties m.p. –70.0°C, b.p. 157.0°C

Chemical name 4,4′-Diaminodiphenylmethane
Chemical Abstract # 101-77-9
Common name DDM
Synonyms *p,p′*-Methylenedianiline
Uses In the synthesis of isocyanates, polyurethane elastomers, as epoxy hardening agent, as curative for neoprene, as antioxidant in footwear
Molecular formula $C_{13}H_{14}N_2$
Molecular weight 198.26
Properties m.p. 93.0°C, b.p. 231.0°C (at 11 mm)

Chemical name 2,4-Diaminophenol hydrochloride
Chemical Abstract # 137-09-7
Common name Amidol
Synonyms
Uses
Molecular formula $C_6H_8N_2O$, 2HCl
Molecular weight 197.08
Properties

Chemical name 1,3-Diamino-2-propanol tetraacetic acid
Chemical Abstract #
Common name DAPTA
Synonyms
Uses As chelating agent
Molecular formula
Molecular weight
Properties

Chemical name	*O,O*-Diethyl-*O*(2-isopropyl-4-methyl-6-pyridiminyl) phosphorothioate
Chemical Abstract #	333-41-5
Common name	Diazinon, AlfaTox, Basudin, Spectracide, Neocidol, Diazide, Diazol
Synonyms	Diethyl-4-(2-isopropyl-6 methylpyrimidinyl) phosphorothionate
Uses	As insecticide in rice paddy fields
Molecular formula	$C_{12}H_{21}N_2O_3PS$
Molecular weight	304.38
Properties	b.p. 84.0°C at 0.002 mm

Chemical name	1,2-5,6-Dibenzanthracene
Chemical Abstract #	53-70-3
Common name	
Synonyms	Dibenz(a,h)anthracene
Uses	
Molecular formula	$C_{22}H_{14}$
Molecular weight	278.36
Properties	m.p. 267.0°C, b.p. 524.0°C

Chemical name	1,2-Dibromo-3-chloropropane
Chemical Abstract #	96-12-8
Common name	Nematox, Nemazon, Nematocide, Nemagon
Synonyms	3-Chloro-1,2-dibromopropane
Uses	As soil fumigant, nematocide, intermediate in organic synthesis
Molecular formula	$C_3H_5Br_2Cl$
Molecular weight	236.35
Properties	b.p. 196.0°C

Chemical name	*n*-Dibutylamine
Chemical Abstract #	111-92-2
Common name	
Synonyms	*n*-Butyl-1-butanamine
Uses	
Molecular formula	$C_8H_{19}N$
Molecular weight	129.28
Properties	m.p. –59.0°C, b.p. 159.0°C

Chemical name	2,6-Di-*tert*-butyl-*p*-cresol
Chemical Abstract #	128-37-0
Common name	BHT,CP-antioxidant
Synonyms	Butylated hydroxytoluene
Uses	As antioxidant for food products, petroleum products, jet fuels, rubber, plastics, and animal feeds
Molecular formula	$C_{15}H_{24}O$
Molecular weight	220.36
Properties	m.p. 70.0°C, b.p. 265.0°C

Chemical name	Di-*n*-butyl phthalate
Chemical Abstract #	84-74-2
Common name	Butyl phthalate
Synonyms	Dibutyl-1,1-benzene dicarboxylate
Uses	In manufacturing plasticizers, cosmetics, as diluent, textile lubricating agent, in insecticides, printing inks, paper coatings, and adhesives; used in rocket propellant fuels
Molecular formula	$C_{16}H_{22}O_4$
Molecular weight	278.38
Properties	m.p. –36.0°C, b.p. 340.0°C

Chemical name	Dibutyl thiourea
Chemical Abstract #	109-46-6
Common name	Pennzone B,Thiate U
Synonyms	*N,N'*-dibutyl thiourea
Uses	
Molecular formula	$C_9H_{20}N_2S$
Molecular weight	188.37
Properties	m.p. 60.0°C

Chemical name	*O*-(2-Chloro-4-nitrophenyl) *O,O*,-dimethyl-phosphorothioate
Chemical Abstract #	2463-84-5
Common name	Dicapthon
Synonyms	Dicaptan
Uses	Nonsystemic insecticide and acaricide
Molecular formula	$C_8H_9ClNO_5PS$
Molecular weight	297.68
Properties	m.p. 53.0°C

Chemical name	2,6-Dicholorobenzonitrile
Chemical Abstract #	1194-65-6
Common name	Dichlobenil, Carsoron, Decabane
Synonyms	Du-sprex
Uses	As herbicide
Molecular formula	$C_7H_3Cl_2N$
Molecular weight	172.01
Properties	m.p. 144.0°C

Chemical name	2,3-Dichloro-1,4-naphthaquinone
Chemical Abstract #	117-80-6
Common name	Diclone, Algistat, Phygon, Sanquinon
Synonyms	2,3-Dicholoro-1,4-naphthalenedione
Uses	As funigicide and algicide
Molecular formula	$C_{10}H_4Cl_2O_2$
Molecular weight	227.04
Properties	m.p. 193.0°C

Chemical name	Dichloroacetic acid
Chemical Abstract #	79-43-6
Common name	Urner's liquid, dicholoroethanoic acid
Synonyms	2,2-Dichloroacetic acid
Uses	
Molecular formula	$C_2H_2Cl_2O_2$
Molecular weight	128.94
Properties	m.p. 5.0–6.0°C, b.p. 194.0°C

Chemical name	*O*-Dicholorobenzene
Chemical Abstract #	95-50-1
Common name	Chloroben, Dizene, Termitkil, Dowtherm E
Synonyms	1,2-Dichlorobenzene
Uses	As pesticide, fumigant, solvent, in metal polishes
Molecular formula	$C_6H_4Cl_2$
Molecular weight	147.01
Properties	m.p. –17.5°C, b.p. 180.0–183.0°C

Chemical name	3,6-Dichloro-*o*-anisic acid
Chemical Abstract #	1918-00-9
Common name	Dicamba, Dianat, Banvel
Synonyms	3,6-Dichloro-2-methoxybenzoic acid
Uses	Herbicide
Molecular formula	$C_8H_6Cl_2O_3$
Molecular weight	221.04
Properties	m.p. 114.0–116.0°C

Chemical name	*m*-Dichlorobenzene
Chemical Abstract #	541-73-1
Common name	
Synonyms	1,3-Dichlorobenzene
Uses	
Molecular formula	$C_6H_4Cl_2$
Molecular weight	147.01
Properties	m.p. –24.8°C, b.p. 173.0°C

Chemical name	*p*-Dichlorobenzene
Chemical Abstract #	106-46-7
Common name	Paradi, Paradow, Paramoth, Parazene, Santochlor
Synonyms	1,4-Dichlorobenzene
Uses	In making moth repellants, air deodorizers, dyes, and pharmaceuticals; as soil fumigant, pesticide
Molecular formula	$C_6H_4Cl_2$
Molecular weight	147.01
Properties	m.p. 53.0°C, b.p. 173.4°C

Chemical name	3,3′-Dicholorobenzidine
Chemical Abstract #	91-94-1
Common name	Curithane C126
Synonyms	3,3′-Dichlorobiphenyl-4,4′-diamine
Uses	In the manufacture of azo pigments, urethane resins; as curing agent for resins
Molecular formula	$C_{12}H_{10}Cl_2N_2$
Molecular weight	253.14
Properties	m.p. 133.0°C

Chemical name	2,5-Dichlorobenzoic acid
Chemical Abstract #	50-79-3
Common name	
Synonyms	
Uses	
Molecular formula	$C_7H_4Cl_2O_2$
Molecular weight	191.01
Properties	m.p. 151.0–154.0°C, b.p. 301.0°C

Chemical name	1,4-Dicholoro-2-butene
Chemical Abstract #	764-41-0
Common name	1,4-DCB
Synonyms	
Uses	
Molecular formula	$C_4H_6Cl_2$
Molecular weight	125.0
Properties	m.p. 3.5°C, b.p. 157.0°C

Chemical name	Dicholorodifluoromethane
Chemical Abstract #	75-71-8
Common name	Freon F-12, Halon, Ucon 12, Refrigerant 12, Propellant 12, Genetron 12
Synonyms	Difluorodichloromethane
Uses	As refrigerant and air conditioner, aerosol propellant, plastics, blowing agent, low-temperature solvent, leak-detecting agent
Molecular formula	CCl_2F_2
Molecular weight	120.91
Properties	m.p. –158.0°C, b.p. –29.8°C

Chemical name	5,5′-Dichloro-2,2′-dihydroxydiphenylmethane
Chemical Abstract #	97-23-4
Common name	Dichlorophen, Antiphen, Preventol GD9 (Bayer)
Synonyms	2,2′-Methylene bis(4-chlorophenol)
Uses	As fungicide, bactericide, herbicide
Molecular formula	$C_{13}H_{10}Cl_2O_2$
Molecular weight	269.0
Properties	m.p. 177.0–178.0°C

Chemical name	1,1-Dicholoroethane
Chemical Abstract #	75-34-3
Common name	Ethyledine chloride
Synonyms	Ethyledine dichloride
Uses	In the manufacture of vinyl chloride, paint and varnish removers, metal degreasers, in organic synthesis
Molecular formula	$C_2H_4Cl_2$
Molecular weight	98.96
Properties	m.p. –97.4°C, b.p. 57.3°C

Chemical name	1,1-Dichloroethylene
Chemical Abstract #	75-35-4
Common name	Sconatex
Synonyms	Vinyledene chloride, 1,1-dichloroethene
Uses	In adhesives, synthetic fibers
Molecular formula	$C_2H_2Cl_2$
Molecular weight	96.94
Properties	m.p. -122.5°C, b.p. 31.9°C

Chemical name	1,2-Dichloroethylene
Chemical Abstract #	540-59-0
Common name	Acetylene dichloride
Synonyms	1,2-Dichloroethane
Uses	As solvent for fats, camphors, phenols, etc.; as refrigerant, additive to dyes and lacquer solutions, low-temperature solvent for heat-sensitive compounds such as caffeine; as constituent in perfumes and thermoplastics, in organic synthesis, used to retard fermentation
Molecular formula	$C_2H_2Cl_2$
Molecular weight	96.95
Properties	m.p. –81.0°C (*cis*) and –50.0°C (*trans*); b.p. 60.0°C (*cis*) and 48.0°C (*trans*)

Chemical name	2,2-Dichloroethyl ether
Chemical Abstract #	111-44-4
Common name	Sym. dichloroethyl ether
Synonyms	1-Chloro-2-(β-chloroethoxy)ethane
Uses	In fumigants and insecticides, processing fats, waxes, greases, cellulose esters; as solvent, constituent in paints, varnishes, and lacquers
Molecular formula	$C_4H_8Cl_2O$
Molecular weight	143.02
Properties	m.p. –50.0°C, b.p. 178.0°C

Chemical name	2,3-Dichloro-1,4-naphthaquinone
Chemical Abstract #	117-80-6
Common name	Algistat, Dichlone, Phygon, Quintar, Uniroyal
Synonyms	2,3-Dichloronaphthaquinone
Uses	Fungicide and algaecide
Molecular formula	$C_{10}H_4Cl_2O_2$
Molecular weight	227.04
Properties	m.p. 193.0°C

Chemical name	1,1-Dichloro-1-nitroethane
Chemical Abstract #	594-72-9
Common name	Ethide
Synonyms	Dichloronitroethane
Uses	Fumigant
Molecular formula	$C_2H_3Cl_2NO_2$
Molecular weight	143.96
Properties	b.p. 124.0°C

Chemical name	2,4-Dichlorophenol
Chemical Abstract #	120-83-2
Common name	2,4-DCP
Synonyms	DCP
Uses	In organic synthesis
Molecular formula	$C_6H_4Cl_2O$
Molecular weight	163.0
Properties	m.p. 45.0°C, b.p. 210.0°C

Chemical name	2,6-Dicholorophenol
Chemical Abstract #	87-65-0
Common name	
Synonyms	
Uses	
Molecular formula	$C_6H_4Cl_2O$
Molecular weight	163.0
Properties	m.p. 65.0–66.0°C, b.p. 218.0–220.0°C

Chemical name	2,4-Dicholorophenoxyacetic acid
Chemical Abstract #	94-75-7
Common name	2,4-D, Amidox, Choloroxone, Ded-Weed, Weed-B-Gon
Synonyms	Weed Tox
Uses	Systemic herbicide and defoliant
Molecular formula	$C_8H_6Cl_2O_3$
Molecular weight	221.04
Properties	m.p. 141.0°C, b.p. 160.0°C

Chemical name	2,4-Dichlorophenoxyacetic acid, butoxyethyl ester
Chemical Abstract #	1929-73-3
Common name	2,4-D(BEE)
Synonyms	Butoxyethyl 2,4-dichlorphenoxyacetate
Uses	Herbicide
Molecular formula	$C_{14}H_{18}Cl_2O_4$
Molecular weight	321.22
Properties	

Chemical name	2,4-Dichlorophenoxyacetic acid, dimethylamine
Chemical Abstract #	2008-39-1
Common name	Bladex G, Demise
Synonyms	Dimethylammonium 2,4-dichlorophenoxyacetate
Uses	Weed killer
Molecular formula	$C_{10}H_{11}Cl_2NO_3$
Molecular weight	264.12
Properties	m.p. 85.0–87.0°C

Chemical name	3(3,4-Dichlorophenyl)-1,1-dimethylurea
Chemical Abstract #	330-54-1
Common name	Dalon, Diurex, Diuron, DMU, Karmex, Herbatox, Telvar
Synonyms	1,1-Dimethyl-3-(3,4-dichlorophenyl)urea
Uses	Herbicide
Molecular formula	$C_9H_{10}Cl_2N_2O$
Molecular weight	233.11
Properties	m.p. 158.0–159.0°C

Chemical name	3-(3,4-Dichlorophenyl)-1-methoxy-1-methylurea
Chemical Abstract #	330-55-2
Common name	Garnitan, Linuron, Lorox, Methoxydiuron
Synonyms	*N*-(3,4-Dichlorophenyl)-*N′*-methyl-*N′*-methoxyurea
Uses	Pesticide, selective herbicide in farming
Molecular formula	$C_9H_{10}Cl_2N_2O_2$
Molecular weight	249.11
Properties	m.p. 93.0–94.0°C

Chemical name	3-(3,4-Dichloro)-1-methyl-1-*n*-butylurea
Chemical Abstract #	555-37-3
Common name	Neburon
Synonyms	*N*-Butyl-*N′*-(3,4-dichlorophenyl)-*N*-methylurea
Uses	Pesticide
Molecular formula	$C_{12}H_{16}Cl_2N_2O$
Molecular weight	275.0
Properties	m.p. 102.0°C

Chemical name	*N*-(3,4-Dichlorophenyl)propionamide
Chemical Abstract #	709-98-8
Common name	Propanil, Stam F-34
Synonyms	3,4-Dichloropropionanilide
Uses	Post-emergence contact herbicide
Molecular formula	$C_9H_9Cl_2NO$
Molecular weight	218.0
Properties	m.p. 91.0°C

Chemical name	1,2-Dichloropropane
Chemical Abstract #	78-87-5
Common name	Propylene chloride
Synonyms	Propylene dichloride
Uses	Intermediate for perchloroethylene and carbon tetrachloride, soil fumigant for nematodes
Molecular formula	$C_3H_6Cl_2$
Molecular weight	112.99
Properties	m.p. –100.0 to –80.0°C, b.p. 96.8°C

Chemical name	1,3-Dichloropropane
Chemical Abstract #	142-28-9
Common name	
Synonyms	Trimethylene dichloride
Uses	
Molecular formula	$C_3H_6Cl_2$
Molecular weight	112.99
Properties	b.p. 125.0°C

Chemical name	1,3-Dichloro-1-propene
Chemical Abstract #	542-75-6
Common name	Telone II
Synonyms	1,3-Dichloro-1-propylene
Uses	Soil fumigant and nematocide
Molecular formula	$C_3H_4Cl_2$
Molecular weight	110.97
Properties	b.p. 104.0°C (*cis*) and 112.0°C (*trans*)

Chemical name	2,3-Dichloro-1-propene
Chemical Abstract #	78-88-6
Common name	2,3-Dichloropropene
Synonyms	2,3-Dichloropropylene
Uses	
Molecular formula	$C_3H_4Cl_2$
Molecular weight	110.97
Properties	

Chemical name	2,2-Dichloropropionic acid, Na
Chemical Abstract #	127-20-8
Common name	Basfapon, Dalapon-Na, Ded-weed, Unipon
Synonyms	Sodium α,α-dichloropropionate
Uses	Selective herbicide
Molecular formula	$C_3H_3Cl_2O_2Na$
Molecular weight	164.95
Properties	Decomposes at 166.0°C

Chemical name	2,2-Dichlorovinyl *O,O*-dimethyl phosphate
Chemical Abstract #	62-73-7
Common name	DDVP, Vapona
Synonyms	Dichlorvos
Uses	Contact and systemic insecticide
Molecular formula	$C_4H_7Cl_2PO_4$
Molecular weight	222.0
Properties	b.p. 84.0°C (at 1.0 mm)

Chemical name	α,α-Dichloro-*m*-xylene
Chemical Abstract #	626-16-4
Common name	*m*-Xylene dichloride
Synonyms	1,3-Bis(chloromethyl)benzene
Uses	
Molecular formula	$C_8H_8Cl_2$
Molecular weight	175.06
Properties	m.p. 34.0–37.0°C, b.p. 250.0–255.0°C

Chemical name	α-Dicyclopentadiene
Chemical Abstract #	77-73-6
Common name	DCPD
Synonyms	Tricyclo-(5,2,1,0)-3,8-decadiene
Uses	
Molecular formula	$C_{10}H_{12}$
Molecular weight	132.2
Properties	m.p. 33.0°C, b.p. 171.0°C

Chemical name	3,4,5,6,9,9-Hexachloro-1a,2,2a,3,6,6a,7,7a-octahydro-2,7,3:6-dimethanonaphth(2,3-*b*)oxirene
Chemical Abstract #	60-57-1
Common name	Dieldrin, Alvit, Heod, Octalox, Quintox
Synonyms	Hexachloroepoxyoctahydro-endo,exo-dimethano-naphthalene
Uses	Insecticide, used in wool-processing industry
Molecular formula	$C_{12}H_8Cl_6$
Molecular weight	380.90
Properties	m.p. 176.0°C

Chemical name	2,2′-Aminodiethanol
Chemical Abstract #	111-42-2
Common name	Dithanolamine, DEA
Synonyms	β,β-Dihydroxydiethylamine
Uses	In liquid detergents; for emulsion paints, cutting oils, shampoos, polishers; in making resins and plasticizers
Molecular formula	$C_4H_{11}NO_2$
Molecular weight	105.14
Properties	m.p. 28.0°C, b.p. 269.1°C

Chemical name	Diethylamine
Chemical Abstract #	109-89-7
Common name	2-Aminopentane
Synonyms	*N,N′*-Diethylamine
Uses	In dyes, floatation agents, corrosion inhibitors, polymerization inhibitors, as solvent
Molecular formula	$C_4H_{11}N$
Molecular weight	73.14
Properties	m.p. –49.0°C, b.p. 57.0°C

Chemical name	2-Diethylaminoethanol
Chemical Abstract #	100-37-8
Common name	Diethylethanolamine
Synonyms	2-Hydroxytriethylamine
Uses	
Molecular formula	$C_6H_{15}NO$
Molecular weight	117.19
Properties	b.p. 163.0°C

Chemical name	*o*-Diethylbenzene
Chemical Abstract #	25340-17-4
Common name	
Synonyms	1,2-Diethylbenzene
Uses	
Molecular formula	$C_{10}H_{14}$
Molecular weight	134.24
Properties	m.p. –20.0°C, b.p. 183.5°C

Chemical name	Diethyleneglycol
Chemical Abstract #	111-46-6
Common name	DEG, carbitol, diglycol, glycol ether
Synonyms	2,2′-Dihydroxyethyl ether
Uses	In making polyurethane and unsaturated polyester resins, as textile softener, in plasticizers and surfactants
Molecular formula	$C_4H_{10}O_3$
Molecular weight	106.12
Properties	m.p. –10.0°C, b.p. 245.0°C

Chemical name	Diethylene glycol monobutyl ether
Chemical Abstract #	112-34-5
Common name	Butyl carbitol, Butyl glycol, Dowanol DB, Polysolv DB
Synonyms	2-(2-Hydroxyethoxy)ethanol
Uses	
Molecular formula	$C_8H_{18}O_3$
Molecular weight	162.26
Properties	m.p. –61.0°C, b.p. 230.6°C

Chemical name	Diethylene glycol monoethyl ether
Chemical Abstract #	111-90-0
Common name	Carbitol, Ethyldigol, Diglycol, Dowanol DE, Dioxitol
Synonyms	2-(2-Ethoxyethoxy)ethanol
Uses	
Molecular formula	$C_{10}H_{14}O_3$
Molecular weight	134.2
Properties	b.p. 202.0°C

Chemical name	Diethylene glycol monomethyl ether
Chemical Abstract #	111-77-3
Common name	Dowanol DM, Methyl carbitol, Poly-Solv DM
Synonyms	2-(2-Methoxyethoxy)ethanol
Uses	
Molecular formula	$C_5H_{12}O_3$
Molecular weight	120.17
Properties	b.p. 194.2°C

Chemical name	Diethylenetriamine
Chemical Abstract #	111-40-0
Common name	
Synonyms	2,2′-Diaminodiethylamine
Uses	
Molecular formula	$C_4H_{13}N_3$
Molecular weight	103.2
Properties	m.p. –39.0°C, b.p. 207.0°C

Chemical name	Diethylenetriamine pentaacetic acid
Chemical Abstract #	
Common name	DTPA
Synonyms	
Uses	Chelating agent
Molecular formula	$C_{14}H_{23}O_{10}N_3$
Molecular weight	393.85
Properties	

Chemical name	Diethyl fumarate
Chemical Abstract #	623-91-6
Common name	
Synonyms	
Uses	
Molecular formula	$C_8H_{12}O_4$
Molecular weight	172.18
Properties	m.p. 1–2.0°C, b.p. 218.0–219.0°C

Chemical name	Diethyl oxalate
Chemical Abstract #	95-92-1
Common name	Ethyloxalate
Synonyms	Diethyl ethanedioate
Uses	As solvent, dye intermediate, and in organic synthesis
Molecular formula	$C_6H_{10}O_4$
Molecular weight	146.16
Properties	m.p. –41.0°C, b.p. 185.0°C

Chemical name	Diethyl phosphite
Chemical Abstract #	762-04-9
Common name	
Synonyms	Diethyl hydrogen phosphite
Uses	
Molecular formula	$C_4H_{11}O_3P$
Molecular weight	138.12
Properties	b.p. 138.0°C

Chemical name	Diethyl phthalate
Chemical Abstract #	84-66-2
Common name	Ethyl phthalate, Estol 1550, Solvanol
Synonyms	1,2-Benzenedicarboxylic acid, diethyl ester
Uses	In the manufacture of plasticizers, plastics, perfumes; used in insecticidal sprays, mosquito repellants, as alcohol denaturant
Molecular formula	$C_{12}H_{14}O_4$
Molecular weight	222.26
Properties	m.p. –40.5°C; b.p. 302.0°C

Chemical name	Diethylstilbestrol
Chemical Abstract #	56-53-1
Common name	Antigestil, DES, Dibestrol, Estrogen, Menostilbeen, Stilkap, Vagestrol
Synonyms	3,4-Bis(*p*-hydroxyphenyl)-3-hexene
Uses	A synthetic estrogen (nonsteroid), in medicine, animal feed, research
Molecular formula	$C_{18}H_{20}O_2$
Molecular weight	268.38
Properties	m.p. 171.0°C

Chemical name	1,3-Diethylthiourea
Chemical Abstract #	105-55-5
Common name	Thiate H, Pennzone E
Synonyms	*n,n′*-Diethylthiocarbamide
Uses	
Molecular formula	$C_5H_{12}N_2S$
Molecular weight	132.25
Properties	m.p. 68.0–71.0°C

Chemical name	N-[(Chlorophenylamino)carbonyl]-2,6-difluoro-benzamide
Chemical Abstract #	35367-38-5
Common name	Diflubenzuron
Synonyms	1-(4-Chlorophenyl)-3-(2,6-difluorobenzoyl)urea
Uses	Insecticide, larvicide
Molecular formula	$C_{14}H_9ClF_2N_2O_2$
Molecular weight	310.68
Properties	m.p. 239.0°C

Chemical name	*N*-(1,1,2,2-Tetrachloroethylthio)cyclohex-4-ene-1,2-dicarboxamide
Chemical Abstract #	2425-06-1
Common name	Captafol, Difolitan
Synonyms	3a,4,7,7a-Tetrahydro-2-[(1,1,2,2-tetrachloroethyl)-1*H*-iso-indole-1,3(2H)-dione
Uses	Agricultural fungicide, especially for potatoes
Molecular formula	$C_{10}H_9Cl_4NO_2S$
Molecular weight	349.09
Properties	m.p. 160.0–161.0°C

Chemical name	Dihydroheptachlor
Chemical Abstract #	14168-01-5
Common name	Dilor
Synonyms	β-DHC
Uses	Insecticide
Molecular formula	$C_{10}H_7Cl_7$
Molecular weight	375.32
Properties	

Chemical name	5,6-Dihydro-2-methyl-N-phenyl-1,4-oxathin-3-carboxamide
Chemical Abstract #	5234-68-4
Common name	Carboxin, DCMO, Vitavax
Synonyms	2,3-Dihydro-5-carboxanilido-6-methyl-1, 4-oxathin
Uses	Systemic plant fungicide, effective against loose smut in cereals
Molecular formula	$C_{12}H_{13}NO_2S$
Molecular weight	235.31
Properties	m.p. 93.0–95.0°C

Chemical name	Diisobutylamine
Chemical Abstract #	110-96-3
Common name	
Synonyms	2-Methyl-*N*-(2-methylpropyl)-1-propanamine
Uses	
Molecular formula	$C_8H_{19}N$
Molecular weight	129.28
Properties	m.p. –70.0°C, b.p. 139.0°C

Chemical name	Diisobutylcarbinol
Chemical Abstract #	108-82-7
Common name	*sec*-Nonyl alcohol
Synonyms	2,6-Dimethyl-4-hepatanol
Uses	As defoamer, in surface-active agents, lubricants, flotation agents
Molecular formula	$C_9H_{20}O$
Molecular weight	144.29
Properties	m.p. –65.0°C, b.p. 173.3°C

Chemical name	Diisobutyl ketone
Chemical Abstract #	108-83-8
Common name	Isovalerone, Valerone, DIBK
Synonyms	2,6-Dimethyl-4-heptanone
Uses	
Molecular formula	$C_9H_{18}O$
Molecular weight	142.24
Properties	m.p. –44.0°C, b.p. 165.0–168.0°C

Chemical name	Diisopropanolamine
Chemical Abstract #	110-97-4
Common name	DIPA
Synonyms	Bis(2-hydroxypropyl)amine
Uses	
Molecular formula	$C_6H_{15}NO_2$
Molecular weight	133.22
Properties	m.p. 42.0°C, b.p. 249.0°C

Chemical name	Diisopropylamine
Chemical Abstract #	108-18-9
Common name	DIPA
Synonyms	*N*-(1-Methylethyl)-2-propanamine
Uses	In organic synthesis
Molecular formula	$C_6H_{15}N$
Molecular weight	101.19
Properties	m.p. –96.3°C, b.p. 83.4°C

Chemical name	Diisopropyl ether
Chemical Abstract #	108-20-3
Common name	DIPE, Isopropyl ether
Synonyms	2-Isopropoxypropane
Uses	Solvent for oils, waxes and resins, paint and varnish removers, in rubber cements
Molecular formula	$C_6H_{14}O$
Molecular weight	102.2
Properties	m.p. –86.0 to –60.0°C, b.p. 69.0°C

Chemical name	*O,O*-Dimethyl-*S*-(*n*-methyl carbamoylmethyl) phosphorothioate
Chemical Abstract #	60-51-5
Common name	Dimethoate, Fosfamid, Cygon, Devigon, Fosfotox, Lurgo, Perfecthion, Racusan, Rebelate, Sinoratox, Trimetion
Synonyms	*N*-Monomethylamide of *O,O*-dimethyl dithiophosphorylacetic acid
Uses	Systemic and contact insecticide
Molecular formula	$C_5H_{12}NO_3PS_2$
Molecular weight	229.28
Properties	m.p. 52.0°C

Chemical name	1,4-Dimethoxybenzene
Chemical Abstract #	150-87-5
Common name	*p*-Dimethoxybenzene
Synonyms	Hydroquinone dimethyl ether
Uses	
Molecular formula	$C_8H_{10}O_2$
Molecular weight	138.18
Properties	m.p. 56.0°C, b.p. 212.6°C

Chemical name	Dimethoxymethane
Chemical Abstract #	109-87-5
Common name	Methylal, Formal
Synonyms	Methylene dimethyl ether
Uses	
Molecular formula	$C_3H_8O_2$
Molecular weight	76.09
Properties	m.p. –104.8°C, b.p. –44.0°C

Chemical name	Dimethylamine
Chemical Abstract #	124-40-3
Common name	
Synonyms	*N*-Methylmethanamine
Uses	As accelerator in vulcanizing rubber, in tanning, in the manufacture of detergent soaps
Molecular formula	C_2H_7N
Molecular weight	45.08
Properties	m.p. –96.0°C, b.p. 7.0°C

Chemical name	4-Dimethylaminoazobenzene
Chemical Abstract #	60-11-7
Common name	Methyl yellow, butter yellow
Synonyms	*N,N*-Dimethyl-4-(phenylazo)benzamine
Uses	Used to test free HCl in gastric juice, in spot test identification of peroxidized fats
Molecular formula	$C_{14}H_{15}N_3$
Molecular weight	225.28
Properties	m.p. 114.0–117.0°C

Chemical name	2,3-Dimethylaniline
Chemical Abstract #	87-59-2
Common name	
Synonyms	2,3-Xylidine
Uses	
Molecular formula	C_8H_9N
Molecular weight	121.18
Properties	m.p. 2.5°C, b.p. 221.0°C

Chemical name	2,4-Dimethylaniline
Chemical Abstract #	95-68-1
Common name	
Synonyms	2,4-Xylidine
Uses	
Molecular formula	C_8H_9N
Molecular weight	121.18
Properties	b.p. 218.0°C

Chemical name 2,6-Dimethylaniline
Chemical Abstract # 87-62-67
Common name
Synonyms 2,6-Xylidine
Uses
Molecular formula C_8H_9N
Molecular weight 121.18
Properties m.p. 10.0–12.0°C, b.p. 214.0°C

Chemical name 1,1-Dimethylhydrazine
Chemical Abstract # 57-14-7
Common name UDMH
Synonyms *N,N*-Dimethylhydrazine
Uses As base in rocket fuel formulations
Molecular formula $C_2H_8N_2$
Molecular weight 60.10
Properties m.p. –58.0°C, b.p. 64.0°C

Chemical name *O,O*-Dimethyl *O-p*-nitrophenylphosphorothioate
Chemical Abstract # 298-00-0
Common name Metaphos, methyl parathion, Metacide, Nitrox 80
Synonyms Phosphorothioic acid, *O,O*-dimethyl-(4-nitrophenyl)ester
Uses Insecticide
Molecular formula $C_8H_{10}NO_5PS$
Molecular weight 263.23
Properties m.p. 37.0–38.0°C

Chemical name Dimethyl phthalate
Chemical Abstract # 131-11-3
Common name Avolin, Palatinol M, Solvanom, Solvarone
Synonyms Dimethyl-1,2-benzene dicarboxylate
Uses As pesticide; in manufacturing of plasticizers, latex, cellulose acetate films, in rocket fuel formulations, used in insect repellants, perfumes, component in plastics, rubber, lacquers, safety glass, molding powders, and coating agents
Molecular formula $C_{10}H_{10}O_4$
Molecular weight 194.20
Properties m.p. 0.0°C, b.p. 283.7°C

Chemical name Dimethyl sulfate
Chemical Abstract # 77-78-1
Common name Methyl sulfate
Synonyms Dimethyl monosulfate
Uses As methylating agent
Molecular formula $C_2H_6O_4S$
Molecular weight 126.14
Properties m.p. –31.8°C, b.p. 188.0°C

Chemical name *O,O*-Dimethyl-(2,2,2-trichloro-1-hydroxyethyl) phosphorothioate
Chemical Abstract # 52-68-6
Common name Chlorfos, Chlorfon, Trichlorphene, Dipterex
Synonyms (2,2,2-Trichloro-1-hydroxyethyl)-phosphonic acid dimethyl ester
Uses Insecticide for the control of flies and roaches, and in anthelmintic compositions for animals
Molecular formula $C_4H_8Cl_3O_4P$
Molecular weight 257.45
Properties m.p. 83.0–84.0°C

Chemical name	*m*-Dinitrobenzene
Chemical Abstract #	99-65-0
Common name	
Synonyms	1,3-Dinitrobenzene
Uses	
Molecular formula	$C_6H_4N_2O_4$
Molecular weight	168.12
Properties	m.p. 89.0°C, b.p. 301.0°C

Chemical name	2,4-Dinitro-6-*sec*-butylphenol
Chemical Abstract #	88-85-7
Common name	Dinoseb, Preemerge, Subitex, Caldon
Synonyms	2-(1-Methylpropyl)-4,6-dinitrophenol
Uses	Herbicide, insecticide, miticide
Molecular formula	$C_{10}H_{12}N_2O_5$
Molecular weight	240.22
Properties	m.p. 38.0–42.0°C

Chemical name	4,6-Dinitro-*o*-cresol
Chemical Abstract #	534-52-1
Common name	Arborol, Dinitrol, Sandolin, winterwash
Synonyms	2-Methyl-4,6-dinitrophenol
Uses	Selective herbicide, insecticide (ovicidal spray for dormant fruit trees)
Molecular formula	$C_7H_6N_2O_5$
Molecular weight	198.15
Properties	m.p. 85.8°C

Chemical name	2,4-Dinitrophenol
Chemical Abstract #	51-28-5
Common name	Aldifen
Synonyms	*o*-Dinitrophenol
Uses	In the manufacture of dyes, as wood preservative, insecticide
Molecular formula	$C_6H_4N_2O_5$
Molecular weight	184.11
Properties	m.p. 112.0–114.0°C

Chemical name	3,5-Dinitro-*o*-toluamide
Chemical Abstract #	148-01-6
Common name	Zoalene, Zoamix
Synonyms	2-Methyl 3,5-dinitrobenzamide
Uses	Coccidiostat, permitted food additive
Molecular formula	$C_8H_7N_3O_5$
Molecular weight	225.18
Properties	m.p. 177.0°C

Chemical name	2,3-Dinitrotoluene
Chemical Abstract #	602-01-7
Common name	2,3-DNT
Synonyms	1-Methyl 1,3-dinitrobenzene
Uses	
Molecular formula	$C_7H_6N_2O_4$
Molecular weight	182.15
Properties	m.p. 59.0–61.0°C

Chemical name	2,4-Dinitrotoluene
Chemical Abstract #	121-14-2
Common name	2,4-Dinitrotoluol
Synonyms	1-Methyl 2,4-dinitrobenzene
Uses	
Molecular formula	$C_7H_6N_2O_4$
Molecular weight	182.15
Properties	m.p. 69.5°C, b.p. 300.0°C

Chemical name	2,6-Dinitrotoluene
Chemical Abstract #	606-20-2
Common name	2,6-DNT
Synonyms	2-Methyl 2,6-dinitrobenzene
Uses	In the manufacture of TNT, urethane polymers, flexible and rigid foams, surface coatings and dyes
Molecular formula	$C_7H_6N_2O_4$
Molecular weight	182.15
Properties	m.p. 64.0–66.0°C

Chemical name	Dinocap
Chemical Abstract #	39300-45-3
Common name	Karathane, Crotothane
Synonyms	A mixture of 2(or 4)-octyl 4,6-(or 2,6-)-dinitrophenyl butenoate in which the octyl is a mixture of 1-methylheptyl, 1-ethylhexyl and 1-propylpentyl
Uses	Acaricide, fungicide
Molecular formula	$C_{18}H_{24}N_2O_6$
Molecular weight	364.39
Properties	b.p. 138.0–140.0°C (at 0.05 mm)

Chemical name	Glycol ethylene ether
Chemical Abstract #	123-91-1
Common name	1,4-Dioxane
Synonyms	1,4-Diethylene dioxide
Uses	Solvent for a wide range of organic products, lacquers, paints, varnishes; in cleaning and detergent preparations, in cosmetics, deodorants, and fumigants
Molecular formula	$C_4H_8O_2$
Molecular weight	88.10
Properties	m.p. 11.8°C, b.p. 101.0°C

Chemical name	2,3-*p*-Dioxanedithiol-*S,S*-bis-(*O,O*-diethyl phosphorothioate
Chemical Abstract #	78-34-2
Common name	Dioxathion, Delnav, Navadel, Hercules 528
Synonyms	Phosphorodithioic acid *S,S'*,-1,4-dioxane-2,3-diyl *O,O,O,O*-tetraethyl ester
Uses	Insecticide, acaricide, miticide
Molecular formula	$C_{12}H_{26}O_6P_2S_4$
Molecular weight	456.54
Properties	m.p. –20.0°C

Chemical name	*N,N*-Dimethyl-2,2-diphenylacetamide
Chemical Abstract #	957-51-7
Common name	Diphenamid
Synonyms	*N,N*-Dimethyl-α-phenylbenzene acetamide
Uses	Herbicide
Molecular formula	$C_{16}H_{17}NO$
Molecular weight	239.3
Properties	m.p. 135.0°C

Chemical name	1,1-Ethylene-2,2'-bipyridylium dibromide
Chemical Abstract #	85-00-7
Common name	Diquat, diquat dibromide, Reglone
Synonyms	6,7-Dihydrodipyrido[1,2a:2',1'-c]pyrazinediium dibromide
Uses	Contact herbicide, defoliant
Molecular formula	$C_{12}H_{12}Br_2N_2$
Molecular weight	344.07
Properties	m.p. 320.0°C (decomposes)

Chemical name	*O,O*-Diethyl-*S*-[2(ethylthio)ethyl] phosphorodithioate
Chemical Abstract #	298-04-4
Common name	Disulfoton, Thiodemeton, Disyston, Solvirex
Synonyms	Phosphorodithioic acid *O,O*-diethyl-*S*-[2-(ethylthio)-ethyl] ester
Uses	Insecticide
Molecular formula	$C_8H_{19}O_2PS_3$
Molecular weight	274.38
Properties	b.p. 108.0°C (at 0.01 mm)

Chemical name	Laurylguanidine acetate
Chemical Abstract #	2439-10-3
Common name	Dodine, Carpine, Cyprex, Melprex
Synonyms	*n*-Dodecylguanidine monoacetate
Uses	Fungicide
Molecular formula	$C_{15}H_{33}N_3O_2$
Molecular weight	287.44
Properties	m.p. 136.0°C

Chemical name	*O,O*-Diethyl-*O*-quinoxalinyl (2)thionophosphate
Chemical Abstract #	13593-03-8
Common name	Ekalux, Quinalphos, Bayrusil, Chinalphos, Sandoz 6538, Wie oben
Synonyms	*O,O*-Diethyl-*O*-(2-chinoxalyl) phosphorothioate
Uses	Insecticide
Molecular formula	$C_{12}H_{15}N_2O_3PS$
Molecular weight	298.32
Properties	

Chemical name	*O,O*-Dimethyl-*S*-[2-(ethylthio)ethyl] phosphorodithioate
Chemical Abstract #	
Common name	Ekatin
Synonyms	Thiometon, Ekatin, Dithiometon
Uses	Insecticide and acaricide
Molecular formula	$C_6H_{15}PS_3O_2$
Molecular weight	245.44
Properties	b.p. 121.0°C (at 1.0 mm)

Chemical name	6,7,8,9,10,10-Hexachloro-1,5,5a,6,9,9a-hexahydro-6,9-methano-2,3,4-benzodioxyanthiepin 3-oxide
Chemical Abstract #	115-29-7
Common name	Endosulfan, Chlorthiepin, Malix, Thiodan, Thionex
Synonyms	1,2,3,4,7,7-Hexachlorobicyclo[2.2.1]-2 heptene-5,6-bisoxymethylenesulfite
Uses	Insecticide
Molecular formula	$C_9H_6Cl_6O_3S$
Molecular weight	406.95
Properties	m.p. 70.0–100.0°C (106.0°C for pure compound)

Chemical name	1,2,3,4,10,10-Hexachloro-6, 7-epoxy-1,4,4a,5,6,7,8,8a-octahydro-1,4-endo-5,8-dimethyl-naphthalene
Chemical Abstract #	72-20-8
Common name	Endrin, Mendrin, Nendrin, Hexadrin
Synonyms	3,4,5,6,9,9-Hexachloro-1a,2,2a,3,6,6a,7,7a-octahydro-2,7:3,6-dimethanonaphth[2,3-b]oxirene
Uses	Formerly as an insecticide; discontinued in U.S.
Molecular formula	$C_{12}H_8Cl_6O$
Molecular weight	380.93
Properties	Decomposes at 245.0°C

Chemical name	1-Chloro-2,3-epoxypropane
Chemical Abstract #	106-89-8
Common name	Epichlorhydrin
Synonyms	Chloromethyloxirane
Uses	Solvent for natural and synthetic resins, gum, cellulose esters and ethers paints and varnishes, nail enamels and lacquers, cement for celluloid
Molecular formula	C_3H_5ClO
Molecular weight	92.53
Properties	m.p. –25.6°C, b.p. 117.0°C

Chemical name	*O*-Ethyl-*O*-*p*-nitrophenylphosphonothioate
Chemical Abstract #	2104-64-5
Common name	EPN
Synonyms	Ethyl-*p*-nitrophenyl benzenethiophosphonate
Uses	Insecticide, acaricide
Molecular formula	$C_{14}H_{14}NO_4PS$
Molecular weight	323.31
Properties	m.p. 36.0°C

Chemical name *S*-Ethyldipropylthiocarbamate
Chemical Abstract # 759-94-4
Common name Eptam, Eradicane, EPTC
Synonyms Dipropylthiocarbamic acid *S*-ethyl ester
Uses Selective herbicide
Molecular formula $C_9H_{19}NOS$
Molecular weight 189.31
Properties b.p. 127.0°C

Chemical name Ethyl alcohol
Chemical Abstract # 64-17-5
Common name Ethanol
Synonyms Methyl carbinol
Uses In the manufacture of several organic chemicals, plastics, plasticizers, lacquers, perfumes, cosmetics, mouthwash products, alcoholic beverages, soaps and cleaning products, dyes, and explosives, and as solvent
Molecular formula C_2H_5OH
Molecular weight 46.07
Properties m.p. –114.0°C, b.p. 78.4°C

Chemical name 2-Aminoethanol
Chemical Abstract # 141-43-5
Common name Ethanolamine
Synonyms 2-Aminoethanol, 2-hydroxyethylamine
Uses In scrubbing CO_2 and SO_2 from natural gas and other gases, in the synthesis of surface-active agents, in polishes, hair-waving solutions, and emulsifiers, as softening agents for hides, and dispersing agent for agricultural chemicals
Molecular formula C_2H_7NO
Molecular weight 61.08
Properties m.p. 11.0°C, b.p. 172.0°C

Chemical name *O,O,O,O*-Tetraethyl-*S,S*-methylene bisphosphorodithioate
Chemical Abstract # 563-12-2
Common name Ethion, Diethion, Niagra 1240, Nialate
Synonyms Bis[*S*-(diethoxyphosphinothioyl)mercapto]methane
Uses Insecticide and acaricide
Molecular formula $C_9H_{22}O_4P_2S_4$
Molecular weight 384.48
Properties m.p. –12.0 to –15.0°C

Chemical name Ethyl acetate
Chemical Abstract # 141-78-6
Common name Vinegar naphtha
Synonyms Acetic acid ethyl ester
Uses In artificial fruit essences, as solvent for nitrocellulose, varnishes, lacquers and airplane dopes, in making smokeless powders, artificial leather, photographic films and plates, artificial silk, perfumes; as cleaning agent for textiles
Molecular formula $C_4H_8O_2$
Molecular weight 88.10
Properties m.p. –83.0°C, b.p. 77.0°C

Chemical name Ethyl acetoacetate
Chemical Abstract # 141-97-9
Common name Acetoacetic ester
Synonyms Ethyl acetylacetonate
Uses In organic synthesis, in flavoring
Molecular formula $C_6H_{10}O_3$
Molecular weight 130.14
Properties m.p.–45.0°C, b.p. 180.8°C

Chemical name Ethyl acrylate
Chemical Abstract # 140-88-5
Common name
Synonyms Ethyl propenoate
Uses In the manufacture of paints, textile and paper coatings, leather finish resins, and adhesives
Molecular formula $C_5H_8O_2$
Molecular weight 100.11
Properties m.p. –75.0°C, b.p. 100.0°C

Chemical name Ethylamine
Chemical Abstract # 75-04-7
Common name Aminoethane
Synonyms Monoethylamine
Uses In resins, as stabilizer for rubber latex, as intermediate for dyes and pharmaceuticals, and in oil refining
Molecular formula C_2H_7N
Molecular weight 45.08
Properties m.p. –82.0°C, b.p. 16.6°C

Chemical name 5-Methyl 3-heptanone
Chemical Abstract # 106-68-3
Common name Ethyl amyl ketone
Synonyms Ethyl *iso*-amylketone
Uses Solvent for nitrocellulose-alkyd, nitrocellulose-laleic, and vinyl resins
Molecular formula $C_8H_{16}O$
Molecular weight 128.21
Properties b.p. 157.0–162.0°C

Chemical name Ethylbenzene
Chemical Abstract # 100-41-4
Common name Ethyl benzol
Synonyms Phenylethane
Uses As resin solvent, in making styrene monomer
Molecular formula C_8H_{10}
Molecular weight 106.16
Properties m.p. –95.0°C, b.p. 136.25°C

Chemical name 2-Ethyl-1-butanol
Chemical Abstract # 97-95-0
Common name *sec*-Hexanol
Synonyms 3-Methylolpentane
Uses
Molecular formula $C_6H_{14}O$
Molecular weight 102.2
Properties m.p. –50.0°C, b.p. 150.0°C

Chemical name Ethylene dibromide
Chemical Abstract # 106-93-4
Common name EDB, glycol dibromide
Synonyms 1,2-Dibromoethane
Uses As anti-knock in gasoline; grain and fruit fumigant
Molecular formula $C_2H_4Br_2$
Molecular weight 187.88
Properties m.p. 9.97°C, b.p. 131.6°C

Chemical name Ethylenediamine
Chemical Abstract # 107-15-3
Common name Diaminoethane
Synonyms 1,2-Ethanediamine
Uses Solvent for casein, albumin, shellac, and sulfur; emulsifier; stabilizer in rubber latex; as inhibitor in antifreeze solutions; in textile lubricants
Molecular formula $C_2H_8N_2$
Molecular weight 60.1
Properties m.p. 8.5°C, b.p. 116.0–117.0°C

Chemical name	Ethylenediamine tetraacetate,Na_4
Chemical Abstract #	64-02-8
Common name	EDTA, tetraNa salt, Chelon 100, Complexone
Synonyms	*N,N′*-1,2-Ethanediylbis[*N*-(carboxymethyl)glycine] tetrasodium salt
Uses	Sequestering agent, chelating agent
Molecular formula	$C_{10}H_{12}N_2Na_4O_8$
Molecular weight	380.20
Properties	m.p. 300.0°C

Chemical name	*N,N*-1,2-Ethanediylbis[*N*-(carboxymethyl)glycine], trisodium salt
Chemical Abstract #	150-38-9
Common name	EDTA, Na_3
Synonyms	(Ethylenedinitrilo) tetraacetic acid, triNa salt
Uses	Chelating agent
Molecular formula	$C_{10}H_{13}N_2Na_3O_8$
Molecular weight	358.20
Properties	m.p. > 300.0°C

Chemical name	Ethylenediamine tetraacetic acid
Chemical Abstract #	60-00-4
Common name	EDTA, Havidote, versene acid
Synonyms	*N,N′*-1,2-Ethanediylbis[*N*-(carboxymethyl)glycine]-(ethylenenitrilo) tetraacetic acid
Uses	Chelating agent in treating lead and heavy metal poisoning of farm animals
Molecular formula	$C_{10}H_{16}N_2O_8$
Molecular weight	292.24
Properties	Decomposes at 240.0°C

Chemical name	Ethylenediamine tetraacetic acid, dihydrate, Ca chelate of the disodium salt
Chemical Abstract #	62-33-9
Common name	EDTA calcium, Ledclair, Mosatil
Synonyms	Calcium disodium ethylenediamine tetraacetate
Uses	Chelating agent in lead poisoning treatment
Molecular formula	$C_{10}H_{22}CaN_2Na_2O_8$
Molecular weight	374.28
Properties	

Chemical name	Ethylene dichloride
Chemical Abstract #	107-06-2
Common name	Dichloroethylene, EDC, ethylene chloride
Synonyms	1,2-Dichloroethane
Uses	In cosmetics, as food additive, and as fumigant
Molecular formula	$C_2H_4Cl_2$
Molecular weight	99.0
Properties	m.p. –35.4°C, b.p. 85.3°C

Chemical name	Ethylene glycol
Chemical Abstract #	107-21-1
Common name	MEG, 1,2-ethanediol
Synonyms	1,2-Dihydroxyethane
Uses	As antifreeze in cooling and heating systems, in hydraulic brake fluids, industrial humectant, in electrolytic condensers, as solvent in paint and plastics industries, in printers' inks, stamp pad inks, ballpoint pen inks; as softening agent for cellophane; as stabilizer for soybean foam used to extinguish oil and gasoline fires; in the synthesis of explosives, alkyd resins (unsaturated), plasticizers, elastomers, synthetic fibers (Terylene, Dacron), and synthetic waxes
Molecular formula	$C_2H_6O_2$
Molecular weight	62.07
Properties	m.p. –17.0 to –12.6°C, b.p. 198.0°C

Chemical name Ethyleneglycol acetate
Chemical Abstract # 542-59-6
Common name Glycol monoacetin
Synonyms 1,2-Ethanediol, monoacetate
Uses
Molecular formula $C_4H_8O_3$
Molecular weight 104.12
Properties b.p. 182.0°C

Chemical name Ethylene glycol diethyl ether
Chemical Abstract # 629-14-1
Common name Ethyl glyme, diethylcellosolve
Synonyms 1,2-Diethoxyethane
Uses
Molecular formula $C_6H_{14}O_2$
Molecular weight 118.20
Properties b.p. 118.2°C

Chemical name 2-Ethoxyethanol
Chemical Abstract # 110-80-5
Common name Ethoxal, cellosolve, ethyl glycol, Dowanol EE, Oxitol
Synonyms Ethylene glycol monoethyl ether
Uses Solvent for nitrocellulose, lacquers, and dopes; in varnish removers, cleaning solutions, dye baths; finishing leather with water pigments; stabilizer of emulsions
Molecular formula $C_4H_{10}O_2$
Molecular weight 90.12
Properties m.p. –70.0°C, b.p. 135.0°C

Chemical name Ethylene glycol monoethyl ether acetate
Chemical Abstract # 111-15-9
Common name Cellosolve acetate
Synonyms 2-Ethoxyethyl acetate
Uses In automobile lacquers to provide high gloss and minimize evaporation
Molecular formula $C_6H_{12}O_3$
Molecular weight 132.16
Properties m.p. –62.0 to –58.0°C, b.p. 156.0°C

Chemical name Ethylene glycol monomethyl ether
Chemical Abstract # 109-86-4
Common name Methylcellosolve, methyl glycol, Dowanol EM, Methyloxitol
Synonyms 2-Methoxyethanol
Uses Solvent for low-viscosity cellulose acetate, natural resins, some synthetic resins; in dyeing leather; sealing moisture-proof cellophane; in nail polishes, quick-drying varnishes and enamels, wood stains
Molecular formula $C_3H_8O_2$
Molecular weight 76.09
Properties m.p. –85.0°C, b.p. 124.0°C

Chemical name Ethylene glycol monomethyl ether acetate
Chemical Abstract # 110-49-6
Common name Methylcellosolve acetate
Synonyms Methoxyethanol acetate
Uses As industrial solvent
Molecular formula $C_5H_{10}O_3$
Molecular weight 118.13
Properties m.p. –65.1°C, b.p. 145.0°C

Chemical name Ethyleneimine
Chemical Abstract # 151-56-4
Common name Aminoethylene, Azirane, Aziridine, Ethylimine
Synonyms Dihydro-1*H*-azirine
Uses
Molecular formula C_2H_5N
Molecular weight 43.08
Properties m.p. –71.0°C, b.p. 55.0–56.0°C

Chemical name Ethylene oxide
Chemical Abstract # 75-21-8
Common name Amprolene, Oxirane, Anproline, ethene oxide, Oxyfume
Synonyms 1,2-Epoxy ethane
Uses As fumigant for foodstuffs and textiles; to sterilize surgical instruments; agricultural fungicide; in organic synthesis and in making acrylonitrile and nonionic surfactants
Molecular formula C_2H_4O
Molecular weight 44.05
Properties m.p. –111.0°C, b.p. 10.7°C

Chemical name Ethyl ether
Chemical Abstract # 60-29-7
Common name Ether, diethyl ether
Synonyms Ethoxyethane
Uses As solvent for waxes, fats, oils, perfumes, alkaloids, and gums; excellent solvent for nitrocellulose; in making gun powder; as primer in gasoline engines; in organic synthesis
Molecular formula $C_4H_{10}O$
Molecular weight 74.14
Properties m.p. –116.2°C, b.p. 34.6°C

Chemical name 2-Ethylhexylamine
Chemical Abstract # 104-75-6
Common name
Synonyms 1-Amino-2-ethylhexane
Uses
Molecular formula $C_8H_{19}N$
Molecular weight 129.28
Properties b.p. 169.2°C

Chemical name *O*-Ethyl-*S*-phenylethylphosphonodithioate
Chemical Abstract # 944-22-9
Common name Fonofos, Dyfonate
Synonyms Ethylphosphonodithioic acid *O*-ethyl *S*-phenyl ester
Uses As soil insecticide
Molecular formula $C_{10}H_{15}OPS_2$
Molecular weight 246.32
Properties b.p. 130.0°C at 0.1 mm

Chemical name Ethyl propionate
Chemical Abstract # 105-37-3
Common name Propionic ether
Synonyms Propionic acid ethyl ester
Uses
Molecular formula $C_5H_{10}O_2$
Molecular weight 102.13
Properties m.p. –73.0°C, b.p. 99.0°C

Chemical name *N*-[(Dichlorofluoromethyl)thio]-*N,N'*-dimethyl-*N*-phenylsulfamide
Chemical Abstract # 1085-98-9
Common name Euparen, Dichlofluanid, Elvaron
Synonyms 1,1-Dichloro-*N*-[dimethylaminosulfonyl]-fluoro-*N*-phenylmethanesulfenamide
Uses Fungicide
Molecular formula $C_9H_{11}Cl_2FN_2O_2S_2$
Molecular weight 333.21
Properties m.p. 105.0–105.6°C

Chemical name	*O,O*-Dimethyl-*O*-(4-nitro-*m*-tolyl) phosphorothioate
Chemical Abstract #	122-14-5
Common name	Fenitrothion, Metathion, Accothion, Sumithion, Cyfen, Folithion
Synonyms	Phosphorothioic acid *O,O*-dimethyl-*O*-(3-methyl-4-nitrophenyl) ester
Uses	Insecticide
Molecular formula	$C_9H_{12}NO_5PS$
Molecular weight	277.25
Properties	b.p. 140.0–145.0°C at 0.1 mm

Chemical name	Diethyl-4-(methylsulfinyl) phenylphosphorothionate
Chemical Abstract #	115-90-2
Common name	Fensulfothion, Dasanit, Terracur P
Synonyms	Phosphorothioic acid, *O,O*-diethyl-*O*-[4(methylsulfinyl)phenyl] ester
Uses	Nematocide, insecticide; especially for the control of nematodes in golf courses and cemeteries; 10% solution is used for the control of onion maggots
Molecular formula	$C_{11}H_{17}O_4PS_2$
Molecular weight	308.35
Properties	b.p. 138.0–141.0°C at 0.01 mm

Chemical name	(+)-α-Cyano-3-phenoxybenzyl-(+)-α-(4-chlorophenyl) isovalerate
Chemical Abstract #	51630-58-1
Common name	Fenvalerate, Belmark, Pydrin, Pyridin, Sumicidin, and Tirade
Synonyms	4-Chloro-α-(1-methylethyl)benzene acetic acid cyano(3-phenoxyphenyl) methyl ester
Uses	Insecticide
Molecular formula	$C_{25}H_{22}ClNO_3$
Molecular weight	419.92
Properties	Decomposes gradually between 150.0 and 300.0°C

Chemical name	9*H*-Fluorene
Chemical Abstract #	86-73-7
Common name	
Synonyms	*o*-Biphenylmethane
Uses	
Molecular formula	C_3H_{10}
Molecular weight	166.21
Properties	m.p. 117.0°C, b.p. 295.0°C

Chemical name	*N*-(Trichloromethylthio)phthalimide
Chemical Abstract #	133-07-3
Common name	Folpet, Phaltan, Thiopal
Synonyms	2-[(Trichloromethyl)thio]-1*H*-isoindole 1,3(2*H*)-dione
Uses	Agricultural fungicide; bactericide for vinyls, paints, and enamels
Molecular formula	$C_9H_4Cl_3NO_2S$
Molecular weight	296.58
Properties	m.p. 177.0°C

Chemical name	Formaldehyde
Chemical Abstract #	50-00-0
Common name	
Synonyms	Methanal
Uses	In the production of phenolic, urea, melamine, and acetal resins; in textiles; used in embalming fluids, fungicides, air fresheners, and cosmetics
Molecular formula	CH_2O
Molecular weight	30.03
Properties	m.p. –92.0°C, b.p. –19.5°C

Chemical name	Formic acid
Chemical Abstract #	64-18-6
Common name	
Synonyms	Methanoic acid
Uses	As decalcifier; reducer in dyeing wool-fast colors; dehairing and plumping hides; tanning; electroplating; coagulating rubber latex
Molecular formula	CH_2OH
Molecular weight	46.02
Properties	m.p. 8.4°C, b.p. 100.5°C

Chemical name	4-(Triphenylmethyl)morpholine
Chemical Abstract #	1420-06-0
Common name	Frescon, Trifenmorph
Synonyms	4-Tritylmorpholine
Uses	As molluscicide
Molecular formula	$C_{23}H_{23}NO$
Molecular weight	329.44
Properties	m.p. 174.0–176.0°C

Chemical name	Furfural
Chemical Abstract #	98-01-1
Common name	Fural, furfuraldehyde
Synonyms	2-Furancarboxaldehyde
Uses	In the manufacture of furfural-phenol plastics; in refining petroleum; as solvent for nitrated cotton, cellulose acetate, and gums; in the manufacture of varnishes; to accelelarate vulcanization; as insecticide, fungicide, and germicide
Molecular formula	$C_5H_4O_2$
Molecular weight	96.08
Properties	m.p. –36.5°C, b.p. 161.8°C

Chemical name	Furfuryl alcohol
Chemical Abstract #	98-00-0
Common name	2-Furancarbinol
Synonyms	2-Furanmethanol
Uses	As solvent; in wetting agents and resins
Molecular formula	$C_5H_6O_2$
Molecular weight	98.10
Properties	m.p. –14.6°C, b.p. 170.0°C

Chemical name	3,4,5-Trihydroxybenzoic acid
Chemical Abstract #	149-91-7
Common name	Gallic acid
Uses	In manufacturing of gallic acid esters, inks; as photographic developers; in tanning; dyeing
Synonyms	
Molecular formula	$C_7H_6O_5$
Molecular weight	170.12
Properties	m.p. 220.0°C (decomposes)

Chemical name	1,3-Propanedicarboxylic acid
Chemical Abstract #	110-94-1
Common name	Glutaric acid
Synonyms	Pentanedioic acid
Uses	
Molecular formula	$C_3H_8O_4$
Molecular weight	132.11
Properties	m.p. 98.0°C; b.p. 303.0°C (decomposes)

Chemical name	1,2,3-Propanetriol
Chemical Abstract #	56-81-5
Common name	Glycerol
Synonyms	Glycerine; glycyl alcohol; trihydroxypropane
Uses	Used as a solvent; humectant; plasticizer; emollient; sweetener; in explosive manufacturing; bacteriostat; ink production; lubricants; liquors; confectioneries
Molecular formula	$C_3H_8O_3$
Molecular weight	92.09
Properties	m.p. 17.8°C, b.p. 291.0°C

Chemical name	2,3-Epoxy-1-propanol
Chemical Abstract #	556-52-5
Common name	Glycidol
Synonyms	Glycide; epihydric alcohol; oxiranemethanol; 3-hydroxypropylene oxide
Uses	
Molecular formula	$C_3H_6O_2$
Molecular weight	74.08
Properties	b.p. 167.0°C (decomposes)

Chemical name	Ethylene glycol diacetate
Chemical Abstract #	111-55-7
Common name	Glycol diacetate
Synonyms	Ethylene diacetate
Uses	Solvent for cellulose esters and resins; used in inks and lacquers; perfume fixative; nondiscoloring plasticizer
Molecular formula	$C_6H_{10}O_4$
Molecular weight	146.14
Properties	m.p. –31.0°C, b.p. 186.0–191.0°C

Chemical name	7-Chloro-2′,4,6-trimethoxy-6′-methylspiro[benzofuran-2(3*H*),1′-[2]cyclohexene] 3,4′-dione
Chemical Abstract #	126-07-8
Common name	Griseofulvin, Amudane, Fulcin, Lamoryl, Polygris, Poncyl, Sporostatin
Synonyms	7-Chloro-4,6-dimethoxycoumaran-3-one-2 spiro-1′-(2- methoxy-6′-methylcyclohex-2′-en-4′-one)
Uses	Fungicide (antibiotic)
Molecular formula	$C_{17}H_{17}ClO_6$
Molecular weight	352.77
Properties	m.p. 220.0°C

Chemical name	Hydroxyacetic acid
Chemical Abstract #	79-14-1
Common name	Glycolic acid
Synonyms	Hydroxyethanoic acid
Uses	Used as cheap organic acid for cleaning, copper brightening, dyeing, electroplating, pickling; and milling of metals; in processing textiles, leather, and metals
Molecular formula	$C_2H_4O_3$
Molecular weight	76.05
Properties	m.p. 80.0°C; b.p. decomposes

Chemical name	Gusathion
Chemical Abstract #	86-50-0
Common name	Gusathion
Synonyms	Active ingredient: Azinphosmethyl (25%); Demeton-*S*-methylsulfone (7.5%)
Uses	Insecticide; acaricide
Molecular formula	
Molecular weight	
Properties	

Chemical name 1,4,5,6,7,8,8-Heptachloro-3a,4,7,7a-tetrahydro-4,7-methano-indene
Chemical Abstract # 76-44-8
Common name Heptachlor
Synonyms Heptachlorodicyclopentadiene; velsocol 104; Drinox; Heptamul
Uses Insecticide for control of cottonboll weevil
Molecular formula $C_{10}H_5Cl_7$
Molecular weight 373.35
Properties m.p. 95.0–96.0°C

Chemical name *n*-Heptane
Chemical Abstract # 142-82-5
Common name
Synonyms
Uses Standard for testing knock in gasoline engines
Molecular formula C_7H_{16}
Molecular weight 100.20
Properties m.p. –91.0°C, b.p. 98.0°C

Chemical name *n*-Heptylalcohol
Chemical Abstract # 111-70-6
Common name 1-Heptanol
Synonyms Enanthic alcohol; 1-hydroxyheptane
Uses
Molecular formula $C_7H_{16}O$
Molecular weight 116.20
Properties m.p. –34.6°C, b.p. 176.0°C

Chemical name Hexachlorobenzene
Chemical Abstract # 118-74-1
Common name Perchlorobenzene
Synonyms Anticarie; Bunt-cure; Julin's carbon chloride
Uses In organic syntheses; fungicide; seed treatment; impregnation of paper
Molecular formula C_6Cl_6
Molecular weight 284.80
Properties m.p. 227.0–229.0°C, b.p. 322.0–326.0°C

Chemical name Hexachlorobutadiene
Chemical Abstract # 87-68-3
Common name
Synonyms
Uses Solvent for natural and synthetic rubber transformer liquid; hydraulic fluids; heat-transfer liquids
Molecular formula C_4Cl_6
Molecular weight 54.09
Properties m.p. –19.0°C, b.p. 210.0–220.0°C

Chemical name γ-Hexachlorocyclohexane
Chemical Abstract # 58-89-9
Common name Lindane
Synonyms Benzenehexachloride; BHC; HCCH; HCH; TBH; Chloran
Uses Medicinal manufacturing; insecticide manufacturing
Molecular formula $C_6H_6Cl_6$
Molecular weight 290.85
Properties m.p. 112.5°C

Chemical name	Hexachlorocyclopentadiene
Chemical Abstract #	77-47-4
Common name	PCL, C-56
Synonyms	Perchlorocyclopentadiene
Uses	Used in the production of insecticides, shockproof plastics, acids, esters, ketones, and fluorocarbons
Molecular formula	C_5Cl_6
Molecular weight	273.0
Properties	m.p. 9.0–10.0°C, b.p. 234.0°C

Chemical name	Hexachloroethane
Chemical Abstract #	67-72-1
Common name	Phenohep, HCE, Egitol, Distopan
Synonyms	Perchloroethane; carbon hexachloride
Uses	Manufacturing of grenades and smoke candles; plasticizers for cellulose esters; accelerator in rubber vulcanizing; moth repellant; medical manufacturing
Molecular formula	C_2Cl_6
Molecular weight	236.72
Properties	b.p. 187.0°C (readily sublimes without melting)

Chemical name	1-Hexadecanol
Chemical Abstract #	36653-82-4
Common name	Ethol; ethal
Synonyms	Cetyl alcohol; *n*-hexadecylalcohol; palmityl alcohol
Uses	
Molecular formula	$C_{16}H_{34}O$
Molecular weight	242.44
Properties	m.p. 49.3°C, b.p. 190.0°C

Chemical name	*n*-Hexaldehyde
Chemical Abstract #	66-25-1
Common name	*n*-Hexanol
Synonyms	Caproaldehyde; *n*-caproic aldehyde; *n*-hexoic aldehyde
Uses	Insecticides; synthetic resins; organic synthesis of plasticizers
Molecular formula	$C_6H_{12}O$
Molecular weight	100.16
Properties	m.p. –56.0°C, b.p. 128.0–131.0°C

Chemical name	*n*-Hexane
Chemical Abstract #	110-54-3
Common name	Gettysolve-B
Synonyms	Hexane
Uses	Used in thermometers instead of mercury; to determine refractive index of minerals
Molecular formula	C_6H_{14}
Molecular weight	86.17
Properties	m.p. –94.3°C, b.p. 68.7°C

Chemical name	*n*-Hexanol
Chemical Abstract #	111-27-3
Common name	1-Hexanol
Synonyms	*n*-Hexyl alcohol; amylcarbinol; pentylcarbinol
Uses	Used in the manufacture of antiseptics, hypnotics; solvents; intermediate for textile and leather finishing; pharmaceuticals
Molecular formula	$C_6H_{14}O$
Molecular weight	102.2
Properties	m.p. –51.6°C, b.p. 157.0°C

Chemical name	2-Hexanol
Chemical Abstract #	26401-20-7
Common name	
Synonyms	Butylmethylcarbinol
Uses	
Molecular formula	$C_6H_{14}O$
Molecular weight	102.17
Properties	m.p. 136.0–140.0°C

Chemical name	3-Hexanol
Chemical Abstract #	544-12-7
Common name	
Synonyms	Ethylpropylcarbinol
Uses	
Molecular formula	$C_6H_{14}O$
Molecular weight	102.17
Properties	b.p. 135.0°C

Chemical name	*N*-Hexylamine
Chemical Abstract #	111-26-2
Common name	
Synonyms	
Uses	
Molecular formula	$C_6H_{15}N$
Molecular weight	101.19
Properties	m.p. –19.0°C, b.p. 132.0°C

Chemical name	Hexylene glycol
Chemical Abstract #	107-41-5
Common name	HG; pinakon
Synonyms	2-Methyl-2,4-pentanediol
Uses	Cosmetics; hydraulic brake fluid; printing inks; emulsifying agent; ice inhibitor in carburetors
Molecular formula	$C_6H_{14}O_2$
Molecular weight	118.17
Properties	m.p. –40.0°C, b.p. 198.0°C

Chemical name	Hydrazine
Chemical Abstract #	302-01-2
Common name	Diamide
Synonyms	Diamine
Uses	Reducing agent; propellant in aerospace operations
Molecular formula	H_4N_2
Molecular weight	32.05
Properties	m.p. 1.4°C, b.p. 113.5°C

Chemical name	Hydrocyanic acid
Chemical Abstract #	74-90-8
Common name	Cyclon, HCN
Synonyms	Prussic acid, hydrogen cyanide
Uses	Rodenticide; insecticide
Molecular formula	HCN
Molecular weight	27.03
Properties	m.p. –13.3°C, b.p. 25.6°C

Chemical name	Hydrogen sulfide
Chemical Abstract #	7783-06-4
Common name	Stink damp
Synonyms	Sulfurated hydrogen; hydrosulfuric acid
Uses	Chemical manufacturing; analytical reagent
Molecular formula	H_2S
Molecular weight	34.08
Properties	m.p. –85.5°C, b.p. –60.4°C

Chemical name	Hydroquinone
Chemical Abstract #	123-31-9
Common name	Quinol
Synonyms	1,4-Dihydroxybenzene; 1,4-benzenediol; hydroquinol; *p*-hydroxyphenol
Uses	Inhibitor of polymerization; photographic developer; dye intermediate; medicine
Molecular formula	$C_6H_6O_2$
Molecular weight	110.12
Properties	m.p. 170.5°C, b.p. 286.2°C

Chemical name	Hydroquinone monosulfonate, sodium
Chemical Abstract #	
Common name	
Synonyms	
Uses	Reducing agent; substitution product of hydroquinone
Molecular formula	$C_6H_7O_5SNa$
Molecular weight	212.12
Properties	

Chemical name	*m*-Hydroxybenzoic acid
Chemical Abstract #	99-06-9
Common name	*m*-Salicylic acid
Synonyms	3-Carboxyphenol
Uses	Pharmaceuticals; intermediate for plasticizers; light stabilizers; petroleum additives
Molecular formula	$C_7H_6O_3$
Molecular weight	138.12
Properties	m.p. 201.0–204.0°C

Chemical name	Hydroxylamine sulfate
Chemical Abstract #	10039-54-0
Common name	Oxammonium sulfate
Synonyms	
Uses	Anti-oxidant in developing solutions
Molecular formula	$H_8N_2O_6S$
Molecular weight	164.14
Properties	m.p. 170.0°C

Chemical name	4-Hydroxymethyl-2,6-di-*tert*-butylphenol
Chemical Abstract #	88-26-6
Common name	Ionox 100
Synonyms	
Uses	Antioxidant
Molecular formula	$C_{15}H_{24}O_2$
Molecular weight	236.4
Properties	m.p. 141.0°C

Chemical name	Polyoxyethylated alkylphenol
Chemical Abstract #	26027-38-3
Common name	Igepals
Synonyms	
Uses	Nonionic emulsifiers
Molecular formula	
Molecular weight	
Properties	

Chemical name	*N*-(Mercaptomethyl)phthalimide *S*-(*O*,*O*-dimethylphosphorodithioate
Chemical Abstract #	732-11-6
Common name	Imidan
Synonyms	Phosmet; PMP
Uses	Insecticide; acaricide
Molecular formula	$C_{11}H_{12}NO_4PS_2$
Molecular weight	317.32
Properties	m.p. 71.9°C

Chemical name	Indene
Chemical Abstract #	95-13-6
Common name	
Synonyms	Indonaphthene
Uses	Paint, coating, and tile manufacturing
Molecular formula	C_9H_8
Molecular weight	116.15
Properties	m.p. –1.8°C, b.p. 181.6°C

Chemical name	Isobutanol
Chemical Abstract #	78-83-1
Common name	Isobutyl alcohol
Synonyms	Isopropylcarbinol; 2-methylpropan-1-ol
Uses	Manufacturing of esters; paint solvent; varnish remover
Molecular formula	$C_4H_{10}O$
Molecular weight	74.12
Properties	m.p. –108.0°C, b.p. 108.0°C

Chemical name	Isobutyl acetate
Chemical Abstract #	110-19-0
Common name	2-Methyl propyl acetate
Synonyms	β-methyl propyl ethanoate
Uses	Flavoring; solvent
Molecular formula	$C_6H_{12}O_2$
Molecular weight	116.16
Properties	m.p. –99.0°C, b.p. 118.0°C

Chemical name	2-Methyl-1-propanamine
Chemical Abstract #	78-81-9
Common name	Isobutylamine
Synonyms	1-Amino-2-methylpropane
Uses	
Molecular formula	$C_4H_{11}N$
Molecular weight	73.14
Properties	m.p. –85.0°C, b.p. 68.0°C

Chemical name	2-Methylpropanoic acid
Chemical Abstract #	79-31-2
Common name	Isobutyric acid
Synonyms	Dimethylacetic acid
Uses	
Molecular formula	$C_4H_8O_2$
Molecular weight	88.10
Properties	m.p. –47.0°C, b.p. 154.4°C

Chemical name	2-Methylpropanal
Chemical Abstract #	78-84-2
Common name	Isobutraldehyde; isobutyric aldehyde
Synonyms	Isobutanal; isobutyl aldehyde
Uses	For synthesizing pantothenic acid, valine, leucine, cellulose esters, perfumes, plasticizers, and gasoline additives
Molecular formula	C_4H_8O
Molecular weight	72.10
Properties	m.p. –65.9°C, b.p. 64.0°C

Chemical name	Isodrin
Chemical Abstract #	465-73-6
Common name	Endo,endo isomer of Aldrin
Synonyms	1,2,3,4,10,10-Hexachloro-1,4,4a,5,8,8a-hexahydro-1,4,5,8-endo,endo-dimethanonaphthalene
Uses	Insecticide
Molecular formula	$C_{12}H_8Cl_6$
Molecular weight	364.90
Properties	m.p. 240.0–242.0°C

Chemical name	Isooctyl alcohol
Chemical Abstract #	26952-21-6
Common name	
Synonyms	Isooctanol
Uses	Solvent
Molecular formula	$C_8H_{18}O$
Molecular weight	130.23
Properties	b.p. 182.0/195.0°C

Chemical name	1,3-Benzenedicarboxylic acid
Chemical Abstract #	121-91-5
Common name	Isophthalic acid
Synonyms	*m*-Phthalic acid
Uses	
Molecular formula	$C_8H_6O_4$
Molecular weight	166.13
Properties	m.p. 345.0–348.0°C, b.p. sublimes without decomposition

Chemical name 7-Ethenyl-1,2,3,4,4a,4b,5,6,7,8,10,10a-dodecahydro-1,4a,7-trimethyl-1-phenanthrenecarboxylic acid
Chemical Abstract # 5835-26-7
Common name Isopimaric acid
Synonyms Miropinic acid
Uses
Molecular formula $C_{20}H_{30}O_2$
Molecular weight 302.44
Properties m.p. 160.0°C

Chemical name 2-Methyl-1,3-butadiene
Chemical Abstract # 78-79-5
Common name Isoprene
Synonyms Methylbivinyl; hemiterpene
Uses In the manufacture of synthetic rubber, butyl rubber, butyl elastomers
Molecular formula C_5H_8
Molecular weight 68.11
Properties m.p. –145.95°C, b.p. 34.07

Chemical name 2-Propanol
Chemical Abstract # 67-63-0
Synonyms Isopropyl alcohol
Common name Dimethylcarbinol; petrohol; avantine
Uses Solvent; in antifreeze composition; fast-drying oils and inks; in body rubs and hand lotions; manufacturing of acetone, glycerol, and isopropyl acetate
Molecular formula C_3H_8O
Molecular weight 60.09
Properties m.p. –88.5°C, b.p. 82.4°C

Chemical name 2-Isopropoxyethanol
Chemical Abstract # 109-59-1
Common name Isopropyl glycol
Synonyms Isopropyl glycol ether; isopropyloxitol
Uses
Molecular formula $C_5H_{12}O_2$
Molecular weight 104.15
Properties b.p. 140.0–144.0°C

Chemical name Isopropyl acetate
Chemical Abstract # 108-21-4
Common name
Synonyms
Uses Solvent for cellulose derivatives, plastics, paints, lacquers, printing inks and fats; used in perfumery
Molecular formula $C_5H_{10}O_2$
Molecular weight 102.1
Properties m.p. –73.0°C, b.p. 89.0°C

Chemical name 2-Propanamine
Chemical Abstract # 75-31-0
Common name Isopropylamine
Synonyms 2-Aminopropane
Uses
Molecular formula C_3H_9N
Molecular weight 59.08
Properties m.p. –101.0°C, b.p. 33.0–34.0°C

Chemical name	2-Phenylpropane
Chemical Abstract #	98-82-8
Common name	Isopropylbenzene
Synonyms	Cumene; (1-methylethyl)benzene
Uses	Manufacture of acetone, phenol, polymerization catalysts; component of motor fuel; solvent
Molecular formula	C_9H_{12}
Molecular weight	120.19
Properties	m.p. –96.0°C, b.p. 152.7°C

Chemical name	α,α-Dimethylbenzyl hydroperoxide
Chemical Abstract #	80-15-9
Common name	Cumine hydroperoxide
Synonyms	Isopropylbenzene hydroxyperoxide
Uses	Production of acetone, phenol; polymerization catalyst
Molecular formula	$C_9H_{12}O_2$
Molecular weight	152.2
Properties	

Chemical name	Isopropyl-*N*-(3-chlorophenyl)carbamate
Chemical Abstract #	101-21-3
Common name	Chlorpropham
Synonyms	CIPC; chloro-PIC; 1-methyl ethyl ester
Uses	Herbicide; plant growth regulator; pesticide
Molecular formula	$C_{10}H_{12}ClNO_2$
Molecular weight	213.68
Properties	m.p. 40.7–41.1°C, b.p. 149.0°C

Chemical name	6-Isopropyl-4-methyl-2-pyrimidinol
Chemical Abstract #	2814-20-2
Common name	
Synonyms	
Uses	
Molecular formula	
Molecular weight	
Properties	

Chemical name	Isopropyl-*N*-phenylcarbamate
Chemical Abstract #	122-42-9
Common name	Propham; isopropylcarbanilate
Synonyms	IPC; phenylcarbamic acid 1-methyl ethyl ester
Uses	Herbicide, applied as spray to the soil
Molecular formula	$C_{10}H_{13}NO_2$
Molecular weight	179.21
Properties	m.p. 90.0°C

Chemical name	Isoquinoline
Chemical Abstract #	119-65-3
Common name	Benzo(c)pyridine
Synonyms	2-Benzazine
Uses	Insecticides; synthesis of dyes; rubber accelerator
Molecular formula	C_9H_7N
Molecular weight	129.15
Properties	m.p. 26.48°C, b.p. 242.0°C

Chemical name	3-Methylbutanoic acid
Chemical Abstract #	503-74-2
Common name	Isovaleric acid
Synonyms	Isopropylacetic acid; delphinic acid
Uses	In flavors; in perfumes; in manufacturing sedatives
Molecular formula	$C_5H_{10}O_2$
Molecular weight	102.13
Properties	m.p. –37.6°C, b.p. 175.0–177.0°C

Chemical name	4-Chloro-α-(4-chlorophenyl)-α-(trichloromethyl) benzenemethanol
Chemical Abstract #	115-32-2
Common name	Kelthane, Dicofol, Acarin, Mitigan
Synonyms	1,1-Bis(*p*-chlorophenyl)-2,2,2 trichloroethanol
Uses	Acaricide, miticidal pesticide
Molecular formula	$C_{14}H_9Cl_5O$
Molecular weight	370.47
Properties	m.p. 77.0–78.0°C

Chemical name Decachlor octahydro-1,3,4-metheno-2*H*-cyclobuta-(c,d)pentalen-2-one
Chemical Abstract # 143-50-0
Common name Kepone
Synonyms Chlordecone
Uses Insecticide; fungicide
Molecular formula $C_{10}Cl_{10}O$
Molecular weight 490.68
Properties Decomposes at 350.0°C

Chemical name 2-Hydroxypropanoic acid
Chemical Abstract # 598-82-3
Common name *dl*-Lactic acid
Synonyms Racemic lactic acid; α-hydroxypropionic acid
Uses Solvent; manufacturing of cheese; in brewing; dehairing, plumping, and decalcifying hides; flux for soft solder
Molecular formula $C_3H_6O_3$
Molecular weight 90.08
Properties m.p. 16.0–18.0°C, b.p. 122.0°C

Chemical name 2-Hydroxypropanenitrile
Chemical Abstract # 78-97-7
Common name Lactonitrile
Synonyms Acetaldehydecyanohydrin; ethylidenecyanohydrin
Uses
Molecular formula C_3H_5NO
Molecular weight 71.08
Properties m.p. –40.0°C, b.p. 183.0°C; 103.0°C (at 50.0 mm)

Chemical name Dodecanoic acid
Chemical Abstract # 143-07-7
Common name Lauric acid, Neofat-12, Wecoline-1295
Synonyms Laurostearic acid
Uses Wetting agent; soaps; detergents; insecticides; cosmetics; food additives
Molecular formula $C_{12}H_{24}O_2$
Molecular weight 200.31
Properties m.p. 44.0°C, b.p. 225.0°C (at 100.0 mm)

Chemical name Sulfuric acid monododecyl ester sodium salt
Chemical Abstract # 151-21-3
Common name Sodium lauryl sulfate
Synonyms Sodium dodecyl sulfate; SDS; Irium
Uses Wetting agent; detergent; electrophoretic separation of proteins and lipids; ingredient in toothpaste
Molecular formula $C_{12}H_{25}NaO_4S$
Molecular weight 288.38
Properties m.p. 204.0–207.0°C

Chemical name Lead acetate
Chemical Abstract # 301-04-2
Common name Sugar of lead
Synonyms Salt of saturn
Uses Dyeing and printing cottons; medicinal; waterproofing; varnishes; insecticides; antifouling paints
Molecular formula $C_4H_6O_4Pb$
Molecular weight 325.28
Properties m.p. 75.0°C; decomposes completely > 200.0°C

Chemical name	Phenylphosphonothioic acid *O*-(4-bromo-2,5-dichlorophenyl) *O*-methyl ester
Chemical Abstract #	21609-90-5
Common name	Leptophos, Phosvel, MBCP
Synonyms	*O*-(4-Bromo-2,5-dichlorophenyl) *O*-methyl phenylphosphonothioate
Uses	Insecticide
Molecular formula	$C_{13}H_{10}BrCl_2O_2PS$
Molecular weight	412.06
Properties	m.p. 55.0–67.0°C

Chemical name	Linear alkyl sulfonate
Chemical Abstract #	68411-30-3
Common name	LAS
Synonyms	
Uses	
Molecular formula	
Molecular weight	
Properties	

Chemical name	Linear plasticizer alcohols
Chemical Abstract #	83968-18-7
Common name	Linevol 79 (composition: C_7 47%, C_8 36%, C_9 17%); Linevol 911 (composition: C_9 19%, C_{10} 48%, C_{11} 33%)
Synonyms	Neoflex
Uses	Manufacturing of plasticizers
Molecular formula	
Molecular weight	
Properties	Linevol 79 boiling range 180.0–214.0°C; Linevol 911 boiling range 225.0–248.0°C

Chemical name	[(Dimethoxyphosphinothioyl)thio]butanedioic acid diethyl ether
Chemical Abstract #	121-75-5
Common name	Malathion, Mercaptothion, Carbofos, Phosphothion
Synonyms	*S*-(1,2-Dicarbethoxyethyl) *O,O*-dimethyldithiophosphate
Uses	Insecticide
Molecular formula	$C_{10}H_{19}O_6PS_2$
Molecular weight	330.36
Properties	m.p. 2.85°C, b.p. 156.0–157.0°C (at 0.7 mm)

Chemical name	(*cis*)-Butenedioic acid
Chemical Abstract #	110-16-7
Common name	Maleic acid, toxilic acid
Synonyms	*cis*-1,2-Ethylenedicarboxylic acid
Uses	Dyeing and finishing of cotton, wool and silk; preservative for oil and fats; manufacturing of artificial resins
Molecular formula	$C_4H_4O_4$
Molecular weight	116.07
Properties	m.p. 130.0–131.0°C, b.p. 135.0°C (decomposes)

Chemical name	2,5-Furandione
Chemical Abstract #	108-31-6
Common name	Maleic anhydride, toxilic anhydride
Synonyms	(*cis*)-Butenedioic anhydride
Uses	Pharmaceuticals; agricultural chemicals; copolymerization reactions; dye intermediates
Molecular formula	$C_4H_2O_3$
Molecular weight	98.06
Properties	m.p. 52.8°C, b.p. 202.0°C

Chemical name 1,2-Dihydro-3,6-pyridazinedione
Chemical Abstract # 123-33-1
Common name Maleic hydrazide; regulox 36
Synonyms Maleic acid hydrazide; MH; Fazor; malazide
Uses Herbicide; tobacco suckering control; in synthesis of pyridazine
Molecular formula $C_4H_4N_2O_2$
Molecular weight 112.09
Properties Decomposes at 260.0°C

Chemical name Hydroxybutanedioic acid
Chemical Abstract # 6915-15-7
Common name Malic acid
Synonyms Hydroxysuccinic acid
Uses In organic synthesis as intermediate
Molecular formula $C_4H_6O_5$
Molecular weight 134.09
Properties m.p. 131.0°C (*dl* form)

Chemical name Propanedinitrile
Chemical Abstract # 109-77-3
Common name Malononitrile
Synonyms Methylene cyanide; dicyanomethane; cyanoacetonitrile
Uses In organic syntheses
Molecular formula $C_3H_2N_2$
Molecular weight 66.06
Properties m.p. 32.0°C, b.p. 218.0°C

Chemical name {[1,2-Ethanediylbis-(carbamodithioato)]-(2-)} manganese
Chemical Abstract # 12427-38-2
Common name Dithane; Maneb; Trimangol 80
Synonyms Manganese ethylene bisbithiocarbamate
Uses Agricultural fungicide
Molecular formula $(C_4H_6MnN_2S_4)_n$
Molecular weight Polymeric salt
Properties

Chemical name 1,3,5-Triazine 2,4,6-triamine
Chemical Abstract # 108-78-1
Common name Melamine, cyanurotriamide
Synonyms 2,4,6-Triamino-*s*-triazine
Uses Melamine resins; leather tanning; organic synthesis
Molecular formula $C_3H_6N_6$
Molecular weight 126.13
Properties m.p. < 250.0°C, b.p. sublimes

Chemical name Mercaptoacetic acid
Chemical Abstract # 68-11-1
Common name Thioglycolic acid
Synonyms Thioranic acid; 2-mercaptoethanoic acid
Uses Used in bacteriology; in manufacturing of thioglycolates
Molecular formula $C_2H_4O_2S$
Molecular weight 92.12
Properties m.p. –16.5°C, b.p. 108.0°C (at 15 mm)

Chemical name 2-Mercaptobenzothiazole
Chemical Abstract # 149-30-4
Common name Captax; MBT
Synonyms Dermacid; Mertax; Thiotax; 2-benzothiazolethiol
Uses Fungicide; rubber vulcanization accelerator
Molecular formula $C_7H_5NS_2$
Molecular weight 167.25
Properties m.p. 180.2–181.7°C (170.0–175.0°C, technical grade)

Chemical name Mercuric acetate
Chemical Abstract # 1600-27-7
Common name Mercuryl acetate
Synonyms Diacetoxymercury
Uses Mercuration of organic compounds; for ethylene absorption
Molecular formula $C_4H_6HgO_4$
Molecular weight 318.70
Properties m.p. 178.0–180.0°C, decomposes when overheated

Chemical name Mercuric thiocyanate
Chemical Abstract # 592-85-8
Common name Mercuric dithiocyanate
Synonyms Mercuric sulfocyanate; mercuric sulfocyanide
Uses Fireworks; intensifier in photography
Molecular formula $C_2HgN_2S_2$
Molecular weight 316.79
Properties Decomposes > 165.0°C

Chemical name 1,3,5-Trimethylbenzene
Chemical Abstract # 108-67-8
Common name Mesitylene
Synonyms Trimethylbenzene
Uses Ultraviolet oxidation stabilizer for plastics
Molecular formula C_9H_{12}
Molecular weight 120.21
Properties m.p. –44.8°C, b.p. 164.7°C

Chemical name 4-Methyl-3-penten-2-one
Chemical Abstract # 141-79-7
Common name Mesityl oxide; MO
Synonyms Methylisobutenyl ketone
Uses Solvent for many gums and resins; in making isobutyl ketone
Molecular formula $C_6H_{10}O$
Molecular weight 98.16
Properties m.p. –59.0°C, b.p. 130.0°C

Chemical name 3,5-Dimethyl-4-(methylthio)phenol methylcarbamate
Chemical Abstract # 2032-65-7
Common name Mesurol, Methiocarb, Mercaptodimethur
Synonyms 4-(Methylthio)-3,5-xylyl methyl carbamate
Uses Insecticide; molluscicide; bird repellent
Molecular formula $C_{11}H_{15}NO_2S$
Molecular weight 225.31
Properties m.p. 121.5°C

Chemical name 2-Methylpropenoic acid
Chemical Abstract # 79-41-4
Common name Methacrylic acid
Synonyms α-Methylacrylic acid
Uses Manufacturing of resins and plastics
Molecular formula $C_4H_6O_2$
Molecular weight 86.09
Properties m.p. 16.0°C, b.p. 163.0°C

Chemical name 2-Methyl-2-propenenitrile
Chemical Abstract # 126-98-7
Common name Methacrylonitrile
Synonyms Isopropene cyanide
Uses In preparation of acids, amides, amines, esters, nitriles, and polymers
Molecular formula C_4H_5N
Molecular weight 76.09
Properties m.p. –35.8°C, b.p. 90.3°C

Chemical name Methanesulfonyl chloride
Chemical Abstract # 594-44-5
Common name Mesylchloride
Synonyms
Uses Flame-resistant products; biological chemicals
Molecular formula CH_3ClO_2S
Molecular weight 114.55
Properties b.p. 164.0°C

Chemical name	Methyl alcohol
Chemical Abstract #	67-56-1
Common name	Methanol; wood alcohol
Synonyms	Carbinol; wood spirit
Uses	Industrial solvent; making formaldehyde and methyl esters; antifreeze; octane booster; fuel; extractant for oils
Molecular formula	CH_4O
Molecular weight	32.04
Properties	m.p. –98°C, b.p. 64.7°C

Chemical name	1,1′-(2,2,2-Trichloroethylidene)-bis(4-methoxybenzene)
Chemical Abstract #	72-43-5
Common name	Methoxychlor; Methoxy DDT; DMDT;
Synonyms	2,2-Bis(*p*-methoxyphenyl)-1,1,1- trichloroethane
Uses	Insecticide
Molecular formula	$C_{16}H_{15}Cl_3O_2$
Molecular weight	345.65
Properties	m.p. 78.0–78.2°C or 86.0–88.0°C

Chemical name	4-Methoxy-4-methyl-2-pentanone
Chemical Abstract #	107-70-0
Common name	Pentoxane; ME-6K
Synonyms	Methoxyhexanone
Uses	
Molecular formula	$C_7H_{15}O_2$
Molecular weight	130.02
Properties	b.p. 159.1°C

Chemical name	1-Methoxy-2-nitrobenzene
Chemical Abstract #	91-23-6
Common name	*o*-Nitrophenylmethyl ether
Synonyms	*o*-Nitroanisole
Uses	Dye manufacturing; organic syntheses
Molecular formula	$C_7H_7NO_3$
Molecular weight	153.13
Properties	m.p. 9.4°C, b.p. 277.0°C

Chemical name	1-Methoxy-4-nitrobenzene
Chemical Abstract #	100-17-4
Common name	*p*-Nitrophenylmethyl ether
Synonyms	*p*-Nitroanisole
Uses	Dye intermediate; organic syntheses
Molecular formula	$C_7H_7NO_3$
Molecular weight	153.13
Properties	m.p. 54.0°C, b.p. 260.0°C

Chemical name	Methyl acetate
Chemical Abstract #	79-20-9
Common name	
Synonyms	
Uses	Solvent for nitrocellulose, acetyl cellulose, resins and oils; used in making artificial leathers
Molecular formula	$C_3H_6O_2$
Molecular weight	74.08
Properties	m.p. –98.0°C, b.p. 56.9°C

Chemical name 2-Propenoic acid methyl ester
Chemical Abstract # 96-33-3
Common name Methylacrylate
Synonyms Acrylic acid methyl ester
Uses Plastic films; synthetic leather industry; paper and textile coatings; for making the hardest resin of acrylate ester series
Molecular formula $C_4H_6O_2$
Molecular weight 86.09
Properties m.p. –76.5°C, b.p. 80.0°C

Chemical name Methanamine
Chemical Abstract # 74-89-5
Common name Methylamine; aminomethane
Synonyms Monomethylamine
Uses Insecticide; fungicide; tanning and dyeing industry; solvent; photographic developer; rocket propellant
Molecular formula CH_5N
Molecular weight 31.06
Properties m.p. –93.5°C, b.p. –6.5°C

Chemical name Bromomethane
Chemical Abstract # 74-83-9
Common name Methyl bromide; embafume
Synonyms Monobromomethane
Uses Extraction of oils from seeds; organic syntheses; insect fumigant
Molecular formula CH_3Br
Molecular weight 94.95
Properties m.p. –93.66°C, b.p. 3.56°C

Chemical name 4-Methylaminophenolsulfate
Chemical Abstract # 55-55-0
Common name Verol, Rhodol, Armol, Genol, Photo-Rex
Synonyms *p*-Hydroxymethylaniline sulfate
Uses Photographic developer; for dyeing furs
Molecular formula $C_{14}H_{20}N_2O_2S$; $(HOC_6H_4NHCH_3)_2H_2SO_4$
Molecular weight 344.28
Properties m.p. 260.0°C with decomposition

Chemical name 2-Methyl-2-butanol
Chemical Abstract # 115-19-5
Common name *tert*-Amyl alcohol; *tert*-pentanol; dimethylethylcarbinol
Synonyms *tert*-Pentyl alcohol
Uses In organic synthesis
Molecular formula $C_5H_{12}O$
Molecular weight 88.15
Properties m.p. –11.9°C, b.p. 101.8°C

Chemical name Chloromethane
Chemical Abstract # 74-87-3
Common name Methyl chloride
Synonyms
Uses Manufacturing of fumigants, organic chemicals, synthetic rubber, silicones; solvent; herbicide; medicine; refrigerants
Molecular formula CH_3Cl
Molecular weight 50.49
Properties m.p. –97.7°C, b.p. –24.0°C

Chemical name	2-Methyl-4-chlorophenoxyacetic acid
Chemical Abstract #	94-74-6
Common name	MCPA; Methoxone; MCP; Agritox; Cornox
Synonyms	(4-Chloro-*o*-toloxy) acetic acid
Uses	Selective weed killer; hormone-type herbicide
Molecular formula	$C_9H_9ClO_3$
Molecular weight	200.63
Properties	m.p. 120.0°C

Chemical name	Methylcyclohexane
Chemical Abstract #	108-87-2
Common name	
Synonyms	Hexahydrotoluene; cyclohexylmethane
Uses	Organic synthesis; solvent
Molecular formula	C_7H_8
Molecular weight	98.18
Properties	m.p. –126.0°C, b.p. 101.0°C

Chemical name	2-Methylcyclohexanone
Chemical Abstract #	583-60-8
Common name	Sexton B
Synonyms	Methylanon
Uses	
Molecular formula	C_7H_8O
Molecular weight	112.17
Properties	m.p. –14.0°C, b.p. 165.1°C

Chemical name	Dichloromethane
Chemical Abstract #	75-09-2
Common name	Methylene chloride
Synonyms	Methylene dichloride; methylene bichloride
Uses	Solvent; degreasing and cleansing fluid; solvent in food processing; organic syntheses; pharmaceuticals
Molecular formula	CH_2Cl_2
Molecular weight	84.93
Properties	m.p. –97.0°C, b.p. 40.0–42.0°C

Chemical name	3,7-Bis(dimethylamino)phenothiazin-5-ium chloride
Chemical Abstract #	61-73-4
Common name	Methylene blue; C.I. basic blue 9
Synonyms	Methylthioninium chloride; tetramethylthionine chloride
Uses	Indicator in reactions; stain in bacteriology; chemical reagent
Molecular formula	$C_{16}H_{18}ClN_3S$
Molecular weight	319.85
Properties	m.p. 190.0°C; decomposes

Chemical name	2-Butanone
Chemical Abstract #	78-93-3
Common name	Methyl ethyl ketone
Synonyms	Ethyl methyl ketone; 2-oxobutane
Uses	Solvent; manufacturing of smokeless powder; cements and adhesives; cleaning fluids; swelling agent for resins
Molecular formula	C_4H_8O
Molecular weight	72.1
Properties	m.p. –86.4°C, b.p. 79.6°C

Chemical name	2-Methylfuran
Chemical Abstract #	534-22-5
Common name	Silvan; sylvan
Synonyms	Methylfuran
Uses	Chemical intermediate
Molecular formula	C_5H_6O
Molecular weight	82.10
Properties	m.p. –88.7°C, b.p. 63.7°C

Chemical name	Methyl iodide
Chemical Abstract #	74-88-4
Common name	
Synonyms	Iodomethane
Uses	In methylations and microscopy; for pyridine testing; as imbedding material for the examination of diatoms
Molecular formula	CH_3I
Molecular weight	141.95
Properties	m.p. –66.1°C, b.p. 42.5°C

Chemical name	Methyl isoamyl ketone
Chemical Abstract #	110-12-3
Common name	
Synonyms	5-Methyl-2-hexanone
Uses	Organic synthesis; solvent
Molecular formula	$C_7H_{14}O$
Molecular weight	114.0
Properties	b.p. 144.0°C

Chemical name	4-Methyl-2-pentanone
Chemical Abstract #	108-10-1
Common name	Hexone; hexanone; MIBK
Synonyms	Isopropyl acetone; methyl isobutyl ketone
Uses	Solvent for gums, resins, paints, varnishes, lacquers; organic syntheses; denaturing of alcohol
Molecular formula	$C_6H_{12}O$
Molecular weight	100.16
Properties	m.p. –85.0°C, b.p. 116.0–119.0°C

Chemical name	2-Methyl-1-butene-3-one
Chemical Abstract #	814-78-8
Common name	
Synonyms	Methyl isopropenyl ketone
Uses	Solvent
Molecular formula	C_5H_8O
Molecular weight	84.06
Properties	m.p. –53.7°C, b.p. 97.7°C

Chemical name	Isothiocyanatomethane
Chemical Abstract #	8066-01-1
Common name	Methyl mustard oil
Synonyms	Methyl isothiocyanate
Uses	Pesticide
Molecular formula	C_2H_3NS
Molecular weight	73.12
Properties	m.p. 35–36.0°C, b.p. 119.0°C

Chemical name	Methanethiol
Chemical Abstract #	74-93-1
Common name	Mercaptomethane
Synonyms	Methyl mercaptan; thiomethyl alcohol
Uses	Jet fuel; pesticides; fungicides; plastics
Molecular formula	CH_4S
Molecular weight	48.11
Properties	m.p. –123.1°C, b.p. 6.0–7.6°C

Chemical name	Dimethylmercury
Chemical Abstract #	22967-92-6
Common name	Methylmercury
Synonyms	Mercury dimethyl
Uses	Inorganic agent
Molecular formula	C_2H_6Hg
Molecular weight	230.66
Properties	b.p. 92°C at 740 mm

Chemical name	Methyl methacrylate
Chemical Abstract #	80-62-6
Common name	Methyl ester
Synonyms	Diakon
Uses	Manufacture of methacrylate resins and plastics
Molecular formula	$C_5H_8O_2$
Molecular weight	100.11
Properties	m.p. –50.0°C, b.p. 101.0°C

Chemical name	*N*-Methylmorpholine
Chemical Abstract #	109-02-4
Common name	
Synonyms	4-Methylmorpholine
Uses	Solvent for resins, waxes, dyes; emulsifier; antioxidant
Molecular formula	$C_5H_{11}NO$
Molecular weight	101.2
Properties	m.p. –65.0°C, b.p. 111.0–115.0°C

Chemical name	1-Methylnaphthalene
Chemical Abstract #	90-12-0
Common name	
Synonyms	α-Methylnaphthalene
Uses	Manufacturing of insecticides and phthalic anhydride; constituent of asphalt and naphtha
Molecular formula	$C_{11}H_{20}$
Molecular weight	142.19
Properties	m.p. –22.0°C, b.p. 244.6°C

Chemical name	*O*-Methylnitrobenzene
Chemical Abstract #	88-72-2
Common name	Methylnitrobenzene
Synonyms	2-Nitrotoluene
Uses	Manufacture of dyes; organic synthesis
Molecular formula	$C_7H_7NO_2$
Molecular weight	137.15
Properties	m.p. –10.0°C, b.p. 222.3°C

Chemical name	*m*-Methylnitrobenzene
Chemical Abstract #	99-08-1
Common name	Methylnitrobenzene
Synonyms	3-Nitrotoluene
Uses	Manufacture of dyes; organic synthesis
Molecular formula	$C_7H_7NO_2$
Molecular weight	137.15
Properties	m.p. 15.1°C, b.p. 231.0°C

Chemical name	*p*-Methylnitrobenzene
Chemical Abstract #	99-99-0
Common name	Methylnitrobenzene
Synonyms	4-Nitrotoluene
Uses	Manufacture of dyes; organic synthesis
Molecular formula	$C_7H_7NO_2$
Molecular weight	137.15
Properties	m.p. 51.3°C, b.p. 238.0°C

Chemical name	1-Methylphenanthrene
Chemical Abstract #	832-69-9
Common name	
Synonyms	
Uses	
Molecular formula	$C_{15}H_{12}$
Molecular weight	192.27
Properties	

Chemical name	Propanoic acid methyl ester
Chemical Abstract #	554-12-1
Common name	Methyl propylate
Synonyms	Methyl propionate
Uses	In organic synthesis
Molecular formula	$C_4H_8O_2$
Molecular weight	88.12
Properties	m.p. –87.0°C, b.p. 79.8°C

Chemical name	*N*-Methyl-2-pyrrolidone
Chemical Abstract #	872-50-4
Common name	M-Pyrol, NMP
Synonyms	1-Methyl-2-pyrrolidone
Uses	
Molecular formula	C_5H_9NO
Molecular weight	99.15
Properties	m.p. –24.0°C, b.p. 202.0°C

Chemical name	2-Hydroxybenzoic acid methyl ester
Chemical Abstract #	119-36-8
Common name	Oil of wintergreen; betula oil; sweet birch oil
Synonyms	Methyl salicylate
Uses	Candy flavoring; perfume industry
Molecular formula	$C_8H_8O_3$
Molecular weight	152.14
Properties	m.p. –8.6°C, b.p. 222.0°C

Chemical name	2-Methylquinoline
Chemical Abstract #	91-63-4
Common name	Quinaldine
Synonyms	*o*-Toluquinoline
Uses	As anesthetic in transport and handling of fish
Molecular formula	$C_{10}H_9N$
Molecular weight	143.18
Properties	b.p. 246.0–247.0°C

Chemical name	6-Methylquinoline
Chemical Abstract #	91-62-3
Common name	*p*-Toluquinoline
Synonyms	*p*-Methyl quinoline
Uses	
Molecular formula	$C_{10}H_9N$
Molecular weight	143.18
Properties	

Chemical name	3-[(Dimethoxyphosphinyl)oxy]-2-butenoic acid methyl ester
Chemical Abstract #	7786-34-7
Common name	Mevinphos; Phosdrin
Synonyms	3-Hydroxycrotonic acid methyl ester dimethyl phosphate
Uses	Contact and systemic insecticide and acaricide
Molecular formula	$C_7H_{13}O_6P$
Molecular weight	224.16
Properties	b.p. 106.0–107.5°C

Chemical name	4-(Dimethylamino)-3,5-dimethylphenol methylcarbamate (ester)
Chemical Abstract #	315-18-4
Common name	Mexacarbate; zectran
Synonyms	Methylcarbamic acid 4-(dimethylamino)-3,5-xylyl ester
Uses	Pesticide
Molecular formula	$C_{12}H_{18}N_2O_2$
Molecular weight	222.29
Properties	m.p. 85.0°C

Chemical name	1,1a,2,2,3,3a,4,5,5,5a,5b,6-Dodeca-chlorooctahydro-1,3,4-metheno-1*H*-cyclobuta(cd)pentalene
Chemical Abstract #	2385-85-5
Common name	Mirex; dechlorane
Synonyms	Hexachloropentadiene dimer
Uses	Insecticide; fire retardant for plastics, rubber, and electrical goods
Molecular formula	$C_{10}Cl_{12}$
Molecular weight	545.59
Properties	Decomposes at 485.0°C

Chemical name	*S*-Ethylhexahydro-1*H*-azepine-1-carbthioate
Chemical Abstract #	2212-67-1
Common name	Molinate; ordram
Synonyms	
Uses	Herbicide
Molecular formula	
Molecular weight	187.30
Properties	b.p. 117.0°C at 10 mm

Chemical name	Monochloroacetic acid
Chemical Abstract #	79-11-8
Common name	
Synonyms	Chloroethanoic acid
Uses	
Molecular formula	$C_2H_3ClO_2$
Molecular weight	94.50
Properties	m.p. α form 63.0°C, β form 55.0–56.0°C, γ form 50.0°C

Chemical name	Monochlorohydroquinone
Chemical Abstract #	615-67-8
Common name	
Synonyms	
Uses	Substitution product; reducing agent in photography
Molecular formula	$C_6H_5ClO_2$
Molecular weight	144.56
Properties	

Chemical name	Monofluoroacetic acid
Chemical Abstract #	144-49-0
Common name	
Synonyms	
Uses	Sodium salt is used as a water-soluble rodenticide
Molecular formula	$C_2H_3FO_2$
Molecular weight	78.04
Properties	

Chemical name	Monomethylhydrazine
Chemical Abstract #	60-34-4
Common name	MMH
Synonyms	Methylhydrazine
Uses	Chemical syntheses; in rocket fuel
Molecular formula	CH_6N_2
Molecular weight	46.07
Properties	m.p. –52.4°C, b.p. 87.5°C

Chemical name	Monosodium methane arsonate
Chemical Abstract #	2163-80-6
Common name	MSMA, Ansar-170, Herb-All, Weed-Hol
Synonyms	Methyl arsenic acid, Na salt
Uses	Herbicide
Molecular formula	CH_4AsO_3,Na
Molecular weight	161.96
Properties	

Chemical name	Tetrahydro-1,4-oxazine
Chemical Abstract #	110-91-8
Common name	Morpholine
Synonyms	Diethyleneimide oxide
Uses	As cheap solvent for resins, waxes, casein, and dyes; morpholine-fatty acid salts are used as surface-active agents and emulsifiers; some compounds are used as corrosion inhibitors, antioxidants, plasticizers, viscosity improvers, insecticides, fungicides, herbicides; as local anesthetics and antiseptics
Molecular formula	C_4H_9NO
Molecular weight	87.12
Properties	m.p. –7.5°C, b.p. 128.9°C

Chemical name	Tetradecanoic acid
Chemical Abstract #	544-63-8
Common name	Crodacid
Synonyms	Myristic acid, 1-tridecanecarboxylic acid
Uses	Lubricants; in soaps, cosmetics, creams; in coatings for anodized aluminum
Molecular formula	$C_{14}H_{28}O_2$
Molecular weight	228.36
Properties	m.p. 58.5°C, b.p. 250.5°C (at 100.0 mm)

Chemical name	1, 2-Ethanediylbiscarbamodithioic acid disodium salt
Chemical Abstract #	142-59-6
Common name	Nabam; Dithane D-14
Synonyms	Ethylenebis(dithiocarbamic acid) disodium salt
Uses	Fungicide for agriculture
Molecular formula	$C_4H_6N_2Na_2S_4$
Molecular weight	256.35
Properties	

Chemical name	Phosphoric acid 1, 2-dibromo-2, 2-dichloroethyl dimethylester
Chemical Abstract #	300-76-5
Common name	Naled; Bromchlophos, Bromex, Dibrom
Synonyms	Dimethyl 1,2-dibromo-2,2-dichloroethyl phosphate
Uses	Acaricide, insecticide
Molecular formula	$C_4H_7Br_2Cl_2O_4P$
Molecular weight	380.79
Properties	m.p. 26.5–27.5°C, b.p. 110°C (at 0.5 mm)

Chemical name	Naphthalene
Chemical Abstract #	91-20-3
Common name	Tar camphor; moth balls
Synonyms	Naphthalin
Uses	Used in the manufacturing of dyes, synthetic resins, smokeless powder, solvents
Molecular formula	$C_{10}H_8$
Molecular weight	128.16
Properties	m.p. 80.1°C, b.p. 217.9°C

Chemical name	1-Naphthol
Chemical Abstract #	90-15-3
Common name	α-Naphthol
Synonyms	1-Naphthalenol
Uses	Manufacturing of dyes; synthetic perfumes; organic syntheses
Molecular formula	$C_{10}H_8O$
Molecular weight	144.18
Properties	m.p. 96.0°C, b.p. 282.5°C

Chemical name	2-Naphthol
Chemical Abstract #	135-19-3
Common name	β-Naphthol
Synonyms	2-Naphthalenol
Uses	Manufacturing of dyes; medicinal organics; perfumes; insecticides; antioxidants for rubber, oils, and fats
Molecular formula	$C_{10}H_8O$
Molecular weight	144.16
Properties	m.p. 121.0–123.0°C, b.p. 285.0–286.0°C

Chemical name	1,4-Naphthoquinone
Chemical Abstract #	130-15-4
Common name	α-Naphthoquinone
Synonyms	1,4-Naphthalenedione
Uses	Reagent
Molecular formula	$C_{10}H_6O_2$
Molecular weight	158.15
Properties	m.p. 126.0°C (sublimes)

Chemical name	1-Naphthylamine
Chemical Abstract #	134-32-7
Common name	α-Naphthylamine
Synonyms	1-Aminonaphthalene
Uses	Dye manufacturing; reagent
Molecular formula	$C_{10}H_9N$
Molecular weight	143.20
Properties	m.p. 50.0°C, b.p. 309.0°C

Chemical name	2-Naphthylamine
Chemical Abstract #	91-59-8
Common name	β-Naphthylamine
Synonyms	2-Naphthalenamine
Uses	Dye manufacturing
Molecular formula	$C_{10}H_9N$
Molecular weight	143.20
Properties	m.p. 115.0°C, b.p. 306.1°C

Chemical name	2,2-Dihydroxy-1*H*-indene-1,3(2*H*)-dione
Chemical Abstract #	485-47-2
Common name	Ninhydrin
Synonyms	2,2-Dihydroxy-1,3-indanedione
Uses	Indicator; chemical intermediate
Molecular formula	$C_9H_6O_4$
Molecular weight	178.14
Properties	m.p. 241.0°C (decomposes)

Chemical name	Nitrilotriacetic acid
Chemical Abstract #	139-13-9
Common name	NTA; TGA
Synonyms	*N,N*-Bis(carboxymethyl)glycine
Uses	Chemical syntheses; chelating agent
Molecular formula	$C_6H_9NO_6$
Molecular weight	191.14
Properties	m.p. 230.0–235.0°C, decomposes at 241.5°C

Chemical name	Nitrilotriacetic acid, monohydrated trisodium salt
Chemical Abstract #	18662-53-8
Common name	Triglycine, Na
Synonyms	*N,N*-Bis(carboxymethyl)glycine trisodium salt
Uses	Chelating and sequestering agent
Molecular formula	$C_6H_6NO_6Na_3H_2O$
Molecular weight	275.12
Properties	

Chemical name	*p*-Nitroaniline
Chemical Abstract #	100-01-6
Common name	Developer-P
Synonyms	*p*-Nitraniline
Uses	Dye manufacturing
Molecular formula	$C_6H_6N_2O_2$
Molecular weight	138.12
Properties	m.p. 148.5°C, b.p. 332.0°C, decomposes

Chemical name	Nitrobenzene
Chemical Abstract #	98-95-3
Common name	Oil of mirbane
Synonyms	Nitrobenzol
Uses	Manufacturing of aniline, dyes, and rubber production; solvents; lubricants
Molecular formula	$C_6H_5NO_2$
Molecular weight	123.11
Properties	m.p. 6.0°C, b.p. 210.0–211.0°C

Chemical name	*m*-Nitrobenzenesulfonate, sodium
Chemical Abstract #	127-68-4
Common name	Ludigol F,60
Synonyms	Nitrobenzene *m*-sulfonate, Na
Uses	
Molecular formula	$C_6H_4NO_5S,Na$
Molecular weight	225.16
Properties	

Chemical name	*o*-Nitrobenzoic acid
Chemical Abstract #	552-16-9
Common name	
Synonyms	2-Nitrobenzoic acid
Uses	A reagent for alkaloids and thorium
Molecular formula	$C_7H_5NO_4$
Molecular weight	167.12
Properties	m.p. 147.5°C

Chemical name	*m*-Nitrobenzoic acid
Chemical Abstract #	121-92-6
Common name	
Synonyms	
Uses	A reagent for alkaloids and thorium
Molecular formula	$C_7H_5NO_4$
Molecular weight	167.12
Properties	m.p. 142.0°C

Chemical name	2-Nitro-*p*-cresol
Chemical Abstract #	119-33-5
Common name	
Synonyms	4-Methyl-2-nitrophenol
Uses	
Molecular formula	$C_7H_7NO_3$
Molecular weight	153.13
Properties	m.p. 32.0°C, b.p. 125.0°C (at 22 mm)

Chemical name	4-Nitro-*m*-cresol
Chemical Abstract #	2581-34-2
Common name	
Synonyms	3-Methyl-4-nitrophenol
Uses	
Molecular formula	$C_7H_7NO_3$
Molecular weight	153.13
Properties	

Chemical name	1,2,3-Propanetriol trinitrate
Chemical Abstract #	55-63-0
Common name	Nitroglycerin; glycerol; trinitrin
Synonyms	Trinitroglycerol; glyceryl trinitrate
Uses	Manufacturing of explosives
Molecular formula	$C_3H_5N_3O_9$
Molecular weight	227.11
Properties	m.p. 13.0°C, explodes at 218.0°C

Chemical name	Nitromethane
Chemical Abstract #	75-52-5
Common name	
Synonyms	Nitrocarbol
Uses	Coating industry; rocket fuel
Molecular formula	CH_3NO_2
Molecular weight	61.04
Properties	m.p. –29.0°C, b.p. 101.2°C

Chemical name	2-Nitrophenol
Chemical Abstract #	88-75-5
Common name	*o*-Nitrophenol
Synonyms	2-Hydroxynitrobenzene
Uses	Chemical manufacturing; indicator
Molecular formula	$C_6H_5NO_3$
Molecular weight	139.12
Properties	m.p. 45.0°C, b.p. 214.5°

Chemical name 3-Nitrophenol
Chemical Abstract # 554-84-7
Common name *m*-Nitrophenol
Synonyms 3-Hydroxynitrobenzene
Uses Chemical manufacturing; indicator
Molecular formula $C_6H_5NO_3$
Molecular weight 139.12
Properties m.p. 97.0°C, b.p. 194.0°C (at 70.0 mm)

Chemical name 4-Nitrophenol
Chemical Abstract # 100-02-7
Common name *p*-Nitrophenol
Synonyms 4-Hydroxynitrobenzene
Uses Organic synthesis; fungicide for leather; indicator
Molecular formula $C_6H_5NO_3$
Molecular weight 139.12
Properties m.p. 114.0°C (sublimes)

Chemical name *n*-Nonyl alcohol
Chemical Abstract # 143-08-8
Common name Nonalol
Synonyms 1-Nonalol
Uses Manufacturing of artificial lemon oil
Molecular formula $C_9H_{20}O$
Molecular weight 144.26
Properties m.p. –5.0°C, b.p. 194.0–213.0°C

Chemical name Nonylphenol
Chemical Abstract # 25154-52-3
Common name Hydroxy NO. 253
Synonyms 2,6-Dimethyl-4-heptylphenol (*o* and *p*)
Uses Surfactant; preparation of oils and resins; fungicides; bactericides; drugs; adhesives
Molecular formula $C_{15}H_{24}O$
Molecular weight Approximately 215
Properties b.p. 293.0–297.0°C

Chemical name Nonylphenol ethoxylate
Chemical Abstract #
Common name
Synonyms
Uses Surfactant
Molecular formula $C_{17}H_{29}O_2$
Molecular weight 265.42
Properties

Chemical name Octane
Chemical Abstract # 111-65-9
Common name
Synonyms *n*-Octane
Uses Organic synthesis; rubber industry; solvent
Molecular formula C_8H_{18}
Molecular weight 114.23
Properties m.p. –56.5°C, b.p. 125.7°C

Chemical name 1-Octanol
Chemical Abstract # 111-87-5
Common name Caprylic alcohol, Alfol-8, Octilin, Sipol L8
Synonyms Heptylcarbinol
Uses Petroleum esters and perfumes manufacturing
Molecular formula $C_8H_{18}O$
Molecular weight 130.26
Properties m.p. –16.0 to –17.0°C, b.p. 194.0–195.0°C

Chemical name 2-Octanol
Chemical Abstract # 123-96-6
Common name Secondary caprylic alcohol
Synonyms Methyl hexyl carbinol
Uses Solvent; perfume and soap manufacturing
Molecular formula $C_8H_{18}O$
Molecular weight 130.23
Properties m.p. –38.6°C, b.p. 178.5°C

Chemical name	Oleic acid
Chemical Abstract #	112-80-1
Common name	Red oil; Emersol 210; Century CD fatty acid
Synonyms	(Z)-9-Octadecenoic acid
Uses	Waterproofing; wool oiling; solvent; pharmaceuticals; in polishing compounds and soft soaps
Molecular formula	$C_{18}H_{34}O_2$
Molecular weight	282.45
Properties	m.p. 6.0°C, b.p. 360.0°C

Chemical name	Sodium oxalate
Chemical Abstract #	62-76-0
Common name	Oxalic acid, disodium salt
Synonyms	Ethanedioic acid disodium salt
Uses	Finishing textiles and leather; standardizing potassium permanganate solution
Molecular formula	$C_2Na_2O_4$
Molecular weight	134.01
Properties	m.p. 250.0–270.0°C, decomposes

Chemical name	Oxalic acid
Chemical Abstract #	142-62-7
Common name	NCI-C55209
Synonyms	Ethanedioic acid
Uses	Printing; dyeing; bleaching; stain removal; reducing agent; photography; engraving; rubber manufacturing;
Molecular formula	$C_2H_2O_4$
Molecular weight	90.04
Properties	m.p. 101.0°C, b.p. 150.0°C (sublimes)

Chemical name	[*S*-(2-Ethylsulfinyl)ethyl] *O,O*-dimethylphosphorothioate
Chemical Abstract #	
Common name	
Synonyms	Demeton *O*-methyl sulfoxide, oxydemetonmethyl
Uses	Insecticide
Molecular formula	$C_6H_{15}O_2PS_2$
Molecular weight	214.32
Properties	m.p. < –10.0°C, b.p. 106.0°C (at 0.01 mm)

Chemical name	Oxydipropionitrile
Chemical Abstract #	1656-48-0
Common name	Ether, bis(2-cyanoethyl)
Synonyms	3,3-Oxydipropionitrile
Uses	
Molecular formula	$C_6H_8N_2O$
Molecular weight	124.16
Properties	m.p. –26.3°C, b.p. 172.0°C (at 10 mm)

Chemical name	Hexadecanoic acid
Chemical Abstract #	57-10-3
Common name	Palmitic acid
Synonyms	*n*-Hexadecylic acid
Uses	
Molecular formula	$C_{17}H_{32}O_2$
Molecular weight	256.42
Properties	m.p. 64.0°C, b.p. 271.5°C (at 100.0 mm)

Chemical name	2,4,6-Trimethyl-1,3,5-trioxane
Chemical Abstract #	123-63-7
Common name	Para-acetaldehyde
Synonyms	Paraldehyde
Uses	Organic compound manufacturing
Molecular formula	$C_6H_{12}O_3$
Molecular weight	132.16
Properties	m.p. 12.6°C, b.p. 124.4°C (at 752 mm)

Chemical name	1.1′-Dimethyl-4,4′-bipyridinium
Chemical Abstract #	4685-14-7
Common name	Paraquat; methyl viologen
Synonyms	*N,N′*-Dimethyl-γ,γ′-dipyridylium
Uses	Herbicide
Molecular formula	$[C_{12}H_{14}N_2]^{2+}$
Molecular weight	186.25
Properties	m.p. 300.0°C, decomposes

Chemical name	Phosphorothioic acid *O,O*-diethyl *O*-(4-nitrophenyl)ester
Chemical Abstract #	56-38-2
Common name	Parathion
Synonyms	*O,O*-diethyl-*O*-*p*-nitrophenyl phosphorothioate
Uses	Insecticide; acaricide
Molecular formula	$C_{10}H_{14}NO_5PS$
Molecular weight	291.28
Properties	m.p. 6.0°C, b.p. 375.0°C

Chemical name	Butylethylcarbamothioic acid *S*-propyl ester
Chemical Abstract #	1114-71-2
Common name	Pebulate; Tillam; Stauffer 2061
Synonyms	*S*-Propyl butylethylthiocarbamate
Uses	Herbicide
Molecular formula	$C_{10}H_{21}NOS$
Molecular weight	203.36
Properties	b.p. 142.0°C (at 20 mm)

Chemical name	Pelargonic acid
Chemical Abstract #	112-05-0
Common name	Nonoic acid; nonylic acid
Synonyms	Nonanoic acid
Uses	In manufacturing hydrotropic salts
Molecular formula	$C_9H_{18}O_2$
Molecular weight	158.23
Properties	m.p. 12.5°C, b.p. 254.0°C

Chemical name	Pentachlorobenzene
Chemical Abstract #	608-93-5
Common name	RCRA Waste # U183
Synonyms	QCB
Uses	Experimental teratogen
Molecular formula	C_6HCl_5
Molecular weight	250.34
Properties	m.p. 85.0–86.0°C, b.p. 275.0–277.0°C

Chemical name	2,4,5,2′,5′-pentachlorobiphenyl
Chemical Abstract #	25429-29-2
Common name	
Synonyms	
Uses	Electrical insulation; fire-resistant and hydraulic fluids; high-temperature lubricants; adhesives; paints; waxes
Molecular formula	$C_{12}H_5Cl_5$
Molecular weight	332.42
Properties	

Chemical name	Pentachloroethane
Chemical Abstract #	76-01-7
Common name	Pentalin
Synonyms	Ethane pentachloride
Uses	Solvent
Molecular formula	C_2HCl_5
Molecular weight	202.31
Properties	m.p. –29.0°C, b.p. 162.0°C

Chemical name	Sodium pentachlorophenate
Chemical Abstract #	131-52-2
Common name	Santobrite; Dowicide G; Pentaphenate
Synonyms	Sodium pentachlorophenoxide
Uses	Insecticide
Molecular formula	C_6Cl_5ONa
Molecular weight	288.30
Properties	

Chemical name	Pentachlorophenol
Chemical Abstract #	87-86-5
Common name	Penta; PCP
Synonyms	Pentachlorol
Uses	Insecticide; herbicide; wood preservation
Molecular formula	C_6HCl_5O
Molecular weight	266.35
Properties	m.p. 191.0°C, b.p. 310.0°C, decomposes

Chemical name	Pentane
Chemical Abstract #	109-66-0
Common name	Pentane
Synonyms	*n*-Pentane; amyl hydride
Uses	Fuel; chemical manufacturing; solvent; blowing agent for plastics; pesticide
Molecular formula	C_5H_{12}
Molecular weight	72.15
Properties	m.p. –130.0°C, b.p. 36.0°C

Chemical name	*n*-Pentanol
Chemical Abstract #	71-41-0
Common name	Pentyl alcohol
Synonyms	*n*-Butylcarbinol; *n*-amyl alcohol
Uses	Solvent; organic synthesis; plastic processing; chemical manufacturing; pharmaceutical preparations
Molecular formula	$C_5H_{12}O$
Molecular weight	88.15
Properties	m.p. –78.9°C, b.p. 138.0°C

Chemical name	2,4-Pentanedione
Chemical Abstract #	123-54-6
Common name	
Synonyms	Acetylacetone; acetylmethane
Uses	Gasoline and lubricant additives; used in fungicides and pesticides
Molecular formula	$C_5H_8O_2$
Molecular weight	100.11
Properties	m.p. –23.2°C, b.p. 139.0°C (at 746 mm)

Chemical name	3-Pentanol
Chemical Abstract #	584-02-1
Common name	Pentanol 3
Synonyms	Diethyl carbinol
Uses	In organic synthesis; solvent; flotation agent; solvent manufacturing; pharmaceuticals
Molecular formula	$C_5H_{12}O$
Molecular weight	88.15
Properties	b.p. 115.6°C

Chemical name	1,5-Pentanediol
Chemical Abstract	111-29-5
Common name	Pentamethylene glycol
Synonyms	1,5-Dihydroxypentane
Uses	Plasticizer, in adhesives and brake fluids
Molecular formula	$C_5H_{12}O_2$
Molecular weight	104.15
Properties	m.p. –18.0°C, b.p. 239.0°C

Chemical name 3-(2,2-Dichloroethenyl)-2,2-dimethylcyclopropanecarboxylic acid (3-phenoxyphenyl)methyl ester
Chemical Abstract # 52645-53-1
Common name Permethrin; FMC 33297; Ambush; Pour-on
Synonyms 3-(Phenoxyphenyl)methyl (±)-*cis,trans*-3-(2,2-dichloroethenyl) 2,2-dimethylcyclopropanecarboxylate
Uses Insecticide
Molecular formula $C_{21}H_{20}Cl_2O_3$
Molecular weight 391.29
Properties m.p. approx. 35.0°C, b.p. 220.0°C at 0.05 mm

Chemical name Phenanthrene
Chemical Abstract # 85-01-8
Common name
Synonyms Phenanthrin
Uses Explosives; dyes; drug syntheses
Molecular formula $C_{14}H_{10}$
Molecular weight 178.22
Properties m.p. 100.0°C, b.p. 340.0°C

Chemical name Phenol
Chemical Abstract # 108-95-2
Common name Carbolic acid
Synonyms Phenylic acid; phenyl hydroxide
Uses Disinfectant; manufacturing of resins, dyes, medical compounds; preservative
Molecular formula C_6H_6O
Molecular weight 94.11
Properties m.p. 41.0°C, b.p. 182.0°C

Chemical name Hydroxybenzenesulfonic acid sodium salt
Chemical Abstract # 98-67-9
Common name Sodium sulfocarbolate
Synonyms Sodium phenolsulfonate
Uses Antiseptic
Molecular formula $C_6H_5NaO_4S$
Molecular weight 196.15
Properties

Chemical name Phenoxyacetic acid
Chemical Abstract # 122-59-8
Common name Phenylium; phenyl ether glycolic acid
Synonyms Phenoxyethanoic acid
Uses Fungicide; used for softening calluses, corns, hard skin surfaces
Molecular formula $C_8H_8O_3$
Molecular weight 152.14
Properties m.p. 98.0°C, b.p. 285.0°C (some decomposition)

Chemical name Phenyl acetate
Chemical Abstract # 122-79-2
Common name Acetylphenol
Synonyms Acetic acid phenyl ester
Uses Laboratory reagent
Molecular formula $C_8H_8O_2$
Molecular weight 136.14
Properties b.p. 195.0–196.0°C

Chemical name 3(Phenyl)-1,1-dimethylurea
Chemical Abstract # 101-42-8
Common name Fenuron; Dybar
Synonyms *N,N*-Dimethyl-*N*-phenylurea
Uses Herbicide
Molecular formula $C_9H_{12}N_2O$
Molecular weight 164.20
Properties m.p. 131.0–133.0°C

Chemical name *p*-Phenylenediamine
Chemical Abstract # 106-50-3
Common name Orsin; Ursol D
Synonyms 1,4-Benzenediamine
Uses Fur dyeing; vulcanization; milk testing
Molecular formula $C_6H_8N_2$
Molecular weight 108.14
Properties m.p. 145.0–147.0°C, b.p. 276.0°C (sublimes)

Chemical name β-phenylethylamine sulfate
Chemical Abstract # 64-04-0 (Phenylethyl amine)
Common name
Synonyms
Uses Complexing agent for silver
Molecular formula $C_8H_{11}ON, {}^1/_2\ H_2SO_4$
Molecular weight 188.22
Properties

Chemical name Phenylmercuric acetate
Chemical Abstract # 62-38-4
Common name Ceresan slaked lime; gallotox; riogen
Synonyms Acetoxyphenyl mercury
Uses Slimicide; fungicide
Molecular formula $C_8H_8HgO_2$
Molecular weight 336.75
Properties m.p. 149.0°C

Chemical name *N*-Phenyl-α-naphthylamine
Chemical Abstract # 90-30-2
Common name Neozone A; aceto pan; PANA
Synonyms Phenylnaphthylamine
Uses Rubber syntheses
Molecular formula $C_{16}H_{13}N$
Molecular weight 219.30
Properties m.p. 62.0°C, b.p. 335.0°C (at 528 mm)

Chemical name *N*-Phenyl-β-naphthylamine
Chemical Abstract # 135-88-6
Common name PBNA; Aceto PBN; Neozone D
Synonyms Anilinonaphthalene
Uses Antioxidant; rubber syntheses
Molecular formula $C_{16}H_{13}N$
Molecular weight 219.30
Properties m.p. 107.0–108.0°C, b.p. 395.5°C

Chemical name *o*-Phenylphenol
Chemical Abstract # 90-43-7
Common name Dowicide 1,Orthoxenol
Synonyms (1,1′-Biphenyl)-2-ol; 2-biphenylol
Uses Disinfectant; fungicide; in rubber industry
Molecular formula $C_{12}H_{10}O$
Molecular weight 170.20
Properties m.p. 55.5–57.5°C, b.p. 280.0–284.0°C

Chemical name 1-Phenyl-3-pyrazolidinone
Chemical Abstract # 92-43-3
Common name Phenidone
Synonyms 1-Phenyl-3-pyrazolidone
Uses Photographic development; reducing agent
Molecular formula $C_9H_{10}N_2O$
Molecular weight 162.19
Properties m.p. 121.0°C

Chemical name Phloroglucinol
Chemical Abstract # 108-73-6
Common name Spasfon-Lyoc; Dilospan
Synonyms 1,3,5-Benzenetriol; 1,3,5-trihydroxybenzene
Uses Dyeing; decalcifier; chemical reagent
Molecular formula $C_6H_6O_3$
Molecular weight 126.11
Properties m.p. 218.0°C; sublimes with decomposition

Chemical name	Carbonic dichloride
Chemical Abstract #	75-44-5
Common name	Phosgene
Synonyms	Carbonyl chloride; chloroformyl chloride
Uses	As a weapon; organic chemical preparation
Molecular formula	CCl_2O
Molecular weight	98.92
Properties	m.p. –118.0°C, b.p. 8.3°C

Chemical name	Phosphamidon (2-chloro-2-diethylcarbomyl-1-methylvinyl dimethyl phosphate)
Chemical Abstract #	13171-21-6
Common name	Ciba 570; Dimecron; ENT 25515
Synonyms	Phosphoric acid 2-chloro-3-(diethylamino)-1-methyl-3-oxo-1-propenyl dimethyl ester
Uses	Insecticide
Molecular formula	$C_{10}H_{19}ClNO_5P$
Molecular weight	299.69
Properties	m.p. –45.0°C, b.p. 162.0°C (at 1.5 mm)

Chemical name	Phorate [*O,O*-diethyl-*S*-(ethylthio)methyl-phosphorothioate]
Chemical Abstract #	298-02-2
Common name	Thimet
Synonyms	Phosphorodithioic acid *O,O*-diethyl *S*-[(ethylthio) methyl] ester
Uses	Insecticide
Molecular formula	$C_7H_{17}O_2PS_3$
Molecular weight	260.40
Properties	b.p. 118.0–120.0°C (at 0.8 mm)

Chemical name	Phorone (diisopropylidene acetone)
Chemical Abstract #	504-20-1
Common name	Phoron
Synonyms	2,6-Dimethyl 4-oxo-2,5-heptadien-4-one
Uses	
Molecular formula	$C_9H_{14}O$
Molecular weight	138.20
Properties	m.p. 28.0°C, b.p. 198.0–199.0°C

Chemical name	Phthalic anhydride
Chemical Abstract #	85-44-9
Common name	Phthalandione
Synonyms	1,3-Isobenzofurandione
Uses	Manufacturing of artificial resins, specialty chemicals, synthetic fibers; used in dyes, pigments, pharmaceuticals, insecticides
Molecular formula	$C_8H_4O_3$
Molecular weight	148.11
Properties	m.p. 130.8°C, b.p. 295.0°C; sublimes

Chemical name	Phygon
Chemical Abstract #	177-80-6
Common name	Dichlone
Synonyms	2,3-Dichloro-1,4-naphthalenedione
Uses	Fungicide; herbicide
Molecular formula	$C_{10}H_4Cl_2O_2$
Molecular weight	227.06
Properties	m.p. 193.0°C, sublimes

Chemical name Picloram (4-amino-3,5,6-trichloro-2-pyridine carboxylic acid)
Chemical Abstract # 1918-02-1
Common name Tordon
Synonyms 4-Amino-3,5,6-trichloropicolinic acid
Uses Herbicide; defoliant
Molecular formula $C_6H_3Cl_3N_2O_2$
Molecular weight 241.48
Properties m.p. 218.0–219.0°C

Chemical name α-Picoline
Chemical Abstract # 109-06-8
Common name 2-Picoline
Synonyms 2-Methylpyridine
Uses Solvent; dye industry
Molecular formula C_6H_7N
Molecular weight 93.12
Properties m.p. –70.0°C, b.p. 129.0°C

Chemical name Picric acid
Chemical Abstract # 88-89-1
Common name Carbazotic acid
Synonyms 2,4,6-Trinitrophenol; nitroxanthic acid
Uses Explosives; tissue fixing; leather industry; electric batteries; copper etching
Molecular formula $C_6H_3N_3O_7$
Molecular weight 229.11
Properties m.p. 122.0–123.0°C, b.p. explodes > 300.0°C

Chemical name Pimaric acid [13α-methyl-13vinylpodocarp-8-(14)-en-15-oic acid]
Chemical Abstract # 127-27-5
Common name Dextropimaric acid
Synonyms 7-Ethenyl-1,2,3,4,4a,4b,5,6,7,9,10,10a-dodecahydro-1,4a,7-trimethyl-1-phenanthrenecarboxylic acid
Uses Used in resins
Molecular formula $C_{20}H_{30}O_2$
Molecular weight 302.44
Properties m.p. 217.0–219.0°C, b.p. 282.0°C at 18 mm

Chemical name Pimelic acid
Chemical Abstract # 111-16-0
Common name Heptanedioic acid
Synonyms 1,5-Pentanedicarboxylic acid
Uses
Molecular formula $C_7H_{12}O_4$
Molecular weight 160.17
Properties m.p. 105.0–106.0°C, b.p. 272.0°C (at 100 mm), sublimes

Chemical name Pivalic acid
Chemical Abstract # 75-98-9
Common name Trimethylacetic acid; Versatic 5
Synonyms 2,2-Dimethylpropanoic acid
Uses
Molecular formula $C_5H_{10}O_2$
Molecular weight 102.13
Properties m.p. 35.5°C, b.p. 164.0°C

Chemical name Polychlorinated biphenyls
Chemical Abstract # 1336-36-3
Common name Polychlorobiphenyls; Aroclor; Fenclor; Montar
Synonyms Chlorinated biphenyls
Uses In electrical capacitors, transformers; also used in lubricants, cutting oils, adhesives, inks, hydraulic fluids, plasticizers
Molecular formula $C_{12}H_{(10-n)}Cl_n$
Molecular weight Variable
Properties b.p. variable (340.0–375.0°C)

Chemical name Polyethylene glycol
Chemical Abstract # 25322-68-3
Common name Macrogol; PEG; Carbowax; Solbase
Synonyms α-Hydro-ω-hydroxypoly-(oxy-1,2-ethanediyl)
Uses Lubricant; paints; finishes; polishes; ceramic industry
Molecular formula $H(OCH_2\text{-}CH_2)_nOH$ (where *n* is greater than or equal to 4
Molecular weight 190-9000
Properties m.p. 4.0–63.0°C

Chemical name Polypropylene glycol
Chemical Abstract # 25322-69-4
Common name Jeffox
Synonyms
Uses Cosmetics; brake and lubricating fluids; greases; rubber processing
Molecular formula $(C_3H_8O_2)_n$
Molecular weight 400-4000
Properties

Chemical name Prometon [2,4-bis(isopropylamino)-6-methoxy-*s*-triazine]
Chemical Abstract # 1610-18-0
Common name Prometone
Synonyms 6-Methoxy-*N,N′*-bis(1-methylethyl)-1,3,5-triazine-2,4-diamine
Uses Nonselective herbicide
Molecular formula $C_{10}H_{19}N_5O$
Molecular weight 225.29
Properties m.p. 91.0–92.0°C

Chemical name Propane
Chemical Abstract # 74-98-6
Common name Propane
Synonyms Dimethyl methane
Uses Fuel; in organic syntheses; refrigerant
Molecular formula C_3H_8
Molecular weight 44.09
Properties m.p. –189.9°C, b.p. –42.0°C

Chemical name *n*-Propyl alcohol
Chemical Abstract # 71-23-8
Common name Optal, ethylcarbinol
Synonyms 1-Propanol
Uses Solvent
Molecular formula C_3H_8O
Molecular weight 60.11
Properties m.p. –127.0°C, b.p. 97.2°C

Chemical name Propargyl alcohol
Chemical Abstract # 107-19-7
Common name Propiolic alcohol
Synonyms 2-Propyn-1-ol; ethynylcarbinol
Uses Organic syntheses
Molecular formula C_3H_4O
Molecular weight 56.06
Properties m.p. –48.0°C to –52.0°C, b.p. 115.0°C

Chemical name	Propylene oxide
Chemical Abstract #	75-56-9
Common name	Propene oxide
Synonyms	Methyloxirane; 1,2-epoxypropane
Uses	Solvent; chemical syntheses; used in lubricant, surfactant and oil emulsifiers preparations
Molecular formula	C_3H_6O
Molecular weight	58.08
Properties	m.p. –112.13°C, b.p. 34.23°C

Chemical name	Propionaldehyde
Chemical Abstract #	123-38-6
Common name	Propanal
Synonyms	Methylacetaldehyde
Uses	Disinfectant; manufacturing of plastics
Molecular formula	C_3H_6O
Molecular weight	58.1
Properties	m.p. –81.0°C, b.p. 49.0°C

Chemical name	Sodium propionate
Chemical Abstract #	137-40-6
Common name	Mycoban
Synonyms	Propionic acid sodium salt
Uses	Fungicide; food preservative
Molecular formula	$C_3H_5NaO_2$
Molecular weight	96.07
Properties	

Chemical name	Propionic acid
Chemical Abstract #	79-09-4
Common name	Ethylformic acid
Synonyms	Propanoic acid
Uses	Fruit flavors; perfume base; production of propionates; manufacturing of ester solvents
Molecular formula	$C_3H_6O_2$
Molecular weight	74.08
Properties	m.p. –21.5°C, b.p. 141.1°C

Chemical name	*n*-Propyl acetate
Chemical Abstract #	109-60-4
Common name	Propyl acetate
Synonyms	Acetic acid *n*-propyl ester
Uses	Solvent; flavor and perfume manufacturing
Molecular formula	$C_5H_{10}O_2$
Molecular weight	102.13
Properties	m.p. –92.0°C, b.p. 102.0°C

Chemical name	Propylamine
Chemical Abstract #	107-10-8
Common name	*n*-Propylamine
Synonyms	1-Propanamine
Uses	Used in dyes, pharmaceuticals, agricultural chemicals, leather finishing resins; rubber manufacturing
Molecular formula	C_3H_9N
Molecular weight	59.11
Properties	m.p. –83.0°C, b.p. 48.0–49.0°C

Chemical name	Vernolate(Dipropylcarbamothioic acid *S*-propyl ester)
Chemical Abstract #	1929-77-7
Common name	Vernam
Synonyms	*S*-Propyldipropylthiocarbamate
Uses	Selective herbicide
Molecular formula	$C_{10}H_{21}NOS$
Molecular weight	203.35
Properties	b.p. 149.0–150.0°C (at 30 mm)

Chemical name	Propylene glycol
Chemical Abstract #	57-55-6
Common name	Methyl glycol
Synonyms	1,2-Propanediol; 1,2-dihydroxypropane
Uses	Antifreeze; manufacturing of synthetic resins; fermentation inhibitor; disinfectant
Molecular formula	$C_3H_8O_2$
Molecular weight	76.09
Properties	m.p. –59.0°C, b.p. 188.2°C

Chemical name	Pyrene
Chemical Abstract #	129-00-0
Common name	β-Pyrine
Synonyms	Benzo(def)phenanthrene
Uses	Experimental tumorigen
Molecular formula	$C_{16}H_{10}$
Molecular weight	202.26
Properties	m.p. 156.0°C, b.p. 404.0°C

Chemical name	Pyrethrin
Chemical Abstract #	97-11-0
Common name	Cinerin I or II; pyrethrin I or II
Synonyms	2-Cyclopentenyl-4-hydroxy-3-methyl-2-cyclo-penten-1-one chrysanthemate
Uses	Insecticide
Molecular formula	$C_{21}H_{28}O_3$
Molecular weight	328.49
Properties	b.p. 170.0°C (at 0.1 mm), decomposes

Chemical name	Pyridine
Chemical Abstract #	110-86-1
Common name	Azine
Synonyms	Azobenzene
Uses	Solvent; waterproofing; manufacturing of rubber chemicals
Molecular formula	C_5H_5N
Molecular weight	79.11
Properties	m.p. –42.0°C, b.p. 115.0°C

Chemical name	Pyrogallol
Chemical Abstract #	87-66-1
Common name	Pyrogallic acid
Synonyms	1,2,3-Benzenetriol
Uses	Photography; metal colloidal solutions; dye manufacturing; hair dyeing; reducer of gold, silver, and mercury salts
Molecular formula	$C_6H_6O_3$
Molecular weight	126.11
Properties	m.p. 131.0–133.0°C, b.p. 309.0°C, decomposes

Chemical name	2-Pyrrolidone
Chemical Abstract #	616-45-5
Common name	α-Pyrrolidone
Synonyms	Butyrolactam
Uses	Insecticide; solvent; inks; high-boiling solvent in petroleum industry; plasticizer for acrylic latexes
Molecular formula	C_4H_7NO
Molecular weight	85.10
Properties	m.p. 25.0°C, b.p. 245.0°C

Chemical name	Quinhydrone
Chemical Abstract #	106-34-3
Common name	Green hydroquinone
Synonyms	2,5-Cyclohexadiene-1,4-dione compound with 1,4-benzenediol (1:1)
Uses	Indicator
Molecular formula	$C_{12}H_{10}O_4$
Molecular weight	218.20
Properties	m.p. 171.0°C, b.p. sublimes with partial decomposition

Chemical name	Resorcinol
Chemical Abstract #	108-46-3
Common name	Resorcin, *m*-hydroxyphenol
Synonyms	1,3-Benzenediol; 1,3-dihydroxybenzene
Uses	Manufacturing of resins, resin adhesives, explosives, adhesives, dyes; tanning; cosmetics; dyeing and printing textiles; as topical antipruritic and antiseptic
Molecular formula	$C_6H_6O_2$
Molecular weight	110.11
Properties	m.p. 109.0–111.0°C, b.p. 280.0°C

Chemical name	Benzalkonium chloride
Chemical Abstract #	
Common name	Roccal; Benirol; Rodalon; Zephinon
Synonyms	Alkyldimethylethylbenzyl ammonium chloride, alkyl[(ethylphenyl)methyl]dimethyl quaternary ammonium chlorides
Uses	Topical antiseptic; bactericide; fungicide; germicide
Molecular formula	Mixture, C_8H_{17} to $C_{18}H_{37}$
Molecular weight	Mixture
Properties	

Chemical name	Phosphorothioic acid, *O,O*-dimethyl *O*-(2,4,5-trichlorophenyl)ester
Chemical Abstract #	299-84-3
Common name	Ronnel, Fenchlorphos, Trolene, Nankor, Viozene
Synonyms	Dimethyl trichlorophenyl thiophosphate
Uses	Systemic insecticide
Molecular formula	$C_8H_8Cl_3O_3PS$
Molecular weight	321.57
Properties	m.p. 41.0°C

Chemical name	Rotenone
Chemical Abstract #	83-79-4
Common name	Cubor; Deril; Extrax; Noxfish; Rotocide; Fish-Tox
Synonyms	[2*R*-(2α,6aα,12aα)]-1,2,12,12a-Tetrahydro-8,9-dimethoxy-2-(1-methylethenyl)-[1]benzopyrano-[3,4-b]furo[2,3-h][1]benzopyran-6(6a*H*)-one
Uses	Selective nonsystemic insecticide, pesticide, and ectoparasiticide
Molecular formula	$C_{23}H_{22}O_6$
Molecular weight	394.45
Properties	m.p. 165.0–166.0°C, b.p. 210.0–220.0°C (at 0.5 mm)

Chemical name	Salicylaldehyde
Chemical Abstract #	90-02-8
Common name	Salicylic aldehyde
Synonyms	2-Hydroxybenzaldehyde
Uses	In perfume industry; gasoline additives; fumigant; analytical chemistry
Molecular formula	$C_7H_6O_2$
Molecular weight	122.12
Properties	m.p. –7.0°C, b.p. 196.0°C

Chemical name	Salicylic acid
Chemical Abstract #	69-72-7
Common name	Keralyt; Verrugon
Synonyms	2-Hydroxybenzoic acid
Uses	Food preservative; reagent in analytical chemistry
Molecular formula	$C_7H_6O_3$
Molecular weight	138.12
Properties	m.p. 158.0°C, b.p. 256.0°C

Chemical name	5-Benzyl-(3-furyl)methyl-*cis,trans*-(+)-2,2-dimethyl-3-(2-methylpropenyl) cyclopropanecarboxylate
Chemical Abstract #	35764-59-1
Common name	SBP-1382, Resmethrin, Pyrethroid
Synonyms	5-Benzyl-3-furylmethyl (+)-*cis*-chrysanthemate
Uses	Insecticide
Molecular formula	$C_{22}H_{26}O_3$
Molecular weight	338.48
Properties	

Chemical name	Strychnine
Chemical Abstract #	57-24-9
Common name	Ro-Dex, Cer-Tox, Kwik-Kil, Mouse-Rid
Synonyms	Strychnidin-10-one
Uses	Pesticide, mainly for rodent control
Molecular formula	$C_{21}H_{22}N_2O_2$
Molecular weight	334.45
Properties	m.p. 268.0–290.0°C (depends on rate of heating), b.p. 270.0°C

Chemical name	Styrene
Chemical Abstract #	100-42-5
Common name	Styrol; styrolene; vinylbenzene
Synonyms	Ethenylbenzene
Uses	Manufacturing of plastics, resins; synthetic rubber
Molecular formula	C_8H_8
Molecular weight	104.14
Properties	m.p. –30.6°C, b.p. 145.0–146.0°C

Chemical name	Octanedioic acid
Chemical Abstract #	505-48-6
Common name	Suberic acid
Synonyms	1,6-Hexanedicarboxylic acid
Uses	In plastic industry
Molecular formula	$C_8H_{14}O_4$
Molecular weight	174.19
Properties	m.p. 140.0–144.0°C, b.p. 279.0°C (at 100 mm)

Chemical name	Succinic acid
Chemical Abstract #	110-15-6
Common name	Amber acid
Synonyms	Butanedioic acid
Uses	Manufacturing of dyes, esters, succinates
Molecular formula	$C_4H_6O_4$
Molecular weight	118.09
Properties	m.p. 185.0–187.0°C, b.p. 235.0°C, decomposes

Chemical name	Succinonitrile
Chemical Abstract #	110-61-2
Common name	Ethylene dicyanide
Synonyms	Butanedinitrile
Uses	An experimental teratogen
Molecular formula	$C_4H_4N_2$
Molecular weight	80.09
Properties	m.p. 57.15°C, b.p. 265.0–267.0°C

Chemical name	Sulfolane
Chemical Abstract #	126-33-0
Common name	Thiophan sulfone
Synonyms	Tetrahydrothiophene 1,1-dioxide
Uses	Solvent for liquid-vapor extraction
Molecular formula	$C_4H_8O_2S$
Molecular weight	120.16
Properties	m.p. 27.4–27.8°C, b.p. 285.0°C

Chemical name	Tannic acid
Chemical Abstract #	1401-55-4
Common name	Tannin
Synonyms	Gallotannic acid
Uses	Tanning; clarification agent; photography
Molecular formula	$C_{76}H_{52}O_{46}$
Molecular weight	1701.23
Properties	m.p. 200.0°C, decomposes at 210.0–215.0°C

Chemical name	*dl*-Tartaric acid
Chemical Abstract #	133-37-9
Common name	Racemic acid; paratartaric acid
Synonyms	2,3-Dihydroxybutanedioic acid
Uses	Photography; ceramics
Molecular formula	$C_4H_6O_6$
Molecular weight	150.09
Properties	m.p. 206.0°C

Chemical name	Tetraethyldithiopyrophosphate(TEDP)
Chemical Abstract #	3689-24-5
Common name	TEDP, Sulfotep, ASP-47
Synonyms	Thiodiphosphoric acid tetraethyl ester
Uses	Insecticide; acaricide
Molecular formula	$C_8H_{20}O_5P_2S_2$
Molecular weight	322.30
Properties	b.p. 136.0–139.0°C

Chemical name	Teepol 715
Chemical Abstract #	42615-29-2
Common name	ABS
Synonyms	Alkyl aryl sulfonate; alkylbenzene sulfonate
Uses	Used in degreasers, moisturizers, detergents
Molecular formula	$R(C_6H_4)SO_3{\cdot}Na$
Molecular weight	Variable
Properties	

Chemical name	Tetraethyl pyrophosphate (TEPP)
Chemical Abstract #	107-49-3
Common name	TEPP; Tetron; Vapotone; Fosvex; Nifos; Bladan
Synonyms	Diphosphoric acid tetraethyl ester
Uses	Insecticide for aphids and mites, rodenticide
Molecular formula	$C_8H_{20}O_7P_2$
Molecular weight	290.20
Properties	Decomposition range 170.0–213.0°C with copious formation of ethylene

Chemical name	Terephthalic acid
Chemical Abstract #	100-21-0
Common name	Tephthol
Synonyms	1,4-Benzenedicarboxylic acid
Uses	Manufacturing of plastic films and sheets
Molecular formula	$C_8H_6O_4$
Molecular weight	166.13
Properties	Sublimes at 402.0°C

Chemical name	1,2,3,4-Tetrachlorobenzene
Chemical Abstract #	634-66-2
Common name	
Synonyms	
Uses	Used in dielectric fluids; chemical syntheses
Molecular formula	$C_6H_2Cl_4$
Molecular weight	215.88
Properties	m.p. 47.5°C, b.p. 245.0°C

Chemical name	1,2,3,5-Tetrachlorobenzene
Chemical Abstract #	634-90-2
Common name	
Synonyms	
Uses	
Molecular formula	$C_6H_2Cl_4$
Molecular weight	215.88
Properties	m.p. 50.0–52.0°C, b.p. 246.0°C

Chemical name	1,2,4,5-Tetrachlorobenzene
Chemical Abstract #	95-94-3
Common name	
Synonyms	RCRA waste number U207
Uses	Chemical syntheses
Molecular formula	$C_6H_2Cl_4$
Molecular weight	215.88
Properties	m.p. 138.0°C, b.p. 245.0°C

Chemical name	2,3,7,8-Tetrachlorodibenzo-*p*-dioxin
Chemical Abstract #	1746-01-6
Common name	Dioxin
Synonyms	TCDD
Uses	
Molecular formula	$C_{12}H_4Cl_4O_2$
Molecular weight	321.96
Properties	m.p. 305.0–306.0°C

Chemical name	1,1,2,2-Tetrachloroethylene
Chemical Abstract #	127-18-4
Common name	Perchloroethylene; Dow-Per; Percosolve
Synonyms	Tetrachloroethylene
Uses	Dry cleaning; degreasing metals; solvent; manufacturing of paint removers, inks, fluorocarbons
Molecular formula	C_2Cl_4
Molecular weight	165.83
Properties	m.p. –23.5°C, b.p. 121.4°C

Chemical name	2,3,4,5-Tetrachlorophenol
Chemical Abstract #	4901-51-3
Common name	
Synonyms	
Uses	
Molecular formula	$C_6H_2Cl_4O$
Molecular weight	231.88
Properties	m.p. 69.0–70.0°C, b.p. 164.0°C (at 23 mm)

Chemical name	2,3,4,6-Tetrachlorophenol
Chemical Abstract #	58-90-2
Common name	
Synonyms	
Uses	Fungicide
Molecular formula	$C_6H_2Cl_4O$
Molecular weight	231.88
Properties	m.p. 69.0–70.0°C, b.p. 164.0°C (at 23 mm)

Chemical name	2,3,5,6-Tetrachlorophenol
Chemical Abstract #	935-95-5
Common name	
Synonyms	
Uses	
Molecular formula	$C_6H_2Cl_4O$
Molecular weight	231.88
Properties	m.p. 114.0–116.0°C

Chemical name	2,3,5,6-Tetrachloropyridine
Chemical Abstract #	2402-79-1
Common name	
Synonyms	
Uses	
Molecular formula	C_5HCl_4N
Molecular weight	216.87
Properties	m.p. 91.0°C, b.p. 251.0°C

Chemical name	Tetraethyl lead
Chemical Abstract #	78-00-2
Common name	TEL
Synonyms	Lead tetraethyl
Uses	Antiknock in gasoline
Molecular formula	$C_8H_{20}Pb$
Molecular weight	323.47
Properties	m.p. –125.0 to 150.0°C, b.p. 198.0 to 202.0°C, (decomposes)

Chemical name	Tetrahydrofuran
Chemical Abstract #	109-99-9
Common name	THF; diethylene oxide
Synonyms	Tetramethylene oxide
Uses	Solvent for high polymers; chemical syntheses
Molecular formula	C_4H_8O
Molecular weight	72.12
Properties	m.p. –108.5°C, b.p. 65.4°C

Chemical name	Tetrahydrofurfuryl alcohol
Chemical Abstract #	97-99-4
Common name	THFA
Synonyms	Tetrahydro-2-furylmethanol; tetrahydro-2-furan carbinol
Uses	Solvent for fats, waxes, and resins; in organic syntheses
Molecular formula	$C_5H_{10}O_2$
Molecular weight	102.15
Properties	m.p. $<$ –80.0°C, b.p. 178.0°C

Chemical name	Tetramethyl lead
Chemical Abstract #	75-74-1
Common name	TML
Synonyms	Tetramethylplumbane
Uses	Antiknock compound in gasoline
Molecular formula	$C_4H_{12}Pb$
Molecular weight	267.35
Properties	m.p. –27.5°C, b.p. 110.0°C, decomposes

Chemical name	Thallium acetate
Chemical Abstract #	563-68-8
Common name	Thallous acetate
Synonyms	Thallium monoacetate
Uses	Ore separation by flotation; in medicine
Molecular formula	$C_2H_3O_2Tl$
Molecular weight	263.42
Properties	m.p. 110.0°C

Chemical name	Thiophanate methyl
Chemical Abstract #	23564-06-9 (for thiophanate)
Common name	Thiophanate
Synonyms	[1,2-Phenylenebis(aminocarbonothioyl)] biscarbamic acid diethyl ester *O,O*-dimethyl analog
Uses	Fungicide
Molecular formula	$C_{12}H_{14}N_4O_4S_2$
Molecular weight	342.42
Properties	m.p. 181.0–182.5°C

Chemical name	Thiourea
Chemical Abstract #	62-56-6
Common name	
Synonyms	Thiocarbamide
Uses	Silver complexing agent; photographic fixing
Molecular formula	CH_4N_2S
Molecular weight	76.12
Properties	m.p. 176.0–178.0°C, b.p. decomposes

Chemical name	Tillam (butylethyl carbamothioic acid *S*-propyl ester)
Chemical Abstract #	1114-71-2
Common name	Pebulate, Stauffer 2061, PEBC
Synonyms	*S*-Propyl butylethyl thiocarbamate
Uses	Herbicide
Molecular formula	$C_{10}H_{21}NOS$
Molecular weight	203.36
Properties	b.p. 142.0°C (at 20 mm)

Chemical name	Toluene
Chemical Abstract #	108-88-3
Common name	Toluol; methacide
Synonyms	Methylbenzene
Uses	Manufacturing of benzene derivatives, medicines, dyes, perfumes, explosives; solvent; gasoline additive
Molecular formula	C_7H_8
Molecular weight	92.13
Properties	m.p. –95.0°C, b.p. 110.6°C

Chemical name	2,4-Toluenediisocyanate
Chemical Abstract #	584-84-9
Common name	2,4-TDI; Nacconate 100
Synonyms	Toluene 2,4-diisocyanate
Uses	Manufacturing of polyurethane foam and other elastomers
Molecular formula	$C_9H_6N_2O_2$
Molecular weight	174.15
Properties	m.p. 19.5–21.5°C, b.p. 251.0°C

Chemical name	2,6-Toluenediisocyanate
Chemical Abstract #	91-08-7
Common name	2,6-TDI
Synonyms	Toluene 2,6-diisocyanate
Uses	Elastomer manufacturing
Molecular formula	$C_9H_6N_2O_2$
Molecular weight	174.15
Properties	

Chemical name	*p*-Toluene sulfonic acid
Chemical Abstract #	104-15-4
Common name	Toluene sulfonic acid
Synonyms	4-Methylbenzenesulfonic acid
Uses	Medicinal and dye manufacturing
Molecular formula	$C_7H_8O_3S$
Molecular weight	172.20
Properties	m.p. 106.0–107.0°C, b.p. 140.0°C (at 20 mm)

Chemical name	*O*-Toluic acid
Chemical Abstract #	118-90-1
Common name	Methylbenzoic acid; toluic acid
Synonyms	*o*-Methylbenzoic acid
Uses	
Molecular formula	$C_8H_8O_2$
Molecular weight	136.14
Properties	m.p. 107.0–108.0°C, b.p. 258.0–260.0°C

Chemical name	*p*-Toluic acid
Chemical Abstract #	99-94-5
Common name	Methylbenzoic acid; toluic acid
Synonyms	*p*-Methylbenzoic acid
Uses	
Molecular formula	$C_8H_8O_2$
Molecular weight	136.14
Properties	m.p. 179.0–180.0°C, b.p. 274.0–275.0°C

Chemical name	*m*-Toluic acid
Chemical Abstract #	99-04-7
Common name	Methylbenzoic acid; toluic acid
ynonyms	*m*-Methylbenzoic acid
Uses	
Molecular formula	$C_8H_8O_2$
Molecular weight	136.14
Properties	m.p. 111.0–113.0°C, b.p. 263.0°C (sublimes)

Chemical name	*o*-Toluidine
Chemical Abstract #	95-53-4
Common name	*o*-Methylaniline
Synonyms	2-Methylbenzamine
Uses	Manufacturing of dyes and organic chemicals; textile printing
Molecular formula	C_7H_9N
Molecular weight	107.15
Properties	m.p. –16.3°C, b.p. 200.0–202.0°C

Chemical name	*m*-Toluidine
Chemical Abstract #	108-44-1
Common name	*m*-Methylaniline
Synonyms	3-Methylbenzamine
Uses	Manufacturing of organic chemicals, dyes
Molecular formula	C_7H_9N
Molecular weight	107.15
Properties	m.p. –50°C, b.p. 203.0–204.0°C

Chemical name *p*-Toluidine
Chemical Abstract # 106-49-0
Common name *p*-Methylaniline
Synonyms 4-Methylbenzamine
Uses Manufacturing of organic chemicals and dyes
Molecular formula C_7H_9N
Molecular weight 107.15
Properties m.p. 45.0°C, b.p. 200.3°C

Chemical name Toxaphene
Chemical Abstract # 8001-35-2
Common name Hercules 3956; Alltox; Geniphene
Synonyms Polychlorocamphene; camphechlor
Uses Insecticide
Molecular formula Approximate formula = $C_{10}H_{10}Cl_8$
Molecular weight Approximately 413.80
Properties m.p. 65.0–90.0°C

Chemical name Tri-*N*-allylamine
Chemical Abstract # 102-70-5
Common name Triallylamine
Synonyms *N,N*-Di-2-propenyl-2-propen-1-amine
Uses
Molecular formula $C_9H_{15}N$
Molecular weight 137.25
Properties m.p. < –70.0°C, b.p. 149.5°C

Chemical name Tri-*n*-butylamine
Chemical Abstract # 102-82-9
Common name Tributylamine
Synonyms Tri-*n*-butylamine
Uses Solvent inhibitor
Molecular formula $C_{12}H_{27}N$
Molecular weight 185.34
Properties m.p. –70.0°C, b.p. 213.0°C

Chemical name Tri-*n*-butylphosphate
Chemical Abstract # 126-73-8
Common name TBP; Celluphos 4
Synonyms Tributylphosphate
Uses Plasticizer for lacquers, plastics, vinyl resins
Molecular formula $C_{12}H_{27}O_4P$
Molecular weight 266.32
Properties m.p. < –80.0°C, b.p. 289.0°C, decomposes

Chemical name *S,S,S*-Tributylphosphorotrithioate
Chemical Abstract # 150-50-5
Common name DEF
Synonyms
Uses Pesticide
Molecular formula $C_{12}H_{27}S_3PO$
Molecular weight 313.59
Properties

Chemical name Trichloroacetic acid
Chemical Abstract # 76-03-9
Common name TCA
Synonyms Trichloroethanoic acid
Uses Herbicide; protein precipitant; decalcifier
Molecular formula $C_2HCl_3O_2$
Molecular weight 163.40
Properties m.p. 57.0°C, b.p. 197.5°C

Chemical name 2,4,6-Trichloroaniline
Chemical Abstract # 634-93-5
Common name Trichloroaniline
Synonyms 2,4,6-Trichlorobenzenamine
Uses Experimental carcinogen
Molecular formula $C_6H_4Cl_3N$
Molecular weight 196.46
Properties m.p. 77.5–78.5°C, b.p. 262.0°C (at 46 mm)

Chemical name	1,2,3-Trichlorobenzene
Chemical Abstract #	87-61-6
Common name	
Synonyms	*vic*-Trichlorobenzene
Uses	Coolant, solvent; used in insecticides
Molecular formula	$C_6H_3Cl_3$
Molecular weight	181.46
Properties	m.p. 52.0°C, b.p. 219.0°C

Chemical name	1,2,4-Trichlorobenzene
Chemical Abstract #	120-82-1
Common name	
Synonyms	*unsym*-Trichlorobenzene
Uses	Solvent; dyes and dye intermediates; dielectric fluid; lubricant; insecticide
Molecular formula	$C_6H_3Cl_3$
Molecular weight	181.46
Properties	m.p. 17.0°C, b.p. 213.0°C

Chemical name	1,3,5-Trichlorobenzene
Chemical Abstract #	108-70-3
Common name	
Synonyms	*sym*-Trichlorobenzene
Uses	Solvent; dielectric fluid; lubricant; insecticide
Molecular formula	$C_6H_3Cl_3$
Molecular weight	181.46
Properties	m.p. 63.4°C, b.p. 208.4°C

Chemical name	3,3,4′-Trichlorocarbanilide
Chemical Abstract #	101-20-2
Common name	
Synonyms	*N*-(3,4-Dichlorophenyl)-*N′*-(4-chlorophenyl)urea
Uses	Bacteriostat in soap; used in plastics
Molecular formula	$C_{13}H_9Cl_3N_2O$
Molecular weight	315.59
Properties	m.p. 250.0°C

Chemical name	1,1,2-Trichloroethane
Chemical Abstract #	79-00-5
Common name	
Synonyms	Vinyl trichloride
Uses	Solvent
Molecular formula	$C_2H_3Cl_3$
Molecular weight	133.40
Properties	m.p. –35.0°C, b.p. 113.0–114.0°C

Chemical name	1,1,1-Trichloroethane
Chemical Abstract #	71-55-6
Common name	Chloroethane
Synonyms	Methylchloroform
Uses	Cleaning of plastics and metals
Molecular formula	$C_2H_3Cl_3$
Molecular weight	133.40
Properties	m.p. –32.0°C, b.p. 74.1°C

Chemical name	Trichloroethylene
Chemical Abstract #	79-01-6
Common name	Petzinol; Algylen; Triasol
Synonyms	Ethylenetrichloride; 1-chloro-2,2-dichloroethylene
Uses	Solvent for fats, waxes, resins, oils, rubber, paints; solvent extraction
Molecular formula	C_2HCl_3
Molecular weight	131.38
Properties	m.p. –84.8°C, b.p. 86.7°C

Chemical name	3,4,6-Trichloro-2-nitrophenol
Chemical Abstract #	82-62-2
Common name	Dowlap
Synonyms	2-Nitro-3,4,6-trichlorophenol
Uses	For controlling lampreys
Molecular formula	$C_6H_2Cl_3NO_3$
Molecular weight	242.44
Properties	m.p. 92.0–93.0°C

Chemical name	2,3,6-Trichlorophenol
Chemical Abstract #	933-75-5
Common name	
Synonyms	
Uses	Pesticide
Molecular formula	$C_6H_3Cl_3O$
Molecular weight	197.44
Properties	m.p. 62.0°C, b.p. 253.0°C

Chemical name	2,4,5-Trichlorophenol
Chemical Abstract #	95-95-4
Common name	
Synonyms	
Uses	Fungicide; bactericide
Molecular formula	$C_6H_3Cl_3O$
Molecular weight	197.44
Properties	m.p. 57.0°C, b.p. 252.0°C

Chemical name	2,4,6-Trichlorophenol
Chemical Abstract #	88-06-2
Common name	Dowicide 2S; Omal
Synonyms	
Uses	Fungicide; bactericide; preservative
Molecular formula	$C_6H_3Cl_3O$
Molecular weight	197.44
Properties	m.p. 69.0°C, b.p. 246.0°C

Chemical name	2,4,5-Trichlorophenoxyacetic acid
Chemical Abstract #	93-76-5
Common name	2,4,5-T, Dacamine
Synonyms	Weedone, Reddox
Uses	Herbicide
Molecular formula	$C_8H_5Cl_3O_3$
Molecular weight	255.48
Properties	m.p. 151.0–153.0°C

Chemical name	2,4,5-Trichlorophenoxyacetic acid, isooctyl esters
Chemical Abstract #	93-76-5 (2,4,5-Trichlorophenoxyacetic acid)
Common name	
Synonyms	
Uses	Herbicide
Molecular formula	$C_{16}H_{19}Cl_3O_3$
Molecular weight	365.78
Properties	

Chemical name	2-(2,4,5-Trichlorophenoxy)propionic acid
Chemical Abstract #	93-72-1
Common name	Fenoprop; Garlon; Silvex, 2,4,5-TP
Synonyms	α-(2,4,5-Trichlorophenoxy) propionic acid
Uses	Herbicide
Molecular formula	$C_9H_7Cl_3O_3$
Molecular weight	269.51
Properties	m.p. 182.0°C

Chemical name	2-(2,4,5-Trichlorophenoxy)propionic acid, butoxy ethyl ester
Chemical Abstract #	93-72-1 [2-(2,4,5-Trichlorophenoxy) propionic acid]
Common name	2,4,5-TP, BEE
Synonyms	
Uses	Herbicide
Molecular formula	
Molecular weight	
Properties	

Chemical name	2-(2,4,5-Trichlorophenoxy)propionic acid, isoctyl esters
Chemical Abstract #	93-72-1 [2-(2,4,5-Trichlorophenoxy)propionic acid]
Common name	Silvex (IOE)
Synonyms	
Uses	Herbicide
Molecular formula	
Molecular weight	
Properties	b.p. 373.0°C

Chemical name	2-(2,4,5-Trichlorophenoxy)propionic acid, propylene glycol butyl ether ester
Chemical Abstract #	6047-17-2
Common name	PBGE; Silvex, PGBEE
Synonyms	Kuron
Uses	Herbicide
Molecular formula	$C_{16}H_{21}Cl_3O_4$
Molecular weight	383.72
Properties	b.p. 327.0°C

Chemical name	2,3,6-Trichlorophenylecetic acid
Chemical Abstract #	85-34-7
Common name	Fenac; Trifene; Chlorfenac
Synonyms	2,3,6-Trichlorbenzene acetic acid
Uses	Herbicide
Molecular formula	$C_8H_5Cl_3O_2$
Molecular weight	239.48
Properties	m.p. 160.0°C

Chemical name	1,2,3-Trichloropropene
Chemical Abstract #	96-19-5
Common name	
Synonyms	
Uses	
Molecular formula	$C_3H_3Cl_3$
Molecular weight	145.41
Properties	b.p. 142.0°C

Chemical name	1,2,3-Trichloropropane
Chemical Abstract #	96-18-4
Common name	Trichlorohydrin
Synonyms	Glyceroltrichlorohydrin
Uses	
Molecular formula	$C_3H_5Cl_3$
Molecular weight	147.43
Properties	m.p. –14.0°C

Chemical name	Tri-*o*-cresylphosphate
Chemical Abstract #	78-30-8
Common name	TCP
Synonyms	Tri-*o*-tolyl phosphate
Uses	Plasticizer in lacquers and varnishes
Molecular formula	$C_{21}H_{21}O_4P$
Molecular weight	368.36
Properties	m.p. –25.0 to –30.0°C, b.p. 420.0°C

Chemical name	Tri-*p*-cresyl phosphate
Chemical Abstract #	1330-78-5
Common name	
Synonyms	Tritolyl phosphate
Uses	Plasticizer in vinyl plastics
Molecular formula	$C_{21}H_{21}O_4P$
Molecular weight	368.36
Properties	b.p. 265.0°C

Chemical name	Triethanolamine
Chemical Abstract #	102-71-6
Common name	TEA; triethylolamine
Synonyms	2,2′,2″-Nitrilotrisethanol
Uses	Manufacturing of herbicides, waxes, polishes, toilet goods, cutting oils; solvent; lubricant production
Molecular formula	$C_6H_{15}NO_3$
Molecular weight	149.19
Properties	m.p. 21.6°C, b.p. 335.4°C

Chemical name	Triethylamine
Chemical Abstract #	121-44-8
Common name	
Synonyms	*N,N*-Diethylethanamine
Uses	Preparation of quaternary ammonium compounds
Molecular formula	$C_6H_{15}N$
Molecular weight	101.22
Properties	m.p. –115.0°C, b.p. 89.0–90.0°C

Chemical name Triethylene glycol
Chemical Abstract # 112-27-6
Common name TEG
Synonyms 2,2′-[1,2-Ethanediylbis(oxy)]-bisethanol
Uses Stabilizer for plastics
Molecular formula $C_6H_{14}O_4$
Molecular weight 150.20
Properties m.p. –4.0 to –7.0°C, b.p. 287.4°C

Chemical name Triethyllead chloride
Chemical Abstract # 1067-14-7
Common name
Synonyms Triethylchloroplumbane
Uses Experimental teratogen
Molecular formula $C_6H_{15}ClPb$
Molecular weight 329.85
Properties

Chemical name Trifluralin
Chemical Abstract # 1582-09-8
Common name Treflan; Triflurex
Synonyms 2,6-Dinitro-*N,N*-dipropyl-4-(trifluoromethyl) benzenamine
Uses Herbicide
Molecular formula $C_{13}H_{16}F_3N_3O_4$
Molecular weight 335.29
Properties m.p. 46.0–47.0°C, b.p. 139.0–140.0°C (at 4.2 mm)

Chemical name Trimethyllead chloride
Chemical Abstract # 1520-78-1
Common name TML
Synonyms Chlorotrimethylplumbane
Uses Experimental reproductive effects
Molecular formula C_3H_9ClPb
Molecular weight 287.76
Properties

Chemical name 2,4,4-Trimethyl-1-pentene
Chemical Abstract # 25167-70-8
Common name Diisobutene
Synonyms Diisobutylene
Uses Organic syntheses
Molecular formula C_8H_{16}
Molecular weight 112.24
Properties b.p. 104.5°C

Chemical name 2,4,6-Trinitroresorcinol
Chemical Abstract # 82-71-3
Common name Styphnic acid
Synonyms 2,4,6-Trinitro-1,3-benzenediol
Uses Organic syntheses; explosive manufacturing
Molecular formula $C_6H_3N_3O_8$
Molecular weight 245.11
Properties m.p. 179.0–180.0°C

Chemical name 2,4,6-Trinitrotoluene
Chemical Abstract # 118-96-7
Common name TNT; Trotyl
Synonyms Trinitrotoluene
Uses Explosive manufacturing
Molecular formula $C_7H_5N_3O_6$
Molecular weight 227.15
Properties m.p. 80.7°C, b.p. 240.0°C (explodes)

Chemical name 1,3,5-Trioxane
Chemical Abstract # 110-88-3
Common name Triformol
Synonyms 1,3,5-Trioxacyclohexane
Uses Explosives manufacturing
Molecular formula $C_3H_6O_3$
Molecular weight 90.08
Properties m.p. 64.0°C, b.p. 115.0°C

Chemical name	Triphenylphosphate
Chemical Abstract #	115-86-6
Common name	
Synonyms	
Uses	Manufacturing of plasticizers; insecticide; gasoline additive; fire retardant
Molecular formula	$C_{18}H_{15}O_4P$
Molecular weight	326.28
Properties	m.p. 50.0°C, b.p. 245.0°C (at 11 mm)

Chemical name	Triphenyltin hydroxide
Chemical Abstract #	76-87-9
Common name	Du-Ter; TPTH; Fenolovo
Synonyms	Hydroxytriphenylstannane
Uses	Pesticide; fungicide
Molecular formula	$C_{18}H_{16}OSn$
Molecular weight	367.03
Properties	m.p. 122.0–123.5°C

Chemical name	*N*-Tritylmorpholine
Chemical Abstract #	1420-06-0
Common name	Frescon; Trifenmorph
Synonyms	4-(Triphenylmethyl)morpholine
Uses	Molluscicide
Molecular formula	$C_{23}H_{23}NO$
Molecular weight	329.44
Properties	m.p. 174.0–176.0°C; resolidifies and melts again at 185.0–187.0°C

Chemical name	Undecylenic acid
Chemical Abstract #	112-38-9
Common name	Sevinon; Declid; Renselin
Synonyms	10-Undecenoic acid
Uses	Herbicide; topical antifungal, weed defoliant
Molecular formula	$C_{11}H_{20}O_2$
Molecular weight	184.27
Properties	m.p. 24.5°C, b.p. 195.6°C

Chemical name	Urea
Chemical Abstract #	57-13-6
Common name	Aquadrate; Hyanit; Ureaphil; Urepearl
Synonyms	Carbamide
Uses	Fertilizer; in animal feeds; in paper industry
Molecular formula	CH_4N_2O
Molecular weight	60.06
Properties	m.p. 132.7°C; decomposes on further heating

Chemical name	*n*-Valeraldehyde
Chemical Abstract #	110-62-3
Common name	Pentanal
Synonyms	*n*-Valeric aldehyde; *n*-amylaldehyde
Uses	Flavoring; rubber industry; resin chemistry
Molecular formula	$C_5H_{10}O$
Molecular weight	86.13
Properties	m.p. –91.0°C, b.p. 103.0°C

Chemical name	*n*-Valeric acid
Chemical Abstract #	109-52-4
Common name	Propylacetic acid
Synonyms	Pentanoic acid
Uses	Perfume industry; lubricant; plasticizer; pharmaceuticals
Molecular formula	$C_5H_{10}O_2$
Molecular weight	102.13
Properties	m.p. –34.5°C, b.p. 186.0–187.0°C

Chemical name	Vanillin
Chemical Abstract #	121-33-5
Common name	Vanillic aldehyde
Synonyms	4-Hydroxy-3-methoxybenzaldehyde
Uses	Flavoring agent; perfumery; food industry; beverage industry
Molecular formula	$C_8H_8O_3$
Molecular weight	152.14
Properties	m.p. 80.0–81.0°C, b.p. 285.0°C

Chemical name	Versatic 10
Chemical Abstract #	71700-95-3
Common name	
Synonyms	C_{10} carbonic acid
Uses	Foam catalyst; manufacturing of paint dryers
Molecular formula	Mixture of saturated, mainly tertiary carbon acids
Molecular weight	
Properties	

Chemical name	Acetic acid vinyl ester
Chemical Abstract #	108-05-4
Common name	Vinyl acetate
Synonyms	Acetic acid ethenyl ester
Uses	In films, and lacquers; foam industry
Molecular formula	$C_4H_6O_2$
Molecular weight	86.09
Properties	m.p. –100.2°C, b.p. 73.0°C

Chemical name	Vinyl chloride
Chemical Abstract #	75-01-4
Common name	MVC; chloro-ethene
Synonyms	Chloroethylene
Uses	Adhesive; refrigerant; plastic industry
Molecular formula	C_2H_3Cl
Molecular weight	62.5
Properties	m.p. –160.0°C, b.p. –13.9°C

Chemical name	Warfarin
Chemical Abstract #	81-81-2
Common name	200 Coumarin; compound 42; Rodex; Co-Rax
Synonyms	4-Hydroxy-3-(3-oxo-1-phenylbutyl)-2*H*-1-benzopyran-2-one
Uses	Rodenticide; anticoagulant
Molecular formula	$C_{19}H_{16}O_4$
Molecular weight	308.32
Properties	m.p. 161.0°C

Chemical name	*o*-Xylene
Chemical Abstract #	95-47-6
Common name	*o*-Xylol
Synonyms	1,2-Dimethylbenzene
Uses	Manufacturing of phthalic acid, terephthalic acid; solvent recovery; aviation fuel; dye and insecticide manufacturing
Molecular formula	C_8H_{10}
Molecular weight	106.16
Properties	m.p. –25.0°C, b.p. 144.0°C

Chemical name	*m*-Xylene
Chemical Abstract #	108-38-3
Common name	*m*-Xylol
Synonyms	1,3-Dimethylbenzene
Uses	Solvent; manufacturing of dyes, polyester fibers; clearing agent in microscopic work
Molecular formula	C_8H_{10}
Molecular weight	106.16
Properties	m.p. –48.0°C, b.p. 139.0°C

Chemical name	*p*-Xylene
Chemical Abstract #	106-42-3
Common name	*p*-Xylol
Synonyms	1,4-Dimethylbenzene
Uses	Solvent; manufacturing of dyes; fibers; fuel additive
Molecular formula	C_8H_{10}
Molecular weight	106.16
Properties	m.p. 13.0–14.0°C, b.p. 138.3°C

Chemical name	2,4-Xylenol
Chemical Abstract #	105-67-9
Common name	*m*-Xylenol
Synonyms	2,4-Dimethylphenol
Uses	In manufacturing of pharmaceuticals, insecticides, fungicides, rubbers, solvents
Molecular formula	$C_8H_{10}O$
Molecular weight	122.16
Properties	m.p. 25.4–26.6°C, b.p. 211.0°C

Chemical name	2,5-Xylenol
Chemical Abstract #	95-87-4
Common name	*p*-Xylenol
Synonyms	2,5-Dimethylphenol
Uses	Manufacturing of resins, rubbers, pesticides, solvents
Molecular formula	$C_8H_{10}O$
Molecular weight	122.16
Properties	m.p. 74.5°C, b.p. 211.5–213.5°C

Chemical name	2,6-Xylenol
Chemical Abstract #	576-26-1
Common name	*vic-m*-Xylenol
Synonyms	2,6-Dimethylphenol
Uses	Manufacturing of resins, solvents
Molecular formula	$C_8H_{10}O$
Molecular weight	122.16
Properties	m.p. 49.0°C, b.p. 203.0°C

Chemical name	3,4-Xylenol
Chemical Abstract #	95-65-8
Common name	*asym-o*-Xylenol
Synonyms	3,4-Dimethylphenol
Uses	Manufacturing of solvents, resins, disinfectants
Molecular formula	$C_8H_{10}O$
Molecular weight	122.16
Properties	m.p. 65.0°C, b.p. 226.9°C

Chemical name	3,5-Xylenol
Chemical Abstract #	108-68-9
Common name	*sym-m*-Xylenol
Synonyms	3,5-Dimethylphenol
Uses	Manufacturing of disinfectants, solvents
Molecular formula	$C_8H_{10}O$
Molecular weight	122.16
Properties	m.p. 64.0°C, b.p. 219.5°C

Chemical name	Zectran
Chemical Abstract #	315-18-4
Common name	Mexacarbate
Synonyms	4-(Dimethylamino)-3,5-dimethylphenol methyl-carbamate
Uses	Pesticide, herbicide
Molecular formula	$C_{12}H_{18}N_2O_2$
Molecular weight	222.29
Properties	m.p. 85.0°C

Chemical name	Zinc stearate
Chemical Abstract #	557-05-1
Common name	
Synonyms	Octadecanoic acid zinc salt
Uses	Antiseptic; dietary supplement; lubricant; plastics
Molecular formula	$C_{36}H_{70}O_4Zn$
Molecular weight	632.33
Properties	m.p. 120.0–130.0°C

Chemical name	Zineb{[1,2-Ethanediylbis[carbamodithioato](2-)] zinc}
Chemical Abstract #	12122-76-7
Common name	Karamate, Parzate, Lonacol
Synonyms	Zinc ethylene bisdithiocarbamate
Uses	Fungicide
Molecular formula	$C_4H_6N_2S_4Zn$
Molecular weight	275.85
Properties	

Chemical name	Zolone
Chemical Abstract #	2310-17-0
Common name	Phosalone
Synonyms	Phosphorodithioic acid *S*-[(6-chloro-2-oxo-3(2H)-benzoxazolyl)methyl] *O,O*-diethyl ester
Uses	Insecticide, acaricide
Molecular formula	$C_{12}H_{15}ClNO_4PS_2$
Molecular weight	367.80
Properties	m.p. 47.5–48.0°C

Chemical name	Phenylmercuric acetate
Chemical Abstract #	62-38-4
Common name	PMA, Phix, Mersolite, Tag fungicide
Synonyms	Acetoxyphenylmercury
Uses	As herbicide and fungicide
Molecular formula	$C_8H_8HgO_2$
Molecular weight	336.75
Properties	m.p. 149.0°C

Chemical name	Mercuric chloride
Chemical Abstract #	7487-94-7
Common name	Corrosive sublimate
Synonyms	Mercury perchloride
Uses	For preserving wood and anatomical specimens; in embalming, browning, and etching steel and iron; tanning leather; electroplating aluminium; as depolarizer for dry batteries; in extracting gold from lead ores; mordant for rabbit and beaver furs; treating seed potatoes; topical antiseptic and disinfectant
Molecular formula	$HgCl_2$
Molecular weight	271.52
Properties	m.p. 277.0°C; volatilizes unchanged at 300.0°C

Chemical name	Sodium selenite
Chemical Abstract #	10102-18-8
Common name	
Synonyms	
Uses	For removing green color from glass during its manufacture
Molecular formula	Na_2O_3Se
Molecular weight	172.95
Properties	

Chemical name	Selenium
Chemical Abstract #	7782-49-2
Common name	
Synonyms	
Uses	As pigment in making colored glasses; as metallic base in making electrodes for arc ights, electrical instruments, and apparatuses; as rectifier in radio and television sets; in photocells, semiconductor fusion mixtures; as vulcanizing agent for rubber
Molecular formula	Se
Molecular weight	78.96
Properties	

Chemical name	Sulfur
Chemical Abstract #	7704-34-9
Common name	Brimstone
Synonyms	Sulphur
Uses	In the manufacture of sulfuric acid, carbon disulfide, insecticides, plastics, enamels, metal-glass cements; in vulcanizing rubber; in the synthesis of dyes; in making gunpowder, matches; in bleaching straw, wool, silk, felt, and linen
Molecular formula	S
Molecular weight	32.064
Properties	m.p. 115.21°C, b.p. 444.6°C

Chemical name	Cadmium chloride
Chemical Abstract #	10108-64-2
Common name	Caddy, Vi-Cad
Synonyms	
Uses	In photography; for dyeing and calico printing; in the vacuum tube industry; in the manufacture of special mirrors; as ice-nucleating agent; as lubricant; as absorber of hydrogen sulfide; as fungicide
Molecular formula	$CdCl_2$
Molecular weight	183.32
Properties	m.p. 568.0°C, b.p. 960.0°C

Chemical name	Arsenic pentoxide
Chemical Abstract #	1303-28-2
Common name	
Synonyms	Arsenic acid anhydride
Uses	In the manufacture of colored glass; in adhesives for metals; as wood preservative; in weed control; as fungicide
Molecular formula	As_2O_5
Molecular weight	229.82
Properties	

Chemical name	Zinc chloride
Chemical Abstract #	7646-85-7
Common name	Butter of zinc
Synonyms	
Uses	As antiseptic, as astringent; in medicine
Molecular formula	$ZnCl_2$
Molecular weight	136.29
Properties	m.p. 290.0°C, b.p. 732.0°C

Chemical name	Zinc sulfate
Chemical Abstract #	7733-02-0
Common name	White vitriol, Medizinc, Optrax
Synonyms	Zincate, Zincomed
Uses	As mordant in calico-printing; preserving wood and skin; clarifying glue; as a reagent in analytical chemistry; in medicine
Molecular formula	$ZnSO_4$
Molecular weight	161.44
Properties	m.p. 100.0°C

Chemical name	Arsenic trioxide
Chemical Abstract #	1327-53-3
Common name	White arsenic
Synonyms	Arsenous acid
Uses	In the manufacture of colored glass, Paris green, enamels; in weed killers; preserving hides; as rodenticide, insecticide; in sheep dips; as textile mordant
Molecular formula	As_2O_3
Molecular weight	197.82
Properties	m.p. 275.0°C, b.p. 465.0°C

Chemical name	Sodium fluoride
Chemical Abstract #	7681-49-4
Common name	Cavi-Trol, Antibulit, Gleem, Duraphat
Synonyms	Karidium, Osteofluor
Uses	As insecticide for roaches and ants; in other pesticide formulations; in making vitreous enamels and glass mixes; in fluxes; in fluoridating drinking water; as disinfectant; prophylactic for dental caries
Molecular formula	NaF
Molecular weight	42.00
Properties	m.p. 993.0°C, b.p. 1704.0°C

Chemical name	Nickel sulfate
Chemical Abstract #	7786-81-4
Common name	Nickelous sulfate
Synonyms	Nickel(2+) sulfate(1:1)
Uses	In nickel plating; as mordant in dyeing and printing fabrics; for blackening zinc and brass
Molecular formula	$NiSO_4$
Molecular weight	154.77
Properties	m.p. 840.0°C (loss of SO_3)

Chemical name	Nickel chloride, hexahydrate
Chemical Abstract #	7791-20-0
Common name	
Synonyms	
Uses	In nickel plating cast zinc; in making special inks
Molecular formula	$NiCl_2$, $6H_2O$
Molecular weight	237.33
Properties	Sublimes

Chemical name	Cobalt chloride
Chemical Abstract #	7646-79-9
Common name	
Synonyms	Cobalt dichloride
Uses	Used in invisible inks; in hygrometers, electroplating; as fertilizer and feed additive; as a foam stabilizer in beer; as absorbent for poisonous gases and ammonia; in making vitamin B_{12}
Molecular formula	$CoCl_2$
Molecular weight	129.85
Properties	m.p. 735.0°C, b.p. 1049.0°C

Chemical name	Sodium arsenite
Chemical Abstract #	7784-46-5
Common name	Atlas "A," Kill-All, Sodanit
Synonyms	Arsenous acid, Na salt
Uses	In making medicinal soaps; as insecticide against termites; as a topical acaricide
Molecular formula	$NaAsO_2$
Molecular weight	129.91
Properties	

Chemical name	Potassium cyanide
Chemical Abstract #	151-50-8
Common name	
Synonyms	Hydrocyanic acid, K salt
Uses	
Molecular formula	KCN
Molecular weight	65.11
Properties	m.p. 634.0°C

Chemical name	Cupric sulfate
Chemical Abstract #	7758-98-7
Common name	Blue stone, blue vitreol, blue copper
Synonyms	Copper sulfate
Uses	In agricultural fungicides, algicides, bactericides, herbicides formulations; preserving hides, tanning leather; as pigment in paints, varnishes; in pyrotechnic compositions
Molecular formula	$CuSO_4$
Molecular weight	159.61
Properties	

Chemical name	Aluminum chloride
Chemical Abstract #	7446-70-0
Common name	Pearsall
Synonyms	Trichloroaluminum
Uses	As preservative, disinfectant for stables; in deodorant and antiseptic formulations
Molecular formula	$AlCl_3$
Molecular weight	133.34
Properties	Sublimes at 181.0°C

Chemical name	Sodium cyanide
Chemical Abstract #	143-33-9
Common name	
Synonyms	
Uses	For silver plating
Molecular formula	AgCN
Molecular weight	133.90
Properties	

6 Aquatic Toxicity Data

Chemical	Species	Toxicity test	24 h	48 h	96 h	Chronic exposure	Ref.
A, Pentachlorophenol	*Ambystoma mexicanum* (3–4 wk old)	LC_{50}		0.30			1
A, Allylamine	*Ambystoma mexicanum*, imm.	LC_{50}		1.8			1
A, Salicylaldehyde	*Ambystoma mexicanum* (3–4 wk after hatch)	LC_{50}		7.0			1
A, *o*-Cresol	*Ambystoma mexicanum*, imm.	LC_{50}		40.0			1
A, Trichloroethylene	*Ambystoma mexicanum* (3–4 wk old)	LC_{50}		48.0			1
A, 1-Heptanol	*Ambystoma mexicanum* (3–4 wk after hatch)	LC_{50}		52.0			1
A, Ethyl propionate	*Ambystoma mexicanum* (3–4 wk after hatch)	LC_{50}		54.0			1
A, Ethylacetate	*Ambystoma mexicanum* (3 wk after hatch)	LC_{50}		150.0			1
A, Benzene	*Ambystoma mexicanum*, imm.	LC_{50}		370.0			1
A, Pyridine	*Ambystoma mexicanum* (3–4 wk old)	LC_{50}		950.0			1
A, *n*-Propanol	*Ambystoma mexicanum* (3–4 wk old)	LC_{50}		4000.0			2
A, Acetone	*Ambystoma mexicanum*, imm.	LC_{50}		20000.0			1
A, Cyanatrine	*Rana temporaria*	LC_{50}			30.0		371
A, Pentachlorophenol	*Xenopus laevis* (3–4 wk old)	LC_{50}		0.260			1
A, Allylamine	*Xenopus laevis*, imm.	LC_{50}		5.0			1
A, Salicylaldehyde	*Xenopus laevis* (3–4 wk after hatch)	LC_{50}		7.7			1
A, *o*-Cresol	*Xenopus laevis*, imm.	LC_{50}		38.0			1
A, 1-Heptanol	*Xenopus laevis* (3–4 wk after hatch)	LC_{50}		44.0			1
A, Trichloroethylene	*Xenopus laevis* (3–4 wk old)	LC_{50}		45.0			1
A, Ethyl propionate	*Xenopus laevis* (3–4 wk after hatch)	LC_{50}		56.0			1
A, Ethyl acetate	*Xenopus laevis* (3 wk after hatch)	LC_{50}		180.0			1
A, Benzene	*Xenopus laevis*, imm.	LC_{50}		190.0			1
A, Pyridine	*Xenopus laevis* (3–4 wk old)	LC_{50}		1400.0			1
A, *n*-Propanol	*Xenopus laevis* (3–4 wk old)	LC_{50}		4000.0			1
A, Quinoline	*Xenopus laevis* (embryo)	LC_{50}			26.3		3
Al, Triethyllead chloride	Algae	EC_{50}				0.1 (as Pb, 6 h)	114
Al, Trimethyllead chloride	Algae	EC_{50}				0.8 (Pb)(6 h)	114
Al, Tetramethyllead chloride	Algae	EC_{50}				1.30 (6 h)	114
Al, Tetraethyllead	Algae	LC_{50}			0.10		114

Al, Picloram	*Algae*	NOEL		1.0 (10 wk)	771
Al, Tetraethyllead	*Ankistrodesmus falcatus*	EC_{50}		0.3 (4 h)	350
Al, Pentachlorophenol	*Chlorella pyrenoidosa*	LC_{100}		0.001	22
Al, *p*-Bromophenol	*Chlorella pyrenoidosa*	LC_{100}		36.0	22
Al, *m*-Bromophenol	*Chlorella pyrenoidosa*	LC_{100}		36.0	22
Al, *p*-Chlorophenol	*Chlorella pyrenoidosa*	LC_{100}		40.0	22
Al, *m*-Chlorophenol	*Chlorella pyrenoidosa*	LD_{100}		40.0	22
Al, *o*-Aminophenol	*Chlorella pyrenoidosa*	LC_{100}		47.0	22
Al, *o*-Bromophenol	*Chlorella pyrenoidosa*	LC_{100}		78.0	22
Al, *o*-Chlorophenol	*Chlorella pyrenoidosa*	LC_{100}		96.0	22
Al, *m*-Aminophenol	*Chlorella pyrenoidosa*	LC_{100}		140.0	22
Al, *p*-Aminophenol	*Chlorella pyrenoidosa*	LC_{100}		140.0	22
Al, *m*-Cresol	*Chlorella pyrenoidosa*	LC_{100}		148.0–171.0	22
Al, 3-Butenoic acid	*Chlorella pyrenoidosa*	LC_{100}		280.0	22
Al, *n*-Butyric acid	*Chlorella pyrenoidosa*	LC_{100}		340.0	22
Al, Isobutyric acid	*Chlorella pyrenoidosa*	TOXIC		345.0	22
Al, Acetic anhydride	*Chlorella pyrenoidosa*	TOXIC		360.0	22
Al, *n*-Butanol	*Chlorella pyrenoidosa*	LC_{100}		8500.0	22
Al, *sec*-Butanol	*Chlorella pyrenoidosa*	LC_{100}		8900.0	22
Al, Isopropanol	*Chlorella pyrenoidosa*	LC_{100}		17,400.0	22
Al, *t*-Butanol	*Chlorella pyrenoidosa*	LC_{100}		24,200.0	22
Al, Ethylene glycol	*Chlorella pyrenoidosa*	LC_{100}		180,000.0	22
Al, Nitrilotriacetic acid	*Chlorella vulgaris*	NOEL	100.0		372
Al, Acetic anhydride	*Chlorella pyrenoidosa*	LC_{100}		360.0	22
Al, *p*-Benzoquinone	*Cyanophyta*	LC_{100}		1.0	8
Al, Butylbenzyl phthalate	*Dunaliella* (cell counts) (S)	LC_{50}	1.0	0.3 (NOAE)	9
Al, Urea	*Entosiphon sulcatum*	TT		29.0	10
Al, *n*-Hexane	*Macrocystis pyrifera*	NOAE		10.0	11
Al, Atrazine	*Microcystis aeruginosa*	TT		0.003	12
Al, Cyclohexylamine	*Microcystis aeruginosa*	TT		0.02	12

Chemical	Species	Toxicity test	24 h	48 h	96 h	Chronic exposure	Ref.
Al, 2-Ethylhexylamine	*Microcystis aeruginosa*	TT				0.02	12
Al, Carbaryl	*Microcystis aeruginosa*	TT				0.03	12
Al, Acrolein	*Microcystis aeruginosa*	TT				0.04	12
Al, Ethylene diamine	*Microcystis aeruginosa*	TT				0.08	12
Al, *m*-Dinitrobenzene	*Microcystis aeruginosa*	TT				0.1	12
Al, Ethyleneimine	*Microcystis aeruginosa*	TT				0.12	12
Al, 2,4-Dinitrotoluene	*Microcystis aeruginosa*	TT				0.13	12
Al, *n*-Butylamine	*Microcystis aeruginosa*	TT				0.14–0.19	12
Al, Acrylic acid	*Microcystis aeruginosa*	TT				0.15	12
Al, 4,6-Dinitrocresol	*Microcystis aeruginosa*	TT				0.15	12
Al, Aniline	*Microcystis aeruginosa*	TT				0.16	12
Al, 2,3-Dinitrotoluene	*Microcystis aeruginosa*	TT				0.22	12
Al, 2,3-Dichloro-1,4-naphthaquinone	*Microcystis aeruginosa*	LC_{100}				0.25	8
Al, Lindane	*Microcystis aeruginosa*	TT				0.30	12
Al, *o*-Toluidine	*Microcystis aeruginosa*	TT				0.31	12
Al, Styphnic acid	*Microcystis* aeruginosa	TT				0.32	12
Al, Allylamine	*Microcystis aeruginosa*	TT				0.35	12
Al, *p*-Nitroaniline	*Microcystis aeruginosa*	TT				0.35	12
Al, Formaldehyde	*Microcystis aeruginosa*	TT				0.39	12
Al, Leadacetate	*Microcystis aeruginosa*	TT				0.45	12
Al, Diallyl phthalate	*Microcystis aeruginosa*	TT				0.65	12
Al, Ethylamine	*Microcystis aeruginosa*	TT				1.3	12
Al, Salicylaldehyde	*Microcystis aeruginosa*	TT				1.60	12
Al, Nitrobenzene	*Microcystis aeruginosa*	TT				1.90	12
Al, 2,4-Dichlorophenol	*Microcystis aeruginosa*	TT				2.0	13
Al, Furfural	*Microcystis* aeruginosa	TT				2.7	12
Al, Benzonitrile	*Microcystis aeruginosa*	TT				3.4	12
Al, 1-Heptanol	*Microcystis aeruginosa*	TT				3.5	12
Al, Butyl phosphate	*Microcystis aeruginosa*	TT				4.1	12

Al, Phenol	*Microcystis aeruginosa*	TT	4.60	12
Al, Furfuryl alcohol	*Microcystis aeruginosa*	TT	5.2	12
Al, Dinoseb	*Microcystis aeruginosa*	TT	5.7	12
Al, *o*-Cresol	*Microcystis aeruginosa*	TT	6.8	12
Al, Isooctanol	*Microcystis aeruginosa*	TT	7.3	12
Al, Phenyl acetate	*Microcystis aeruginosa*	TT	7.50	12
Al, 2,4-Pentanedione	*Microcystis aeruginosa*	TT	8.50	12
Al, Diethyl oxalate	*Microcystis aeruginosa*	TT	9.0	12
Al, *n*-Hexanol	*Microcystis aeruginosa*	TT	12.0	12
Al, *m*-Cresol	*Microcystis aeruginosa*	TT	13.0	12
Al, Ethyl acrylate	*Microcystis aeruginosa*	TT	14.0	12
Al, Styphnic acid	*Microcystis aeruginosa*	TT	15.0	14
Al, Diethyl phthalate	*Microcystis aeruginosa*	TT	15.0	12
Al, Diethanolamine	*Microcystis aeruginosa*	TT	16.0	12
Al, *m*-Nitrophenol	*Microcystis aeruginosa*	TT	17.0	12
Al, *n*-Pentanol	*Microcystis aeruginosa*	TT	17.0	12
Al, Butyraldehyde	*Microcystis aeruginosa*	TT	19.0	12
Al, Aniline	*Microcystis aeruginosa*	LD_{50}	20.0	12
Al, Benzaldehyde	*Microcystis aeruginosa*	TT	20.0	12
Al, *o*-Nitrophenol	*Microcystis aeruginosa*	TT	27.0	12
Al, Pyridine	*Microcystis aeruginosa*	TT	28.0	12
Al, Cyclopentanol	*Microcystis aeruginosa*	TT	28.0	12
Al, Benzylchloride	*Microcystis aeruginosa*	TT	30.0	12
Al, 2-Nitro-*p*-cresol	*Microcystis aeruginosa*	TT	32.0	12
Al, 3-Hexanol	*Microcystis aeruginosa*	TT	32.0	12
Al, 2-Hexanol	*Microcystis aeruginosa*	TT	32.0	12
Al, 2,4-Dinitrophenol	*Microcystis aeruginosa*	TT	33.0	12
Al, Ethylbenzene	*Microcystis aeruginosa*	TT	33.0	15
Al, Vinyl acetate	*Microcystis aeruginosa*	TT	35.0	12
Al, Butyl cellosolve	*Microcystis aeruginosa*	TT	35.0	12

Chemical	Species	Toxicity test	24 h	48 h	96 h	Chronic exposure	Ref.
Al, Ethyl-*sec*-amyl ketone	*Microcystis aeruginosa*	TT				40.0	12
Al, Cyclohexyl acetate	*Microcystis aeruginosa*	TT				46.0	12
Al, Triethanolamine	*Microcystis aeruginosa*	TT				47.0	12
Al, Trinitrotoluene	*Microcystis aeruginosa*	TT				50.0	12
Al, Cyclohexanone	*Microcystis aeruginosa*	TT				52.0	12
Al, *o*-Dichlorobenzene	*Microcystis aeruginosa*	TT				53.0	12
Al, Butyldigol	*Microcystis aeruginosa*	TT				53.0	12
Al, Benzoic acid	*Microcystis aeruginosa*	TT				55.0	12
Al, *p*-Nitrophenol	*Microcystis aeruginosa*	TT				56.0	12
Al, Trichloroethylene	*Microcystis aeruginosa*	TT				63.0	12
Al, Cyclopentanone	*Microcystis aeruginosa*	TT				63.0	12
Al, *n*-Amyl acetate	*Microcystis aeruginosa*	TT				63.0	12
Al, Styrene	*Microcystis aeruginosa*	TT				67.0	12
Al, Picric acid	*Microcystis aeruginosa*	TT				72.0	12
Al, Ethylenediaminetetra aceticacid	*Microcystis aeruginosa*	TT				76.0	16
Al, Chloral	*Microcystis aeruginosa*	TT				78.0	12
Al, Citric acid	*Microcystis aeruginosa*	TT				80.0	12
Al, Acetic acid	*Microcystis aeruginosa*	TT				90.0	12
Al, Methyl cellosolve	*Microcystis aeruginosa*	TT				100.0	12
Al, *n*-Butanol	*Microcystis aeruginosa*	TT				100.0	12
Al, Toluene	*Microcystis aeruginosa*	TT				105.0	12
Al, Ethylene dichloride	*Microcystis aeruginosa*	TT				105.0	12
Al, Chlorobenzene	*Microcystis aeruginosa*	TT				120.0	12
Al, Cyclohexene	*Microcystis aeruginosa*	TT				160.0	12
Al, Chloroform	*Microcystis aeruginosa*	TT				185.0	12
Al, Isobutryic acid	*Microcystis aeruginosa*	TT				205.0	12
Al, Tetrhydrofuran	*Microcystis aeruginosa*	TT				225.0	12
Al, Trichloroacetic acid	*Microcystis aeruginosa*	TT				250.0	12
Al, *n*-Propanol	*Microcystis aeruginosa*	TT				255.0	12

Al, *n*-Butylacetate	*Microcystis aeruginosa*	TT		280.0	12
Al, Isobutanol	*Microcystis aeruginosa*	TT		290.0	12
Al, *sec*-Butanol	*Microcystis aeruginosa*	TT		312.0	12
Al, *n*-Butyric acid	*Microcystis aeruginosa*	TT		318.0	12
Al, *t*-Butyl acetate	*Microcystis aeruginosa*	TT		420.0	12
Al, Nitrilotriacetic acid	*Microcystis aeruginosa*	TT		510.0	12
Al, Acetonitrile	*Microcystis aeruginosa*	TT		520.0	12
Al, *n*-Propyl acetate	*Microcystis aeruginosa*	TT		530.0	12
Al, Diacetone alcohol	*Microcystis aeruginosa*	TT		530.0	12
Al, Acetone	*Microcystis aeruginosa*	TT		530.0	12
Al, Ethyl acetate	*Microcystis aeruginosa*	TT		550.0	12
Al, 1,4-Dioxane	*Microcystis aeruginosa*	TT		575.0	12
Al, Ethyl butyrate	*Microcystis aeruginosa*	TT		700.0	12
Al, Isopropanol	*Microcystis aeruginosa*	TT		1000.0	12
Al, Benzene	*Microcystis aeruginosa*	TT		1400.0	12
Al, Isopropyl acetate	*Microcystis aeruginosa*	TT		1400.0	12
Al, Ethanol	*Microcystis aeruginosa*	TT		1450.0	12
Al, Ethanolamine	*Microcystis aeruginosa*	TT		1600.0	12
Al, Diethylene glycol	*Microcystis aeruginosa*	TT		1700.0	12
Al, Ethylene glycol	*Microcystis aeruginosa*	TT		2000.0	12
Al, Glycerol	*Microcystis aeruginosa*	TT		2900.0	12
Al, Triethylene glycol	*Microcystis aeruginosa*	TT		3600.0	12
Al, Acetamide	*Microcystis aeruginosa*	TT		6200.0	12
Al, Nitrilotriacetic acid	*Microcystis aeruginosa*	NOEL	100.0		372
Al, Butylbenzyl phthalate	*Microcystis* (cell counts)(S)	LC_{50}	1000.0	560.0 (NOAE)	9
Al, Epichlorhydrin	*Microcystis aeruginosa*	TT		6.0	12
Al, Butylbenzyl phthalate	*Navicula* (cell counts)(S)	LC_{50}	0.6	0.3 (NOAE)	9
Al, 2,3-Dichloro-1,4-naphthaquinone	*Plectonema promelas*	LC_{100}		0.25	8
Al, *m*-Cresol	*Scenedesmus aeruginosa*	TT		15.0	10
Al, Ethylene diamine tetraacetic acid	*Scenedesmus quadricauda*	TT		11.0	10

Chemical	Species	Toxicity test	24 h	48 h	96 h	Chronic exposure	Ref.
Al, Lauryl sulfate	*Scenedesmus quadricauda*	TT				0.02	10
Al, Formaldehyde	*Scenedesmus*	LC_{100}				0.3–0.5	19
Al, 2-Ethylhexylamine	*Scenedesmus quadricauda*	TT				0.36	10
Al, Ethyleneimine	*Scenedesmus quadricauda*	TT				0.37	10
Al, Cyclohexylamine	*Scenedesmus quadricauda*	TT				0.51	10
Al, *n*-Butylamine	*Scenedesmus quadricauda*	TT				0.53	10
Al, *m*-Dinitrobenzene	*Scenedesmus quadricauda*	TT				0.7	10
Al, Ethanolamine	*Scenedesmus quadricauda*	TT				0.75	10
Al, 2,3-Dinitrotoluene	*Scenedesmus quadricauda*	TT				0.83	10
Al, Hydroquinone	*Scenedesmus quadricauda*	TT				0.93	10
Al, Triethylamine	*Scenedesmus*	LD_0				1.0	8
Al, 1,3,5-Trioxane	*Scenedesmus*	LD_0				1.0	8
Al, Succinic acid	*Scenedesmus*	NOAE				1.0	8
Al, Trinitrotoluene	*Scenedesmus quadricauda*	TT				1.6	10
Al, Triethanolamine	*Scenedesmus quadricauda*	TT				1.80	10
Al, Allyl amine	*Scenedesmus quadricauda*	TT				2.2	10
Al, Ethyl amine	*Scenedesmus quadricauda*	TT				2.3	10
Al, Formaldehyde	*Scenedesmus quadricauda*	TT				2.5	10
Al, 2,4-Dinitrotoluene	*Scenedesmus quadricauda*	TT				2.7	10
Al, 2,4-Pentanedione	*Scenedesmus quadricauda*	TT				2.70	10
Al, Chloral	*Scenedesmus quadricauda*	TT				2.8	10
Al, Diallyl phthalate	*Scenedesmus quadricauda*	TT				2.9	10
Al, Phenyl acetate	*Scenedesmus quadricauda*	TT				3.0	10
Al, Tri-*n*-butyl phosphate	*Scenedesmus quadricauda*	TT				3.20	10
Al, 2,4-Dichlorophenol	*Scenedesmus quadricauda*	TT				3.6	10
Al, 2-Nitro-*p*-cresol	*Scenedesmus quadricauda*	TT				3.80	10
Al, Hydroquinone	*Scenedesmus*	LC_0				4.0	20
Al, Diethyl amine	*Scenedesmus*	LC_{100}				4.0	21
Al, Quinhydrone	*Scenedesmus*	LD_0				4.0	8

Al, *o*-Nitrophenol	*Scenedesmus quadricauda*	TT	4.30	10
Al, Salicylaldehyde	*Scenedesmus quadricauda*	TT	4.90	10
Al, 2,4-dimethyl aniline	*Scenedesmus quadricauda*	TT	5.0	10
Al, Styphnic acid	*Scenedesmus capricornutum*	TT	5.0	14
Al, Cyclohexyl acetate	*Scenedesmus quadricauda*	TT	5.3	10
Al, Epichlorhydrin	*Scenedesmus quadricauda*	TT	5.4	10
Al, *p*-Cresol	*Scenedesmus*	LC_0	6.0	8
Al, *p*-Benzoquinone	*Scenedesmus*	LC_{100}	6.0	8
Al, Benzylamine	*Scenedesmus*	LC_{100}	6.0	8
Al, *p*-Aminophenol	*Scenedesmus*	LC_{100}	6.0	8
Al, Catechol	*Scenedesmus*	LD_0	6.0	8
Al, Allyl chloride	*Scenedesmus quadricauda*	TT	6.3	10
Al, *o*-Toluidine	*Scenedesmus quadricauda*	TT	6.30	10
Al, Nitroglycerine	*Scenedesmus*	LD_0	6.50	2
Al, 4-Nitro-*m*-cresol	*Scenedesmus quadricauda*	TT	6.80	10
Al, 6-Nitro-*m*-cresol	*Scenedesmus quadricauda*	TT	7.0	10
Al, *p*-Nitrophenol	*Scenedesmus quadricauda*	TT	7.40	10
Al, Phenol	*Scenedesmus quadricauda*	TT	7.50	10
Al, Ethyl acetoacetate	*Scenedesmus quadricauda*	TT	7.6	10
Al, *m*-Nitrophenol	*Scenedesmus quadricauda*	TT	7.60	10
Al, *o*-Toluidine	*Scenedesmus*	LD_0	8.0	8
Al, Pyrogallol	*Scenedesmus*	LD_0	8.0	8
Al, *m*-Toluidine	*Scenedesmus*	LD_0	8.0	8
Al, Nitrilotriacetic acid	*Scenedesmus quadricauda*	TT	8.30	10
Al, Aniline	*Scenedesmus quadricauda*	TT	8.3	10
Al, Isooctanol	*Scenedesmus quadricauda*	TT	8.5	10
Al, Trinitrotoluene	*Scenedesmus capricornutum*	TT	9.0	12
Al, Ethylene glycol acetate	*Scenedesmus quadricauda*	TT	9.0	10
Al, Aniline	*Scenedesmus*	LC_{100}	10.0	2
Al, *p*-*tert*-Butylphenol	*Scenedesmus*	LD_0	10.0	8

Chemical	Species	Toxicity test	24 h	48 h	96 h	Chronic exposure	Ref.
Al, *p*-Toluidine	*Scenedesmus*	LD_0				10.0	8
Al, Diethyl phthalate	*Scenedesmus quadricauda*	TT				10.0	10
Al, *o*-Cresol	*Scenedesmus quadricauda*	TT				11.0	10
Al, *p*-Nitroaniline	*Scenedesmus quadricauda*	TT				11.0	10
Al, 2,6-Dinitrotoluene	*Scenedesmus quadricauda*	TT				12.0	10
Al, 4,6-Dinitro-*o*-cresol	*Scenedesmus quadricauda*	TT				13.0	10
Al, Ethyl propionate	*Scenedesmus quadricauda*	TT				14.0	12
Al, Ethyl acetate	*Scenedesmus quadricauda*	TT				15.0	10
Al, 2,4-Dinitrophenol	*Scenedesmus quadricauda*	TT				16.0	10
Al, 1-Heptanol	*Scenedesmus quadricauda*	TT				17.0	10
Al, Propargyl alcohol	*Scenedesmus quadricauda*	TT				18.0	10
Al, Acrylic alcohol	*Scenedesmus quadricauda*	TT				18.0	10
Al, *p*-Chlorophenol	*Scenedesmus*	LC_{100}				20.0	22
Al, *p*-Nitroaniline	*Scenedesmus*	LD_0				20.0	8
Al, *o*-Diethylbenzene	*Scenedesmus quadricauda*	TT				20.0	10
Al, *n*-Butyl acetate	*Scenedesmus quadricauda*	TT				21.0	10
Al, Furfuryl alcohol	*Scenedesmus quadricauda*	TT				25.0	10
Al, *n*-Propyl acetate	*Scenedesmus quadricauda*	TT				26.0	10
Al, *m*-Nitrophenol	*Scenedesmus*	LD_0				28.0	8
Al, 4,4′-Diaminodiphenylmethane	*Scenedesmus*	LC_{100}				30.0	8
Al, Diisopropyl ether	*Scenedesmus*	NOAE				30.0	8
Al, *n*-Hexanol	*Scenedesmus quadricauda*	TT				30.0	10
Al, Furfural	*Scenedesmus quadricauda*	TT				31.0	10
Al, Nitrobenzene	*Scenedesmus quadricauda*	TT				33.0	10
Al, Benzaldehyde	*Scenedesmus quadricauda*	TT				34.0	10
Al, 4,6-Dinitro-*o*-cresol	*Scenedesmus*	LC_{100}				36.0	8
Al, *o*-Nitrophenol	*Scenedesmus*	LD_0				36.0	8
Al, *m*-Cresol	*Scenedesmus*	LC_0				40.0	22
Al, 2,4-Dinitrophenol	*Scenedesmus*	LC_{100}				40.0	8

Al, 3,5-Xylenol	*Scenedesmus*	LD_0	40.0	8
Al, Nitrobenzene	*Scenedesmus*	LD_0	40.0	8
Al, 3,4-Xylenol	*Scenedesmus*	LD_0	40.0	22
Al, 2,4-Xylenol	*Scenedesmus*	LD_0	40.0	8
Al, Allylthiourea	*Scenedesmus quadricauda*	TT	41.0	10
Al, *p*-Nitrophenol	*Scenedesmus*	LD_0	42.0	8
Al, Diethanolamine	*Scenedesmus quadricauda*	TT	44.0	10
Al, Ethyl butyrate	*Scenedesmus quadricauda*	TT	47.0	10
Al, Benzyl chloride	*Scenedesmus quadricauda*	TT	50.0	10
Al, Ethyl-*sec*-amyl ketone	*Scenedesmus quadricauda*	TT	53.0	10
Al, *o*-Chlorophenol	*Scenedesmus*	LC_{100}	60.0	8
Al, Resorcinol	*Scenedesmus*	LD_0	60.0	8
Al, 3-Hexanol	*Scenedesmus quadricauda*	TT	63.0	10
Al, 2-Hexanol	*Scenedesmus quadricauda*	TT	72.0	10
Al, Benzonitrile	*Scenedesmus quadricauda*	TT	75.0	10
Al, Isobutryic acid	*Scenedesmus quadricauda*	TT	80.0	10
Al, *n*-Amyl acetate	*Scenedesmus quadricauda*	TT	80.0	10
Al, Butyraldehyde	*Scenedesmus quadricauda*	TT	83.0	10
Al, *sec*-Butanol	*Scenedesmus quadricauda*	TT	95.0	10
Al, Formic acid	*Scenedesmus*	LC_{100}	100.0	8
Al, Triethanolamine	*Scenedesmus*	LD_0	100.0	8
Al, Nitroglycol	*Scenedesmus*	LD_0	100.0	8
Al, Benzotrichloride	*Scenedesmus quadricauda*	TT	100.0	10
Al, *o*-Dichlorobenzene	*Scenedesmus quadricauda*	TT	100.0	10
Al, *o*-Chlorotoluene	*Scenedesmus quadricauda*	TT	100.0	10
Al, Pyridine	*Scenedesmus quadricauda*	TT	120.0	10
Al, Quinoline	*Scenedesmus*	LD_0	140.0	8
Al, Ethylbenzene	*Scenedesmus quadricauda*	TT	160.0	10
Al, Isopropyl acetate	*Scenedesmus quadricauda*	TT	165.0	10
Al, *n*-Butyric acid	*Scenedesmus*	LC_{100}	200.0	8

Chemical	Species	Toxicity test	24 h	48 h	96 h	Chronic exposure	Ref.
Al, Phloroglucinol	*Scenedesmus*	LD_0				200.0	8
Al, Trichloroacetic acid	*Scenedesmus quadricauda*	TT				200.0	10
Al, Styrene	*Scenedesmus quadricauda*	TT				200.0	10
Al, Picric acid, 50%	*Scenedesmus*	LD_0				240.0	12
Al, Dimethyl amine	*Scenedesmus*	LC_{100}				250.0	8
Al, Cyclopentanol	*Scenedesmus quadricauda*	TT				255.0	10
Al, *n*-Pentanol	*Scenedesmus quadricauda*	TT				260.0	10
Al, Isobutanol	*Scenedesmus quadricauda*	TT				350.0	10
Al, Vinyl acetate	*Scenedesmus quadricauda*	TT				370.0	10
Al, Cyclohexanone	*Scenedesmus quadricauda*	TT				370.0	10
Al, Chlorobenzene	*Scenedesmus quadricauda*	TT				390.0	10
Al, Toluene	*Scenedesmus quadricauda*	TT				400.0	10
Al, 1,1,2-Tricloroethane	*Scenedesmus quadricauda*	TT				430.0	10
Al, Citric acid	*Scenedesmus quadricauda*	TT				640.0	10
Al, Ethylene dicholoride	*Scenedesmus quadricauda*	TT				710.0	10
Al, *n*-Butanol	*Scenedesmus quadricauda*	TT				875.0	10
Al, Butyl cellosolve	*Scenedesmus quadricauda*	TT				900.0	10
Al, Cyclohexanone	*Scenedesmus*	LC_0				1000.0	8
Al, 2-Butanone oxime	*Scenedesmus*	LC_{100}				1000.0	8
Al, Glutaric acid	*Scenedesmus*	NOAE				1000.0	8
Al, Ethylene glycol	*Scenedesmus*	NOAE				1000.0	8
Al, Butyldigol	*Scenedesmus quadricauda*	TT				1000.0	10
Al, Trichloroethylene	*Scenedesmus quadricauda*	TT				1000.0	10
Al, Chloroform	*Scenedesmus quadricauda*	TT				1100.0	10
Al, Benzene	*Scenedesmus quadricauda*	TT				1400.0	10
Al, Benzoic acid	*Scenedesmus quadricauda*	TT				1630.0	10
Al, Isopropanol	*Scenedesmus quadricauda*	TT				1800.0	10
Al, Cyclopentanone	*Scenedesmus quadricauda*	TT				1900.0	10
Al, *n*-Butyric acid	*Scenedesmus quadricauda*	TT				2600.0	10

Al, Diethylene glycol	*Scenedesmus quadricauda*	TT			2700.0	10
Al, Diacetone alcohol	*Scenedesmus quadricauda*	TT			3000.0	10
Al, *n*-Propanol	*Scenedesmus quadricauda*	TT			3100.0	10
Al, Acetic anhydride	*Scenedesmus quadricauda*	TT			3400.0	10
Al, *t*-Butyl acetate	*Scenedesmus quadricauda*	TT			3700.0	10
Al, Acetic acid	*Scenedesmus quadricauda*	TT			4000.0	10
Al, Ethanol	*Scenedesmus quadricauda*	TT			5000.0	10
Al, Acetonitrile	*Scenedesmus quadricauda*	TT			7300.0	10
Al, Acetone	*Scenedesmus quadricauda*	TT			7500.0	10
Al, Glycerol	*Scenedesmus quadricauda*	TT			10000.0	10
Al, Urea	*Scenedesmus quadricauda*	TT			10000.0	10
Al, Triethylene glycol	*Scenedesmus quadricauda*	TT			10000.0	10
Al, Acetamide	*Scenedesmus quadricauda*	TT			10000.0	10
Al, Methyl cellosolve	*Scenedesmus quadricauda*	TT			10000.0	10
Al, Ethylene glycol	*Scenedesmus quadricauda*	TT			10000.0	10
Al, *n*-Butyl acetate	*Scenedesmus*	TL_m	320.0			18
Al, Benzyl alcohol	*Scenedesmus*	TL_m		640.0		18
Al, Hydroquinone	*Selenastrum capricornutum*	NOAE			0.1	20
Al, 1-Phenyl-3-pyrazolidone	*Selenastrum capricornutum*	NOAE			0.1	20
Al, Hydroxylamine sulfate	*Selenastrum capricornutum*	NOAE			0.10	20
Al, *t*-Butylamineborane	*Selenastrum capricornutum*	NOAE			0.1	20
Al, Hydroquinone	*Selenastrum capricornutum*	NOAE			0.4	20
Al, Ethylene diamine	*Selenastrum quadricauda*	TT			0.85	10
Al, Diethylene triamine penta-acetic acid	*Selenastrum capricornutum*	LC_{100}			1.0	20
Al, Hydroxylamine sulfate	*Selenastrum capricornutum*	LC_{100}			1.0	20
Al, *t*-Butylamine borane	*Selenastrum capricornutum*	LC_{100}			1.0	20
Al, Nitrilotriacetic acid	*Selenastrum capricornutum*	NOAE			1.0	20
Al, Hydroquinone sulfonate, Na	*Selenastrum capricornutum*	NOAE			1.0	20
Al, 1-Phenyl-3-pyrazolidone	*Selenastrum capricornutum*	NOAE			1.0	20
Al, Pyrrolidone	*Selenastrum capricornutum*	NOAE			1.0	20

Chemical	Species	Toxicity test	24 h	48 h	96 h	Chronic exposure	Ref.
Al, Ethylene diamine tetraacetic-acid, 3Na	*Selenastrum capricornutum*	NOAE				1.0	20
Al, Diethanolamine	*Selenastrum capricornutum*	NOAE				1.0	20
Al, 1,3-Diamino-2-propanol tetra-acetic acid	*Selenastrum capricornutum*	NOAE				1.0	20
Al, Diethylene glycol	*Selenastrum capricornutum*	NOEL				1.0	20
Al, 1,3-Diamino-2-propanol tetra-acetic acid	*Selenastrum capricornutum*	LC_{100}				10.0	20
Al, 1-Phenyl-3-pyrazolidone	*Selenastrum capricornatum*	NOAE				10.0	20
Al, Nitrilotriacetic acid	*Selenastrum capricornutum*	NOAE				10.0	20
Al, Pyrrolidone	*Selenastrum capricornutum*	NOAE				10.0	20
Al, Hydroquinone sulfonate, Na	*Selenastrum capricornutum*	NOAE				10.0	20
Al, β-Phenylethylaminesulfate	*Selenastrum capricornutum*	NOAE				10.0	20
Al, Diethanolamine	*Selenastrum capricornutum*	NOAE				10.0	20
Al, Citrazinic acid	*Selenastrum capricornutum*	NOAE				10.0	20
Al, Thiourea	*Selenastrum capricornutum*	NOAE				10.0	20
Al, Diethylene glycol	*Selenastrum capricornutum*	NOEL				10.0	2
Al, Ethylene diamine tetraacetic-acid, 3Na	*Selenastrum capricornutum*	TT				10.0	20
Al, Diethanolamime	*Selenastrum capricornutum*	LC_{100}				100.0	20
Al, Hydroquinonesulfonate, Na	*Selenastrum capricornutum*	LC_{100}				100.0	20
Al, β-Phenylethylamine sulfate	*Selenastrum capricornutum*	LC_{100}				100.0	20
Al, Thiourea	*Selenastrum capricornutum*	LC_{100}				100.0	20
Al, Ethylenediamine	*Selenastrum capricornutum*	NOAE				100.0	20
Al, Pyrrolidone	*Selenastrum capricornutum*	NOAE				100.0	20
Al, EDTA, ammonium ferric salt	*Selenastrum capricornutum*	NOEL				100.0	20
Al, Diethylene glycol	*Selenastrum capricornutum*	NOEL				100.0	20
Al, Citrazinic acid	*Selenastrum capricornutum*	TT				100.0	20
Al, Butyl benzyl phthalate	*Selenastrum* (cell counts)(S)	LC_{50}			0.4	0.1(NOAE)	9

Al, Butyl benzyl phthalate	*Skeletonema* (cell counts)(S)	LC_{50}	0.6	0.1(NOAE)	9
Ba, Ethylene dichloride	Bacteria	MUTAG (weak)			36
Ba, Adipic acid	Bacteria, algae, protozoa	NOAE		100.0	8
Ba, Formaldehyde	*Escherichia coli*	LC_{100}		1.0	19
Ba, Picric acid	*Escherichia coli*	NOAE		1.0	2
Ba, *n*-Pentanol	*Escherichia coli*	NOAE		1.0	8
Ba, Catechol	*Escherichia coli*	LD_0		1.6	8
Ba, Quinhydrone	*Escherichia coli*	LD_0		5.0	8
Ba, *p*-Aminophenol	*Escherichia coli*	LC_{100}		8.0–10.0	22
Ba, Hydroquinone	*Escherichia coli*	LD_0		50.0	8
Ba, *p*-Benzoquinone	*Escherichia coli*	LC_{100}		55.0	8
Ba, 4,6-Dinitro-*o*-cresol	*Escherichia coli*	LC_{100}		100.0	8
Ba, 2,4-Dinitrophenol	*Escherichia coli*	LC_{100}		100.0	8
Ba, 3,5-Xylenol	*Escherichia coli*	LD_0		100.0	8
Ba, 3,4-Xylenol	*Escherichia coli*	LD_0		100.0	22
Ba, *p*-Nitrophenol	*Escherichia coli*	LD_0		100.0	22
Ba, *p*-*tert*-Butyl phenol	*Escherichia coli*	LD_0		100.0	8
Ba, *p*-Nitroaniline	*Escherichia coli*	NOAE		100.0	8
Ba, Toluene	*Escherichia coli*	LD_0		200.0	2
Ba, Ethylene diamine	*Escherichia coli*	LD_{50}		200.0	20
Ba, *m*-Nitrophenol	*Escherichia coli*	LD_0		300.0	8
Ba, 2,4-Xylenol	*Escherichia coli*	LD_0		500.0	8
Ba, Nitrobenzene	*Escherichia coli*	LD_0		600.0	8
Ba, *n*-Butyl acetate	*Escherichia coli*	LC_0		1000.0	2
Ba, Diethyl amine	*Escherichia coli*	LC_0		1000.0	21
Ba, Resorcinol	*Escherichia coli*	LD_0		1000.0	8
Ba, *o*-Nitrophenol	*Escherichia coli*	LD_0		1000.0	8
Ba, Phloroglucinol	*Escherichia coli*	LD_0		1000.0	8
Ba, Triethyl amine	*Escherichia coli*	NOAE		1000.0	8
Ba, Benzyl alcohol	*Escherichia coli*	NOAE		1000.0	18

Chemical	Species	Toxicity test	24 h	48 h	96 h	Chronic exposure	Ref.
Ba, Quinoline	*Escherichia coli*	NOAE				1000.0	3
Ba, *o*-Toluidine	*Escherichia coli*	NOAE				1000.0	8
Ba, Formic acid	*Escherichia coli*	NOAE				1000.0	8
Ba, Benzyl amine	*Escherichia coli*	NOAE				1000.0	8
Ba, *p*-Toluidine	*Escherichia coli*	NOAE				1000.0	8
Ba, *m*-Toluidine	*Escherichia coli*	NOAE				1000.0	8
Ba, Eulan	*Escherichia coli*	NOAE				5000.0	2
Ba, Acrolein	*Pseudomonas putida*	TT				0.21	12
Ba, Ethylene diamine	*Pseudomonas putida*	TT				0.85	10
Ba, *o*-Nitrophenol	*Pseudomonas putida*	TT				0.90	10
Ba, 1,3,5-Trioxane	*Pseudomonas*	LD_0				1.0	8
Ba, Chloral	*Pseudomonas putida*	TT				1.6	10
Ba, Lead acetate	*Pseudomonas putida*	TT				1.80	12
Ba, Nitroglycerine	*Pseudomonas*	LD_0				2.0	8
Ba, *p*-Nitroaniline	*Pseudomonas putida*	TT				4.0	10
Ba, 2-Nitro-*p*-cresol	*Pseudomonas putida*	TT				4.0	10
Ba, *p*-Nitrophenol	*Pseudomonas putida*	TT				4.0	10
Ba, Benzyl chloride	*Pseudomonas putida*	TT				4.8	10
Ba, Lindane	*Pseudomonas putida*	TT				5.0	12
Ba, Ethyleneimine	*Pseudomonas putida*	TT				5.5	10
Ba, Vinyl acetate	*Pseudomonas putida*	TT				6.0	10
Ba, 4-Nitro-*m*-cresol	*Pseudomonas putida*	TT				6.0	10
Ba, 2,4-Dichlorophenol	*Pseudomonas putida*	TT				6.0	10
Ba, *m*-Nitrophenol	*Pseudomonas putida*	TT				7.0	10
Ba, Nitrobenzene	*Pseudomonas putida*	TT				7.0	10
Ba, 6-Nitro-*m*-cresol	*Pseudomonas putida*	TT				7.0	10
Ba, 2,4-Dimethyl aniline	*Pseudomonas putida*	TT				8.0	10
Ba, 2,3-Dinitrotoluene	*Pseudomonas putida*	TT				9.0	10
Ba, Salicylaldehyde	*Pseudomonas putida*	TT				10.0	10

Ba, Atrazine	*Pseudomonas putida*	TT	10.0	12
Ba, Diethanolamine	*Pseudomonas putida*	TT	10.0	10
Ba, Benzonitrile	*Pseudomonas putida*	TT	11.0	10
Ba, Ethylbenzene	*Pseudomonas putida*	TT	12.0	10
Ba, *m*-Dinitrobenzene	*Pseudomonas putida*	TT	14.0	10
Ba, Formaldehyde	*Pseudomonas putida*	TT	14.0	10
Ba, 4,4-Diaminodiphenylmethane	*Pseudomonas*	LC_{100}	15.0	8
Ba, *o*-Dichlorobenzene	*Pseudomonas putida*	TT	15.0	10
Ba, *o*-Chlorotoluene	*Pseudomonas putida*	TT	15.0	10
Ba, Furfural	*Pseudomonas putida*	TT	16.0	10
Ba, *o*-Toluidine	*Pseudomonas putida*	TT	16.0	10
Ba, 4,6-Dinitro-*o*-cresol	*Pseudomonas putida*	TT	16.0	10
Ba, Chlorobenzene	*Pseudomonas putida*	TT	17.0	10
Ba, Cyclohexene	*Pseudomonas putida*	TT	17.0	10
Ba, *p*-Chlorophenol	*Pseudomonas*	LC_{100}	20.0	22
Ba, *o*-Diethylbenzene	*Pseudomonas putida*	TT	20.0	10
Ba, Ethyl *sec*-amyl ketone	*Pseudomonas putida*	TT	25.0	10
Ba, 2,6-Dinitrotoluene	*Pseudomonas putida*	TT	26.0	10
Ba, Ethyl amine	*Pseudomonas putida*	TT	29.0	10
Ba, Toluene	*Pseudomonas putida*	TT	29.0	10
Ba, *o*-Chlorophenol	*Pseudomonas*	LC_{100}	30.0	8
Ba, *o*-Cresol	*Pseudomonas putida*	TT	33.0	10
Ba, Ethyl acetoacetate	*Pseudomonas putida*	TT	33.0	10
Ba, Dinoseb	*Pseudomonas putida*	TT	40.0	12
Ba, Acrylic acid	*Pseudomonas putida*	TT	41.0	10
Ba, Carbaryl	*Pseudomonas putida*	TT	50.0	12
Ba, *m*-Cresol	*Pseudomonas putida*	TT	53.0	10
Ba, Acrylonitrile	*Pseudomonas putida*	TT	53.0	12
Ba, Epichlorhydrin	*Pseudomonas putida*	TT	55.0	10
Ba, 2,4-Dinitrotoluene	*Pseudomonas putida*	TT	57.0	10

Chemical	Species	Toxicity test	24 h	48 h	96 h	Chronic exposure	Ref.
Ba, Hydroquinone	*Pseudomonas putida*	TT				58.0	10
Ba, *n*-Hexanol	*Pseudomonas putida*	TT				62.0	10
Ba, 2-Hexanol	*Pseudomonas putida*	TT				63.0	10
Ba, Isooctanol	*Pseudomonas putida*	TT				63.0	10
Ba, Phenol	*Pseudomonas putida*	TT				64.0	10
Ba, Trichloroethylene	*Pseudomonas putida*	TT				65.0	10
Ba, 1-Heptanol	*Pseudomonas putida*	TT				67.0	10
Ba, 2,4-Pentanedione	*Pseudomonas putida*	TT				67.0	10
Ba, Styrene	*Pseudomonas putida*	TT				72.0	10
Ba, *t*-Butyl acetate	*Pseudomonas putida*	TT				78.0	10
Ba, 2-Ethylhexylamine	*Pseudomonas putida*	TT				82.0	10
Ba, Cyclohexyl acetate	*Pseudomonas putida*	TT				83.0	10
Ba, Benzene	*Pseudomonas putida*	TT				92.0	10
Ba, 1,1,2-Trichloroethane	*Pseudomonas putida*	TT				93.0	10
Ba, Butyraldehyde	*Pseudomonas putida*	TT				100.0	10
Ba, Trinitrotoluene	*Pseudomonas putida*	TT				100.0	10
Ba, Styphnic acid	*Pseudomonas putida*	TT				100.0	12
Ba, Butyl phosphate	*Pseudomonas putida*	TT				100.0	12
Ba, Benzotrichloride	*Pseudomonas putida*	TT				100.0	10
Ba, Tri-*n*-butyl phosphate	*Pseudomonas putida*	TT				100.0	12
Ba, Diallyl phthalate	*Pseudomonas putida*	TT				100.0	10
Ba, 3-Hexanol	*Pseudomonas putida*	TT				105.0	10
Ba, Ethylene diamine tetraacetic acid	*Pseudomonas putida*	TT				105.0	10
Ba, Allyl chloride	*Pseudomonas putida*	TT				115.0	10
Ba, 2,4-Dinitrophenol	*Pseudomonas putida*	TT				115.0	10
Ba, Phenyl acetate	*Pseudomonas putida*	TT				115.0	10
Ba, *n*-Butyl acetate	*Pseudomonas putida*	TT				115.0	10
Ba, Succinic acid	*Pseudomonas*	NOAE				125.0	8
Ba, Diisopropyl ether	*Pseudomonas*	NOAE				125.0	8

Ba, Glutaric acid	*Pseudomonas*	NOAE	125.0	8
Ba, Chloroform	*Pseudomonas putida*	TT	125.0	10
Ba, Aniline	*Pseudomonas putida*	TT	130.0	10
Ba, Benzaldehyde	*Pseudomonas putida*	TT	132.0	10
Ba, Ethylene dichloride	*Pseudomonas putida*	TT	135.0	10
Ba, Ethyl butyrate	*Pseudomonas putida*	TT	140.0	10
Ba, Allyl thiourea	*Pseudomonas*	TT	140.0	10
Ba, *n*-Amyl acetate	*Pseudomonas putida*	TT	145.0	10
Ba, Propargyl alcohol	*Pseudomonas putida*	TT	150.0	10
Ba, *n*-Propyl acetate	*Pseudomonas putida*	TT	170.0	10
Ba, Cyclopentanone	*Pseudomonas putida*	TT	175.0	10
Ba, Cyclohexanone	*Pseudomonas putida*	TT	180.0	10
Ba, Furfuryl alcohol	*Pseudomonas putida*	TT	180.0	10
Ba, Nitroglycol	*Pseudomonas*	LD_0	190.0	8
Ba, Isopropyl acetate	*Pseudomonas putida*	TT	190.0	10
Ba, Isobutyl acetate	*Pseudomonas putida*	TT	200.0	10
Ba, *n*-Pentanol	*Pseudomonas putida*	TT	220.0	10
Ba, Ethylene glycol	*Pseudomonas*	LC_{100}	250.0	2
Ba, Cyclopentanol	*Pseudomonas putida*	TT	250.0	10
Ba, Butyl digol	*Pseudomonas putida*	TT	255.0	10
Ba, Ethyl propionate	*Pseudomonas putida*	TT	270.0	10
Ba, Ethyl acrylate	*Pseudomonas putida*	TT	270.0	12
Ba, Isobutanol	*Pseudomonas putida*	TT	280.0	10
Ba, Lauryl sulfate	*Pseudomonas putida*	TT	290.0	10
Ba, Triethylene glycol	*Pseudomonas putida*	TT	320.0	10
Ba, Pyridine	*Pseudomonas putida*	TT	340.0	10
Ba, Diethyl phthalate	*Pseudomonas putida*	TT	400.0	10
Ba, Cyclohexamine	*Pseudomonas putida*	TT	420.0	10
Ba, Benzoic acid	*Pseudomonas putida*	TT	480.0	10
Ba, Cyclohexanone	*Pseudomonas*	LC_{100}	500.0	8

Chemical	Species	Toxicity test	24 h	48 h	96 h	Chronic exposure	Ref.
Ba, *sec*-Butanol	*Pseudomonas putida*	TT				500.0	10
Ba, Tetrahydrofuran	*Pseudomonas putida*	TT				580.0	10
Ba, 2-Butanoneoxime	*Pseudomonas*	LC_{100}				630.0	8
Ba, *n*-Butanol	*Pseudomonas putida*	TT				650.0	10
Ba, Ethyl acetate	*Pseudomonas putida*	TT				650.0	10
Ba, Acetonitrile	*Pseudomonas putida*	TT				680.0	10
Ba, Butyl cellosolve	*Pseudomonas putida*	TT				700.0	10
Ba, Allyl amine	*Pseudomonas putida*	TT				700.0	10
Ba, *n*-Butyl amine	*Pseudomonas putida*	TT				800.0	10
Ba, Diacetone alcohol	*Pseudomonas putida*	TT				825.0	10
Ba, *n*-Butyric acid	*Pseudomonas putida*	TT				875.0	10
Ba, Ethylene glycol acetate	*Pseudomonas putida*	TT				875.0	10
Ba, Dimethyl amine	*Pseudomonas*	NOAE				1000.0	8
Ba, Trichloroacetic acid	*Pseudomonas putida*	TT				1000.0	10
Ba, Picric acid, 50%	*Pseudomonas putida*	TT				1020.0	12
Ba, Isopropanol	*Pseudomonas putida*	TT				1050.0	10
Ba, Acetic anhydride	*Pseudomonas putida*	TT				1150.0	10
Ba, Acetone	*Pseudomonas putida*	TT				1700.0	10
Ba, 1,4-Dioxane	*Pseudomonas putida*	TT				2700.0	12
Ba, *n*-Propanol	*Pseudomonas putida*	TT				2700.0	10
Ba, Acetic acid	*Pseudomonas putida*	TT				2850.0	10
Ba, Ethanolamine	*Pseudomonas putida*	TT				6300.0	10
Ba, Ethanol	*Pseudomonas putida*	TT				6500.0	10
Ba, Diethylene glycol	*Pseudomonas putida*	TT				8000.0	10
Ba, Triethanolamine	*Pseudomonas*	LD_0				10,000.0	8
Ba, Eulan	*Pseudomonas fluorescens*	NOAE				10,000.0	2
Ba, Triethanolamine	*Pseudomonas putida*	TT				10,000.0	10
Ba, Methyl cellosolve	*Pseudomonas putida*	TT				10,000.0	10

Ba, Diethyl oxalate	*Pseudomonas putida*	TT			10,000.0	12
Ba, Citric acid	*Pseudomonas putida*	TT			10,000.0	10
Ba, Acetamide	*Pseudomonas putida*	TT			10,000.0	10
Ba, Nitrilotriacetic acid	*Pseudomonas putida*	TT			10,000.0	10
Ba, Urea	*Pseudomonas putida*	TT			10,000.0	10
Ba, Ethylene glycol	*Pseudomonas putida*	TT			10,000.0	10
Ba, Glycerol	*Pseudomonas putida*	TT			10,000.0	10
Ba, Nitrilotriacetic acid	*Pseudomonas fluorescens*	NOEL		100.0		372
Ba, Glycidol	*Salmonella* test	MUTAG				6
Ba, Ethylneimine	*Salmonella* test	MUTAG				6
C, Sodium arsenite	*Acartia clausi* (S)	LC_{50}			0.508 (h?)	374
C, Nitrilotriacetic acid, Na, H_2O	*Acartia clausi* (S)	TL_{50}			1350.0 (72 h)	37
C, Toxaphene	*Acartia tonsa* (S)	LC_{50}		0.0000072		38
C, Mercuric salt	*Acartia tonsa* (20C)	LC_{50}		0.001		377
C, Tributyltin oxide	*Acartia tonsa* (1/2S, 20C)	LC_{50}		0.001	0.00055 (144 h)	378
C, Diazinon	*Acartia tonsa* (S)	LC_{50}		0.00257		38
C, Toxaphene	*Acartia tonsa*	LC_{50}		0.011		376
C, Azodrin	*Acartia tonsa* (S)	LC_{50}		0.240		38
C, Dimethyl parathion	*Acartia tonsa* (S)	LC_{50}		0.89		38
C, Nickel (2+)	*Acartia clausi* (S)	LC_{50}		2.85		375
C, Sodium selenite	*Allorchestes compressa* (F, 20°C, pH 7.5)	LC_{50}		4.77		448
C, Diflubenzuron	*Artemia salina*	LC_{100}			0.010 (3 d)	382
C, Captan	*Artemia salina*	NOAE			10.0 (h?)	383
C, Dimethyl phthalate	*Artemia salina* embryo (S, 26°C, pH 8.3)	NOAE			50.0 (72 h)	384
C, Phenol	*Artemia salina*	TL_m			56.0 (25–50 h)	51
C, Aflatoxin B1	*Artemia salina*	LC_{60}	0.0005			372
C, Aflatoxin B1	*Artemia salina*	LC_{90}	0.001			372
C, 1-Chloronaphthalene	*Artemia salina* (L, semiS, 19°C, pH 8.6)	EC_{50}	0.78			386
C, 1-Chloronaphthalene	*Artemia salina* (L, syn. seaW, S, 19°C, pH 8.7)	EC_{50}	0.91			387
C, Endosulfan	*Artemia salina*	LC_{100}	2.50			380

Chemical	Species	Toxicity test	24 h	48 h	96 h	Chronic exposure	Ref.
C, Demeton	*Artemia salina* (S, 35°C, pH 8.1)	LC_0	10.0				589
C, 1,3,5-Trichlorobenzene	*Artemia salina*, female (S, 22°C)	LC_{100}	10.0				394
C, 1-Methylethyl benzene	*Artemia salina* (S, 19–21°C)	LC_{50}	13.7				385
C, Ethylbenzene	*Artemia salina* (19–21°C)	LC_{50}	15.4				385
C, *n*-Butylamine	*Artemia salina* (S)	TL_m	30.0				52
C, Sodium arsenite	*Artemia* sp. (syn. seaW, S)	LC_{50}	32.1				395
C, Zinc sulfate $7H_2O$	*Artemia salina* (S, 23–27°C, syn. seaW)	LC_{50}	63.2				395
C, Styrene	*Artemia salina*	TL_m	68.0				50
C, Isopropylbenzene	*Artemia salina* (S, 24.5°C)	LC_{50}	110.0				389
C, 1,1-Dichloroethane	*Artemia salina*	TL_m	320.0				51
C, Ethyl acetate	*Artemia salina* (S, 19°C, larvae, pH 8.6)	EC_{50}	346.0				386
C, Dow Corning, 193	*Artemia salina* (S, seaW, 25°C, dark)	NOAE	500.0				392
C, 2-Butoxyethanol	*Artemia salina* (S, 24.5°C, syn. seaW)	LC_{50}	1000.0				388
C, Ethyl acetate	*Artemia salina* (S, 25°C)	LC_{50}	1590.0				388
C, *n*-Propanol	*Artemia salina* (S, 24.5°C)	LC_{50}	4200.0				388
C, 2-Methoxyethanol	*Artemia salina* (S, 24.5°C)	LC_{50}	10,000.0				388
C, 1,2-Ethanediol	*Artemia salina* (S, 24°C, pH 7)	LC_{50}	20,000.0				390
C, Tetraethyllead	*Artemia salina* (24 h old)	LC_0		0.025			40
C, Tetraethyllead	*Artemia salina* (24 h old)	LC_{50}		0.085			40
C, Tetramethyllead	*Artemia salina* (24 h old)	LC_0		0.18			40
C, Tetramethyllead	*Artemia salina* (24 h old)	LC_{50}		0.25			40
C, Tetraethyllead	*Artemia salina* (24 h old)	LC_{100}		0.26			40
C, Tetramethyllead	*Artemia salina* (24 h old)	LC_{100}		0.67			40
C, α-Hexachlorocyclohexane	*Artemia salina* (3 d old)	LC_{50}		1.4			41
C, *n*-Nitrosodimethylamine	*Artemia salina* (egg, 27°C, pH 7–8)	NOAE		10.0			383
C, Benzene	*Artemia salina*	TL_m	66.0	21.0			22
C, Acetic acid	*Artemia salina*	TL_m	42.0	32.0			51
C, 2-Chloroethanol	*Artemia salina* (S, 24°C, pH 7)	LC_{50}		680.0			390
C, Ethylene oxide	*Artemia salina* (S, 24°C, pH 7)	LC_{50}		745.0			390

C, Ammonium salt of a saturated carboxylic acid dispersant	*Artemia salina*	LD_{50}		1000.0			42
C, Acetone	*Artemia salina*	TL_m	2100.0	2100.0			51
C, α-Hexachlorocyclohexane	*Artemia salina* (1/2S, 24°C, adult)	LC_{50}			0.5		393
C, Hexachlorocyclohexane	*Artemia salina* (3 d old)	LC_{50}			0.5		41
C, Fenoprop, butoxyethyl ester (BEE)	*Asellus brevicaudus*	LC_{50}		40.0			43
C, Hydrocyanic acid	*Asellus aquaticus* (F, 18°C, pH 8)	NOAEL				0.029 (115 d)	398
C, Hydrocyanic acid	*Asellus communis* (F, 18°C, pH 8)	NOAEL				0.041 (98 h)	398
C, Hydrocyanic acid	*Asellus aquaticus* (F, 18°C, pH 8)	LC_{100}				0.432 (115 d)	398
C, Parathion	*Asellus aquaticus* (I, artif. FW, 18°C)	LC_0				1.0 (21 d)	399
C, Hydrocyanic acid	*Asellus communis* (F, 18°C, pH 8)	TC_{LO}				1.895 (10–12 d)	398
C, Parathion	*Asellus aquaticus* (I, artif. freshW, 18°C)	LC_{50}				4.80 (21 d)	399
C, Methyl parathion	*Asellus aquaticus* (I, 18°C)	LC_{50}				7.5 (18 d)	399
C, Photo-aldrin	*Asellus*	LC_{50}	0.04				44
C, Photo-heptachlor	*Asellus*	LC_{50}	0.060				44
C, Aldrin	*Asellus*	LC_{50}	0.08				44
C, Endosulfan	*Asellus aquaticus*	LC_{100}	0.100				396
C, Heptachlor	*Asellus*	LC_{50}	0.10				44
C, Dichlone	*Asellus brevicaudus*	LC_{50}		0.200			43
C, Trifluralin	*Asellus brevicaudus*	LC_{50}		0.20			43
C, Fenoprop, propylene glycol butyl ether ester (PGBEE)	*Asellus brevicaudus*	LC_{50}		0.50			43
C, Potassium cyanide	*Asellus aquaticus* (S, 20°C)	LC_{10}		0.84			400
C, Potassium cyanide	*Asellus aquaticus* (S, 20°C)	LC_{50}		2.68			400
C, 2,4-D, butoxyethanol ester	*Asellus brevicaudus*	LC_{50}		3.2			43
C, Vernam	*Asellus brevicaudus*	LC_{50}		5.6			43
C, Dichlobenil	*Asellus brevicaudus*	LC_{50}		34.0			43
C, Amitrol	*Asellus brevicaudus*	NOAE		100.0			43
C, Simazine	*Asellus brevicaudus*	NOAE		100.0			43
C, Trifene	*Asellus brevicaudus*	NOAE		100.0			43

Chemical	Species	Toxicity test	24 h	48 h	96 h	Chronic exposure	Ref.
C, Diphenamid	*Asellus brevicaudus*	NOAE		100.0			403a
C, Dicamba	*Asellus brevicaudus*	NOEL		100.0			43
C, 2,4-D (diethylamine) salt	*Asellus brevicaudus*	NOEL		100.0			43
C, Diphenamid	*Asellus brevicaudus*	NOEL		100.0			43
C, *n*-Propanol	*Asellus aquaticus* (S, 20°C, pH 8.4)	LC_{50}		2500.0			401
C, Dichlorodiphenyltrichloroethane (DDT)	*Asellus brevicaudus*	LC_{50}			0.004		45
C, Dieldrin	*Asellus brevicaudus*	LC_{50}			0.005		45
C, Lindane	*Asellus brevicaudus*	LC_{50}			0.010		45
C, Dichlorodiphenyldichloroethane (DDD)	*Asellus brevicaudus*	LC_{50}			0.010		45
C, Azinphosmethyl	*Asellus brevicaudus*	LC_{50}			0.021		45
C, Binapacryl	*Asellus brevicaudus* (S, 16°C, pH 7.4)	LC_{50}	0.130		0.029		404
C, Hydrogen sulfide	*Asellus*	TL_m			0.111		46
C, Naled	*Asellus brevicaudus*	LC_{50}			0.23		45
C, Carbaryl	*Asellus brevicaudus*	NOEL			0.240		69
C, Parathion	*Asellus brevicaudus*	LC_{50}			0.6		45
C, Parathion	*Asellus brevicaudus* (S, 21°C, pH 7.1)	LC_{50}	5.60		0.60		405
C, Carbophenothion	*Asellus brevicaudus*	LC_{50}			1.100		45
C, Baytex	*Asellus brevicaudus*	LC_{50}			1.8		45
C, Hydrogen cyanide	*Asellus communis* (10–12 d)	LTC			1.90		47
C, Hydrogen cyanide	*Asellus communis*	LC_{50}			2.29		47
C, Hydrocyanic acid	*Asellus communis* (F, 18°C, pH 8)	LC_{50}			2.295		398
C, Malathion	*Asellus brevicaudus*	LC_{50}			3.0		45
C, Aldrin	*Asellus brevicaudus*	LC_{50}			8.0		43
C, Dimethylamine	*Asellus aquaticus*	LC_0			43.0		397
C, Dimethylamine	*Asellus aquaticus*	LC_{50}	146.0	84.0	62.0	65.0 (72 h)	397
C, Aniline	*Asellus intermedius* (S, 21°C, pH 6.5–8.5)	LC_{50}			100.0		408
C, Dicamba	*Asellus brevicaudus* (S, 15°C, pH 7.5)	LC_{50}			100.0		402

C, Diphenamid	*Asellus brevicaudus* (S, 15°C, pH 7.5)	LC_{50}			100.0		403b
C, Phosmet	*Asellus* (F, 20°C, pH 7.5)	LC_{50}			0.044		409
C, Diuron	*Asellus* (S, 15°C, pH 7.2–7.5)	LC_{50}			15.5		410
C, Piperonyl butoxide	*Asellus* (S, 15°C, pH 7.5)	LC_{50}			0.012		411
C, Phosmet	*Asellus* (S, 20°C, pH 7.5)	LC_{50}			0.090		409
C, Isodrin/photiosdrin, toxicity ratio	*Aselus* sp.	TR	0.64				44, 258, 300
C, Endrin	*Aselus brevicaudus*	LC_{50}			0.0015		45
C, Di-Na arsenate	*Bosmina congirostris* (S, 17°C, 6.8, < 24 h)	EC_{50}			0.85		412
C, Aminocarb	*Caecidolea racovitzai* (1/2S, 20°C, pH 7.1)	LC_{50}			12.0		413
C, Aminocarb	*Caecidolea racovitzai* (1/2S, 12°C, pH 7.4)	LC_{50}			31.6		413
C, Naphthalene	*Calanus finmarchicus* (S, 15°C)	LC_{71}			3.8		414
C, Carbaryl	*Callianassa californiensis*, adult	TL_m	0.130				407
C, Carbaryl	*Callianassa californiensis* (larva)	LC_{50}		0.03			415
C, Carbaryl	*Callianassa californiensis*	TL_m		0.030			407
C, Mirex	*Callinectes sapidus* (F, juv., 17–27°C)	LC_{19}				0.000060 (28 d)	428
C, Mirex	*Callinectes sapidus* (larva, 25°C)	LC_{50}				0.001–0.01 (5 d)	425
C, Mirex	*Callinectes sapidus* (0.5S, larva, 25°C)	LC_{100}				0.001 (20 d)	429
C, Mirex	*Callinectes sapidus* (0.5S, larva, 25°C)	LC_0				0.001 (5 d)	449
C, Aroclor 1254	*Callinectes sapidus*	NOAE				0.005 (20 d)	421
C, Chlorpyrifos	*Callinectes sapidus*	LC_{100}	0.240				420
C, Eptam, technical grade	*Callinectes sapidus*	LC_{50}	20.0				25
C, Fenthion	*Callinectes sapidus* (F, 28°C)	LC_{50}		0.004			417
C, Endosulfan	*Callinectes sapidus*	EC_{50}		0.035			417
C, Toxaphene	*Callinectes sapidus*	EC_{50}		0.330			419
C, Chlordane	*Callinectes sapidus* (juv.)	LC_{50}		0.480			416
C, Chlorodecane	*Callinectes sapidus* (F)	LC_{20}		1.00			422
C, Mirex	*Callinectes sapidus* (F, juv., 31°C)	LC_{20}		2.0			417
C, Naphthalene	*Callinectes sapidus* (F, 22–23°C)	LC_{50}		2.30			401
C, DDT	*Callinectes sapidus* (S)	TL_m			0.019		55
C, DDT	*Callinectes sapidus* (S)	TL_m			0.035		55

Chemical	Species	Toxicity test	24 h	48 h	96 h	Chronic exposure	Ref.
C, Chlorodecane	*Callinectes sapidus* (F)	LC_{50}			0.21		48
C, Kepone	*Callinectes sapidus* (F, 19°C)	LC_{50}			0.210		48
C, Kepone	*Callinectes sapidus* (F)	LC_{50}			0.210		48
C, Kepone	*Callinectes sapidus* (F)	LC_{50}			0.210		56
C, Toxaphene	*Callinectes sapidus* (S)	TL_m			0.37		2
C, 3-Chloro-4-methylbenzene-amine, HCl	*Callinectes sapidus*	TL_m			16.0		406
C, Dichromic acid, K_2	*Callinectes sapidus* (seaW, 22°C, 6.5)	LC_{50}			98.0		420
C, Chlordane	*Cancer magister*	NOAE				0.000015 (70 d)	427
C, Chlordane	*Cancer magister* (larva)	LC_{50}				0.00015 (37 d)	427
C, Fenthion	*Cancer magister* (0.5S, 10–12°C)	LC_0				0.002 (72 h)	431
C, Fenthion	*Cancer magister* (0.5S, 10–12°C)	LC_{100}				0.010 (72 h)	431
C, Propanil	*Cancer magister* (juv. F, 12.6°C, pH 7.9)	LC_0				0.08 (80 d)	434
C, Chlordane	*Cancer magister* (S)	LC_{50}				0.220 (h?)	424
C, Propanil	*Cancer magister* (juv. F, 12.6°C, pH 7.9)	LC_{50}				0.80 (30 d)	434
C, Propanil	*Cancer magister* (juv. F, 12.6°C, pH 7.9)	LC_{95}				0.80 (80 d)	434
C, Nitrilotriacetic acid, Na,H_2O	Cancer magister (S)	TL_{50}				1650.0 (72 h)	37
C, Carbaryl	*Cancer magister* (juvenile, male)	EC_{50}	0.600				407
C, Propanil	*Cancer magister* (embryo, S)	NOAE	10.0				436
C, Mercuric chloride	*Cancer magister* (Stage 1, S, 20°C, pH 8)	LC_{50}			0.0066		440
C, Carbaryl	*Cancer magister* (S, larva, 9–11°C)	LC_{50}			0.010		439
C, Cupric sulfate	*Cancer magister* (larva, S, 15°C, pH 8.1)	LC_{50}			0.049		433
C, Propanil	*Cancer magister* (larva, F, 12.6°C, pH 7.9)	LC_0			0.08 (70 d)		434
C, Carbaryl	*Cancer magister* (adult)	LC_{50}			0.180		430
C, Arsenic pentoxide	*Cancer magister* (larvae, 20°C, pH 8.2)	LC_{50}			0.232		433
C, Carbaryl	*Cancer magister* (juvenile)	LC_{50}			0.280		430
C, Propanil	*Cancer magister* (adult, F, 12.6°C, pH 7.9)	LC_0			0.40 (85 d)		434
C, Zinc sulfate	*Cancer magister* (S, 16°C, pH 7.5–8.3)	LC_{50}			0.456		435
C, Propanil	*Cancer magister* (larva, F, 12.6°C, pH 7.9)	LC_{90}			0.80 (20 d)		434

C, Propanil	*Cancer magister* (larva, F, 12.6°C, pH 7.9)	LC_{100}			0.80 (55 d)		434
C, Dichromic acid, K_2	*Cancer magister* (1 stage, S, 16°C, pH 8)	LC_{50}			3.44		433
C, Propanil	*Cancer magister* (adult, F, 12.6°C, pH 7.9)	LC_{50}			4.0 (85 d)		434
C, Nickel sulfate $6H_2O$	*Cancer magister* (Stage 1, L, S, 15°C, pH 8.1)	LC_{50}			4.36		433
C, *o*-Xylene	*Cancer magister* (0.5S, 16°C, larva)	LC_{50}			6.0		432
C, *o*-Xylene	*Cancer magister* (stage 1)	LC_{50}			6.0		57
C, Propanil	*Cancer magister* (larva, F, 12.6°C, pH 7.9)	LC_{50}			7.3		434
C, Captan	*Cancer magister* (zoae 1, juv. + adult)	LC_{50}			10.0		441
C, *m*-Xylene	*Cancer magister* (I stage zoae, 0.5S)	LC_{50}			12.0		432
C, Ethylbenzene	*Cancer magister* (larva, $^1/_2$S, 13°C)	LC_{50}		40.0	13.0		437
C, Propanil	*Cancer magister* (adult, 0.5S, 13°C, pH 7.5)	LC_{50}			26.0		434
C, Toluene	*Cancer magister* (larvae stg. 1)	LC_{50}			28.0		57
C, Benzene	*Cancer magister* (larvae)	LC_{50}			108.0		57
C, Tri-*n*-butyltin oxide	*Carcinides maenas* (larva, 0.5S, 15°C)	LC_{50}				0.10 (3.4 d)	'445
C, Sodium fluoride	*Carcinides maenas* (S, 15°C, pH 7.9–8.3)	LC_0				30.0 (90 d)	447
C, Tributyl tinoxide	*Carcinides maenas*		0.010				442
C, Cupric sulphate	*Carcinides maenas* (larva, S, 15°C)	LC_{50}		0.60			438
C, Dichlobenil	*Carcinides maenas* (0.5S, 15°C)	LC_{50}		10.0			443
C, Simazine	*Carcinides maenas* (15°C, 0.5S, adult)	LC_{50}		100.0			444
C, Hexachlorocyclohexane	*Carcinides maenas* ($^1/_2$S, 15°C, adult)	LC_{50}		100.0			443
C, Cupric sulphate	*Carcinides maenas* (adult, S, 15°C)	LC_{50}		109.0			443
C, Nickel sulphate	*Carcinides maenas* (S, 15°C)	LC_{50}		255.0			443
C, Diflubenzuron	*Carcinides mediterraneus*	LC_{100}			0.02		450
C, Cobalt chloride	*Carcinides maenas* (S, 15°C, pH 7.8, I stage zoae)	LC_{50}			22.7		446
C, Cobalt chloride	*Carcinides maenas* (S, 15°C, pH 7.8, 2nd stg)	LC_{50}			227.0		446
C, Bifenthrin	*Ceriodaphnia dubia*	LC_{50}		0.00007			1083
C, Cyfluthrin	*Ceriodaphnia dubia*	LC_{50}		0.00014			1083
C, Tralomethrin	*Ceriodaphnia dubia*	LC_{50}		0.00026			1083
C, Lambda Cyhalothrin	*Ceriodaphnia dubia*	LC_{50}		0.0003			1083
C, Permethrin	*Ceriodaphnia dubia*	LC_{50}		0.00055			1083

Chemical	Species	Toxicity test	24 h	48 h	96 h	Chronic exposure	Ref.
C, Zinc sulphate	*Ceriodaphnia dubia* (pH 6.0–6.5)	LC_{50}		>0.530			1084
C, Zinc sulphate	*Ceriodaphnia dubia* (pH 7.0–7.5)	LC_{50}		0.360			1084
C, Zinc sulphate	*Ceriodaphnia dubia* (pH 8.0–8.5)	LC_{50}		0.095			1084
C, Cupric nitrate	*Ceriodaphnia dubia* (pH 6.0–6.5)	LC_{50}		>0.095			1084
C, Cupric nitrate	*Ceriodaphnia dubia* (pH 7.0–7.5)	LC_{50}		0.028			1084
C, Cupric nitrate	*Ceriodaphnia dubia* (pH 8.0–8.5)	LC_{50}		0.200			1084
C, Nickel chloride	*Ceriodaphnia dubia* (pH 6.0–6.5)	LC_{50}		>0.200			1084
C, Nickel chloride	*Ceriodaphnia dubia* (pH 7.0–7.5)	LC_{50}		0.140			1084
C, Nickel chloride	*Ceriodaphnia dubia* (pH 8.0–8.5)	LC_{50}		0.013			1084
C, Lead chloride	*Ceriodaphnia dubia* (pH 6.0–6.5)	LC_{50}		0.280			1084
C, Lead chloride	*Ceriodaphnia dubia* (pH 7.0-7.5)	LC_{50}		>2.70			1084
C, Lead chloride	*Ceriodaphnia dubia* (pH 8.0-8.5)	LC_{50}		>2.70			1084
C, Cadmium nitrate	*Ceriodaphnia dubia* (pH 6.0-6.5)	LC_{50}		0.560			1084
C, Cadmium nitrate	*Ceriodaphnia dubia* (pH 7.0-7.5)	LC_{50}		0.350			1084
C, Cadmium nitrate	*Ceriodaphnia dubia* (pH 8.0-8.5)	LC_{50}		0.120			1084
C, Hypochlorite ion (free Cl_2)	*Ceriodaphnia dubia* (S, fed)	LC_{50}	0.08				1085
C, Hypochlorite ion (free Cl_2)	*Ceriodaphnia dubia* (S, unfed)	LC_{50}	0.048				1085
C, Hypochlorite ion (free Cl_2)	*Ceriodaphnia dubia* (F)	LC_{50}	0.006				1085
C, Hypochlorous acid (free Cl_2)	*Ceriodaphnia dubia* (S, fed)	LC_{50}	0.14				1085
C, Hypochlorous acid (free Cl_2)	*Ceriodaphnia dubia* (S, unfed)	LC_{50}	0.035				1085
C, Hypochlorous acid (free Cl_2)	*Ceriodaphnia dubia* (F)	LC_{50}	0.005				1085
C, Monochloramine	*Ceriodaphnia dubia* (S, fed)	LC_{50}	<0.02				1085
C, Monochloramine	*Ceriodaphnia dubia* (S, unfed)	LC_{50}	0.012				1085
C, Monochloramine	*Ceriodaphnia dubia* (F)	LC_{50}	0.016				1085
C, Dichloramine	*Ceriodaphnia dubia* (S, fed)	LC_{50}	0.02				1085
C, Dichloramine	*Ceriodaphnia dubia* (S, unfed)	LC_{50}	0.016				1085
C, Dichloramine	*Ceriodaphnia dubia* (F)	LC_{50}	0.027				1085
C, Sodium arsenite	*Ceridaphnia reticulata* (<24 h, S, 24°C, 8.33)	EC_{50}		1.269			451
C, 1,3-Pentadiene, *trans*	*Chaetogammarus marinus* (15°C, pH 8)	NOAE			10.0		452

C, 1,3-Pentadiene, *cis*	*Chaetogammarus marinus* (15°C, pH 8)	NOAE			10.0		452
C, 1,3-Pentadiene, *trans*	*Chaetogammarus marinus* (15°C, pH 8)	LC_{50}	86.0		18.0		452
C, 1,3-Pentadiene, *cis*	*Chaetogammarus marinus* (15°C, pH 8)	LC_{50}	92.0		35.0		452
C, Chlorpyrifos	*Chlamydotheca arcuata*	LC_{50}		0.20			420
C, *m*-Xylene	Crab (larvae stage 1)	LC_{50}			12.0		57
C, *o*-Cresol	*Crangon septemspinosa* (adult, S, 10°C)	LC_{50}				14.2 (59 h)	456
C, Cresol	*Crangon septemspinosa* (adult, 1/2S, 10°C)	LC_{50}				16.3 (21 h)	456
C, 2,6-Dichlorophenol	*Crangon septemspinosa* (0.5S, 10°C)	LC_{50}				19.1 (52 h)	456
C, Phenol	*Crangon crangon*, 15°C	LC_{50}				40.0 (6 h)	59
C, Phenol	*Crangon crangon*, 15°C	LC_{50}				80.0 (3 h)	59
C, Phenol	*Crangon crangon*, 15°C	LC_{50}				120.0 (1 h)	59
C, Phenol	*Crangon crangon*, 15°C	LC_{50}				400.0 (27 min)	59
C, Phenol	*Crangon crangon*, 15°C	LC_{50}				1000.0 (9 min)	59
C, phenol	*Crangon crangon*, 15°C	LC_{50}				5600.0 (3 min)	59
C, 4-Propylenebenzene sulfonate	*Crangon crangon* (SW, 15°C)	LC_{50}				32,000.0 (72 h)	59
C, 4-Propylenebenzene sulfonate	*Crangon crangon* (SW, 15°C)	LC_{50}				56,000.0 (9 min)	59
C, Dioxathion	*Crangon septemspinosa*	LC_{50}	0.307				453
C, Mixed crude oils	*Crangon crangon* (S, 14–16°C)	LC_{98}	1.0ml/l				459
C, Acrylonitrile	*Crangon crangon* (S, 15°C)	LC_{50}	10.0–33.0				443
C, Hexachlorocyclohexane	*Crangon crangon* (1/2S, adult, 15°C)	LC_{50}		0.001			443
C, Endosulfan	*Crangon crangon*	LC_{50}		0.010			458
C, Kepone	*Crangon crangon* (adult, F, 30°C)	LC_{50}		0.085			417
C, Kelthane	*Crangon franciscorum*	LC_{50}		0.089			58
C, Kelthane	*Crangon franciscorum*	LC_{50}	1.29	0.59		0.1 (100 h)	94
C, Thiometon	*Crangon crangon* (0.5S, 15°C)	LC_{50}		1.0			443
C, Allyl alcohol	*Crangon crangon* (15°C, semiS)	LC_{50}		1.0			443
C, Aroclor	*Crangon crangon* (S)	LC_{50}		3.0–10.0			443
C, Dichlobenil	*Crangon crangon* (0.5S, 15°C)	LC_{50}		3.3			443
C, Swedish EDC tar	*Crangon crangon* (S, 16°C)	LC_{50}		5.0			573
C, Methylchlorophenoxyacetic acid	*Crangon crangon* (15°C, S)	LC_{50}		10.0			443

Chemical	Species	Toxicity test	24 h	48 h	96 h	Chronic exposure	Ref.
C, Cresol	*Crangon crangon* (1/2S, 15°C, adult)	LC_{50}		10.0			443
C, Simazine	*Crangon crangon* (15°C, 0.5S)	LC_{50}		100.0			443
C, Sodium fluoride	*Crangon crangon*	LC_{50}		300.0			443
C, Fenuron	*Crangon crangon* (0.5S, 15°C)	LC_{50}		600.0			454
C, Endosulfan	*Crangon* sp. (S, 10–20°C, no sediment)	LC_{50}			0.0002		457
C, DDT (*p,p′*)	*Crangon septemspinosa* (S)	LC_{50}			0.0006		60
C, Tributyltinoxide	*Crangon crangon* (larvae, 1/2S)	LC_{50}			0.0015		442
C, Endrin	*Crangon septemspinosa* (S)	LC_{50}			0.0017		60
C, *p*-Xylene	*Crangon franciscorum* (16°C)	LC_{50}			0.002		455
C, Dimethyl parathion	*Crangon septemspinosa* (S)	LC_{50}			0.002		38
C, Dichlorvos	*Crangon septemspinosa* (S)	LC_{50}			0.004		60
C, Dichlorvos	*Crangon septemspinosa*	LC_{50}			0.004		453
C, Lindane	*Crangon septemspinosa* (S)	LC_{50}			0.005		453
C, Lindane	*Crangon septemspinosa* (S)	LC_{50}			0.005		60
C, Endosulfan	*Crangon* sp. (S, 10–20°C, sediment)	LC_{50}			0.0069		457
C, Dieldrin	*Crangon septemspinosa* (S)	LC_{50}			0.007		60
C, Hexachlorobenzene	*Crangon septemspinosa* (semiS, 20°C)	LC_0			0.0072		457
C, Heptachlor	*Crangon septemspinosa* (S)	LC_{50}			0.008		60
C, Aldrin	*Crangon septemspinosa* (S)	LC_{50}			0.008		60
C, Aroclor	*Crangon septemspinosa* (0.5S, 20°C)	LC_{50}			0.012		457
C, Malathion	*Crangon septemspinosa* (S)	LC_{50}			0.033		60
C, Dioxathion	*Crangon septemspinosa* (S)	LC_{50}			0.038		60
C, Tributyl tin oxide	*Crangon crangon* (adult, 1/2S)	LC_{50}			0.041		442
C, Mixed crude oils	*Crangon crangon* (0.5S, 14–16°C)	LC_{50}			0.140 ml/l		459
C, Aminocarb	*Crangon septemspinosa* (S)	LC_{50}			0.2		481
C, Nonylphenol	*Crangon septemspinosa* (S)	LC_{50}			0.40		481
C, Aminocarb, liquid formulation	*Crangon septemspinosa* (S)	LC_{50}			0.55		481
C, *o*-Xylene	*Crangon franciscorum*	LC_{50}			1.30		61
C, Methyl parathion	*Crangon septemspinosa* (20°C)	LC_{50}			2.0		453

C, *p*-Xylene	*Crangon franciscorum*	LC_{50}			2.0		61
C, Toluene	*Crangon franciscorum*	LC_{50}			3.70		455
C, *m*-Xylene	*Crangon franciscorum*	LC_{50}			3.7		61
C, Toluene	*Crangon franciscorum*	LC_{50}			4.3		61
C, *o*-Chlorophenol	*Crangon crangon* (0.5S, 10°C)	LC_{50}			5.30		456
C, Acrylonitrile	*Crangon crangon* (S, 15°C, pH 8)	LC_{50}			6.0		457
C, Acrylonitrile	*Crangon crangon*	LC_{50}	25.0	20.0	6.0	3600.0 (1 h)	59
C, Thallium	*Crangon crangon* (0.5S, 15°C)	LC_{50}			10.0		443
C, Diquat dibromide	*Crangon crangon*	LC_{50}			10.0		463
C, 3-Chloro-4-4-methylbenzenamine, HCl	*Crangon* sp.	TL_m			10.8		406
C, 2,3,4,6-Tetrachlorophenol	*Crangon septemspinosa* (0.5S, 10°C)	LC_{50}			11.8		456
C, Benzene	*Crangon franciscorum*	LC_{50}			20.0		61
C, Phenol	*Crangon crangon*, 15°C	LC_{50}	40.0	30.0	25.0	30.0 (72 h)	59
C, Bromoform	*Crangon septemspinosa*	LC_{50}			26.0		465
C, *p*-Nitrophenol	*Crangon septemspinosa*	LC_{100}			26.4		456
C, 2-Nitrophenol	*Crangon septemspinosa*	LC_{100}			32.9		456
C, Malathion	*Crangon septemspinosa* (21°C)	LC_{50}			33.0		453
C, Ethylene dichloride	*Crangon crangon*, Seawater, 15°C	LC_{50}	75.0	65.0	65.0		59
C, 1,2-Dichloroethane	*Crangon crangon* (F, 15°C, pH 8)	LC_{50}			85.0		464
C, 2-Butoxyethanol	*Crangon crangon*	LC_{50}		800.0	775.0		462
C, Trichlorpyr	*Crangon* sp.	LC_{50}			895.0		461
C, 4-Propylenebenzene sulfonate	*Crangon crangon* (SW, 15°C)	LC_{50}	56,000.0	42,000.0	18,000.0		59
C, Creosote	*Crangon*, 10°C, aerated and renewed	LC_{50}			0.13		466
C, Creosote	*Crangon*, 20°C, aerated and renewed	LC_{50}			0.11		466
C, Hydrogen sulfide	*Crangon*, sp.	TL_m			1.07		46
C, Malathion	*Crassostrea virginica* (larvae, S)	TL_m				2.660 (14 d)	39
C, Eptam, technical	*Crassostrea virginica*	EC_{50}				5.0	25
C, Chlordane	*Crassostrea virginica*	EC_{50}			0.0062		369
C, Dieldrin	*Crassostrea virginica* (S)	TL_m		0.0640			64
C, Nabam	Crassostrea virginica (S, larvae)	TL_m		0.50			30

Chemical	Species	Toxicity test	24 h	48 h	96 h	Chronic exposure	Ref.
C, Malathion	*Crassostrea virginica* (eggs)	TL_m		9.07			39
C, Lindane	*Crassostrea virginica* (egg, S)	TL_m		9.10			64
C, Heptachlor	*Crassostrea virginica* (F)	EC_{50}			1500.0		65
C, *N*-Nitrosodimethylamine	*Crayfish* (injection, 20°C, pH 7.8)	LD_{50}				2250.0 mg/kg (12 d)	468
C, *N*-Nitrosodimethylamine	*Crayfish* (injection, adult, 20°C)	LD_{50}				230.0 mg/kg (12 d)	468
C, Endosulfan	*Cyclops strenuus*	NOAE				0.100 (26 h)	396
C, Aflatoxin B1	*Cyclops fuscus* (19°C)	LC40	0.001				590
C, Aflatoxin B1	*Cyclops fuscus* (19°C)	LC80	0.004				590
C, Chlorpyrifos	*Cyclops fuscus* (S)	LC_{100}	0.050				420
C, Calcium chloride	*Cyclops fuscus*	LC_{50}			0.340		471
C, Dichlone	*Cypridopsis vidua*	LC_{50}		0.012			43
C, Fenoprop, PGBEE	*Cypridopsis vidua*	LC_{50}		0.20			43
C, Vernam	*Cypridopsis vidua*	LC_{50}		0.24			66
C, Trifluralin	*Cypridopsis vidua*	LC_{50}		0.250			43
C, 2,4-D-PGBEE	*Cypridopsis vidua* (21°C, S)	LC_{50}		0.32			469
C, 2,4-D-BEE	*Cypridopsis vidua* (21°C, S)	LC_{50}		1.8			469
C, 2,4-D-BEE	*Cypridopsis vidua*	LC_{50}		1.8			43
C, Simazine	*Cypridopsis vidua*	LC_{50}		3.2			43
C, Fenoprop	*Cypridopsis vidua* (21°C, S)	LC_{50}		4.9			469
C, Fenoprop-BEE	*Cypridopsis vidua*	LC_{50}		4.9			43
C, Dichlobenil	*Cypridopsis vidua*	LC_{50}		7.8			43
C, 2,4-D-Dimethylamine	*Cypridopsis vidua* (21°C, S)	LC_{50}		8.0			469
C, 2,4-D-Diethylamine, salt	*Cypridopsis vidua*	LC_{50}		8.0			43
C, Amitrol	*Cypridopsis vidua*	LC_{50}		32.0			43
C, Diphenamid	*Cypridopsis vidua*	LC_{50}		50.0			43
C, Diphenamid	*Cypridopsis vidua* (S, 21°C, pH 7.5)	EC_{50}		51.0			404
C, Dicamba	*Cypridopsis vidua*	LC_{50}		100.0			470

C, Trifene	*Cypridopsis vidua*	NOAE		100.0			43
C, Dicamba	*Cypridopsis vidua*	NOEL		100.0			43
C, Calcium chloride	*Cypridopsis* sp.	LC_{50}			0.190		471
C, Chlorpyrifos	*Cyprinotus incongruens*	LC_{50}	0.20				420
C, Parathion	*Daphnia magna* (F, 18°C, instar 1)	LC_{50}				0.00014 (3 wk)	472
C, Resmethrin, *trans*, racemic, *cis*	*Daphnia pulex*	NOAE				0.0001 (3 h)	373
C, Parathion	*Daphnia magna* (F, 18°C, instar 1)	LC_{50}				0.00028 (1 wk)	472
C, Parathion	*Daphnia magna* (F, 18°C, instar 1)	LC_{50}				0.00025 (2 wk)	472
C, Diazinon	*Daphnia magna*	NOEL				0.00026 (21 d)	78
C, Parathion	*Daphnia magna* (S, 18°C, pH 7.9)	LC_{50}				0.0003 (3 wk)	473
C, Malathion	*Daphnia magna*	NOEL				0.0006 (21 d)	78
C, Premethrin, racemic, *trans, cis*	*Daphnia pulex*	NOAE				0.001 (3 h)	373
C, Fenothrin, (+)*trans*	*Daphnia pulex*	NOAE				0.001 (3 h)	373
C, Fenothrin, (+)*cis*	*Daphnia pulex*	NOAE				0.001 (3 h)	373
C, Fenothrin, racemic	*Daphnia pulex*	NOAE				0.001 (3 h)	373
C, Carbaryl	*Daphnia magna*	NOEL				0.005 (63 d)	90
C, Fenitrothion	*Daphnia*	LC_{50}				0.0092 (3 h)	81
C, Aminocarb (17%)	*Daphnia magna* (S, 21°C, pH 7.5, instar 1)	EC_{50}				0.010 (h?)	482
C, Tetramethrin, *trans/cis*	*Daphnia pulex*	NOAE				0.01 (3 h)	373
C, Phygon	*Daphnia magna*	IC_{50}				0.014	28
C, Dichlone	*Daphnia magna*	IC_{50}				0.014	28
C, Carbendazim	*Daphnia magna*	EC_{50}				0.021 (8 d)	474
C, Aroclor 1248	*Daphnia magna* (S)	LC_{50}				0.025 (3 wk)	483
C, Aroclor 1254	*Daphnia magna* (S)	LC_{50}				0.031 (3 wk)	483
C, Aroclor 1260	*Daphnia magna* (S)	LC_{50}				0.036 (3 wk)	483
C, Aroclor 1262	*Daphnia magna* (S)	LC_{50}				0.043 (3 wk)	483
C, Hydroquinone	*Daphnia magna*	LC_{50}				0.05	20
C, Fenothrin, (–)*trans*	*Daphnia pulex*	NOAE				0.050 (3 h)	373
C, Fenothrin, (–)*cis*	*Daphnia pulex*	NOAE				0.050 (3 h)	373
C, Tetramethrin, racemic	*Daphnia pulex*	NOAE				0.05 (3 h)	373

Chemical	Species	Toxicity test	24 h	48 h	96 h	Chronic exposure	Ref.
C, Resmethrin, *trans, cis*	*Daphnia pulex*	NOAE				0.05 (3 h)	373
C, Aroclor 1268	*Daphnia magna* (S)	LC_{50}				0.053 (3 wk)	476
C, Aroclor 1242	*Daphnia magna* (S)	LC_{50}				0.067 (3 wk)	476
C, Aroclor 1232	*Daphnia magna* (S)	LC_{50}				0.072 (3 wk)	476
C, Triphenyl phosphate	*Daphnia magna*	NOAE				0.1 (3 wk)	480
C, Pentachlorophenol	*Daphnia magna*	LC_{50}				0.170 (21 d)	479
C, Aroclor 1221	*Daphnia magna* (S)	LC_{50}				0.180 (3 wk)	476
C, Cyantrine	*Daphnia pulex*	LC_{10}				0.2 (3 d)	489
C, 4,4′-Diaminophenylmethane	*Daphnia*	LC_{100}				0.25	8
C, Butylbenzyl phthalate	*Daphnia magna*	MATC				0.26–0.76	9
C, Diethylene glycol	*Daphnia magna*	LC_{50}				0.3–1.0	20
C, *p*-Benzoquinone	*Daphnia*	LC_{100}				0.4	82
C, Aniline	*Daphnia magna*	LC_{100}				0.4	291
C, Quinhydrone	*Daphnia*	LD_0				0.40	8
C, Hydroquinone sulfonate, Na	*Daphnia magna*	LC_{50}				0.47 (h?)	20
C, Lindane,TG	*Daphnia*	LC_{50}				0.516	83
C, Hydroquinone	*Daphnia*	LC_0				0.60	8
C, *p*-Aminophenol	*Daphnia magna*	LC_{100}				0.6	8
C, Phloroglucinol	*Daphnia*	LD_0				0.60	8
C, *p*-Toluidine	*Daphnia*	LD_0				0.60	8
C, *m*-Toluidine	*Daphnia*	LD_0				0.60	8
C, Sodium arsenite	*Daphnia magna* (<24 h old, 1/2S, 21°C, pH 7.2–8.1)	NOAE				0.633 (28 d)	477
C, Acridine	*Daphnia magna*	LC_{100}				0.7	8
C, *t*-Butylamineborane	*Daphnia magna*	LC_{50}				0.7	20
C, Resorcinol	*Daphnia*	LD_0				0.80	8
C, Ethylene diamine	*Daphnia magna*	LC_{50}				0.88	20
C, Arsenic trioxide	*Daphnia magna* (S, 16°C, pH 6.9–7.3)	NOAE				0.955 (14 d)	477
C, Hydroxylamine sulfate	*Daphnia magna*	LC_{50}				1.20	20

C, Sodium arsenite	*Daphnia magna* (<24 h old, 1/2S, 21°C, pH 7.2–8.1)	LC_{100}			1.320 (28 d)	477
C, Diethanolamine	*Daphnia magna*	LC_{50}			1.4	20
C, Thiourea	*Daphnia magna*	LC_{50}			1.80	20
C, Formaldehyde	*Daphnia*	LC_{100}			2.0	19
C, Cyantrine	*Daphnia pulex*	LC_{25}			2.0 (3 d)	489
C, Picloram	*Daphnia*	NOEL			1.0 (10 wk)	771
C, Chlordane	*Daphnia*	EC_{50}			0.097	83
C, *p,p*-DDT	*Daphnia*	LC_{50}		0.0011		83
C, *p,p*-DDT (chlorinated emulsion)	*Daphnia*	LC_{50}		0.0017		83
C, *p,p*-DDT (dechlorinated emulsion)	*Daphnia*	LC_{50}		0.101		83
C, DDT (dechlorinated)	*Daphnia*	LC_{50}		0.170		83
C, 2,5-Dichlorobenzene sulfonate, Na	*Daphnia magna*	TL_m			1468.0 (100 h)	74
C, Ethyl-*p*-nitrophenylthionobenzene phosphate	*Daphnia magna*	LC_{50}	0.00032			725
C, β-Phenylethylamine sulfate	*Daphnia magna*	LC_{50}			2.6	20
C, Pyrrolidone	*Daphnia magna*	LC_{50}			3.40	20
C, Crotonaldehyde	*Daphnia magna*	EC_{50}			3.9 (h?)	490
C, Catechol	*Daphnia*	LD_0			4.0	8
C, Propanil	*Daphnia magna*	IC_{50}			4.8	28
C, 1,3,5-Trioxane	*Daphnia*	LD_0			5.0	8
C, Triphenyl phosphate	*Daphnia magna*	EC_{50}			5.60 (2 d)	480
C, 2,4-Dinitrophenol	*Daphnia*	LC_{100}			6.0	8
C, Boric acid	*Daphnia magna* (S, 19°C, pH 7.1–8.7)	NOAE			6.0 (21 d)	491
C, Lindane,1% chlorinated dust	*Daphnia*	LC_{50}			6.442	83
C, 4,6-Dinitro-*o*-cresol	*Daphnia*	LC_{100}			8.0	8
C, *o*-Toluidine	*Daphnia*	LD_0			8.0	8
C, *p-tert*-Butylphenol	*Daphnia*	LD_0			8.0	8
C, Ethylene diamine	*Daphnia*	LD_0			8.0	8
C, Dichlobenil	*Daphnia magna*	IC_{50}			9.8	28
C, 1-Phenyl-3-pyrazolidone	*Daphnia magna*	LC_{50}			10.0	20

Chemical	Species	Toxicity test	24 h	48 h	96 h	Chronic exposure	Ref.
C, Diethylene triamine pentaacetic acid	*Daphnia magna*	LC_{50}				10.0–100.0	2
C, *p*-Cresol	*Daphnia*	LC_0				12.0	8
C, Lindane,1% dechlorinated dust	*Daphnia*	LC_{50}				13.054	83
C, *p*-Nitrophenol	*Daphnia*	LD_0				14.0	8
C, *o*-Cresol	*Daphnia*	LC_0				16.0	8
C, Phenol	*Daphnia*	LD_0				16.0	51
C, 3,4-Xylenol	*Daphnia*	LD_0				16.0	8
C, 3,5-Xylenol	*Daphnia*	LD_0				16.0	8
C, Pyrogallol	*Daphnia*	LD_0				18.0	8
C, Lindane, dechlorinated, technical	*Daphnia*	LC_{50}				19.342	83
C, *p*-Nitroaniline	*Daphnia*	LD_0				24.0	8
C, *m*-Nitrophenol	*Daphnia*	LD_0				24.0	8
C, Allethrin, (+)*trans*	*Daphnia pulex*	LC_{50}				25.0–50.0 (3 h)	373
C, Fenothrin, (+)*trans*	*Daphnia pulex*	LC_{50}				25.0–50.0 (3 h)	373
C, Resmethrin, *trans* + *cis*	*Daphnia pulex*	LC_{50}				25.50 (3 h)	373
C, Nitroglycerine	*Daphnia*	LD_0				26.0	8
C, Boric acid	*Daphnia magna* (S, 19°C, pH 7.1–8.7)	LC_{14}				27.0 (21 d)	491
C, *m*-Cresol	*Daphnia*	LC_0				28.0	8
C, Nitrobenzene	*Daphnia*	LD_0				28.0	8
C, Citrazinic acid	*Daphnia magna*	LC_{50}				32.0	20
C, Ethylamine	*Daphnia*	NOAE				40.0	8
C, Quintozene	*Daphnia*	LC_{50}				40.0 (3 h)	492
C, Permethrin, racemic + *trans* + *cis*	*Daphnia pulex*	LC_{50}				50.0 (3 h)	373
C, Tetramethrin, *trans/cis*	*Daphnia pulex*	LC_{50}				50.0 (3 h)	373
C, Allethrin, racemic, *cis*	*Daphnia pulex*	LC_{50}				50.0 (3 h)	373
C, Tetramethrin, racemic	*Daphnia pulex*	LC_{50}				50.0 (3 h)	373
C, Resmethrin, *trans*, racemic, *cis*	*Daphnia pulex*	LC_{50}				50.0 (3 h)	373
C, Fenothrin, racemic	*Daphnia pulex*	LC_{50}				50.0 (3 h)	373

C, Fenothrin, (–)*cis*	*Daphnia pulex*	LC_{50}	50.0 (3 h)	373
C, Fenothrin, (+)*cis*	*Daphnia pulex*	LC_{50}	50.0 (3 h)	373
C, Fenothrin, (–)*trans*	*Daphnia pulex*	LC_{50}	50.0 (3 h)	373
C, Quinoline	*Daphnia*	LD_0	52.0	3
C, Boric acid	*Daphnia magna* (S, 19°C, pH 7.1–8.7)	LC_{32}	53.0 (21 d)	491
C, Benzylamine	*Daphnia*	LC_{100}	60.0	8
C, *n*-Butyric acid	*Daphnia*	LC_{100}	60.0	8
C, Toluene	*Daphnia*	LD_0	60.0	8
C, *o*-Nitrophenol	*Daphnia*	LD_0	60.0	8
C, Citric acid	*Daphnia magna*	LC_0	80.0 (chr. exp.)	219
C, Picric acid, 50%	*Daphnia*	LD_0	88.0	2
C, Diethylamine	*Daphnia*	LC_{100}	100.0	21
C, Citric acid	*Daphnia magna*	LC_{100}	100.0	8
C, Nitrilotriacetic acid	*Daphnia magna*	LC_{50}	100.0	20
C, 1,3-Diamino-2-propanol tetra-acetic acid (DAPTA)	*Daphnia magna*	LC_{50}	100.0	20
C, EDTA, triNa salt	*Daphnia magna*	LC_{50}	100.0	20
C, Trichloroethylene	*Daphnia*	NOAE	100.0	8
C, Phenol	*Daphnia*	TL_m	100.0 (25–50 h)	51
C, Boric acid	*Daphnia magna* (S, 19°C, pH 7.1–8.7)	LC_{100}	106.0 (21 d)	491
C, Formic acid	*Daphnia*	LC_{100}	120.0	8
C, Citric acid	*Daphnia magna*	LC_{100}	120.0 (chr. exp.)	219
C, Ammonium chloride	*Daphnia magna*	TL_m	139.0 (100 h)	74
C, Benzoic acid	*Daphnia magna*	immob	146.0	84
C, Nitrilotriacetic acid	*Daphnia magna* (0.5S, 15°C)	LC_{50}	150.0 (21 d)	495
C, Nitroglycol	*Daphnia*	LD_0	190.0	8
C, Triethylamine	*Daphnia*	LD_0	200.0	8
C, Dimethylamine	*Daphnia magna*	EC_{50}	286.0 (h?)	490
C, 1,2-Dichloroethane	*Daphnia magna* (S, 20°C, pH 8)	EC_0	385.0 (24 wk)	494
C, *n*-Pentanol	*Daphnia*	LC_{100}	440.0	8
C, Triethyl Phosphate	*Daphnia magna*	NOAE	500.0 (h?)	493

Chemical	Species	Toxicity test	24 h	48 h	96 h	Chronic exposure	Ref.
C, 1,2-Dichloroethane	*Daphnia magna* (S, 20°C, pH 8)	EC_{50}				540.0 (24 wk)	494
C, Trichloroethylene	*Daphnia*	LD_{100}				600.0 (40 h)	8
C, 1,2-Dichloroethane	*Daphnia magna* (S, 20°C, pH 8)	EC_{100}				683.0 (24 wk)	494
C, Triethyl phosphate	*Daphnia magna*	EC_{50}				950.0 (h?)	493
C, 4-Nitrochlorobenzene-2-sulfonate, Na	*Daphnia magna*	TL_m				1474.0 (100 h)	74
C, Triethyl phosphate	*Daphnia magna*	EC_{100}				1500.0 (h?)	493
C, *m*-Nitrobenzene sulfonate, Na	*Daphnia magna*	TL_m				2235.0 (100 h)	74
C, Triethanolamine	*Daphnia*	LD_0				2500.0	8
C, Eulan	*Daphnia pulex*	NOAE				5000.0	2
C, *m*-Nitrobenzene sulfonate, Na	*Daphnia magna*	TL_m				6017.0 (72 h)	77
C, Thiram	*Daphnia magna* (S)	EC_0	0.00005				493
C, Thiram	*Daphnia magna* (S)	EC_{50}	0.00006				493
C, Thiram	*Daphnia magna* (S)	EC_{100}	0.0008				493
C, Lindane	*Daphnia magna* (S)	EC_0	0.003				493
C, Endrin	*Daphnia magna* (S)	EC_0	0.004				493
C, Pentachlorobenzene	*Daphnia magna* (S)	EC_0	0.006				493
C, Isobenzan	*Daphnia magna* (S)	EC_0	0.008				493
C, Pentachlorobenzene	*Daphnia magna* (S)	EC_{100}	0.01				493
C, Chlorpyrifos	*Daphnia* sp.	LC_{50}	0.016				497
C, Isobenzan	*Daphnia magna* (S)	EC_{100}	0.020				493
C, Aldrin	*Daphnia magna* (S)	EC_0	0.03				493
C, Endrin	*Daphnia magna* (S)	EC_{50}	0.03				493
C, Lindane	*Daphnia magna* (S)	EC_{50}	0.04				493
C, Aldrin	*Daphnia magna* (S)	EC_{50}	0.05				493
C, Dieldrin	*Daphnia magna* (S)	EC_0	0.06				493
C, Endrin	*Daphnia magna* (S)	EC_{100}	0.06				493
C, Furfural	*Daphnia magna*	LC_{50}	0.062			0.034 (72 h)	498
C, Hexachloro-1,3-butadiene	*Daphnia magna* (S)	EC_0	0.08				493

C, Heptachlor	*Daphnia magna* (S)	EC_0	0.09	493
C, Hexachlorobenzene	*Daphnia magna* (S)	EC_0	0.1	493
C, Aldrin	*Daphnia magna* (S)	EC_{100}	0.1	493
C, Dinoseb acetate	*Daphnia magna*	LC_{50}	0.10	591
C, Heptachlor	*Daphnia magna* (S)	EC_{100}	0.2	493
C, Hexachloro-1,3-butadiene	*Daphnia magna* (S)	EC_{100}	0.2	493
C, Dieldrin	*Daphnia magna* (S)	EC_{100}	0.2	493
C, Heptachlor	*Daphnia magna* (S)	EC_{50}	0.2	493
C, Benz(a)acridine	*Daphnia pulex*	LC_{50}	0.449	72
C, Isodrin/photoisodrin	*Daphnia pulex*, toxicity ratio	TR	0.45	44, 258, 300
C, Lindane	*Daphnia magna* (S)	EC_{100}	0.5	493
C, Hexachloro-1,3-butadiene	*Daphnia magna* (S)	EC_{50}	0.5	493
C, Hydrazine	*Daphnia pulex* (S, 20°C, pH 7.2, juv.)	LC_{50}	0.51	499
C, 2,4,5-Trichlorophenol	*Daphnia magna* (S)	EC_0	0.8	493
C, Linuron	*Daphnia magna*	LC_0	1.0	496
C, Fenethcarb	*Daphnia magna*	LC_{50}	1.0	67
C, Pentachlorophenol	*Daphnia magna* (S)	EC_0	1.1	493
C, 2,4,5-Trichlorophenol	*Daphnia magna* (S)	EC_{50}	1.2	493
C, Heptachlor/photoheptachlor toxicity ratio	*Daphnia magna*	TR	1.24	44,300
C, Diuron	*Daphnia magna*	LC_{50}	1.4	67
C, 2,4,5-Trichlorophenol	*Daphnia magna* (S)	EC_{100}	1.8	493
C, 1,2,4-Trichlorobenzene	*Daphnia magna* (S)	EC_0	2.0	493
C, Acridine	*Daphnia pulex*	LC_{50}	2.92	72
C, 1,2,4-Trichlorobenzene	*Daphnia magna* (S)	EC_{50}	4.0	493
C, Linuron	*Daphnia magna*	LC_{50}	4.0	496
C, Pentachlorophenol	*Daphnia magna* (S)	EC_{100}	4.5	493
C, Monolinuron	*Daphnia magna*	LC_0	5.0	496
C, Linuron	*Daphnia magna*	LC_{100}	6.0	496
C, Di-Na arsenate	*Daphnia magna* (24 h, S, 20°C, pH 8.2)	EC_0	6.2	494

Chemical	Species	Toxicity test	24 h	48 h	96 h	Chronic exposure	Ref.
C, 4,6-Dinitro-*o*-cresol	*Daphnia magna*	LC_{50}	6.6				494
C, Tetrachloroethylene	*Daphnia magna* (S)	EC_0	7.0				493
C, Diquat dibromide	*Daphnia magna* (21°C)	EC_{50}	7.10				500
C, Monolinuron	*Daphnia magna*	LC_{50}	7.5				496
C, Carbon tetrachloride	*Daphnia magna* (S)	EC_0	9.0				493
C, *o*-Chlorophenol	*Daphnia magna* (S)	EC_0	10.0				493
C, 1,2,4-Trichlorobenzene	*Daphnia magna* (S)	EC_{100}	10.0				493
C, Monolinuron	*Daphnia magna*	LC_{100}	10.0				496
C, Di-Na arsenate	*Daphnia magna* (24 h, S, 20°C, pH 8.2)	EC_{50}	13.0				494
C, 2,4-Dinitrophenol	*Daphnia magna*	LC_{50}	19.0				501
C, Epichlorohydrin	*Daphnia magna* (S, 20–22°C, pH 7.7)	LC_0	20.0				501
C, Tetrachloroethylene	*Daphnia magna* (S)	EC_{50}	22.0				493
C, Caproic acid	*Daphnia magna*	TL_m	22.0				75
C, *o*-Chlorophenol	*Daphnia magna* (S)	EC_{50}	23.0				493
C, Nitrobenzene	*Daphnia magna* (<24 h, S, 20°C, pH 8.2)	EC_0	26.0				494
C, Carbon tetrachloride	*Daphnia magna* (S)	EC_{50}	28.0				493
C, Epichlorohydrin	*Daphnia magna* (S, 20–22°C, pH 7.7)	LC_{50}	30.0				501
C, Disodium arsenate	*Daphnia magna* (24 h, S, 20°C, pH 8.2)	EC_{100}	33.0				494
C, *p*-Nitrophenol	*Daphnia magna*	LC_{50}	35.0				501
C, *m*-Nitrophenol	*Daphnia magna*	LC_{50}	39.0				501
C, Isoquinoline	*Daphnia pulex*	LC_{50}	39.9				72
C, Epichlorohydrin	*Daphnia magna* (S, 20–22°C, pH 7.7)	LC_{100}	44.0				501
C, Acetic acid	*Daphnia magna*	TL_m	47.0				76
C, Dimethylamine	*Daphnia magna* (½S)	EC_{50}	48.0				503
C, Nitrobenzene	*Daphnia magna* (<24 h old, S, 20°C, pH 8.2)	EC_{50}	50.0				494
C, Chloroform	*Daphnia magna* (S)	EC_0	62.0				493
C, *o*-Chlorophenol	*Daphnia magna* (S)	EC_{100}	64.0				493
C, Tetrachloroethylene	*Daphnia magna* (S, 20°C, pH 8)	EC_0	65.0				494
C, 1,2-Dichloroethane	*Daphnia magna* (S)	EC_0	67.0				493

C, 1,2-Dichloroethane	*Daphnia magna* (S, 22°C, pH 6.7–8.1)	NOAE	68.0	502
C, 1-Methylethylbenzene	*Daphnia magna* (S, 20°C, pH 8.2)	EC_0	83.0	494
C, 1-Methylethylbenzene	*Daphnia magna* (S, 20°C, pH 7.8–8)	EC_{50}	91.0	494
C, 1-Methylethylbenzene	*Daphnia magna* (S, 20°C, pH 8.2)	EC_{100}	100.0	494
C, Nitrobenzene	*Daphnia magna* (<24 h old, S, 20°C, pH 8.2)	EC_{100}	100.0	494
C, *o*-Xylene	*Daphnia magna*	TL_m	100.0	75
C, Propionic acid	*Daphnia magna*	TL_m	130.0	75
C, Ethylbenzene	*Daphnia magna* (S, 20°C, pH 8.2)	NOAE	137.0	494
C, Picric acid	*Daphnia magna*	LC_{50}	145.0	501
C, Tetrachloroethylene	*Daphnia magna* (S, 20°C, pH 8)	EC_{50}	147.0	494
C, Carbon tetrachloride	*Daphnia magna* (S)	EC_{100}	159.0	493
C, Hydroxyethane diphosphoric acid	*Daphnia*	LC_{50}	165.0	505
C, Hydroxyethane diphosphoric acid	*Daphnia*	LC_{100}	180.0	505
C, Ethylbenzene	*Daphnia magna* (S, 20°C, pH 8.2)	EC_{50}	184.0	494
C, Ethylbenzene	*Daphnia magna* (S, 20°C, pH 8.2)	EC_{100}	200.0	494
C, 2-Nitrophenol	*Daphnia magna*	LC_{50}	210.0	494
C, Ammonium salt of carboxylic acid, a dispersant	*Daphnia magna*	LC_{50}	280.0	70
C, Chloroform	*Daphnia magna* (S)	EC_{50}	290.0	493
C, Picloram	*Daphnia*	LC_0	380.0	504
C, Triethyl phosphate	*Daphnia magna* (S)	EC_0	500.0	493
C, Chloroform	*Daphnia magna* (S)	EC_{100}	500.0	493
C, Picloram	*Daphnia*	LC_{95}	530.0	504
C, 1,2-Dichloroethane	*Daphnia magna* (S)	EC_{50}	600.0	493
C, 1,2-Dichloroethane	*Daphnia magna* (S, 22°C, pH 7.7)	LC_0	850.0	501
C, Triethylphosphate	*Daphnia magna* (S)	EC_{50}	950.0	493
C, 1,2-Dichloroethane	*Daphnia magna* (S)	EC_{100}	1075.0	493
C, 2-Butoxyethanol	*Daphnia magna* (22°C, pH 7.7, <24 h old)	LC_{10}	1140.0	501
C, Tetrahydrofuran	*Daphnia magna* (20°C, pH 8.2)	EC_0	1250.0	494
C, Isobutyl alcohol	*Daphnia magna*	EC_{50}	1250.0	494
C, 1,2-Dichloroethane	*Daphnia magna* (S, 22°C, pH 7.7)	LC_{50}	1350.0	501

Chemical	Species	Toxicity test	24 h	48 h	96 h	Chronic exposure	Ref.
C, Dioxane	*Daphnia*	EC_{50}	1400.0				507
C, Triethanol amine	*Daphnia*	LC_0	1400.0				505
C, Triethyl phosphate	*Daphnia magna* (S)	EC_{100}	1500.0				493
C, 2-Butoxyethanol	*Daphnia magna* (22°C, pH 7.7, <24 h old)	LC_{50}	1720.0				501
C, 1,2-Dichloroethane	*Daphnia magna* (S, 22°C, pH 7.7)	LC_{100}	1820.0				501
C, *n*-Butyl alcohol	*Daphnia magna*	EC_{50}	1880.0				508
C, *s*-Butyl alcohol	*Daphnia magna*	EC_{50}	2300.0				509
C, 2-Butoxyethanol	*Daphnia magna* (22°C, pH 7.7, <24 h old)	LC_{100}	2500.0				501
C, *n*-Propanol	*Daphnia magna*	EC_0	3336.0				501
C, *e*-Caprolactam	*Daphnia magna* (S)	EC_{100}	4000.0				493
C, *n*-Propanol	*Daphnia magna*	EC_{50}	4415.0				501
C, *n*-Propanol	*Daphnia magna*	EC_{100}	5909.0				501
C, Tetrahydrofuran	*Daphnia magna* (20°C, pH 8.2)	EC_{50}	5930.0				494
C, Dioxane	*Daphnia magna*	EC_{100}	6210.0				494
C, *e*-Caprolactam	*Daphnia magna* (S)	EC_{50}	7000.0				493
C, *e*-Caprolactam	*Daphnia magna* (S)	EC_{100}	8000.0				493
C, Dioxane	*Daphnia magna*	EC_{50}	8450.0				494
C, Tetrachloroethylene	*Daphnia magna* (S)	EC_{100}	9880.0				493
C, Dioxane	*Daphnia magna*	EC_{100}	10,000.0				494
C, Tetrahydrofuran	*Daphnia magna* (20°C, pH 8.2)	EC_{100}	10,000.0				494
C, Dichlorvos	*Daphnia pulex* (15.5°C)	EC_{50}		0.000066			506
C, 2,2-Dichlorovinyl dimethyl phosphate (DDVP)	*Daphnia pulex*	LC_{50}		0.00007			69
C, Dichlorvos	*Daphnia pulex*	LC_{50}		0.00007			510
C, Chlorfos	*Daphnia pulex*	LC_{50}		0.00018			45
C, Methyl parathion	*Daphnia magna* (1 instar, S, 21°C, pH 7.5, softW)	LC_{50}		0.00014			511
C, Tralomethrin	*Daphnia magna*	LC_{50}		0.00015			1083
C, Cyfluthrin	*Daphnia magna*	LC_{50}		0.00017			1083

C, Azinophos-methyl	*Daphnia magna*	LC_{50}		0.0002		510
C, Bifenthrin	*Daphnia magna*	LC_{50}		0.00032		1083
C, DDT	*Daphnia pulex*	LC_{50}		0.00036		69
C, Parathion	*Daphnia magna*	LC_{50}	0.0008	0.00037		67,68
C, DDT	*Daphnia magna*	LC_{50}	0.0044	0.00036		68
C, Parathion	*Daphnia pulex*	LC_{50}		0.0004		510
C, Parathion	*Daphnia magna* (F, 15°C, pH 7.5, 1 instar)	LC_{50}		0.0006		511
C, Parathion	*Daphnia pulex*	LC_{50}		0.00060		69
C, Parathion	*Daphnia pulex* (S, 22°C)	LC_{50}		0.00076		512
C, Baytex	*Daphnia pulex*	LC_{50}		0.00080		69
C, Diazinon	*Daphnia pulex*	LC_{50}		0.00090		69
C, Diazinon	*Daphnia pulex*	LC_{50}		0.0009		510
C, Co-Ral	*Daphnia magna*	LC_{50}		0.001		45
C, lambda-Cyhalothrin	*Daphnia magna*	LC_{50}		0.00104		1083
C, Diazinon	*Daphnia magna*	LC_{50}		0.00122		73
C, Permethrin	*Daphnia magna*	LC_{50}		0.00125		1083
C, Malathion	*Daphnia pulex*	LC_{50}		0.0018		69
C, Malathion	*Daphnia pulex*	LC_{50}		0.0018		510
C, Malathion	*Daphnia magna*	LC_{50}	0.0009	0.0018	0.003 (1 wk)	67, 68, 80
C, Azinphosethyl	*Daphnia pulex*	LC_{50}		0.0032		69
C, DDD	*Daphnia pulex*	LC_{50}		0.0032		69
C, Chlorothion	*Daphnia magna*	LC_{50}		0.0045		235
C, Methyl parathion	*Daphnia magna*	LC_{50}		0.0048		510
C, Phosmet (TG, 95%)	*Daphnia magna* (1 instar, S, 21°C, pH 7.5)	EC_{50}		0.0056		409
C, Carbaryl	*Daphnia pulex*	NOEL		0.0064		235
C, Methomyl, (98% TG)	*Daphnia magna* (S, 20°C, pH 7.5)	LC_{50}		0.0088		484
C, Phosphamidon	*Daphnia pulex*	LC_{50}		0.0088		69
C, Phosmet (50% pdrHW)	*Daphnia magna* (1 instar, S, 21°C, pH 7.5)	EC_{50}		0.009–0.014		409
C, Cartap	*Daphnia magna* (18–22°C, pH 7.1)	LC_{50}		0.010		514
C, Zectran	*Daphnia pulex*	LC_{50}		0.01		69

Chemical	Species	Toxicity test	24 h	48 h	96 h	Chronic exposure	Ref.
C, Toxaphene	*Daphnia* serrulatus	LC_{50}		0.01			515
C, Cartap	*Daphnia magna* (adult)	LC_{50}	0.012	0.010			516
C, Toxaphene	*Daphnia pulex*	LC_{50}		0.015			71
C, Toxaphene	*Daphnia pulex* (1st instar, 15.5°C)	LC_{50}		0.015			515
C, Toxaphene	*Daphnia magna*	LC_{50}	0.094	0.015			67,68
C, Aminocarb	*Daphnia magna* (S, 17°C, pH 7.4)	EC_{50}		0.019			517
C, Endrin	*Daphnia pulex*	LC_{50}		0.020			69
C, Allethrin	*Daphnia pulex*	LC_{50}		0.021			69
C, Pyrethrum	*Daphnia pulex* (15.5°C)	EC_{50}		0.025			510
C, Dichlone	*Daphnia magna*	LC_{50}		0.025			45
C, Pyrethrum	*Daphnia pulex*	LC_{50}		0.025			518
C, Pyrethrin	*Daphnia pulex*	LC_{50}		0.025			69
C, Aldrin	*Daphnia magna*	LC_{50}	0.03	0.028			67,68
C, Chlordane	*Daphnia pulex*	LC_{50}		0.029			69
C, Chlordane	*Daphnia pulex*	LC_{50}		0.029			506
C, Acrolein	*Daphnia magna* (<24 h old, S, 22°C, pH 7–9)	NOAE		0.034			502
C, Heptachlor	*Daphnia pulex*	LC_{50}		0.042			69
C, Cadmium chloride	*Daphnia pulex* (20°C)	LC_{50}	0.147	0.042			1082
C, Cadmium chloride	*Daphnia pulex* (26°C)	LC_{50}	0.063	0.006			1082
C, Aldicarb	*Daphnia laevis* (S, 21–23°C, pH 6.9)	EC_{50}		0.051			519
C, Acrolein	*Daphnia magna* (S)	LC_{50}		0.057			520
C, Endosulfan	*Daphnia magna*	LC_{50}	0.240	0.060			67,68
C, Aldicarb	*Daphnia laevis* (juv., S, 23°C, pH 6.9)	EC_{50}		0.065			519
C, Acrolein	*Daphnia magna* (<24 h old, S, 22°C, pH 7–9)	LC_{50}	0.23	0.083			502
C, 1,3-Dichloropropene	*Daphnia magna* (1 instar, 21°C, pH 7.5)	LC_{50}		0.090			402
C, Rotenone	*Daphnia pulex*	LC_{50}		0.10			69
C, Mirex	*Daphnia pulex* (17°C, pH 7.2–7.5)	LC_{50}		0.100			485
C, 2-Ethylhexyldiphenyl phosphate	*Daphnia magna* (S, 23°C, pH 7–8)	NOAE		0.10			522

C, *p-t*-Butylphenyldiphenyl phosphate	*Daphnia magna* (S, 23°C, pH 7–8.5)	NOAE		0.10			522
C, Isodecyldiphenyl phosphate	*Daphnia magna* (S, 23°C, pH 7–8.5)	NOAE		0.10			522
C, Endosulfan	*Daphnia pulex* (10°C)	LC_{50}	0.178	0.132		0.048 (120 h)	524
C, 2-Ethylhexyldiphenyl phosphate	*Daphnia*	LC_{50}		0.15			523
C, 2-Ethylhexyldiphenyl phosphate	*Daphnia magna* (S, 23°C, pH 7–8.5)	EC_{50}	0.60	0.15			522
C, Hydrazine	*Daphnia pulex* (S, 20°C, pH 7.2, juv.)	EC_{50}		0.16			499
C, Fenoprop-PGBEE	*Daphnia magna*	LC_{50}		0.18			43
C, Propyl glycol butyl ester	*Daphnia magna* (21°C, S)	LC_{50}		0.180			469
C, Endosulfan	*Daphnia magna*	LC_{50}		0.20			396
C, Aldicarb	*Daphnia* laevis (S, 21°C, pH 6.9)	LC_{50}		0.209			519
C, Pentachlorophenol	*Daphnia magna*	EC_{50}		0.240			528
C, Trifluralin	*Daphnia pulex*	EC_{50}		0.240			506
C, Trifluralin	*Daphnia magna*	LC_{50}		0.240			69
C, Trifluralin	*Daphnia pulex*	LC_{50}		0.240			510
C, Dieldrin	*Daphnia pulex*	LC_{50}		0.250			69
C, *p-t*-Butylphenyldiphenyl phosphate	*Daphnia magna* (16–30°C, pH 8.4,field)	EC_{50}		0.261			525
C, *p-t*-Butylphenyldiphenyl phosphate	*Daphnia magna* (S, 23°C, pH 7–8.5)	EC_{50}	1.4	0.30			522
C, Aminocarb	*Daphnia magna* (juv., S, 17°C, pH 7.4)	EC_{50}		0.320			517
C, *p-t*-Butylphenyldiphenyl phosphate	*Daphnia magna* (16–30°C, pH 8.4, laboratory)	EC_{50}		0.343			525
C, DDT	*Daphnia pulex* (S, 15.6°C, pH 7.1)	LC_{50}		0.40			528
C, Aminocarb, oil fraction	*Daphnia pulex* (S)	LC_{50}		0.40			460
C, Lindane	*Daphnia pulex*	LC_{50}		0.460			69
C, Carbendazim	*Daphnia magna* (S, 20°C)	LC_{50}		0.46			474
C, Lindane	*Daphnia pulex* (1 instar, 15°C, pH 7.5)	LC_{50}		0.460			527
C, Lindane	*Daphnia magna*	LC_{50}	1.25	0.46			67,68
C, Isodecyldiphenyl phosphate	*Daphnia magna* (S, 23°C, pH 7–8.5)	EC_{50}	0.79	0.48			522
C, Pentachlorophenate, Na	*Daphnia pulex* (20°C)	LC_{50}	0.64	0.48			1082

Chemical	Species	Toxicity test	24 h	48 h	96 h	Chronic exposure	Ref.
C, Pentachlorophenate, Na	*Daphnia pulex* (26°C)	LC_{50}	0.62	0.47			1082
C, Lindane	*Daphnia magna* (S)	LC_{50}		0.485			529
C, Anilazine	*Daphnia magna*	LC_{50}		0.490			530
C, Trifluralin	*Daphnia magna*	LC_{50}		0.560			43
C, Ethyl parathion	*Daphnia pulex* (15.5°C)	EC_{50}		0.60			506
C, 1-Methylethylbenzene	*Daphnia magna* (S, 21–25°C, pH 6–7)	LC_{50}		0.6			531
C, Benomyl	*Daphnia magna* (20°C)	LC_{50}		0.64			474
C, Mirex	*Daphnia magna* (1 instar, S, 17°C, pH 7.5)	EC_{50}		1.0			485
C, Mirex	*Daphnia pulex* (1 instar, S, 17°C, pH 7.5)	EC_{50}		1.0			485
C, Simazine	*Daphnia magna*	LC_{50}		1.0			43
C, Triphenyl phosphate	*Daphnia magna* (S)	LC_{50}		1.0			533
C, Mirex	*Daphnia magna* (1 instar, 17°C, pH 7.2–7.5)	LC_{50}		1.0			521
C, *N*-Nitrosodiphenylamine	*Daphnia magna* (S, 23°C, pH 6.7–8.1)	NOAE		1.0			502
C, Vernam	*Daphnia magna*	LC_{50}		1.1			43
C, Aminocarb, water-soluble fraction	*Daphnia pulex* (S)	LC_{50}		1.10			460
C, 2-Chlorotoluene-5-sulfonate	*Daphnia magna*, adult	TL_m	3.3 (25 h)	1.3 (50 h)			74
C, Diuron	*Daphnia pulex* (1st instar, S, 15°C)	EC_{50}		1.4			532
C, Diuron	*Daphnia pulex*	LC_{50}		1.4			69
C, Diuron	*Daphnia pulex*	LC_{50}		1.40			534
C, 1-Chloronaphthalene	*Daphnia magna* (23°C, pH 7.4–9.4)	LC_{50}	3.6–10.0	1.6			502
C, Formaldehyde	*Daphnia magna*	LC_{50}		2.0			520
C, Fenoprop-PGBEE (propylene glycol butyl ether ester)	*Daphnia pulex*	LC_{50}		2.0			69
C, Fenoprop, butoxyethyl ester	*Daphnia magna*	LC_{50}		2.1			43
C, Ethylbenzene	*Daphnia magna* (6 wk old, S, 21–25°C, pH 7)	LC_{50}		2.1			531
C, Fenoprop[2-(2,4,5-trichloro-phenoxy) propionic acid]	*Daphnia magna* (S, 21°C)	LC_{50}		2.10			470
C, Sodium arsenite	*Daphnia pulex* (<24 h old, S, 24°C, pH 8.3)	EC_{50}		2.366			451
C, Atrazine	*Daphnia magna* (S)	LC_{10}		3.0			535

C, Methomyl, 24%	*Daphnia magna* (S, 20°C, pH 7.5)	LC_{50}		3.20		484
C, Paraquat	*Daphnia pulex*	LC_{50}		3.7		69
C, Dichlobenil	*Daphnia pulex*	LC_{50}		3.7		70
C, Butyl benzyl phthalate (sanitizer)	*Daphnia magna* (S)	EC_{50}		3.7	1.0 (NOAE)	9
C, 2,4-D (diethylamine) salt	*Daphnia magna*	LC_{50}		4.0		43
C, Trifene	*Daphnia pulex*	LC_{50}		4.5		69
C, Sodium arsenite	*Daphnia magna* (<24 h, S, 24°C, pH 8.3)	EC_{50}		4.501		451
C, Propham	*Daphnia pulex* (1 instar. S, 15°C)	EC_{50}		5.0–13.0		532
C, 2,4-D butoxyethyl ester	*Daphnia magna*	LC_{50}		5.6		43
C, 1,3-Dichloropropene	*Daphnia magna*	LC_{50}		6.150		538
C, Atrazine	*Daphnia magna* (S)	LC_{50}		6.9		542
C, Disodium arsenate, H_20	*Daphnia magna* (24 h, S, 17°C, pH 8.2)	LC_{50}		7.4		543
C, Bis(2-choloroethyl)ether	*Daphnia magna* (21–23°C, pH 7–8)	NOAE		7.8		502
C, *N*-Nitrosodiphenylamine	*Daphnia magna* (S, 23°C, pH 6.7–8.1)	LC_{50}	46.0	7.8		502
C, Phosphamidon	*Daphnia pulex* (15.5°C)	EC_{50}		8.8		506
C, Dichlobenil	*Daphnia magna*	LC_{50}		10.0		43
C, Isopropyl-*N*-phenyl carbamate	*Daphnia pulex*	LC_{50}		10.0		69
C, Tetrachloroethylene	*Daphnia magna* (S, 22°C, pH 6.7–8.1)	NOAE		10.0		502
C, Acetone	*Daphnia magna*	TL_m	10.0	10.0		76,77
C, 2,2-Dichloropropionic acid, Na	*Daphnia pulex*	LC_{50}		11.0		69
C, 2,2-Dichloropropionic acid (Dalapon)	*Daphnia pulex* (1st instar, S, 15°C, pH 7.5)	LC_{50}		11.0		537
C, 1,1-Dichloroethylene	*Daphnia magna* (S, 16–18°C, pH 7.9)	LC_{50}		11.6		544
C, Sodium decyl sulfate	*Daphnia pulex* (20°C)	LC_{50}	18.4	12.6		1083
C, Sodium decyl sulfate	*Daphnia pulex* (26°C)	LC_{50}	13.9	10.2		1083
C, Methiophenate	*Daphnia magna* (19–21°C)	LC_{50}		16.0		474
C, Tetrachloroethylene	*Daphnia magna* (S, 22°C, pH 6.7–8.1)	LC_{50}		18.0		502
C, Diethyl phosphite	*Daphnia*	TL_m		25.0		8
C, Aldrin	*Daphnia pulex*	LC_{50}		28.0		69
C, Sulfur (H_2SO_4)	*Daphnia* sp. (pH 5.0)	LC_{100}	29.0	29.0	29.0 (72 h)	545
C, Amitrol	*Daphnia magna*	LC_{50}		30.0		43

Chemical	Species	Toxicity test	24 h	48 h	96 h	Chronic exposure	Ref.
C, Benzene	*Daphnia magna*	LC_{50}		31.2			531
C, Quinoline	*Daphnia magna* (juv., S, 20°C, pH 7.8)	LC_{50}		34.5			546
C, *n*-Butyl acetate	*Daphnia*	TL_m		44.0			18
C, *n*-Valeric acid	*Daphnia magna*	TL_m		45.0			74
C, Dimethylamine	*Daphnia magna* (½S)	LC_{50}		50.0			541
C, Propionic acid	*Daphnia magna*	TL_m		50.0			77
C, Trichloroethylene	*Daphnia pulex* (S)	EC_{50}		51.0			486
C, Pentabromocyclohexane	*Daphnia magna* (S)	LC_{50}		52.7			547
C, Diphenamid	*Daphnia magna*	LC_{50}		56.0			43
C, Diphenamid	*Daphnia magna* (1st instar, S, 21°C, pH 7.5)	EC_{50}		58.0			403b
C, *n*-Butyric acid	*Daphnia magna*	TL_m		61.0			77
C, Ethylbenzene	*Daphnia magna* (<24 h, old, S, 23°C, pH 8.1)	LC_{50}	77.0	75.0			502
C, Diethylene glycol dinitrate	Daphnid	LC_{50}		90.1			1088
C, Trichloroethylene	*Daphnia pulex* (S)	EC_{50}		100.0			548
C, Dicamba	*Daphnia magna* (1st instar, 21°C, pH 7.5)	LC_{50}		100.0			402
C, Trifene	*Daphnia magna*	NOAE		100.0			43
C, Dicamba	*Daphnia magna*	NOEL		100.0			43
C, Dicamba	*Daphnia*	LC_{50}		111.0			536
C, Isophorone	*Daphnia magna* (S)	LC_{50}		117.0			539
C, 1,2-Dichloroethane	*Daphnia magna* (S, 20°C, pH 7–7.7, unfed)	EC_{50}		160.0			549
C, 1,2-Dichloroethane	*Daphnia magna* (S, 20°C, pH 7–7.7, fed)	EC_{50}		180.0			549
C, Disopropyl amine	*Daphnia magna* (½S)	NOAE		180.0		1.0 (2 wk)	480
C, Disopropyl amine	*Daphnia magna* (½S)	EC_{50}	187.0	180.0			480
C, Ethylene oxide	*Daphnia magna* (17°C, S)	LC_{50}		212.0			390
C, 1,2-Dichloroethane	*Daphnia magna* (S, 22°C, pH 6.7–8.1)	LC_{50}	250.0	220.0			502
C, Boric acid	*Daphnia magna* (S, 20°C)	LC_{50}		226.0		53.2 (21 d)	491
C, *Bis*(-2-chloroethyl)ether	*Daphnia magna* (21–23°C, pH 7–8)	LC_{50}	340.0	240.0			502
C, 1,2-Dichloroethane	*Daphnia magna* (S, 20°C, pH 7–7.7, unfed)	LC_{50}		270.0			549
C, Toluene	*Daphnia magna*	EC_{50}		313.0			550

C, 1,2-Dichloroethane	*Daphnia magna* (S, 20°C, pH 7–7.7, fed)	LC_{50}		320.0			549
C, Benzene	*Daphnia magna* (S)	LC_{50}		356.0			548
C, Benzyl alcohol	*Daphnia*	TL_m		360.0			18
C, *n*-Amyl acetate	*Daphnia magna*	TL_m		440.0			18
C, Nitrilotriacetic acid (NTA)	*Daphnia magna* (0.5 S, 15°C)	LC_{50}		560.0			495
C, 2-Butoxyethanol	*Daphnia magna*	EC_{50}		1054.0			551
C, Trichloroacetic acid	*Daphnia magna*	LC_{50}		2000.0			73
C, *n*-Propanol	*Daphnia pulex*	LC_{50}		3025.0			548
C, Ethylene urea	*Daphnia magna* (20°C, pH 8.1)	LC_{50}		5600.0		3200.0 (14 d)	552
C, *n*-Propanol	*Daphnia cucullata*	LC_{50}		5820.0			548
C, *n*-Propanol	*Daphnia magna*	LC_{50}		7080.0			548
C, Trimethylpropane	*Daphnia magna*	LC_{50}		13,000.0			556
C, Pentaerythritol	*Daphnia magna* (7.6°C)	LC_{50}		33,600.0			556
C, Naled	*Daphnia pulex*	LC_{50}			0.00035		69
C, Parathion	*Daphnia magna* (F, 18°C, 1 instar, Juv.)	LC_{50}	0.0027	0.001	0.00062		472
C, Parathion	*Daphnia magna* (S, 22°C, 1 instar)	LC_{50}	0.00321	0.00127	0.00065		472
C, Rotenone	*Daphnia* sp.	LC_{100}			0.026		557
C, Chlordane	*Daphnia magna*	LC_{50}			0.028		553
C, Chlordane	*Daphnia magna*	LC_{50}			0.0284		427
C, Heptachlor	*Daphnia magna*	LC_{50}			0.052		558
C, Endosulfan	*Daphnia magna*	LC_{50}			0.0529		85
C, Heptachloroepoxide	*Daphnia magna*	LC_{50}			0.120		558
C, Diuron	*Daphnia magna*	LC_{50}			0.20		554
C, Aniline	*Daphnia magna* (S, 21°C, pH 6.5–7.8)	LC_{50}			0.21		408
C, Triethyl phosphate	*Daphnia*	LC_{50}			0.330		555
C, 2-Chlorotoluene 5-sulfonate	*Daphnia magna*, young	TL_m	0.8 (25 h)	0.6 (50 h)	0.4 (100 h)		74
C, Pentabromochlorocyclohexane	*Daphnia magna* (S)	NOAE			1.0		547
C, Captan	*Daphnia magna*(S)	LC_{50}			1.3		558
C, Propoxur	*Daphnia magna*	LC_{50}	17.83	3.99	1.45		559
C, Sodium arsenite	*Daphnia magna* (<24 h old, not fed, S, 15°C, pH 8.1)	EC_{50}	2.160	1.500	1.500		477

Chemical	Species	Toxicity test	24 h	48 h	96 h	Chronic exposure	Ref.
C, Fenoprop-K	*Daphnia magna*	LC_{50}			2.0		469
C, EDTA, ammonium ferric salt	*Daphnia magna*	LC_{50}			2.8		20
C, 4-chloro-*m*-Cresol	*Daphnia pulex*	LC_{50}			3.1		560
C, Aminocarb	*Daphnia magna*	LC_{50}			3.34		561
C, Sodium arsenite	*Daphnia magna* (<24 h old, not fed, S, 15°C, pH 8.1)	EC_{100}			4.190		477
C, Sodium arsenite	*Daphnia magna* (<24 h old, fed, S, 15°C, pH 8.1)	EC_{50}	7.25	4.63	4.34		477
C, 2,4-Dinitrophenol	*Daphnia magna* (S)	LC_{50}			4.71		563
C, Tetrabromophthalic anhydride	*Daphnia magna*	NOAE			5.6		564
C, Acrylonitrile	*Daphnia magna* (S)	LC_{50}			7.55		565
C, Pentabromochlorocyclohexane	*Daphnia magna* (S)	LC_{100}			10.0		547
C, Cyantrine	*Daphnia longispina*	LC_{50}			15.4		371
C, Ammonium chloride	*Daphnia magna*	TL_m	202.0	161.0	50.0	67.0 (72 h)	77
C, Anthraquinone "α" sulfonic acid	*Daphnia magna*	TL_m	186.0	186.0	50.0	12.0 (100 h)	2
C, Nitrilotriacetic acid	*Daphnia magna*	NOEL			100.0		372
C, Ammonium sulfite	*Daphnia magna*	TL_m	299.0	273.0	203.0 (100 h)		74
C, Methylene chloride	*Daphnia magna* (S, 22°C, pH 7.4)	LC_{50}			220.0		502
C, Ammonium sulfate	*Daphnia magna*	TL_m	423.0	433.0	292.0		74
C, Triethyl phosphate	*Daphnia magna*	NOAE			500.0		493
C, 2,5-Dichlorobenzene sulfonate, Na	*Daphnia magna*	TL_m	4931.0	4931.0	938.0	2490.0 (72 h)	77
C, 4-Nitrochlorobenzene 2-sulfonate, Na	*Daphnia magna*	TL_m	4698.0	3483.0	948.0	948.0 (72 h)	77
C, Triethyl phosphate	*Daphnia magna*	EC_{50}			950.0		493
C, Phenol sulfonate, Na	*Daphnia magna*	TL_m	13510.0	13,510.0	1471.0	3494.0 (72 h)	77
C, Triethyl phosphate	*Daphnia magna*	EC_{100}			1500.0		493
C, *p*-Chlorobenzenesulfonic acid	*Daphnia magna*	TL_m	8600.0	7659.0	2150.0	3964.0 (72 h)	77
C, Butyl sulfonate, Na	*Daphnia magna*	TL_m	8000.0	8000.0	2700.0	5400.0 (72 h)	77

C, *m*-Nitrobenzene sulfonate, Na	*Daphnia magna*	TL_m	8665.0	8665.0	5067.0		77
C, Phenol	*Daphnia* (adult)	TL_m				21.0 (25–50 h)	51
C, *dl*-Lactic acid	*Daphnia* (26–72 h)	LD_0				170.0	8
C, Dieldrin	*Daphnia* (F)	LC_{100}			0.230 (h?)		566
C, Phenol	*Daphnia* (young)	TL_m				7.0 (25–50 h)	51
C, Triphenyl phosphate	Daphnids	LC_0				0.136 (28 d)	533
C, Diazinon (TG)	Daphnids	EC_{50}		0.0020			87
C, 2,4,6-Trichlorophenyl-4′-nitro phenyl ether	Daphnids	LC_{50}		40.0			86
C, Diethylthiophosphate	Daphnids	EC_{50}		100.0			87
C, Glyphosate	Daphnids (S, 22°C)	LC_{50}		3.0			562
C, Glyphosate	*Gammarus fasciatus*	LC_{50}	100.0	62.0	43.0		562
I, 2-Ethylhexyldiphenyl phosphate	*Diptera*, larvae	LC_{50}		0.50			523
C, *o*-Cresol	*Elasmopus pectnicrus* (adult, $^1/_2$S, 23°C)	LC_{50}	16.0	11.8	10.2		568
C, Tetrachloroethylene	*Eliminus modestus* (larva, S)	LC_{50}		3.5			570
C, Ethylene dichloride	*Eliminus modestus* (S)	LC_{50}		186.0			570
C, Toluene	*Eualus* spp.(12°C, S)	TL_m			14.7		88
C, Toluene	*Eualus* spp.(8°C, S)	TL_m			20.2		88
C, Toluene	*Eualus* spp. (4°C, S)	TL_m			21.4		88
C, Naphthalene	*Eurytemora affinis* (semi S, 10°C)	NOAE				0.050 (10 d)	571
C, Nitrilotriacetic acid, Na, H_2O	*Eurytemora affinis* (S)	TL_{50}				1250.0 (72 h)	37
C, Kepone	Eurytemora affinis (S, 20°C)	LC_{50}		0.040			569
C, Parathion	*Gammarus fasciatus* (F, 20°C, pH 7.4, wellW)	LC_{50}				0.00007 (43 d)	472
C, Parathion	*Gammarus fasciatus* (F, 19°C, pH 7.4, wellW)	LC_{50}				0.00009 (35 d)	472
C, Malathion	*Gammarus pseudolimnaeus*	LC_{50}				0.000023 (30 d)	91
C, Malathion	*Gammarus pseudolimnaeus*	NOEL				0.000008 (30 d)	91
C, Azinphosmethyl	*Gammarus pseudolimnaeus*	NOAE				0.00010 (30 d)	91
C, Diazinon	*Gammarus pseudolimnaeus*	LC_{50}				0.00027 (30 d)	91
C, Diazinon	*Gammarus pseudolimnaeus*	NOEL				0.00020 (30 d)	91
C, Fenvalerate	*Gammarus pseudolimnaeus* (F, 15°C, pH 8)	LC_{100}				0.00093 (5 h)	575
C, Endrin	*Gammarus fasciatus*	LC_{50}				0.0009 (120 h)	45

Chemical	Species	Toxicity test	24 h	48 h	96 h	Chronic exposure	Ref.
C, Permethrin	*Gammarus* sp.	LC_{100}				0.001 (1 h)	576
C, Tributyltin oxide	*Gammarus oceanicus* (0.5S, 10–15°C)	LC_{50}				0.003 (10 d)	577
C, Tributyltin oxide	*Gammarus oceanicus* (0.5S, 10–15°C)	LC_{100}				0.003 (8 wk)	578
C, Hydrocyanic acid	*Gammarus pseudolimnaeus* (F,18°C, pH 8)	NOAEL				0.004 (98 h)	398
C, Aroclor 1242	*Gammarus pseudolimnaeus* (F, 15°C, pH 7.1)	LC_{50}				0.005 (10 d)	580
C, Hydrocyanic acid	*Gammarus pseudolimnaeus* (F)	NOAEL				0.016 (83 d)	398
C, Carbaryl	*Gammarus pulex*	LC_{50}				0.029 (98 h)	90
C, Chloropyrifos	*Gammarus pulex*	LC_{95}				0.05–1.0 (1 h)	574
C, Aroclor 1248	*Gammarus pseudolimnaeus* (S, 15°C, pH 7.1)	LC_{50}				0.052 (4 d)	580
C, Hydrocyanic acid	*Gammarus pseudolimnaeus* (F)	LC_{100}				0.064 (83 d)	398
C, Hydrocyanic acid	*Gammarus pseudolimnaeus* (F)	TC_{LO}				0.074 (10–12 d)	398
C, Atrazine	*Gammarus fasciatus* (S)	LC_{0}				0.14 (30 d)	542
C, Atrazine	*Gammarus fasciatus* (S)	LC_{0}				0.49 (119 d)	542
C, *m*-Cresol	Gammarus pulex	PL				0.7	8
C, Pentachlorophenol	*Gammarus pseudolimnaeus*	LC_{50}				0.86 (30 d)	581
C, Abate	*Gammarus pulex*	LC_{95}				1.0 (1 h)	574
C, Aroclor 1254	*Gammarus oceanicus*	LC_{50}				1.0 (35.3 d)	579
C, Sodium arsenite	*Gammarus pseudolimnaeus* (F, 19°C, pH 8.1)	EC_{50}				1.990 (43 h)	477
C, Acetic acid	*Gammarus pulex*	TT				6.0	8
C, Aroclor 1254	*Gammarus oceanicus*	LC_{50}				10.0 (100 h + 5.2 d)	579
C, Aroclor 1254	*Gammarus oceanicus*	LC_{50}				100.0 (100 h)	579
C, Furfural	*Gammarus pulex*	TT				800.0	8
C, Formic acid	*Gammarus pulex*	LC_{100}				2500.0	8
C, Acetone	*Gammarus pulex*	TL_{m}				5500.0	8
C, Coumaphos	*Gammarus lacustris*	LC_{50}	0.00032				582
C, Chlorpyrifos	*Gammarus lacustris* (S, 21°C, pH 7.1)	LC_{50}	0.00047–0.0012				583
C, Dimethoate	*Gammarus lacustris*	LC_{50}	0.0009				585
C, Phorate	*Gammarus fasciatus*	LC_{50}	0.006				45

C, Endosulfan	*Gammarus lacustris*	LC_{50}	0.0092				585
C, Chlordane	*Gammarus lacustris*	LC_{50}	0.016				584
C, Dioxathion	*Gammarus lacustris* (S)	LC_{50}	0.038				586
C, Temephos	*Gammarus* sp.	LC_{100}	0.1				587
C, Isodrin/photoisodrin, toxicity ratio (TR)	*Gammarus* sp.	TR	0.43				44,48, 49
C, Monocrotophos	*Gammarus fasciatus* (S, 15°C, pH 7.1)	LC_{50}	1.0			0.26	404
C, Heptachlor/photoheptachlor, toxicity, ratio, TR	*Gammarus* sp.	TR	1.45				44,49
C, Endothal	*Gammarus lacustris*	LC_{50}	2.0				585
C, Trifluralin	*Gammarus lacustris*	LC_{50}	8.80				585
C, Endothal	*Gammarus lacustris*	LC_{50}	100.0				585
C, Coumaphos	*Gammarus lacustris*	LC_{50}		0.00014			592
C, Dichlorvos	*Gammarus lacustris*	LC_{50}		0.0001			510
C, Azinophos-methyl	*Gammarus lacustris*	LC_{50}		0.0003			510
C, Malathion	*Gammarus lacustris*	LC_{50}		0.0018			510
C, Endrin	*Gammarus lacustris*	LC_{50}		0.003			66
C, Parathion	*Gammarus lacustris*	LC_{50}		0.006			510
C, Pyrethrin	*Gammarus fasciatus*	LC_{50}			0.011		45
C, Baytex	*Gammarus pulex*	LC_{50}		0.014			90
C, Pyrethrum	*Gammarus lacustris* (15.5°C)	LC_{50}		0.018			510
C, Allethrin	*Gammarus lacustris* (15.5°C)	LC_{50}		0.020			510
C, Toxaphene	*Gammarus fasciatus*	LC_{50}		0.022			515
C, Carbophenothion	*Gammarus lacustris*	LC_{50}		0.028			593
C, Carbophenothion	*Gammarus lacustris*	LC_{50}	0.045	0.028			585
C, Lindane	*Gammarus pulex*	LC_{50}		0.03			89
C, Lindane	*Gammarus pseudolimnaeus* (S)	LC_{50}		0.039			529
C, Tetraethyl pyrophosphate (TEPP)	*Gammarus lacustris*	LC_{50}		0.052			510
C, Trichlorfon	*Gammarus lacustris*	LC_{50}		0.06			510
C, Toxaphene	*Gammarus lacustris*	LC_{50}		0.07			515
C, Chlordane	*Gammarus lacustris*	LC_{50}		0.080			594

Chemical	Species	Toxicity test	24 h	48 h	96 h	Chronic exposure	Ref.
C, Diazinon	*Gammarus lacustris*	LC_{50}		0.500			510
C, Dioxathion	*Gammarus lacustris* (softW)	LC_{50}		0.690			510
C, Atrazine	*Gammarus fasciatus* (S)	LC_{10}		2.4			535
C, Trifluralin	*Gammarus lacustris*	LC_{50}		5.60			510
C, Atrazine	*Gammarus fasciatus* (S)	LC_{50}		5.7			535
C, Propham	*Gammarus fasciatus* (S, 15°C)	EC_{50}		10.0–34.0			532
C, Quinoline	*Gammarus minus* (adult, S, 24°C, pH 7.8)	LC_{50}		40.9			595
C, Picloram	*Gammarus lacustris* (S, 25°C, pH 7.5)	LC_{50}		50.0			585
C, Trifene	*Gammarus fasciatus*	NOAE		100.0			43
C, Simazine	*Gammarus fasciatus*	NOAE		100.0			43
C, Diphenamid	*Gammarus fasciatus*	NOAE		100.0			403
C, Dicamba	*Gammarus fasciatus*	NOEL		100.0			43
C, 2,4-D (diethylamine) salt	*Gammarus fasciatus*	NOEL		100.0			43
C, *n*-Propanol	*Gammarus pulex* (S, 20°C, pH 8.4)	LC_{50}		1000.0			401
C, Coumaphos	*Gammarus fasciatus* (S, 21°C, pH 7.5)	LC_{50}			0.000074		596
C, Co-Ral	*Gammarus lacustris*	LC_{50}			0.00007		66
C, Fenvalerate	*Gammarus pseudolimnaeus* (F, 15°C, pH 8)	LC_{50}	0.00013		0.00003		575
C, Azinphos-methyl	*Gammarus fasciatus*	LC_{50}			0.00010		45
C, Co-Ral	*Gammarus fasciatus*	LC_{50}			0.00015		45
C, Chloropyrifos	*Gammarus lacustris*	LC_{50}			0.00011		66
C, Azinphos-methyl	*Gammarus lacustris*	LC_{50}			0.00015		66
C, Chlorpyrifos	*Gammarus lacustris* (21°C)	LC_{50}	0.00075	0.0004	0.00011		583
C, Anilazine	*Gammarus fasciatus* (15°C)	LC_{50}			0.00027		597
C, Chloropyrifos	*Gammarus fasciatus*	LC_{50}			0.00032		45
C, 2,2-Dichlorovinyl-dimethyl phosphate	*Gammarus fasciatus*	LC_{50}			0.00040		45
C, Parathion	*Gammarus fasciatus* (F, 22°C, 7.4, wellW)	LC_{50}	0.0021	0.00062	0.00043		472
C, Dichlorvos	*Gammarus lacustris*	LC_{50}			0.0005		585
C, Phorate	*Gammarus fasciatus*	LC_{50}			0.0006		598

C, DDD	*Gammarus lacustris*	LC_{50}			0.00064		66
C, Malathion	*Gammarus fasciatus*	LC_{50}			0.00076	0.0005 (120 h)	45
C, DDD	*Gammarus fasciatus*	LC_{50}			0.00086		45
C, DDT	*Gammarus fasciatus*	LC_{50}			0.0008		45
C, Methoxychlor	*Gammarus lacustris* (S, 21°C, pH 7.5)	LC_{50}			0.0008		586
C, Malathion	*Gammarus lacustris*	LC_{50}			0.001		45
C, DDT	*Gammarus lacustris*	LC_{50}			0.001		66
C, Parathion	*Gammarus fasciatus* (S, 21°C, pH 7.5, softW)	LC_{50}			0.0013		511
C, Parathion	*Gammarus fasciatus* (S, 21°C, pH 7.1, FW)	LC_{50}	0.0032		0.0013		405
C, Ethion	*Gammarus lacustris*	LC_{50}			0.0018		66
C, Phosmet	*Gammarus fasciatus* (S, 15°C, pH 7.5)	LC_{50}			0.002		409
C, Parathion	*Gammarus fasciatus*	LC_{50}			0.0021	0.0016 (120 h)	45
C, Parathion	*Gammarus fasciatus* (S, 21°C, pH 7.4, wellW)	LC_{50}	0.006	0.004	0.0021		405
C, Phosphamidon	*Gammarus lacustris*	LC_{50}			0.0028		66
C, Parathion	*Gammarus lacustris*	LC_{50}			0.0035		66
C, Parathion	*Gammarus lacustris* (S, 21°C, pH 7.1)	LC_{50}	0.012	0.006	0.0035		601
C, Methyl parathion	*Gammarus fasciatus* (15°C, pH 7.2–7.5)	LC_{50}			0.0038		511
C, Phorate	*Gammarus fasciatus*(S, 15°C, pH 7.5)	LC_{50}			0.004		599
C, Carbophenothion	*Gammarus lacustris*	LC_{50}			0.0052		582
C, Carbophenothion	*Gammarus lacustris*	LC_{50}			0.0052		66
C, Endosulfan	*Gammarus fasciatus*	LC_{50}			0.0058		66
C, Endosulfan	*Gammarus lacustris* (S, 25°C, pH 7.5)	LC_{50}			0.0058		600
C, Toxaphene	*Gammarus fasciatus*	LC_{50}			0.006		45
C, Endosulfan	*Gammarus fasciatus*	LC_{50}			0.006		585
C, Ethyl-*p*-nitrophenylthionobenzene phosphate (EPN)	*Gammarus fasciatus* (S, 15°C)	LC_{50}			0.0068		602
C, Ethyl-*p*-nitrophenylthionobenzene phosphate (EPN)	*Gammarus fasciatus*	LC_{50}			0.007		45
C, Allethrin	*Gammarus fasciatus*	LC_{50}			0.008		45
C, Allethrin	*Gammarus fasciatus*	LC_{50}			0.008		518
C, Baytex	*Gammarus lacustris*	LC_{50}			0.0084		66

Chemical	Species	Toxicity test	24 h	48 h	96 h	Chronic exposure	Ref.
C, Dioxathion	*Gammarus fasciatus*	LC_{50}			0.0086		45
C, Dioxathion	*Gammarus fasciatus*	LC_{50}			0.0086		603
C, Phorate	*Gammarus lacustris*	LC_{50}			0.009		66
C, Phorate	*Gammarus lacustris* (S)	LC_{50}	0.024	0.014	0.009		585
C, Ethion	*Gammarus fasciatus*	LC_{50}			0.0094		45
C, Carbophenothion	*Gammarus fasciatus* (S, 16°C, pH 7.5)	LC_{50}			0.01–0.10		604
C, Aroclor 1242	*Gammarus fasciatus* (F)	LC_{50}			0.01		92
C, Lindane	*Gammarus pseudolimnaeus* (S)	LC_{50}			0.01		529
C, Aroclor 1242	*Gammarus pseudolimnaeus* (F, 15°C, pH 7.1)	LC_{50}			0.010		580
C, Ciodrin	*Gammarus fasciatus*	LC_{50}			0.011		66
C, Pyrethrum	*Gammarus fasciatus*	LC_{50}			0.011		518
C, Allethrin	*Gammarus lacustris*	LC_{50}			0.011		66
C, Allethrin	*Gammarus lacustris*	LC_{50}			0.011		518
C, Pyrethrin	*Gammarus lacustris*	LC_{50}			0.012		45
C, Aminocarb	*Gammarus lacustris*	LC_{50}			0.012		66
C, Pyrethrum	*Gammarus lacustris* (15.5°C)	LC_{50}			0.012		518
C, Aminocarb	*Gammarus fasciatus* (S, 21°C, pH 7.2–7.5)	LC_{50}			0.012		475
C, Naled	*Gammarus fasciatus*	LC_{50}			0.014		45
C, Ciodrin	*Gammarus lacustris*	LC_{50}			0.015		66
C, Ethyl-*p*-nitrophenylthiono benzene phosphate (EPN)	*Gammarus lacustris*	LC_{50}			0.015		66
C, Phosphamidon	*Gammarus fasciatus*	LC_{50}			0.016		45
C, Carbaryl	*Gammarus lacustris*	NOEL			0.016		66
C, Disulfoton	*Gammarus fasciatus*	LC_{50}			0.021		45
C, Chlordane	*Gammarus lacustris*	LC_{50}			0.026		66
C, Toxaphene	*Gammarus lacustris*	LC_{50}			0.026		66
C, Carbaryl	*Gammarus fasciatus*	NOEL			0.026		45
C, Demeton	*Gammarus fasciatus*	LC_{50}			0.027		45
C, Picloram	*Gammarus fasciatus* (S, 25°C, pH 7.5)	LC_{50}			0.027		411

C, Heptachlor	*Gammarus lacustris*	LC_{50}			0.029	66
C, Aminocarb	*Gammarus pseudolimnaeus* (S, 17°C, pH 7.4)	LC_{50}			0.029	517
C. Baygon	*Gammarus lacustris*	LC_{50}			0.034	66
C, Propoxur	*Gammarus lacustris* (S, 21°C, pH 7.5)	LC_{50}			0.034	606
C, Propoxur	*Gammarus lacustris* (S, 21°C, pH 7.5)	LC_{50}	0.066	0.050	0.034	582
C, Zectran	*Gammarus fasciatus*	LC_{50}			0.04	45
C, Heptachlor	*Gammarus fasciatus*	LC_{50}			0.040	45
C, Chlordane	*Gammarus fasciatus*	LC_{50}			0.040	45
C, Chlorfos	*Gammarus lacustris*	LC_{50}			0.040	66
C, Zectran	*Gammarus lacustris*	LC_{50}			0.046	66
C, Lindane	*Gammarus lacustris*	LC_{50}			0.048	66
C, Propoxur	*Gammarus fasciatus*	LC_{50}			0.050	605
C, Baygon	*Gammarus fasciatus*	LC_{50}			0.050	45
C, 2,2-Dichlorovinyldimethyl phosphate (DDVP)	*Gammarus lacustris*	LC_{50}			0.050	66
C, Aminocarb	*Gammarus pseudolimnaeus* (S, 10°C, pH 7.5)	LC_{50}			0.050	475
C, Disulfoton	*Gammarus lacustris*	LC_{50}			0.052	66
C, Aroclor 1248	*Gammarus fasciatus* (S)	LC_{50}			0.052	92
C, Aramite	*Gammarus fasciatus* (S, 21°C, pH 7.5)	LC_{50}			0.060	609
C, Aramite	*Gammarus lacustris* (S)	LC_{50}	0.350	0.100	0.060	582
C, 2,3,4-Trichlorobiphenyl	*Gammarus fasciatus* (S)	LC_{50}			0.07	92
C, Disulfoton	*Gammarus lacustris*	LC_{50}	0.110		0.070	585
C, Abate	*Gammarus lacustris*	LC_{50}			0.082	66
C, Dichlone	*Gammarus fasciatus*	LC_{50}			0.100	43
C, 4,4′-Dichlorobiphenyl	*Gammarus fasciatus* (S)	LC_{50}			0.1	92
C, *S,S,S*-Tributylphosphorothioate	*Gammarus lacustris*	LC_{50}			0.10	66
C, Baytex	*Gammarus fasciatus*	LC_{50}			0.110	45
C, Naled	*Gammarus lacustris*	LC_{50}			0.11	66
C, Chlorpyrifos	*Gammarus lacustris*	LC_{50}			0.110	403a
C, Tetradifon	*Gammarus fasciatus* (S)	LC_{50}	0.370	0.140	0.110	585
C, Tetradifon	*Gammarus fasciatus* (S, 21°C, pH 7.5)	LC_{50}			0.111	607

Chemical	Species	Toxicity test	24 h	48 h	96 h	Chronic exposure	Ref.
C, 2,4-Dichlorobiphenyl	*Gammarus fasciatus* (S)	LC_{50}			0.12		92
C, Calcium chloride	*Gammarus pulex* (intermolt, 10°C)	LC_{50}		0.68	0.12		608
C, Mevinphos	*Gammarus lacustris*	LC_{50}			0.130		585
C, Aminocarb	*Gammarus pseudolimnaeus* (S, 17°C, pH 7.4)	LC_{50}			0.145		610
C, Methyl mercuric chloride	*Gammarus duebeni* (S, 14–16°C)	LC_{50}			0.150		611
C, Diuron	*Gammarus lacustris*	LC_{50}			0.160		534
C, Diuron	*Gammarus lacustris*	LC_{50}			0.160		66
C, Hydrocyanic acid	*Gammarus pseudolimnaeus* (F, 18°C)	LC_{50}			0.169		398
C, Hydrogen cyanide	*Gammarus pseudolimnaeus*	LC_{50}			0.17		47
C, Diazinon	*Gammarus lacustris*	LC_{50}			0.200		66
C, 2,4,5-Trichlorphenol	*Gammarus fasciatus* (S)	LC_{50}			0.21		92
C, Fenoprop, butoxyethyl ester	*Gammarus fasciatus*	LC_{50}			0.250		43
C, Dioxathion	*Gammarus lacustris*	LC_{50}			0.270		603
C, Dioxathion	*Gammarus lacustris*	LC_{50}			0.27		66
C, 2,4-D butoxyethyl ester	*Gammarus lacustris*	LC_{50}			0.440		66
C, Dieldrin	*Gammarus lacustris*	LC_{50}			0.460		66
C, Dieldrin	*Gammarus fasciatus*	LC_{50}			0.600		45
C, Ethyl parathion	*Gammarus lacustris* (21°C)	LC_{50}			0.60		585
C, Diuron	*Gammarus fasciatus*	LC_{50}			0.700		43
C, Diuron	*Gammarus fasciatus*	LC_{50}			0.70		534
C, Isopropyldiphenyl phosphate	*Gammarus pseudolimnaeus* (S, 20°C, pH 7.4)	LC_{50}	3.1		0.7		614
C, Difolitan	*Gammarus lacustris*	LC_{50}			0.8		66
C, Fenoprop, propylene glycol butyl ethyl ester (PGBEE)	*Gammarus fasciatus*	LC_{50}			0.84		43
C, Hydrogen sulfide	*Gammarus*	TL_m			0.84		46
C, Sodium arsenite	*Gammarus pseudolimnaeus* (F, 19°C, pH 8.1)	EC_{50}			0.874	1.020 (64 h)	477
C, Trifluralin	*Gammarus fasciatus*	LC_{50}			1.0		43
C, Mirex	*Gammarus pseudolimnaeus* (S, 17°C, pH 7.5)	LC_{50}			1.0		521
C, Methomyl	*Gammarus pseudolimnaeus* (S)	LC_{50}			1.050		513

C, Balan	*Gammarus fasciatus*	LC_{50}		1.1	43
C, Dichlone	*Gammarus lacustris*	LC_{50}		1.1	66
C, Sodium arsenite	*Gammarus pseudolimnaeus* (F, 19°C, pH 8.1)	EC_{100}		1.340	477
C, Dinoseb	*Gammarus fasciatus*	LC_{50}		1.8	43
C, Vernam	*Gammarus lacustris*	LC_{50}		1.8	66
C, Malathion	*Gammarus lacustris* (21°C)	LC_{50}		1.80	585
C, Altosid-SR10	*Gammarus aequieaudu*, adult male	LC_{50}		1.95	612
C, Altosid-SR10	*Gammarus aequieaudu*, adult male	LC_{50}		2.15	612
C, Trifluralin	*Gammarus lacustris*	LC_{50}		2.2	66
C, Aroclor 1254	*Gammarus fasciatus* (S)	LC_{50}		2.4	613
C, 2,4-D, isooctyl ester	*Gammarus lacustris*	LC_{50}		2.4	66
C, Aroclor 1254	*Gammarus pseudolimnaeus* (S, 15°C, pH 7.1)	LC_{50}		2.40	580
C, Rotenone	*Gammarus lacustris*	LC_{50}		2.6	66
C, Nitrofen	*Gammarus fasciatus* (S, 15°C, pH 7.4)	LC_{50}	4.0	3.10	404
C, Dexon	*Gammarus lacustris*	LC_{50}		3.7	66
C, Dicamba	*Gammarus lacustris*	LC_{50}		3.9	66
C, Dicamba	*Gammarus lacustris*	LC_{50}		3.90	585
C, 2,4-D, butoxy ethyl ester	*Gammarus fasciatus*	LC_{50}		5.9	43
C, Phosphamidon	*Gammarus lacustris* (21°C)	LC_{50}		8.8	585
C, Dichlobenil	*Gammarus fasciatus*	LC_{50}		10.0	43
C, Pebulate	*Gammarus fasciatus*	LC_{50}		10.0	43
C, Pebulate	*Gammarus fasciatus*	LC_{50}		10.0	470
C, Isopropyl-*N*-phenyl carbamate	*Gammarus lacustris*	LC_{50}		10.0	66
C, Paraquat	*Gammarus lacustris*	LC_{50}		11.0	66
C, Dichlobenil	*Gammarus lacustris*	LC_{50}		11.0	66
C, Trifene	*Gammarus lacustris*	LC_{50}		12.0	66
C, Simazine	*Gammarus lacustris* (21°C)	LC_{50}		13.0	588
C, Simazine	*Gammarus lacustris*	LC_{50}		13.0	66
C, Vernam	*Gammarus fasciatus*	LC_{50}		13.0	43
C, Ethyl-*p*-nitrophenylthiono benzene phosphate (EPN)	*Gammarus lacustris* (21°C)	LC_{50}		15.0	615

Chemical	Species	Toxicity test	24 h	48 h	96 h	Chronic exposure	Ref.
C, Propanil	*Gammarus fasciatus*	LC_{50}			16.0		43
C, Propanil	*Gammarus fasciatus*	LC_{50}			16.0		585
C, Isopropyl-*N*-phenyl carbamate	*Gammarus fasciatus*	LC_{50}			19.0		43
C, Eptam	*Gammarus fasciatus*	LC_{50}			23.0		43
C, Picloram	*Gammarus lacustris*	LC_{50}			27.0		66
C, Trichlorfon	*Gammarus lacustris* (21°C)	LC_{50}			40.0		585
C, Aniline	*Gammarus fasciatus* (S, 21°C, pH 6.5–7.8)	LC_{50}			100.0		408
C, Dicamba	*Gammarus fasciatus* (S, 15°C, pH 7.5)	LC_{50}			100.0		402
C, Diphenamid	*Gammarus fasciatus* (S, 15°C, pH 7.5)	LC_{50}			100.0		403b
C, Diazinon	*Gammarus lacustris* (21°C)	LC_{50}			200.0		585
C, *n*-Nitrosodimethylamine	*Gammarus limnaeus*	LC_{50}			330.0		567
C, *n*-Nitrosodimethylamine	*Gammarus limnaeus* (20°C, pH 8.2)	LC_{50}			500.0		567
C, Aldrin	*Gammarus fasciatus*	LC_{50}			4300.0		43
C, Aldrin	*Gammarus lacustris*	LC_{50}			9800.0		66
C, 1,2-Dichloroethane	*Globius minutus* (F, 15°C, pH 8)	LC_{50}			185.0		464
C, Ethylene dichloride	*Globius minutus* (seaW, 15°C)	LC_{50}			185.0		59
C, Bis(2-ethylhexyl)phthalate	Harpacticoid copepod (S, 20–22°C)	LC_{50}			1000.0		616
C, Carbandazim	Harpacticoid copepod (20–22°C)	LC_{50}			0.3–0.5		617
C, Carbaryl	*Hemigiapsis oregonensis* (female)	EC_{50}	0.270				407
C, Aldrin	*Pegurus longicarpus* (S)	LC_{50}			0.033		60
C, Rotenone	Hirudiidae	LC_{100}			0.100		557
C, Toxaphene	*Hollandale* sp. (prior history of exposure)	LC_{50}	0.229				515
C, Tributyltin oxide	*Homarus americanus* (larva, 0.5S, 20°C)	LC_0				0.001 (24 d)	445
C, Mercuric chloride	*Homarus americanus* (F, 10–20°C)	NOAE				0.006 (30 d)	618
C, Dinoseb	*Homarus americanus* (larvae, S)	LC_{100}				0.0075	97
C, LaRosa crude, water-soluble fr.	*Homarus americanus* (S, 22°C)	NOAE				0.05 ml/l (5 d)	619
C, Cupric sulfate	*Homarus americanus* (F, 5–13°C, pH 8)	LC_{50}				0.056 (13 d)	620
C, Cobalt chloride	*Homarus vulgaris* (larvae, 1 stage, S, 22°C, pH 7.8)	LC_{50}				0.227 (216 h)	446

C, Dinoseb	*Homarus americanus* (S)	LC_{100}				0.30	97
C, Cobalt chloride	*Homarus vulgaris* (larvae II stage, S, 22°C, pH 7.8)	LC_{50}				0.454 (216 h)	446
C, Cobalt chloride	*Homarus vulgaris* (larvae III stage, S, 22°C, ph 7.8)	LC_{50}				0.454 (216 h)	446
C, Isobutyl alcohol	*Homarus americanus*	LC_{50}				2.0–3.0 ml/kg (h?)	621
C, Nitrilotriacetic acid,Na,H_2O	*Homarus americanus* (1st larval stage, S)	LC_{100}				100.0 (168 h)	37
C, Nitrilotriacetic acid,Na,H_2O	*Homarus americanus* (S)	TL_{50}				3150.0 (168 h)	37
C, Tributyltin oxide	*Homarus americanus* (larva, 0.5S, 20°C)	LC_{100}	0.020			0.005 (6 d)	445
C, Cupric sulfate	*Homarus americanus* (S, 15°C, larva, 1-3 d old)	LC_{50}		0.1–0.33			438
C, Decamethrin	*Homarus americanus* (10°C)	LC_{50}			0.0000014		623
C, Permethrin(*cis*)	*Homarus americanus* (10°C, solution renewed every 48 h)	LC_{50}			0.0004		623
C, Fenitrothion	*Homarus americanus*	LC_{50}			0.001		620
C, Permethrin (*cis, trans*)	*Homarus americanus* (10°C, solution renewed every 48 h)	LC_{50}			0.007		623
C, Venezuelan crude oil	*Homarus americanus* (S, 20–25°C)	LC_{50}			0.02 ml/l		622
C, Dinoseb	*Homarus americanus* (6–10°C)	LC_{50}			0.30	1.00 (1 h)	624
C, Nitrilotriacetic acid	*Homarus americanus* (1 larval stage, S, 15°C)	LC_{100}				100.0 (7 d)	626
C, Nitrilotriacetic acid,Na,H_2O	*Homarus americanus* (S)	TL_{50}			3800.0		37
C, Diquat dibromide	*Hyalella azteca*	LC_{50}	0.58	0.12	0.048		627
C, Calcium chloride	*Hyalella azteca* (23°C)	LC_{50}			0.085		471
C, Chlordane	*Hyalella azteca*	LC_{50}			0.097		553
C, Mirex	*Hyalella azteca* (S, 23–25°C)	LC_{54}				0.001 (600 h)	628
C, Diquat	*Hyalella azteca*	LC_{50}			0.048		98
C, Dichlobenil	*Hyalella azteca*	LC_{50}			8.5		98
I, DDT	*Ischnura verticalis* (S, 15.6°C, pH 7.1)	LC_{50}			0.056		580
C, Aroclor 1254	*Lagodon rhomboides* (F, seaW)	LC_{50}				0.005 (12 d)	613
C, Swedish EDC tar	*Leander adspersus*	LC_{50}		8.0			572
C, Phenyl mercuric chloride	*Limnoria tripunctata* (S, pH 6.8–7.8)	LC_{50}				1.5 (100 h)	629
C, *p*-Dichlorobenzene	*Limnoria tripunctata* (S, pH 6.8–7.8)	LC_{50}				48.4 (100 h)	630

Chemical	Species	Toxicity test	24 h	48 h	96 h	Chronic exposure	Ref.
C, Acrylonitrile	*Limnoria tripunctata* (S, pH 6.8–7.8)	LC_{50}				50.0 (100 h)	631
C, Benzidine	*Limnoria tripunctata* (S, pH 6.8–7.8)	LC_{50}				50.0 (100 h)	632
C, *p*-Phenylenediamine	*Limnoria tripunctata* (S, pH 6.8–7.8)	LC_{50}				50.0 (100 h)	633
C, *o*-Phenylenediamine	*Limnoria tripunctata* (S, pH 6.8–7.8)	LC_{50}				100.0 (100 h)	633
C, Anthracene	*Limnoria tripunctata* (S)	LC_{50}				100.0 (100 h)	634
C, Fluoranthene	*Limnoria tripunctata* (S, pH 6.8–7.8)	LC_{50}				100.0 (100 h)	635
C, *n*-Phenylenediamine, HCl	*Limnoria tripunctata* (S, pH 6.8–7.8)	LC_{50}				7100.0 (100 h)	633
C, Ethyl acetate	*Limnoria tripunctata* (S, pH 7.8)	LC_{50}				10000.0 (100 h)	636
C, Fenitrothion	*Macrobrachium mipponensis*	LC_{50}		0.002			637
C, Monocrotophos	*Macrobrachium lamerii* (1/2S, 28°C, pH 7.7)	LC_{50}	2.11	1.84	1.59	1.69 (72 h)	638
C, Monocrotophos	*Macrobrachium lamerii* (S, 24–26°C)	LC_{50}	5.37	4.46	1.91	3.11 (72 h)	639
C, Cobalt chloride	*Maia squinado*, (S, 15°C, pH 7.8, I zoea)	LC_{50}			4.54		446
C, 1,2-Dibromo-3-chloropropane	*Mercenaria mercenaria* (larvae, S)	TL_m				0.78 (12 d)	64
C, Griseofulvin	*Mercenaria mercenaria* (larvae)	TL_m				1.0	64
C, Nabam	*Mercenaria mercenaria* (S, larvae, 12 d old)	TL_m				1.75	30
C, Dicapthon	*Mercenaria mercenaria* (larva, S)	TL_m				5.74 (12 d)	64
C, Lindane	*Mercenaria mercenaria* (larvae, S)	TL_m				10.0 (12 d)	64
C, *o*-Dichlorobenzene	*Mercenaria mercenaria* (larvae)	LC_{50}				100.0 (288 h)	96
C, Griseofulvin	*Mercenaria mercenaria* (egg)	TL_m		0.25			64
C, Nabam	*Mercenaria mercenaria* (S, egg)	TL_m		0.50			30
C, Dicapthon	*Mercenaria mercenaria* (egg, S)	TL_m		3.34			64
C, Lindane	*Mercenaria mercenaria* (egg, S)	TL_m		10.0			64
C, 1,2-Dibromo-3-chloropropane	*Mercenaria mercenaria* (egg, S)	TL_m		10.0			64
C, *o*-Dichlorobenzene	*Mercenaria mercenaria* (egg)	LC_{50}		100.0			96
C, *p*-Chlorophenol	*Mesidotea entomon* (F, 5°C, pH 7.7)	LC_{50}			59.7		640
C, Abate	*Metapenaeus monoceros* (S)	LC_{50}				0.045 (72 h)	639
C, Phosphate (effluent)	*Munida gregaria* (semi S, 19°C, pH 3.45)	LC_{100}				100.0 (1 h)	641
C, Sodium fluoride	*Munida gregaria* (semi S, 19°C, pH 8)	LC_{50}				100.0 (259 h)	641
I, Hydrogen sulfide	*Musca domestica* (inhalation)	LC_{50}				380 mg/m^3 (960 min)	46

I, Hydrogen sulfide	*Musca domestica* (inhalation)	LC_{50}			1500 mg/m^3 (7 min)	46
C, Sodium arsenite	*Mysidopsis bahia* (<24 h, lifecycle test, F, 23°C, pH 8.2)	EC_{L0}			0.893 (36 d)	642
C, Kepone	*Mysidopsis bahia* (juv., IF, 25–28°C)	LC_{50}			0.0014 (19 d)	643
C, Mercuric chloride	*Mysidopsis bahia*	NOAEL			0.00082–0.00165	644
C, Aldicarb	*Mysidopsis bahia* (22°C, juv.)	LC_{35}			0.0015 (14 d)	652
C, Fluoranthene	*Mysidopsis bahia* (S, life cycle exposure)	NOAE			0.016	645
C, Cyanide	*Mysidopsis bahia* (24 h, F, 24°C, pH 7.8–8.2)	EC_{L0}			0.070 (29 d)	642
C, Dichromic acid, K_2	*Mysidopsis bahia* (S, life cycle exp.)	NOAE			0.132	653
C, Octachloronaphthalene	*Mysidopsis bahia*	LC_{50}			500.0	646
C, No. 2 fuel oil,water-soluble fraction	*Mysidopsis almyra* (S, 18–22°C)	LC_{50}	0.9			656
C, Butylbenzyl phthalate	*Mysidopsis bahia* (S)	LC_{50}	0.9		0.4 (NOAE)	9
C, No. 2 fuel oil	*Mysidopsis almyra* (S, 18–22°C)	LC_{50}	1.3			656
C, Kuwait crude, water-soluble fraction	*Mysidopsis almyra* (S, 18–22°C)	LC_{50}	6.6			656
C, Kuwait crude oil	*Mysidopsis almyra* (S)	LC_{50}	18.0			657
C, Chlorpyrifos	*Mysidopsis bahia* (F, juv. 27°C)	LC_{50}		0.000035		658
C, Chlorpyrifos	*Mysidopsis bahia* (S, juv. 27°C)	LC_{50}		0.000056		654
C, Diflubenzuron	*Mysidopsis bahia*	LC_{50}		0.0021	0.00124 (21 d)	100
C, Lindane	*Mysidopsis bahia* (F)	LC_{50}		0.00628		101
C, Kepone	*Mysidopsis bahia* (juv., F, 25–28°C)	LC_{50}		0.010		643
C, Kepone	*Mysidopsis bahia*	LC_{50}		0.0101	0.0014 (19 d)	422
C, Aldicarb	*Mysidopsis bahia* (S, 28°C, pH 6.8–7.4)	LC_{50}		0.013		647
C, Aldicarb	*Mysidopsis bahia* (F, 22°C, pH 8.4)	LC_{50}		0.016		652
C, Aldicarb	*Mysidopsis bahia* (F, 23°C, pH 8.1–8.4)	LC_{50}		0.016		647
C, Fluoranthene	*Mysidopsis bahia* (S)	LC_{50}		0.040		649
C, Cyanide	*Mysidopsis bahia* (24 h, F, 25°C, pH 8.2)	LC_{50}		0.113		642
C, Selenium	*Mysidopsis bahia* (S)	NOAEL		0.143		648
C, Triphenyl phosphate	*Mysidopsis bahia* (S)	LC_{50}		0.18		533
C, Butylbenzyl phthalate	*Mysidopsis bahia* (S, 20°C)	LC_0		0.4		655

Chemical	Species	Toxicity test	24 h	48 h	96 h	Chronic exposure	Ref.
C, 1,2,4-Trichlorobenzene	*Mysidopsis bahia* (S)	LC_{50}			0.45		650
C, Nickel (2+)	*Mysidopsis bahia* (F, 21°C)	LC_{50}			0.508		659
C, Selenium	*Mysidopsis bahia* (S)	LC_{50}			0.60		648
C, 1,3-Dichloropropene	*Mysidopsis bahia* (S)	LC_{50}			0.790		651
C, Butylbenzylphthalate	*Mysidopsis bahia* (S, 20°C)	LC_{50}			0.90		655
C, Arsenic	*Mysidopsis bahia* (24 h old, F, 25°C, pH 8.2)	LC_{50}			1.74		642
C, *p*-Dichlorobenzene	*Mysidopsis bahia* (S)	LC_{50}			1.99		660
C, Dichromicacid, DiK	*Mysidopsis bahia* (S)	LC_{50}			2.0		653
C, Diflubenzuron	*Mysidopsis bahia*	LC_{50}			2.10		1078
C, 2,4,5-Trichlorophenol	*Mysidopsis bahia* (S)	LC_{50}			3.83		661
C, 2,4-Dinitrophenol	*Mysidopsis bahia* (S)	LC_{50}			4.85		662
C, *p*-Nitrophenol	*Mysidopsis bahia* (S)	LC_{50}			7.17		662
C, Tetrachloroethylene	*Mysidopsis bahia* (S)	LC_{50}			10.20		663
C, Isophorone	*Mysidopsis bahia* (S)	LC_{50}			12.9		539
C, Chlorobenzene	*Mysidopsis bahia* (S)	LC_{50}			16.40		650
C, Picric acid	*Mysidopsis bahia* (S)	LC_{50}			19.70		662
C, 2,3,5,6-Tetrachlorophenol	*Mysidopsis bahia* (S)	LC_{50}			21.9		661
C, *p*-Chlorophenol	*Mysidopsis bahia* (S)	LC_{50}			29.7		661
C, Toluene	*Mysidopsis bahia*	LC_{50}			56.3		550
C, Sodium arsenite	*Nitrocra spinipes* (adult, fecundity test, F, 22°C	EC_{50}				2.0 (13 d)	665
C, Toluene	*Nitrocra spinipes*	LC_{50}	24.2–74.2				664
C, Tributyltin oxide	*Nitrocra spinipes* (S, 21°C, pH 7.8)	LC_{50}			0.002		666
C, Tributyltin fluoride	*Nitrocra spinipes* (S, 21°C, pH 7.8)	LC_{50}			0.002		666
C, Mercuric chloride	*Nitrocra spinipes* (S, 12.2°C, pH 8)	LC_{50}			0.23		670
C, Pentachlorophenol	*Nitrocra spinipes* (S, 21°C, pH 7.8)	LC_{50}			0.27		666
C, Dichlobenil	*Nitrocra spinipes* (S, 21°C, pH 7.8)	LC_{50}			0.270		666
C, Zinc chloride	*Nitrocra spinipes* (S, 20°C, pH 8.0)	LC_{50}			1.45		670
C, Dibutylphthalate	*Nitrocra spinipes* (S, 20–22°C)	LC_{50}			1.7		666

C, Trixylenyl phosphate	*Nitrocra spinipes* (S, pH 7.8)	LC_{50}	1.9		616
C, 1,2,4-Trichlorobenzene	*Nitrocra spinipes* (S, 20–22°C)	LC_{50}	2.60		616
C, Disodium arsenate	*Nitrocra spinipes*	LC_{50}	3.0		665
C, Sodium arsenite	*Nitrocra spinipes* (adult, S, 22°C)	LC_{50}	3.20		672
C, Cobalt chloride	*Nitrocra spinipes* (S, 20°C, pH 8)	LC_{50}	4.5		670
C, Disodium arsenate	*Nitrocra spinipes* (adult, S, 22°C)	LC_{50}	4.99		672
C, *p*-Chlorophenol	*Nitrocra spinipes* (S, 22°C, pH 7.8)	LC_{50}	21.0		666
C, Dimethyl phthalate	*Nitrocra spinipes* (S, 10°C)	LC_{50}	62.0		666
C, Dimethyl phthalate	*Nitrocra spinipes* (S, 20–22°C)	LC_{50}	74.0		616
C, Chlormequat chloride	*Nitrocra spinipes* (S, adult, 22°C, pH 7.8)	LC_{50}	80.0		666
C, Mecoprop-K	*Nitrocra spinipes* (S, 21°C, pH 7.8)	LC_{50}	87.0		666
C, Motor fuels	*Nitrocra spinipes* (S, 20–22°C, pH 7.8)	LC_{50}	150.0–250.0		671
C, Triethyl phosphate	*Nitrocra spinipes* (S, 10°C, pH 7.8)	LC_{50}	950.0		669
C, Methyl-*t*-butyl ether	*Nitrocra spinipes* (S, 20–22°C, pH 7.8)	LC_{50}	1000.0		616
C, Trimethylpropane	*Nitrocra spinipes* (S, 10°C, pH 7.8)	LC_{50}	1000.0		616
C, Trimethylol propane	*Nitrocra spinipes*	LC_{50}	1000.0		616
C, *n*-Butylalcohol	*Nitrocra spinipes* (S, adult)	LC_{50}	1900.0		667
C, *n*-Propanol	*Nitrocra spinipes* (S, 21°C, pH 7.9)	LC_{50}	2300.0		668
C, Methyl-*t*-butyl ether	*Nitrocra spinipes* (S, 20–22°C, pH 7.8)	LC_{50}	10,000.0		671
C, Tributyltin fluoride	*Orchestiodea californiana* (S, 18°C)	LC_0		0.00050 (9 d)	673
C, Tributyltin fluoride	*Orchestiodea californiana* (S, 18°C)	LC_{100}		0.0150 (9 d)	673
C, Tributyltin oxide	*Orchestoidea californiana* (0.5S, 18°C)	LC_0		0.0005 (9 d)	673
C, Tributyltin oxide	*Orchestoidea californiana* (0.5S, 18°C)	LC_{47}		0.006 (9 d)	673
C, Tributyltin oxide	*Orchestoidea californiana* (0.5S, 18°C)	LC_{80}		0.010 (9 d)	673
C, Tributyltin fluoride	*Orchestoidea californiana* (S, 18°C)	LC_{87}		0.010 (9 d)	673
C, Tributyltin oxide	*Orchestoidea californiana* (0.5S, 18°C)	LC_{93}		0.015 (9 d)	673
C, Aroclor 1242	*Orconectes nais* (S, 15°C, pH 7.1)	LC_{50}		0.030 (7 d)	580
C, Aroclor 1242	*Orconectes nais*	LC_{50}		0.030 (7 d)	674
C, Aroclor 1242	*Orconectes nais* (S)	LC_{50}		0.03 (7 d)	92
C, Aroclor 1254	*Orconectes nais*	LC_{50}		0.080 (7 d)	674

Chemical	Species	Toxicity test	24 h	48 h	96 h	Chronic exposure	Ref.
C, Aroclor 1254	*Orconectes nais* (S, 15.6°C, pH 7.1)	LC_{50}				0.080 (7 d)	580
C, Aroclor 1254	*Orconectes nais* (S, 15°C, pH 7.1)	LC_{50}				0.100 (7 d)	580
C, Aroclor 1254	*Orconectes nais* (S)	LC_{50}				0.1 (7 d)	613
C, Dinoseb	*Orconectes rusticus* (8–12°C)	NOAE				1.0 (215 h)	624
C, Dinoseb	*Orconectes rusticus* (8-12°C)	NOAE				10.0 (144 h)	624
C, Ethyl parathion	*Orconectes rusticus*	LC_{100}		0.0001			675
C, Dichlone	*Orconectes nais*	LC_{50}		3.2			43
C, Dichlobenil	*Orconectes nais*	LC_{50}		22.0			43
C, Vernam	*Orconectes nais*	LC_{50}		24.0			43
C, Trifluralin	*Orconectes nais*	LC_{50}		50.0			43
C, Fenoprop, butoxyethyl ester	*Orconectes nais*	LC_{50}		60.0			43
C, Fenoprop	*Orconectes rusticus* (S, 15.5°C)	LC_{50}		60.0			469
C, Dicamba	*Orconectes nais*	LC_{50}		100.0			470
C, 2,4-D	*Orconectes nais* (S, 15.5°C)	LC_{50}		100.0			470
C, Simazine	*Orconectes nais*	NOAE		100.0			43
C, Fenoprop, propylene glycol butyl ether ester (PGBEE)	*Orconectes nais*	NOAE		100.0			43
C, Amitrol	*Orconectes nais*	NOAE		100.0			43
C, Diphenamid	*Orconectes nais*	NOAE		100.0			43
C, Diphenamid	*Orconectes nais*	NOAE		100.0			403a
C, 2,4-D(diethylamine)salt	*Orconectes nais*	NOEL		100.0			43
C, 2,4-D, butoxyethyl ester	*Orconectes nais*	NOEL		100.0			43
C, Dicamba	*Orconectes nais*	NOEL		100.0			43
C, Parathion	*Orconectes nais* (S, 21°C, pH 7.2–7.5, freshwater)	LC_{50}			0.00004		511
C, Parathion	*Orconectes nais*	LC_{50}			0.00004		45
C, Parathion	*Orconectes nais* (S, wellW, 21°C, pH 7.4, juv.)	LC_{50}	0.0001		0.000036		405
C, DDT	*Orconectes nais* (S, 15.6°C, pH 7.1)	LC_{50}			0.00024		676
C, DDT	*Orconectes nais*	LC_{50}			0.00024		45

C, Permethrin	*Orconectes* sp.	LC_{50}			0.00062		677
C, Endrin	*Orconectes nais*	LC_{50}			0.0032		45
C, Heptachlor	*Orconectes nais*	LC_{50}			0.0078		45
C, Carbaryl	*Orconectes nais*	NOEL			0.0086		45
C, Methyl parathion	*Orconectes nais* (15°C, pH 7.2–7.5)	LC_{50}			0.015		511
C, Parathion	*Orconectes nais* (S, wellW, 21°C, pH 7.4, adult)	LC_{50}	0.086		0.015		405
C, Phorate	*Orconectes nais*	LC_{50}			0.050		45
C, Baytex	*Orconectes nais*	LC_{50}			0.050		45
C, Phorate	*Orconectes nais*(S)	LC_{50}			0.050		598
C, Malathion	*Orconectes nais*	LC_{50}			0.180		45
C, Dieldrin	*Orconectes nais*	LC_{50}			0.740		45
C, Naled	*Orconectes nais*	LC_{50}			1.80		45
C, Leptophos	*Orconectes nais* (1 instar, 12°C, pH 7.5)	LC_{50}			7.0		678
C, Phosphamidon	*Orconectes nais*	LC_{50}			7.5		45
C, Methyl parathion	*Oziotelphusa senex senex* (S, 30°C, pH 7.3)	LC_{50}	3.0	1.0			679
C, Zinc chloride	*Pagurus longicarpus* (S, 20°C, pH 8)	NOEL				0.1 (168 h)	680
C, Zinc chloride	*Pagurus longicarpus* (S, 20°C, pH 8)	LC_{50}				0.2 (168 h)	680
C, Chromic acid, K_2	*Pagurus longicarpus* (S, 20°C, pH 8)	NOAE				1.0 (168 h)	680
C, Chromic acid, K_2	*Pagurus longicarpus* (S, 20°C, pH 8)	LC_{50}				2.7 (168 h)	680
C, Nitrilotriacetic acid, Na, H_2O	*Pagurus longicarpus* (S)	TL_{50}				1800.0 (168 h)	37
C, Dioxathion	*Pagurus longicarpus*	LC_{50}	0.30				453
C, Lindane	*Pagurus longicarpus* (S)	LC_{50}			0.005		60
C, DDT	*Pagurus longicarpus* (S)	LC_{50}			0.006		60
C, Dimethyl parathion	*Pagurus longicarpus* (S)	LC_{50}			0.007		38
C, Mercuric chloride	*Pagurus longicarpus* (S, 20°C, pH 8)	NOAE			0.01		680
C, Endrin	*Pagurus longicarpus* (S)	LC_{50}			0.012		60
C, Dieldrin	*Pagurus longicarpus* (S)	LC_{50}			0.018		60
C, 2,2-Dichlorovinyl-dimethyl phosphate)	*Pagurus longicarpus* (S)	LC_{50}			0.045		60
C, Mercuric chloride	*Pagurus longicarpus* (S, 20°C, pH 8)	LC_{50}			0.05		680

Chemical	Species	Toxicity test	24 h	48 h	96 h	Chronic exposure	Ref.
C, Heptachlor	*Pagurus longicarpus* (S)	LC_{50}			0.055		60
C, Dioxathion	*Pagurus longicarpus* (S)	LC_{50}			0.082		60
C, Malathion	*Pagurus longicarpus* (S)	LC_{50}			0.083		60
C, Zinc chloride	*Pagurus longicarpus* (S, 20°C, pH 8)	LC_{50}			0.40		680
C, Chromic acid, K_2	*Pagurus longicarpus* (20°C, pH 8, S)	NOAE			5.0		680
C, Chromic acid, K_2	*Pagurus longicarpus* (18–22°C, pH 8.2)	LC_{50}			10.0		680
C, Nitrilotriacetic acid,Na,H_2O	*Pagurus longicarpus* (S)	TL_{50}			5500.0		37
C, Dioxathion	*Palaemonetes pugio*	LC_{50}	0.50				453
C, Chlordane	*Palaemonetes pugio*	LC_{50}			0.0048		369
C, Baytex	*Palaemonetes macrodactylus* (F)	TL_{50}	0.003				102
C, Baytex	*Palaemonetes macrodactylus* (S)	TL_{50}	0.0053				69
C, Chlorpyrifos	*Palaemonetes macrodactylus* (IF, 13–18°C)	LC_{50}			0.00001		682
C, Chloropyrifos	*Palaemonetes macrodactylus* (F)	TL_{50}			0.00001		99
C, DDT (77%)	*Palaemonetes macrodactylus* (F)	TL_{50}			0.00017		99
C, Chloropyrifos	*Palaemonetes macrodactylus* (S)	TL_{50}			0.00025		99
C, Aldrin	*Palaemonetes macrodactylus* (S)	TL_{50}			0.00074		104
C, DDT (77%)	*Palaemonetes macrodactylus* (S)	TL_{50}			0.00086		103
C, DDD	*Palaemonetes macrodactylus* (F)	TL_{50}			0.0025		104
C, Aldrin	*Palaemonetes macrodactylus* (IF)	TL_{50}			0.003		104
C, Fenthion	*Palaemonetes macrodactylus* (S, 14°C)	LC_{50}			0.0053		683
C, Dieldrin	*Palaemonetes macrodactylus* (F)	LC_{50}			0.0069		105
C, Carbaryl	*Palaemonetes macrodactylus* (IF)	TL_{50}			0.007		104
C, DDD	*Palaemonetes macrodactylus* (S)	TL_{50}			0.0083		104
C, Lindane	*Palaemonetes macrodactylus* (F)	TL_{50}			0.0092		99
C, Carbaryl	*Palaemonetes macrodactylus*	TL_{50}			0.012		99
C, Lindane	*Palaemonetes macrodactylus* (S)	TL_{50}			0.0125		99
C, Dieldrin	*Palaemonetes macrodactylus* (S)	LC_{50}			0.0169		99
C, Toxaphene	*Palaemonetes macrodactylus* (S)	TL_{50}			0.0203		2
C, Abate	*Palaemonetes macrodactylus* (F)	TL_{50}			0.249		99

C, Pentachlorophenate, Na	*Palaemonetes pugio* (L. premolt)	LC_{50}	0.5	0.4	0.4	0.4 (72 h)	106
C, Abate	*Palaemonetes macrodactylus* (S)	TL_{50}			2.550		99
C, Pentachlorophenate, Na	*Palaemonetes pugio* (intermolt)	LC_{50}	4.2	3.5	2.6	3.30 (72 h)	106
C, Pentachlorophenate, Na	*Palaemonetes pugio* (E. premolt)	LC_{50}	5.9	3.6	2.7	3.1 (72 h)	106
C, Cobalt chloride	*Palaemonetes serratus* (lar, S, 15°C, pH 7.8)	LC_{50}			22.7		446
C, Cobalt chloride	*Palaemonetes serratus* (L, 15°C, S, pH 7.8)	LC_{50}			227.0	0.045 (216 h)	446
C, DDT(*p,p′*)	*Palaemonetes vulgaris* (S)	LC_{50}			0.002		60
C, Mirex	*Palaemonetes pugio* (F, adult, 17–27°C)	LC_{41}				0.000060 (28 d)	428
C, Endrin	*Palaemonetes kadiakensis*	LC_{50}				0.0004	45
C, Chlorpyrifos	*Palaemonetes pugio* (F, 20°C)	NOAE				0.001 (1 h)	685
C, Azinphosmethyl	*Palaemonetes kadiakensis*	LC_{50}				0.0012 (120 h)	45
C, DDT	*Palaemonetes kadiakensis* (F, 15°C, pH 7.1)	LC_{50}				0.0013 (5 d)	580
C, DDT	*Palaemonetes kadiakensis* (S, 15°C, pH 7.1)	LC_{50}				0.001 (5 d)	580
C, Baytex	*Palaemonetes kadiakensis*	LC_{50}				0.0015 (20 d)	45
C, Tributyltin oxide	*Palaemonetes pugio* (F, 25°C, pH 7–7.7)	NOAE				0.0023 (40 min)	686
C, Aroclor 1254	*Palaemonetes pugio* (0.5S, 21°C, larva)	LC_{10}				0.0032 (26 d)	687
C, Aroclor 1254	*Palaemonetes kadiakensis* (F)	LC_{50}				0.003 (7 d)	613
C, Aroclor 1254	*Palaemonetes kadiakensis*	LC_{50}				0.003 (7 d)	674
C, Aroclor 1254	*Palaemonetes kadiakensis*	LC_{50}				0.003 (7 d)	580
C, Baytex	*Palaemonetes kadiakensis*	LC_{50}				0.005 (120 h)	45
C, Aroclor 1254	*Palaemonetes pugio* (0.5S, 21°C, larva)	LC_{100}				0.0156 (11 d)	687
C, Aroclor 1254	*Palaemonetes pugio* (0.5S, 21°C, larva)	LC_{75}				0.0156 (6 d)	687
C, Dibutyl phthalate	*Palaemonetes pugio holthuis*	LC_{50}				0.1–1.0 (17 d)	107
C, Mirex	*Palaemonetes pugio* (S, 23–25°C)	LC_{50}				0.190 (120 h)	625
C, Pentachlorophenol	*Palaemonetes pugio* (25°C)	EC_{50}				0.60 (28 d)	684
C, Dibutylphthalate	*Palaemonetes pugio* (0.5S, larva, 23°C, seaW)	NOAE				1.0	688
C, Bis(2-ethylhexyl)phthalate	*Palaemonetes pugio* (0.5S, 23°C, larvae)	NOAE				1.0	688
C, Dimethyl phthalate	*Palaemonetes pugio* (0.5S, 23°C, larvae)	NOAE				10.0	688
C, *o*-Dimethyl phthalate	*Palaemonetes pugio* (larvae)	LC_{50}				100.0 (8 d)	107
C, Toluene-2,6-isocyanate	*Palaemonetes pugio*	LC_0				508.0	93

Chemical	Species	Toxicity test	24 h	48 h	96 h	Chronic exposure	Ref.
C, Nitrilotriacetic acid, Na, H_2O	*Palaemonetes vulgaris* (S)	TL_{50}				1800.0 (168 h)	37
C, Chlorpyrifos	*Palaemonetes pugio* (F, 12°C)	LC_{50}	0.00032				420
C, Chlorpyrifos	*Palaemonetes pugio* (F, 20°C)	LC_{50}	0.0032				685
C, Toxaphene	*Palaemonetes kadiakensis* (15.5°C)	LC_{50}	0.0209				689
C, Heptachlor	*Palaemonetes kadiakensis*	LC_{50}	0.046				690
C, Naphthalene	*Palaemonetes pugio* (S)	LC_{50}	2.6				691
C, Dibutyl phthalate	*Palaemonetes pugio holthuis*	LC_{50}	10.0–50.0				107
C, Toluene	*Palaemonetes pugio*	LC_{50}	17.2–38.1				664
C, Toxaphene	*Palaemonetes macrodactylus*	LC_{50}		0.0037			515
C, Toxaphene	*Palaemonetes pugio*	LC_{50}		0.0052			515
C, Toxaphene	*Palaemonetes kadiakensis*	LC_{50}		0.006			515
C, Malathion	*Palaemonetes vulgaris* (S)	LC_{50}		0.082			60
C, Heptachlor	*Palaemonetes vulgaris* (25–30°C)	LC_{100}		0.400			453
C, Heptachlor	*Palaemonetes vulgaris* (10°C)	LC_{25}		0.400			453
C, Heptachlor	*Palaemonetes vulgaris* (20°C)	LC_{50}		0.400			453
C, Heptachlor	*Palaemonetes vulgaris* (15°C)	NOAE		0.400			453
C, Dichlone	*Palaemonetes kadiakensis*	LC_{50}		0.450			43
C, Trifluralin	*Palaemonetes kadiakensis*	LC_{50}		1.20			43
C, 2,4-D, butoxyethyl ester	*Palaemonetes kadiakensis*	LC_{50}		1.4			43
C, 2,4-D, butoxyethyl ester	*Palaemonetes kadiakensis* (21°C, S)	LC_{50}		1.40			469
C, Vernam	*Palaemonetes kadiakensis*	LC_{50}		1.9			43
C, 2,4-D, propylene glycol butyl ether ester (PGBEE)	*Palaemonetes kadiakensis* (21°C, S)	LC_{50}		2.70			469
C, Fenoprop, propylene glycol butyl ether ester (PGBEE)	*Palaemonetes kadiakensis*	LC_{50}		3.20			43
C, Fenoprop, butoxy ethyl esters	*Palaemonetes kadiakensis*	LC_{50}		8.0			43
C, Dichlobenil	*Palaemonetes kadiakensis*	LC_{50}		9.0			43
C, Diphenamid	*Palaemonetes kadiakensis*	LC_{50}		58.0			403a
C, Diphenamid	*Palaemonetes kadiakensis*	LC_{50}		58.0			43

C, 2,4-D, dimethylamine	*Palaemonetes kadiakensis* (21°C, S)	LC_{50}		100.0			469
C, Trifene	*Palaemonetes kadiakensis*	NOAE		100.0			43
C, Simazine	*Palaemonetes kadiakensis*	NOAE		100.0			43
C, Dicamba	*Palaemonetes kadiakensis*	NOEL		100.0			43
C, 2,4-D(diethylamine)salt	*Palaemonetes kadiakensis*	NOEL		100.0			43
C, Endrin	*Palaemonetes macrodactylus* (IF)	LC_{50}			0.00012		99
C, Ethyl-*p*-nitrophenylthionobenzene phosphate (EPN)	*Palaemonetes kadiakensis*	LC_{50}			0.00056		45
C, DDD	*Palaemonetes kadiakensis*	LC_{50}			0.00068		45
C, Ethyl-*p*-nitrophenylthionobenzene phosphate	*Palaemonetes kadiakensis* (S, 21°C, pH 7.5)	LC_{50}			0.0006		602
C, Heptachlor (mixture)	*Palaemonetes vulgaris* (F)	LC_{50}			0.00106		693
C, Carbophenothion	*Palaemonetes kadiakensis*	LC_{50}			0.0012		45
C, Carbophenothion	*Palaemonetes pugio*	LC_{50}			0.0012		604
C, Endosulfan	*Palaemonetes pugio* (F)	LC_{50}			0.00131		247
C, Parathion	*Palaemonetes kadiakensis*	LC_{50}			0.0015		45
C, Parathion	*Palaemonetes kadiakensis* (S, wellW, 21°C)	LC_{50}	0.01	0.005	0.0015	0.015 (120 h)	405
C, Heptachlor	*Palaemonetes kadiakensis*	LC_{50}			0.0018		45
C, Endrin	*Palaemonetes vulgaris* (S)	LC_{50}			0.0018		60
C, DDT	*Palaemonetes kadiakensis*	LC_{50}			0.0023		45
C, Dimethyl parathion	*Palaemonetes vulgaris* (S)	LC_{50}			0.003		38
C, Endosulfan	*Palaemonetes macrodactylus* (F)	TL_{50}			0.0034		99
C, Chlordane	*Palaemonetes pugio*	LC_{50}			0.004		692
C, Chlordane	*Palaemonetes kadiakensis*	LC_{50}			0.004	0.0025 (120 h)	45
C, Toxaphene	*Palaemonetes pugio*	LC_{50}			0.0044		2
C, Toxaphene	*Palaemonetes pugio* (F)	LC_{50}			0.0044		108
C, Lindane	*Palaemonetes pugio* (F)	LC_{50}			0.00444		101
C, Toxaphene	*Palaemonetes pugio*	LC_{50}			0.0044		694
C, Endrin	*Palaemonetes macrodactylus* (S)	LC_{50}			0.0047		99
C, Carbaryl	*Palaemonetes kadiakensis*	NOEL			0.0056		45
C, Ethion	*Palaemonetes kadiakensis*	LC_{50}			0.0057		45

Chemical	Species	Toxicity test	24 h	48 h	96 h	Chronic exposure	Ref.
C, Xylene	*Palaemonetes pugio* (S, 21°C, pH 8.1)	LC_{50}			0.0074		695
C, Aroclor 1254	*Palaemonetes pugio* (0.5S, 21°C, larva)	LC_{50}			0.0078		687
C, Aldrin	*Palaemonetes vulgaris* (S)	LC_{50}			0.009		60
C, Lindane	*Palaemonetes vulgaris* (S)	LC_{50}			0.010		60
C, Lindane	*Palaemonetes vulgaris* (S)	LC_{50}			0.010		453
C, Malathion	*Palaemonetes kadiakensis*	LC_{50}			0.012	0.009 (120 h)	45
C, Heptachlor	*Palaemonetes macrodactylus* (S)	TL_{50}			0.0145		99
C, 2,2-Dichlorovinyldimethyl phosphate	*Palaemonetes vulgaris* (S)	LC_{50}			0.015		60
C, Endosulfan	*Palaemonetes macrodactylus* (S)	TL_{50}			0.0171		99
C, Toxaphene	*Palaemonetes kadiakensis*	LC_{50}			0.028		45
C, Malathion	*Palaemonetes macrodactylus* (IF)	TL_{50}			0.0337		99
C, Disulfoton	*Palaemonetes kadiakensis*	LC_{50}			0.038		45
C, Dieldrin	*Palaemonetes vulgaris* (S)	LC_{50}			0.05		60
C, Mercuric acetate	*Palaemonetes pugio* (S, 23°C, pH 8.3–8.7)	LC_{50}			0.06		696
C, Halowax 1099	*Palaemonetes pugio* (semiS, 19–24°C, seaW, juv.)	LC_{50}			0.069		697
C, Halowax 1013	*Palaemonetes pugio* (semiS, 19–24°C, seaW)	LC_{50}			0.074		697
C, Malathion	*Palaemonetes macrodactylus* (S)	TL_{50}			0.0815		99
C, Zectran	*Palaemonetes kadiakensis*	LC_{50}			0.083	0.025 (20 d)	454
C, Naled	*Palaemonetes kadiakensis*	LC_{50}			0.09		45
C, Mercuric thiocyanate	*Palaemonetes pugio* (S, 23°C, pH 8.3–8.7)	LC_{50}			0.09		696
C, Halowax 1099	*Palaemonetes pugio* (semiS, 19–24°C, seaW)	LC_{50}			0.090		697
C, Kepone	*Palaemonetes pugio*, (F)	LC_{50}			0.120		56
C, Kepone	*Palaemonetes pugio* (F, 20°C)	LC_{50}			0.121		418
C, Kepone	*Palaemonetes vulgaris*	LC_{50}			0.121		418
C, Chlorodecane	*Palaemonetes vulgaris* (F)	LC_{50}			0.121		422
C, Halowax 1014	*Palaemonetes pugio*	LC_{50}			0.25		423
C, Dioxathion	*Palaemonetes vulgaris* (S)	LC_{50}			0.285		60

C, Halowax 1000	*Palaemonetes pugio* (semiS, 19–24°C, seaW)	LC_{50}			0.325		697
C, Pentachlorophenate, Na	*Palaemonetes pugio* (0.5S, 19–21°C)	LC_{50}			0.436		699
C, Heptachlor	*Palaemonetes vulgaris* (S)	LC_{50}			0.440		60
C, Halowax 1000	*Palaemonetes pugio* (semiS, 19–24°C, seaW, juv.)	LC_{50}			0.440		697
C, Heptachlor	*Palaemonetes vulgaris* (S)	LC_{50}			0.440		453
C, Pentachlorophenate, Na	*Palaemonetes pugio* (F)	LC_{50}			0.515		109
C, Pentachlorophenol	*Palaemonetes pugio* (25°C)	LC_{50}			0.515		699
C, Pentachlorophenate, Na	*Palaemonetes pugio* (S, larv)	LC_{50}			0.649		110
C, Pentachlorophenol	*Palaemonetes pugio* (S, 25°C)	LC_{50}			0.649		698
C, Pentachlorophenol	*Palaemonetes pugio* (larvae)	LC_{50}			0.649		698
C, 2,4,6-Trichlorophenol	*Palaemonetes pugio* (0.5S, 20°C)	LC_{50}			1.21		700
C, Naphthalene	*Palaemonetes pugio* (S, 19–21°C, pH 8.1)	LC_{50}			2.35		701
C, Dibutyl phthalate	*Palaemonetes pugio*	LC_{50}			6.0		702
C, Dimethyl phthalate	*Palaemonetes pugio*	LC_{50}			7.0		703
C, *o*-Xylene	*Palaemonetes pugio*	LC_{50}			7.4		75
C, *o*-Dichlorobenzene	*Palaemonetes pugio*	LC_{50}	14.3	10.3	9.4		93
C, Toluene	*Palaemonetes pugio*	LC_{50}			9.50		95
C, Toluene	*Palaemonetes vulgaris*	LC_{50}			9.5		550
C, *o*-Dichlorobenzene	*Palaemonetes pugio* (S, 23°C, pH 8.3–8.7)	LC_{50}			10.00		704
C, Dibutyl phthalate	*Palaemonetes pugio* (0.5S, larva, 23°C, seaW)	LC_{100}			10.0	1.0 (72 h)	688
C, Chromic acid, K_2	*Palaemonetes pugio* (S, seaW, 22°C, pH 8.2)	LC_{50}			18.1		681
C, Dieldrin	*Palaemonetes kadiakensis*	LC_{50}			0.020		45
C, Ethylene dichloride	*Palaemonetes serratus*	NOAE			25.0		113
C, Benzene	*Palaemonetes pugio*	LC_{50}			27.0		95
C, Diphenamid	*Palaemonetes pugio* (S, 21°C, pH 7.5)	LC_{50}			32.0		403b
C, 2,4-Dichlorophenol	*Palaemonetes pugio* (0.5S, 10°C)	LC_{50}			40.3		700
C, Resorcinol	*Palaemonetes pugio*	LC_{50}	170.0	78.0	42.0		93
C, Aldrin	*Palaemonetes kadiakensis*	LC_{50}			50.0		43
C, Dicamba	*Palaemonetes kadiakensis* (S, 21°C)	LC_{50}			56.0		402
C, *p*-Dichlorobenzene	*Palaemonetes pugio* (S, 23°C, pH 6.8–7.8, seaW)	LC_{50}			60.0		704

Chemical	Species	Toxicity test	24 h	48 h	96 h	Chronic exposure	Ref.
C, *p*-Dichlorobenzene	*Palaemonetes pugio*	LC_{50}		129.0	69.0		93
C, Ammonium fluoride	*Palaemonetes pugio*	LC_{50}	160.0	93.0	75.0		93
C, Benzoyl chloride	*Palaemonetes pugio*	LC_{50}			180.0		93
C, Heptachlor	*Palaemonetes vulgaris*	EC_{50}			1060.0		65
C, Nitrilotriacetic acid, Na, H_2O	*Palaemonetes vulgaris* (S)	TL_{50}			4100.0		37
C, Chlordane	*Palaemonetes* sp. (F)	LC_{50}				0.004 (h?)	427
C, Amitrol	*Palaemonetes kadiakensis*	NOAE		100.0			43
C, Abate	*Penaeus monodon*	LC_{50}				0.045 (72 h)	639
C, Picloram	*Penaeus aztecus*	NOAE		1.0			706
C, *o*-Xylene	*Penaeus californicus* (16°C)	LC_{50}			0.0013		455
C, *m*-Xylene	*Penaeus californicus* (16°C)	LC_{50}			0.0037		455
C, Mercuric chloride	*Pandalus montagui* (S, 15°C)	LC_{50}		0.075			708
C, Ethylbenzene	*Pandalus* sp. (S, 16°C)	LC_{50}	0.0022		0.00049		455
C, Naphthalene	*Pandalus goniurus* (S, 12°C)	TL_m			0.971		88
C, Naphthalene	*Pandalus goniurus* (S, 8°C)	TL_m			1.02		88
C, Naphthalene	*Pandalus goniurus* (S, 4°C)	TL_m			2.16		88
C, Naphthalene	*Pandalus platyceros* (F, 9–11°C)	LC_{100}				0.008 (36 h)	709
C, Cupric sulfate	*Pandalus montagui* (S, 15°C)	LC_{50}		0.14			708
C, Creosote	*Panulirus japonicus* (larvae, 20°C)	LC_{50}			0.02		466
C, Creosote	*Panulirus japonicus* (adult, 10°C)	LC_{50}			1.76		466
C, Zinc chloride	*Paragrapsus quadridentatus* (F, 17°C, pH 8)	LC_{50}				10.5 (120 h)	710
C, Cupric sulfate, $5H_2O$	*Paragrapsus quadridentatus* (S, 17°C, larvae)	LC_{50}			0.17		705
C, Zinc chloride	*Paragrapsus quadridentatus* (F, 15°C)	LC_{50}			1.23		710
C, Kuwait crude oil	*Paragrapsus quadridentatus* (0.5S)	LC_{50}			1555.0		707
C, Naphthalene	*Parhyale hawaiensis* (S, closed bottle)	LC_0	1.0				111
C, Naphthalene	*Parhyale hawaiensis* (S, open bowl)	LC_0	4.0				111
C, Naphthalene	*Parhyale hawaiensis* (S, closed bottle)	LC_{50}	6.5				111
C, Naphthalene	*Parhyale hawaiensis* (S, closed bottle)	LC_{80}	7.5				111
C, Naphthalene	*Parhyale hawaiensis* (S, open bowl)	LC_{50}	15.0				111

C, Naphthalene	*Parhyale hawaiensis* (S, open bowl)	LC_{95}	20.0			111
C, Nickel chloride $6H_20$	*Pegurus longicarpus* (S, 20°C, pH 7.8)	LC_0		35.0		680
C, Nickel chloride $6H_20$	*Pegurus longicarpus* (S, 20°C, pH 7.8)	LC_{50}		47.0		680
C, Cutrine	*Penaeus californiensis* (F)	LC_{50}		1000.0		370
C, Mirex	*Penaeus duorarum* (F, juv., 17–27°C)	LC_{81}			0.00006 (28 d)	428
C, Mirex	*Penaeus duorarum* (14–17°C)	LC_0			0.0001 (35 d)	711
C, Mirex	*Penaeus duorarum* (F, juv., 17°C)	LC_{11}			0.0001 (3 wk)	712
C, Mirex	*Penaeus duorarum* (F, juv., 17°C)	LC_{36}			0.0001 (5 wk)	712
C, DDT,TG	*Penaeus duorarum* (F)	TL_{50}			0.00012 (28 d)	112
C, Aroclor 1254	*Penaeus duorarum* (F, seaW)	LC_{50}			0.00094 (15 d)	613
C, Aroclor	*Penaeus duorarum* (F, 20°C, juv.)	LC_{50}			0.00094 (15 d)	713
C, Aroclor 1254	*Penaeus duorarum* (juv.)	LC_{50}			0.00094 (15 d)	580
C, Mirex	*Penaeus duorarum* (F, juv., 17°C)	LC_{100}			0.001 (11 d)	712
C, Mirex	*Penaeus duorarum* (14–17°C)	LC_{100}			0.001 (11 d)	711
C, Mercuric chloride	*Penaeus setiferus* (0.5S, 25°C)	NOAE			0.001 (60 d)	715
C, Mirex	*Penaeus duorarum* (F, juv., 17°C)	LC_{25}			0.001 (7 d)	712
C, Aroclor	*Penaeus duorarum* (F, 20°C)	LC_{50}			0.0035 (35 d)	713
C, Aroclor 1254	*Penaeus duorarum* (F, seaW)	LC_{50}			0.0035 (35 d)	613
C, Aroclor 1254	*Penaeus duorarum* (F, 27°C, juv.)	LC_{72}			0.005 (10 d)	716
C, Aroclor 1254	*Penaeus duorarum* (16°C)	LC_{72}			0.005 (20 d)	716
C, Aroclor 1254	*Penaeus duorarum*	LC_{75}			0.005 (20 d)	674
C, Chlordane	*Penaeus duorarum* (F)	LC_{50}		0.0004		369
C, Mirex	*Penaeus duorarum* (F, juv.)	LC_{100}			0.1 (22 d)	712
C, Fluoride	*Penaeus indicus*	NOAE			5.5 (113 d)	717
C, Cupric sulfate	*Penaeus californicus* (S, 27°C, juv.)	LC_{50}			250.0 (1 h)	718
C, Nitrilotriacetic acid,H_2O-Na	*Penaeus setiferus* (F)	LC_{78}			1000.0 (22 d)	626
C, Nitrilotriacetic acid, Na, H_2O	*Penaeus setiferus* (F)	LC_{78}			1000.0 (22 d)	37
C, Dieldrin	*Penaeus duorarum*	LC_{70}	0.001			714
C, Chlorpyrifos	*Penaeus* sp. (saltW, aeration, S)	LC_0	0.010			583
C, Polychlorinated biphenyl	*Penaeus* sp. (16°C)	NOAE	0.010			716

Chemical	Species	Toxicity test	24 h	48 h	96 h	Chronic exposure	Ref.
C, Hexachlorocyclohexane, mixed isomers	*Penaeus setiferus* (S, 26°C, pH 8.2)	LC_{50}	0.035				719
C, Chlorpyrifos	*Penaeus* sp. (saltW, aeration 48 h earlier)	LC_{0}	0.060				583
C, Chlorpyrifos	*Penaeus* sp. (saltW, aeration, S)	LC_{100}	0.060				583
C, Polychlorinated biphenyl	*Penaeus* sp. (16°C)	LC_{80}	0.100				716
C, Chlorpyrifos	*Penaeus* sp. (saltW, aeration 48 h earlier)	LC_{100}	0.120				583
C, Hexachlorocyclohexane, mixed isomers	*Penaeus aztecus* (S, 17–22°C, pH 8.2)	LC_{50}	0.40				719
C, Naphthalene	*Penaeus aztecus* (S)	LC_{50}	2.5				691
C, Fenthion	*Penaeus duorarum* (F, 30°C)	LC_{50}		0.00006			417
C, Chlorpyrifos	*Penaeus* sp. (juv.)	LC_{50}		0.0002			720
C, Chlorpyrifos	*Penaeus aztecus* (F, 27°C)	LC_{50}	0.0015	0.0003			583
C, Endosulfan	*Penaeus aztecus*	EC_{50}		0.0004			417
C, Lindane	*Penaeus aztecus* (F)	EC_{50}		0.0004			417
C, Aroclor 1254	*Penaeus duorarum* (16°C)	LC_{50}		0.001			716
C, Chlorpyrifos	*Penaeus duorarum* (F, 12°C)	LC_{50}	0.0028	0.0024			420
C, Toxaphene	*Penaeus aztecus*	LC_{50}		0.0027			515
C, Chlorpyrifos	*Penaeus duorarum* (F, 25°C)	LC_{50}	0.0032	0.0032			420
C, Toxaphene	*Penaeus duorarum*	LC_{50}		0.0042			515
C, Chlordane	*Penaeus aztecus*	LC_{50}		0.0044			416
C, Aroclor 1254	*Penaeus duorarum* (16°C)	LC_{0}		0.010			716
C, Aroclor	*Penaeus duorarum* (F, 16°C, juv.)	LC_{0}		0.010			716
C, Polychlorinated biphenyl	*Penaeus* sp. (16°C)	NOAE		0.010			716
C, Propoxur	*Penaeus duorarum*	EC_{50}	0.064	0.041			721
C, Demeton	*Penaeus duorarum* (26°C, F)	LC_{50}		0.063			722a
C, Arochlor 1254	*Penaeus duorarum*	LC_{100}		0.100			722b
C, Aroclor 1254	*Penaeus duorarum* (16°C)	LC_{100}		0.100			716
C, Aroclor 1254	*Penaeus duorarum* (F, 16°C, juv.)	LC_{100}		0.100			716
C, Polychlorinated biphenyl	*Penaeus* sp. (16°C)	LC_{100}		0.100			716

C, Acrolein	*Penaeus aztecus* (S)	LC_{50}		0.1		520
C, Chlorpyrifos	*Penaeus aztecus* (F, 29°C)	LC_{50}	0.32	0.20		583
C, Cupric sulfate	*Penaeus aztecus* (larva, S, 15°C)	LC_{50}		0.33		438
C, Eptam, technical grade	*Penaeus aztecus*	LC_{50}		0.63		25
C, Diuron	*Penaeus aztecus* (15°C, F, adult)	LC_0		1.0		723
C, Fenuron	*Penaeus aztecus* (F)	LC_{10}		1.0		724
C, Tillam	*Penaeus aztecus*	LC_{50}		1.0		23
C, Mirex	*Penaeus duorarum* (F, adult, 25°C)	LC_{50}		1.2		417
C, Nickel sulfate	*Penaeus duorarum* (15°C)	LC_{50}		13.9		708
C, Nickel sulfate	*Penaeus aztecus* (15°C, S)	LC_{50}		125.0		708
C, Endosulfan	*Penaeus duorarum* (F)	LC_{50}			0.00004	245
C, Heptachlor epoxide	*Penaeus duorarum* (F)	LC_{50}			0.00004	65
C, Lindane	*Penaeus duorarum* (F)	LC_{50}			0.00017	727
C, Lindane	*Penaeus duorarum* (F)	LC_{50}			0.00017	101
C, Ethyl-*p*-nitrophenylthiono-benzene phosphate	*Penaeus duorarum*	LC_{50}			0.00029	725
C, Hexachlorocyclohexane, mixed isomers	*Penaeus duorarum* (F, 21–27°C)	LC_{50}			0.00034	727
C, Lindane	*Penaeus duorarum* (F)	LC_{50}			0.00034	101
C, Chlordane	*Penaeus duorarum*	LC_{50}			0.0004	692
C, Chlordane	*Penaeus duorarum*	LC_{50}			0.0004	728
C, Dieldrin	*Penaeus duorarum*	LC_{50}			0.0007	714
C, Toxaphene	*Penaeus duorarum* (75% oxygen saturated)	LC_{12}			0.001	729
C, Toxaphene	*Penaeus duorarum* (35 ppthousand salinity)	LC_{15}			0.001	729
C, Toxaphene	*Penaeus duorarum* (100% oxygen saturated)	LC_{15}			0.001	729
C, Toxaphene	*Penaeus duorarum* (35°C)	LC_{50}			0.001	730
C, Toxaphene	*Penaeus duorarum* (25°C)	LC_{50}			0.001	730
C, Toxaphene	*Penaeus duorarum* (25°C)	LC_{50}			0.001	729
C, Toxaphene	*Penaeus duorarum* (50 ppthousand salinity)	LC_{67}			0.001	729
C, Toxaphene	*Penaeus duorarum* (35°C)	LC_{90}			0.001	729
C, Toxaphene	*Penaeus duorarum*	LC_{50}			0.0014	2

Chemical	Species	Toxicity test	24 h	48 h	96 h	Chronic exposure	Ref.
C, Toxaphene	*Penaeus duorarum*	LC_{50}			0.0014		729
C, Toxaphene	*Penaeus duorarum* (F)	LC_{50}			0.0014		2
C, Leptophos	*Penaeus duorarum*	LC_{50}			0.00188		731
C, Toxaphene	*Penaeus duorarum*	LC_{50}			0.0018		730
C, Arochlor 1016	*Penaeus aztecus*	LC_{50}			0.0105		722
C, Aldicarb	*Penaeus duorarum* (F, 22°C, pH 8.3)	LC_{50}			0.012		652
C, Arochlor 1016	*Penaeus duorarum*	LC_{50}			0.0125		722
C, Mercuric chloride	*Penaeus setiferus* (0.5S, 25°C)	LC_{50}			0.020		715
C, Hexachlorobenzene	*Penaeus duorarum* (F, 29–30°C)	LC_{33}			0.025		733
C, Heptachlor	*Penaeus duorarum* (F)	LC_{50}			0.03		693
C, Aldicarb	*Penaeus setiferus* (S, 24–26°C)	LC_{50}			0.072		647
C, Aldicarb	*Penaeus setiferus* (S, 25°C)	LC_{50}			0.072		726
C, Hexachloronaphthalene	*Penaeus aztecus* (F)	LC_{50}			0.075		423
C, Mirex	*Penaeus duorarum* (F)	LC_{0}			0.1		712
C, Arochlor 1016	*Penaeus duorarum*	LC_{100}			0.100		722
C, Pentachlorophenate, Na	*Penaeus aztecus* (F)	LC_{50}			0.195		109
C, Pentachlorophenol	*Penaeus aztecus*	LC_{50}			0.195		698
C, Lindane	*Penaeus aztecus* (S)	TL_{m}			0.4		89
C, Nickel chloride, $6H_2O$	*Penaeus merguiensis* (0.5S, 35°C)	LC_{50}			2.8		732
C, Benzyl chloride	*Penaeus setiferus*	LC_{50}	7.1	4.4	3.9		93
C, Thallium acetate	*Penaeus aztecus*	LC_{50}			10.0		53
C, Nickel chloride,$6H_2O$	*Penaeus merguiensis* (0.5S + 3.6% salt, 20°C)	LC_{50}			10.0		732
C, Cupric sulfate	*Penaeus aztecus* (adult, 0.5S, 15°C)	LC_{50}			19.0		708
C, Nickel chloride $6H_2O$	*Penaeus merguiensis* (0.5S + 2% salt, 20°C)	LC_{50}			21.0		732
C, Chloroform	*Penaeus duorarum*	LC_{50}			81.5		629
C, Flouride	*Penaeus indicus*	NOAE			100.0		717
C, Flouride	*Penaeus mondon*	NOAE			100.0		717
C, Butyl cellosolve	*Penaeus aztecus*	LC_{50}		800.0	775.0		54
C, Heptachlor	*Penaeus duorarum* (F)	EC_{50}			1100.0		65

C, Isopropanal	*Penaeus aztecus*	LC_{50}		1400.0	1150.0		54
C, Nitrilotriacetic acid, Na, H_2O	*Penaeus setiferus* (S)	LC_{90}			5000.0		37
C, Nitrilotriacetic acid, H_2O, Na	*Penaeus setiferus* (S)	LC_{90}			5000.0		626
C, Rotenone	Planaria	LC_{100}			0.500		557
C, Swedish EDC tar	*Pleuronectes platessa*	LC_{50}		9.0			572
C, Fluoride	*Penaeus* sp.	NOAE				5.50 (113 d)	734
C, Chlorpyrifos	*Procambarus* sp. (S, 26°C)	LC_{50}			0.002		420
C, Mirex	*Procambarus hayi* (S, pH 7.8)	LD_{65}				0.0001 (48 h + 4 d)	735
C, Mirex	*Procambarus hayi* (S, pH 7.8)	LD_{71}				0.0005 (48 h + 4 d)	735
C, Mirex	*Procambarus blandingi* (S)	LD_0				0.001 (144 h)	736
C, Mirex	*Procambarus blandingi* (S)	LD_{95}				0.001 (144 h + 5 d)	735
C, Methyl parathion	*Procambarus clarkii* (S, 27°C, pH 7.6)	LC_{50}				0.010 (5 d)	737
C, Chlorpyrifos	*Procambarus clarkii* (S, 22–26°C)	LC_{50}				0.041 (36 h)	420
C, Toxaphene	*Procambarus simulans*	LC_{50}	0.45				689
C, Mirex	*Procambarus hayi* (S, pH 7.8)	LD_{19}		0.0001			735
C, Mirex	*Procambarus hayi* (S, pH 7.8)	LD_{12}		0.0005			735
C, Methyl parathion	*Procambarus acutus* (S, pH 8.7, no preexposure)	LC_{50}		0.0024			738
C, Methyl parathion	*Procambarus acutus* (S, pH 8.7, preexposure)	LC_{50}		0.0034			738
C, Methyl parathion	*Procambarus clarkii* (S, 16–32°C, pH 7.6)	LC_{50}	0.05	0.04		0.04 (72 h)	739
C, Toxaphene	*Procambarus acutus*	LC_{50}		0.067			515
C, Permethrin	*Procambarus clarkii*	LC_{50}			0.00039		677
C, Methyl parathion	*Procambarus acutus* (pH 8.4)	LC_{50}			0.003		740
C, Toxaphene	*Procambarus simulans*	LC_{50}	0.45	0.29	0.21	0.24 (72 h)	741
C, Carbofuran	*Procambarus* sp. (26°C, pH 9–11)	LC_{50}			0.500		742
C, Calcium chloride	*Procambarus* sp.	LC_{50}			5.0		471
C, Captan	*Procambarus clarkii* (17–20°C, pH 8.4)	LC_{50}			15,630.0		743
C, Toxaphene	*Rangia cuneata*	LC_{50}	940.0	699.0	460.0	480.0 (72 h)	741

Chemical	Species	Toxicity test	24 h	48 h	96 h	Chronic exposure	Ref.
C, Trichlorpyr	*Ranina* sp.	LC_{50}			1000.0		461
C, Tributyltin oxide	*Rhithropanopeus harrisii* (0.5S, 25°C, larva)	LC_{0}				0.0005 (15 d)	746
C, Tributyltin oxide	*Rhithropanopeus harrisii* (0.5S, 25°C, larva)	LC_{63}				0.0250 (15 d)	747
C, Tributyltin oxide	*Rhithropanopeus harrisii* (larvae, semisS, 25°C)	LC_{50}				0.032 (12 d)	748
C, Kepone	*Rhithropanopeus harrisii* (larvae, S, 25°C)	LC_{95}				0.095 (20 d)	745
C, Halowax 1099	*Rhithropanopeus harrisii* (larvae, S, 25°C)	LC_{85}				0.100 (23 d)	646
C, Chloronaphthalene	*Rhithropanopeus harrisii* (larva, 1/2S, 25°C)	LC_{23}				0.300 (21 d)	749
C, Naphthalene	*Rhithropanopeus harrisii* (S, 30°C, L)	LC_{0}				0.500	750
C, Dibutyl phthalate	*Rhithropanopeus harrisii* (S, 24–26°C, L)	NOAE				1.0 (LC exp.)	749
C, Dimethyl phthalate	*Rhithropanopeus harrisii* (S, 26°C, L)	NOAE				1.0	751
C, Toxaphene	*Rhithropanopeus harrisii* (Stage I zoea)	LC_{50}			0.044		689
C, Toxaphene	*Rhithropanopeus harrisii* (Stage II zoea)	LC_{50}			0.288–0.315		689
C, Toxaphene	*Rhithropanopeus harrisii* (Stage III zoea)	LC_{50}			0.290–0.430		689
C, Arsenic trioxide	*Scylla serrata* (26.5–29.5°C)	LC_{0}			7.50		752
C, Arsenic trioxide	*Scylla serrata* (26.5-29.5°C)	LC_{50}	33.0		17.0		752
C, Arsenic trioxide	*Scylla serrata* (26.5–29.5°C)	LC_{100}			32.0		752
C, Toxaphene	*Sesarma cinnerum* (stage I zoea)	LC_{50}			0.054		753
C, Toxaphene	*Sesarma cinnerum* (stage II zoea)	LC_{50}			0.076		753
C, Toxaphene	*Sesarma cinnerum* (stage III zoea)	LC_{50}			0.74		753
C, Toxaphene	*Sesarma cinnerum* (stage IV zoea)	LC_{50}			6.80		753
C, 1,2-Dichloropropane	Shrimp	TL_{m}		100.0			754
C, Tetraethyl lead	Shrimp	LC_{50}			0.020		114
C, Tetramethyl lead	Shrimp	LC_{50}			0.11		114
C, Triethyllead chloride	Shrimp	LC_{50}			5.8(Pb)		114
C, Trimethyllead chloride	Shrimp	LC_{50}			8.8(Pb)		114
C, 1,2-Dichloroethane	Shrimp, (F, 15°C, pH 8)	LC_{50}			85.0		464
C, Lindane	*Simocephalus serrulatus* (S)	LC_{50}		0.520			506
C, Carbophenothion	*Simocephalus serrulatus*	LC_{50}		0.000009			755

C, 2,2-Dichlorovinyldimethyl phosphate	*Simocephalus serrulatus*	LC_{50}	0.00026	69
C, Parathion	*Simocephalus serrulatus*	LC_{50}	0.00037	69
C, Methyl parathion	*Simocephalus serrulatus* (S, 15°C, pH 7.5)	LC_{50}	0.00037	511
C, Chlorfos	*Simocephalus serrulatus*	LC_{50}	0.00032	45
C, Parathion	*Simocephalus serrulatus* (S, FW, 21°C, juv.)	LC_{50}	0.00047	511
C, Baytex	*Simocephalus serrulatus*	LC_{50}	0.00062	69
C, Diazinon	*Simocephalus serrulatus*	LC_{50}	0.0014	69
C, Malathion	*Simocephalus serrulatus*	LC_{50}	0.0035	69
C, Azinphosethyl	*Simocephalus serrulatus*	LC_{50}	0.004	69
C, DDD	*Simocephalus serrulatus*	LC_{50}	0.0045	69
C, Phosphamidon	*Simocephalus serrulatus*	LC_{50}	0.0066	69
C, Carbaryl	*Simocephalus serrulatus*	NOEL	0.0076	69
C, Toxaphene	*Simocephalus serrulatus*	LC_{50}	0.010	69
C, Toxaphene	*Simocephalus serrulatus* (21.5°C)	LC_{50}	0.010	515
C, Zectran	*Simocephalus serrulatus*	LC_{50}	0.013	69
C, Toxaphene	*Simocephalus serrulatus* (15.5°C)	LC_{50}	0.019	515
C, Chlordane	*Simocephalus serrulatus*	LC_{50}	0.020	69
C, Chlordane	*Simocephalus serrulatus* (15.5°C)	LC_{50}	0.020	594
C, Chlordane	*Simocephalus serrulatus* (21.5°C)	LC_{50}	0.024	506
C, Endrin	*Simocephalus serrulatus*	LC_{50}	0.026	69
C, Pyrethrin	*Simocephalus serrulatus*	LC_{50}	0.042	69
C, Pyrethrum	*Simocephalus serrulatus*	LC_{50}	0.042	756
C, Heptachlor	*Simocephalus serrulatus*	LC_{50}	0.047	69
C, Heptachlor	*Simocephalus serrulatus* (S, 16°C)	LC_{50}	0.047	757
C, Allethrin	*Simocephalus serrulatus*	LC_{50}	0.056	69
C, Heptachlor	*Simocephalus serrulatus* (S, 21°C)	LC_{50}	0.080	757
C, Dieldrin	*Simocephalus serrulatus*	LC_{50}	0.190	69
C, Rotenone	*Simocephalus serrulatus*	LC_{50}	0.19	69
C, Trifluralin	*Simocephalus serrulatus*	EC_{50}	0.450	506
C, Trifluralin	*Simocephalus serrulatus*	LC_{50}	0.450	69

Chemical	Species	Toxicity test	24 h	48 h	96 h	Chronic exposure	Ref.
C, Lindane	*Simocephalus serrulatus*	LC_{50}		0.520			69
C, Diuron	*Simocephalus serrulatus*	LC_{50}		2.00			534
C, Diuron	*Simocephalus serrulatus*	LC_{50}		2.0			69
C, Fenoprop propylene glycol butyl ether ester	*Simocephalus serrulatus*	LC_{50}		2.4			69
C, Paraquat	*Simocephalus serrulatus*	LC_{50}		4.0			69
C, Dichlobenil	*Simocephalus serrulatus*	LC_{50}		5.8			70
C, Trifene	*Simocephalus serrulatus*	LC_{50}		6.6			69
C, Isopropyl-*N*-phenyl carbamate	*Simocephalus serrulatus*	LC_{50}		10.0			69
C, 2,2-Dichloropropionic acid, Na	*Simocephalus serrulatus*	LC_{50}		16.0			69
C, 2,2-Dichloropropionic acid, Na	*Simocephalus serrulatus* (S, 15°C)	LC_{50}		16.0			506
C, Aldrin	*Simocephalus serrulatus*	LC_{50}		23.0			69
C, Naled	*Simocephalus serrulatus*	LC_{50}			0.0011		69
C, DDT	*Simocephalus serrulatus*	LC_{50}			0.0025		69
C, Diuron	*Simocephalus* (1st instar, S, 15°C)	EC_{50}		2.0			532
C, Phosmet	*Streptocephalus seali* (F, 21°C, pH 7.5)	LC_{50}		0.170			758
C, Benzene	*Tigriopus californicus*	LC_{50}			450.0		759
C, Flouride	*Tylediplax blephariskios*	NOAE			100.0		717
C, Kepone	*Uca pugilator* (adult, 1/2S, 24°C)	LC_{50}				0.0300 (7 d)	761
C, Aroclor	*Uca pugilator* (0.5S, 25°C)	LC_0				0.100 (3 wk)	762
C, Mirex	*Uca pugilator* (F, 29°C)	LC_{73}				50.0 (14 d)	712
C, Nitrilotriacetic acid, Na, H_2O	*Uca pugilator* (F)	LC_{46}				1000.0 (45 d)	626
C, Nitrilotriacetic acid, Na, H_2O	*Uca cramulata* (F)	LC_{46}				1000.0 (46 d)	37
C, Kepone	*Uca pugilator*	LC_{50}	1.60	1.345			760
C, Aroclor	*Uca pugilator* (0.5S, 25°C, larva)	LC_{50}			0.010		762
C, Kepone	*Uca cramulata*	LC_{50}	1.6		1.47	0.32(NOEL)	2
C, Dicamba	*Uca pugilator*	LC_{50}			180.0		763
C, Nitrilotriacetic acid, Na, H_2O	*Uca pugilator* (S)	LC_{25}			10,000.0		626
C, Nitrilotriacetic acid, Na, H_2O	*Uca cramulata* (S)	LC_{25}			10,000.0		37

C, Carbaryl	*Upogebia pugettensis*	TL_m		0.040			407
C, Diphenamid	*Gammarus fasciatus*	NOAE		100.0			43
F, Zinc chloride	*Adrichetta forsteri* (F, 25°C, pH 8)	LC_{50}			11.50		764
F, Tributyltin oxide	*Agonus cataptaractus* (semiS)	LC_{50}			0.016		765
F, DC-Z1219 Silane	*Agonus cataptaractus* (S, 14–16°C, pH 5.1)	LC_{50}			0.14		766
F, DC-Z1216 Silane	*Agonus cataptaractus* (S, 14–16°C, pH 4–8)	LC_{50}			0.14		766
F, DC-Z1221 Silane	*Agonus cataptaractus* (S, 16°C, pH 4–8)	LC_{50}			0.20		766
F, Pyridine	*Alburnus alburnus*	LD_0				100.0	8
F, 2,4-D	*Alburnus alburnus* (larvae)	LC_{50}				111.0 (12 h)	64
F, 2,4-D	*Alburnus alburnus* (embryo)	LC_{50}				159.0 (12 h)	64
F, Methyl-*t*-butyl ether	*Alburnus alburnus* (S, 10°C, pH 7.8)	LC_{50}	1700.0				671
F, 2,4-D	*Alburnus alburnus* (embryo)	LC_{50}	129.0	13.0		64.0 (36 h)	64
F, 2,4-D	*Alburnus alburnus* (larvae)	LC_{50}	71.0	52.0		62.0 (36 h)	64
F, Tributyltin fluoride	*Alburnus alburnus* (S, 10°C, pH 7.8)	LC_{50}			0.006–0.008		765
F, Tributyltin oxide	*Alburnus alburnus* (S, 10°C, pH 7.8)	LC_{50}			0.015		765
F, Carbendazim	*Alburnus alburnus* (10°C)	LC_{50}			3.0–4.0		768
F, Dimethyl phthalate	*Alburnus alburnus* (S, 10°C, pH 7.8)	LC_{50}			110.0		384
F, Mecoprop-K	*Alburnus alburnus* (S, 10°C, pH 7-8)	LC_{50}			115.0		666
F, Methyl-*t*-butyl ether	*Alburnus alburnus*	LC_{50}			1000.0		616
F, Trimethylol propane	*Alburnus alburnus*	LC_{50}			1000.0		616
F, Isobutyl alcohol	*Alburnus alburnus* (10°C)	LC_{50}			1000.0-3000.0		667
F, Triethyl phosphate	*Alburnus alburnus* (S, 10°C, pH 7.8)	LC_{50}			2100.0		666
F, Triethyl phosphate	*Alburnus alburnus* (S, 10°C, pH 7.8)	LC_{50}			2100.0–2400.0		666
F, *n*-Butyl alcohol	*Alburnus alburnus* (S, 10°C)	LC_{50}			2250.0–2400.0		667
F, *n*-Propanol	*Alburnus alburnus* (brackish W, S, 10°C, pH 8)	LC_{50}			3800.0		668
F, Chromic acid, Na_2	*Adrichetta forsteri* (18.5°C, pH 8.0)	NOAE			17.9		769
F, Chromic acid, Na_2	*Adrichetta forsteri*, (juv.)	LC_{50}			31.2		769
F, Chromium(3+) nitrate	*Adrichetta forsteri* (S)	LC_{50}			53.0		769
F, Fluoride	*Ambassis safga*	NOAE			100.0		717
F, Carbaryl	*Amiurus* sp.	LC_{50}			20.0		122

Chemical	Species	Toxicity test	24 h	48 h	96 h	Chronic exposure	Ref.
F, "α" Caprolactam	*Amiurus nebulosus*	NOAE				1000.0 (30 d)	8
F, "α" Caprolactam	*Amiurus nebulosus*	LC_{100}				5000.0 (18 h)	8
F, "α" Caprolactam	*Amiurus nebulosus*	LC_{100}				10,000.0 (10 h)	8
F, Carbaryl	*Amiurus nebulosus*	LC_{50}			15.8		122
F, Rotenone (5%)	*Amia calva* (S)	LC_{50}			0.030		28
F, Methyl parathion	*Anguilla rostrata* (S, 20°C, pH 7.2)	LC_{50}	42.6	37.2	6.3		772
F, Abate	*Anguilla japonica*	LC_{50}				7.5 (72 h)	639
F, Ammonium salt of carboxylic acid, a dispersant	*Anguilla anguilla*	LD_{50}		300.0			42
F, Endrin	*Anguilla rostrata* (S)	LC_{50}			0.0006		115
F, Dieldrin (100%)	*Anguilla rostrata* (S)	LC_{50}			0.0009		115
F, DDT(*p,p′*)	*Anguilla rostrata* (S)	LC_{50}			0.004		115
F, Aldrin	*Anguilla rostrata* (S)	LC_{50}			0.005		115
F, Heptachlor	*Anguilla rostrata* (S)	LC_{50}			0.01		115
F, Aldrin	*Anguilla rostrata*	LC_{50}			0.016		117
F, Chlordecone	*Anguilla rostrata* (juv., 19°C)	LC_{50}	0.164	0.085	0.035		767
F, Lindane	*Anguilla rostrata* (S)	LC_{50}			0.056		115
F, Malathion	*Anguilla rostrata* (S)	LC_{50}			0.082		115
F, Malathion	*Anguilla rostrata* (S)	LC_{50}			0.500		2
F, 2,2-Dichlorovinyldimethyl phosphate	*Anguilla rostrata* (S)	LC_{50}			1.8		115
F, Dimethyl parathion	*Anguilla rostrata* (S)	LC_{50}			6.3		116
F, Dimethyl parathion	*Anguilla rostrata* (S)	LC_{50}			16.9		117
F, 2,4,5-T	*Anguilla rostrata* (S)	LC_{50}			43.7		117
F, 2,4-D	*Anguilla rostrata* (S)	LC_{50}			300.6		64
F, Dimetilan	*Aphanius fasciatus* (20°C)	LC_{50}		0.07436	0.07436		773
F, Formaldehyde	Aquatic organisms, sensitive	TL_m				10.0	118
F, Phenol	*Arctopsyche grandis*	TL_m	61.0	56.0	0.001		22
F, Rotenone	*Argulus* sp.	LC_{100}				0.025 (3 d)	557

F, Chromic acid, Na_2	*Atherinamosa microstoma* (18.7°C, pH 8.1)	NOAE			19.3		774
F, Chromic acid, Na_2	*Atherinamosa microstoma* (18.7°C, pH 8.1)	LC_{50}			31.6		774
F, Parathion (paramar 50)	*Barbus machecola* (fry, 26–30°C, pH 7.5)	LC_{50}		2.0			775
F, Aldicarb	*Barbus conchonius* (juv., S, 14–22°C, pH 7.16)	LC_{50}		3.296		0.623 (72 h)	776
F, Aldicarb	*Barbus conchonius* (SW, 14–22°C, pH 7.16)	LC_{50}			0.459		776
F, Aldicarb	*Barbus conchonius* (S, softW, 14–22°C, pH 7.16)	LC_{50}		3.30	0.46	0.62 (72 h)	776
F, Aldicarb	*Barbus conchonius* (S, hardW, 14–22°C, pH 7.16)	LC_{50}		8.99	2.42	2.39 (72 h)	776
F, Aldicarb	*Barbus conchonius* (juv., HW, S, 7.41°C, pH 7.16)	LC_{50}		8.99	2.42	2.39 (72 h)	776
F, Toxaphene	Bass	LC_{50}			0.002		122
F, DDT	Bass	LC_{50}			0.002		122
F, Lindane	Bass	LC_{50}			0.032		122
F, Malathion	Bass	LC_{50}			0.290		122
F, Baytex	Bass	LC_{50}			1.5		122
F, Toluene	Bass	LC_{50}			7.3		61
F, Zectran	Bass	LC_{50}			14.7		122
F, Cobalt chloride	*Blennius pholis* (S, 15°C, pH 7.8)	LC_{50}			454.0–680.0	227.0 (216 h)	777
F, Ammonium chloride	*Lepomis macrochirus*	TL_m	725.0		725.0		74
F, Captan	*Brachydanio rerio* (larva, S)	LC_{50}				0.67 (90 min)	778
F, 4-Chloro-*m*-eresol	*Brachydanio rerio* (F)	LC_{50}	1.0-3.5			10.0–35.0 (2–6 h)	779
F, Dimethoate	*Brachydanio rerio*	TL_m		940.0		259.0 (72 h)	780
F, Pentachlorophenol	*Brachydanio rerio* (F, 25°C)	LC_{50}		1.24	1.13	1.08 (10 d)	155
F, Lauryl sulfate	*Brachydanio rerio* (F, 25°C)	LC_{50}		8.81	7.97	7.97 (10 d)	155
F, Epichlorhydrin	*Brachydanio rerio* (S, pH 7.5)	LC_0			26.5		781
F, Phenol	*Brachydanio rerio* (25°C, F)	LC_{50}		30.9	29.0	29.0 (10 d)	155
F, Epichlorhydrin	*Brachydanio rerio* (S, pH 7.5)	LC_{50}			30.5		781
F, Epichlorhydrin	*Brachydanio rerio* (S, pH 7.5)	LC_{100}			31.0		781
F, Tordon 22K	*Brachydanio rerio* (F, 15°C)	LC_{50}			35.5	35.5	155
F, Endrin	*Brevoortia patronus* (F)	LC_{50}	0.0008				241

Chemical	Species	Toxicity test	24 h	48 h	96 h	Chronic exposure	Ref.
F, Chlordecane	*Brevoortia tyrannus* (F, 28.2°C)	LC_{50}			0.0174		782
F, Toxaphene	*Campostorna anomalum* (S, 28.8°C, pH 7.9)	LC_{50}	0.009	0.0078	0.005		783
F, Toxaphene	*Campostorna anomalum* (S, pH 11.6)	LC_{50}	0.062	0.027	0.014		783
F, Toxaphene	*Campostorna anomalum* (S, 17°C, pH 7.9)	LC_{50}	0.054	0.044	0.032		783
F, Polyethylene glycol	*Carassius auratus*	LD_{50}	5000.0				62
F, Sulfuric acid	*Carassius auratus* (pH 3.9)	LC_{100}			0.0013 m/L		784
F, Dobane 83	*Carassius auratus*	LC_0				0.001	62
F, Phenylmercuric acetate	*Carassius auratus*	MTL				0.0078 (1 wk)	786
F, Dobane 055	*Carassius auratus*	LC_0				0.01	62
F, Dobane JN	*Carassius auratus*	LC_0				0.01	62
F, Endrin	*Carassius carassius*, (juv., S)	LC_{50}				0.028 (72 h)	162
F, Chlorodane	*Carassius auratus*	NOEL				0.100–0.200 (87 h)	785
F, Pentachlorophenol	*Carassius auratus* (F, 25°C)	TL_m				0.189 (336 h)	144
F, Pentachlorophenol	*Carassius auratus* (F, 25°C)	TL_m				0.253 (120 h)	144
F, 2,6-Di-*t*-butyl-4-methyl phenol	*Carassius auratus*	LC_0				0.4	62
F, Dobanol 45	*Carassius auratus*	LC_0				0.7	62
F, Dobanol 25	*Carassius auratus*	LC_0				0.8	62
F, Hydrogen sulfide	*Carassius auratus* (long-term exp., hardW)	LD_0				1.0	219
F, Picloram	*Carassius auratus*	NOEL				1.0 (10 wk)	771
F, Tris-1,3-dichloro-2-propyl phosphate	*Carassius auratus* (20°C)	LC_0				1.0 (168 h)	787
F, Tris-(2,3-dibromopropyl) phosphaate	*Carassius auratus*	LC_{100}				1.0 (5 d)	788
F, 2,3-Dibromopropyl phosphate	*Carassius auratus* (S)	LC_{100}				1.0 (5 d)	214
F, Cyanogen chloride	*Carassius auratus*	LC_{100}				1.0 (6–8 h)	215
F, Dutrex 729 HP	*Carassius auratus*	LC_0				1.4	62
F, Dutrex 719 UK	*Carassius auratus*	LC_0				1.4	62
F, Dobanol 23	*Carassius auratus*	LC_0				2.0-4.0	62
F, Dutrex 726 UK	*Carassius auratus*	LC_0				4.9	62

F, Hydrogen sulfide	*Carassius auratus*	LC_{100}		5.0 (200 h)	84
F, Dutrex 217 UK	*Carassius auratus*	LC_0		5.8	62
F, Acetic acid	*Carassius auratus* (long-term exp., hardW)	LC_0		10.0	219
F, Palmitic acid, Na	*Carassius auratus*	LC_{100}		11.0	213
F, Stearic acid, Na	*Carassius auratus*	LC_{100}		14.0	213
F, Monuron	*Carassius auratus*	LC_0		14.9 (5 d)	789
F, Toluene	*Carassius auratus* (F, 17–19°C)	TL_m		15.0 (720 h)	218
F, *m*-Nitrophenol	*Carassius auratus* (juv., 19°C)	LC_{53}		24.0 (8 h)	790
F, Lead sulfate	*Carassius auratus*	LC_{100}		25.0 (4 h)	791
F, Tris-treated fabric-soaked water	*Carassius auratus* (20°C)	LC_{100}		30.0 (3 h)	793
F, 2-Nitrophenol	*Carassius auratus*	LC_{38}		33.30 (8 h)	796
F, Monuron	*Carassius auratus*	LC_0		59.5 (35 d)	789
F, Acetic acid	*Carassius auratus*	LC_{100}		423.0 (20 h)	215
F, *dl*-Lactic acid	*Carassius auratus* (pH 4.6)	LC_{100}		430.0 (d?)	220
F, *dl*-Lactic acid	*Carassius auratus* (pH 4)	LC_{100}		654.0 (6–43 h)	220
F, Propanediol	*Carassius auratus* (S, 20°C, pH 6–8)	LC_{50}		5000.0 (50 h)	795
F, Colloidal sulphur	*Carassius auratus*	LC_{100}		16,000.0 (5 h)	796
F, Endosulfan	*Carassius auratus*	LC_{100}	0.0010		792
F, Endosulfan	*Carassius auratus*	LC_{50}	0.008		792
F, Acrolein	*Carassius auratus*	LD_{50}	0.08		62
F, Pentachlorophenol	*Carassius auratus*	LC_{50}	0.27		212
F, 2,3,4,6-Tetrachlorophenol	*Carassius auratus*	LC_{50}	0.75		212
F, Allyl alcohol	*Carassius auratus*	LD_{50}	1.0		62
F, 2,4,5-Trichlorophenol	*Carassius auratus* (pH 7.3)	LC_{50}	1.7		212
F, Propylene trimer	*Carassius auratus*	LD_{50}	3.0		62
F, Linevol 911	*Carassius auratus*	LD_{50}	3.0		2
F, *p*-*t*-Butyltoluene	*Carassius auratus*	LD_{50}	3.0		62
F, 2,4,4-Trimethyl-1-pentene	*Carassius auratus*	LD_{50}	3.0		62
F, 1,5,9-Cyclododecatriene	*Carassius auratus*	LD_{50}	4.0		62
F, *n*-Hexane	*Carassius auratus* (ASTM D1345)	LD_{50}	4.0		62

Chemical	Species	Toxicity test	24 h	48 h	96 h	Chronic exposure	Ref.
F, *n*-Heptane	*Carassius auratus* (ASTM D1345)	LD_{50}	4.0				62
F, Hydrogen sulfide	*Carassius auratus*	LC_{100}	4.3				84
F, Tris-1,3-dichloro-2-propyl phosphate	*Carassius auratus* (20°C)	LC_{100}	5.0				787
F, 2,4-Dichlorophenol	*Carassius auratus*	LC_{50}	7.8				212
F, *p*-Chlorophenol	*Carassius auratus*	LC_{50}	9.0				212
F, Linevol 79	*Carassius auratus*	LD_{50}	9.0				2
F, Linuron	*Carassius auratus*	LC_0	10.0				496
F, 2,4,6-Trichlorophenol	*Carassius auratus*	LC_{50}	10.0				212
F, Allyl chloride	*Carassius auratus* (softW)	LD_{50}	10.0				62
F, *o*-Xylene	*Carassius auratus* (ASTM D1345)	LD_{50}	13.0				62
F, 2-Ethylhexyl glycidyl ether	*Carassius auratus*	LD_{50}	14.0				62
F, Cyclooctadiene	*Carassius auratus* (ASTM D1345)	LD_{50}	14.0				62
F, Linuron	*Carassius auratus*	LC_{50}	15.0				496
F, Tropilidene	*Carassius auratus* (F)	LD_{50}	15.0				62
F, *o*-Chlorophenol	*Carassius auratus*	LC_{50}	16.0				212
F, *m*-Xylene	*Carassius auratus* (ASTM D1345)	LD_{50}	16.0				62
F, *p*-Xylene	*Carassius auratus* (ASTM D1345)	LD_{50}	18.0				62
F, Linuron	*Carassius auratus*	LC_{100}	20.0				496
F, *p*-Cresol	Carassius carassius	LC_{50}	21.0				183
F, 3,4-Xylenol	Carassius carassius	TL_m	21.0				184
F, Epichlorhydrin	*Carassius auratus* (S, 20°C, pH 6–8)	LC_{50}	23.0				795
F, Epichlorhydrin	*Carassius auratus*	LD_{50}	23.0				62
F, Monolinuron	*Carassius auratus*	LC_0	25.0				496
F, *m*-Cresol	*Carassius carassius*	TL_m	25.0				183
F, Phenol	*Carassius carassius*	TL_m	25.0				183
F, Styrene	*Carassius auratus* (ASTM D1345)	LD_{50}	26.0				62
F, Picloram	*Carassius auratus* (26.6°C)	LC_{50}	27.0				771
F, 2,4-Xylenol	*Carassius carassius*	TL_m	30.0				184

F, *o*-Cresol	*Carassius carassius*	TL_m	30.0	183
F, Benzene	*Carassius auratus*	LD_{50}	46.0	62
F, Phenol	*Carassius auratus* (ASTM-D1345)	LD_{50}	46.0	62
F, Monolinuron	*Carassius auratus*	LC_{50}	50.0	496
F, 3,5-Xylenol	*Carassius carassius*	TL_m	53.0	183
F, Toluene	*Carassius auratus* (ASTM D1345)	LD_{50}	58.0	62
F, Phorone	*Carassius auratus* (ASTM-D-1345)	LD_{50}	60.0	62
F, Monolinuron	*Carassius auratus*	LC_{100}	75.0	496
F, Ethylene oxide	*Carassius auratus* (S, 20°C, pH 6–8)	LC_{50}	90.0	795
F, Ethylene oxide	*Carassius auratus*	LD_{50}	90.0	62
F, Picloram, triethanolamine salt	*Carassius auratus* (26.6°C)	LC_{50}	90.6	771
F, Ethylbenzene (80%)	*Carassius auratus* (S, juv., softW, 25°C, pH 7.5)	LC_{50}	94.0	797
F, Propylene oxide	*Carassius auratus* (S, 20°C, pH 6–8)	LC_{50}	170.0	795
F, Diisopropyl ether	*Carassius auratus* (ASTM D1345)	LD_{50}	380.0	62
F, Ethanolamine	*Carassius auratus* (pH 7, ASTM D1345)	LD_{50}	500.0	2
F, Ammonium chloride	*Carassius carassius*	TL_m	640.0	74
F, Diethanolamine	*Carassius auratus* (pH 9.6)	LD_{50}	800.0	2
F, Donax	*Carassius auratus*	LD_{50}	1050.0	62
F, Butyl cellosolve	*Carassius auratus*	LD_{50}	1650.0	62
F, Isobutyl alcohol	*Carassius auratus* (S, 21°C)	LC_{50}	2600.0	795
F, Butyldigol	*Carassius auratus* (ASTM D1345)	LD_{50}	2700.0	62
F, Triethanolamine (85% TEA)	*Carassius auratus* (pH 10)	LD_{50}	3500.0	62
F, *sec*-Butyl alcohol	*Carassius auratus* (S, 21°C)	LC_{24}	4300.0	795
F, *sec*-Butanol	*Carassius auratus*	LD_{50}	4300.0	62
F, Pivalic acid	*Carassius auratus* (pH 7)	LD_{50}	4500.0	62
F, Sulfolane	*Carassius auratus* (ASTM 1345)	LD_{50}	4800.0	62
F, Ethylene glycol	*Carassius auratus*	LC_{50}	5000.0	62
F, Glycerol	*Carassius auratus*	LC_{50}	5000.0	62
F, Diethylene glycol	*Carassius auratus*	LC_{50}	5000.0	62
F, *tert*-Butyl alcohol	*Carassius auratus* (20°C)	LC_{50}	5000.0	798

Chemical	Species	Toxicity test	24 h	48 h	96 h	Chronic exposure	Ref.
F, Diethylene glycol (special grade)	*Carassius auratus*	LC_{50}	5000.0				62
F, Diacetone alcohol	*Carassius auratus*	LD_{50}	5000.0				62
F, Acetone	*Carassius auratus*	LD_{50}	5000.0				62
F, Methyl cellosolve	*Carassius auratus*	LD_{50}	5000.0				62
F, Isopropanol	*Carassius auratus*	LD_{50}	5000.0				62
F, Cellosolve	*Carassius auratus*	LD_{50}	5000.0				62
F, 2-Isopropoxyl ethanol	*Carassius auratus*	LD_{50}	5000.0				62
F, Ethyldigol	*Carassius auratus* (ASTM D1345)	LD_{50}	5000.0				62
F, Triethylene glycol	*Carassius auratus* (ASTM D1345)	LD_{50}	5000.0				62
F, Hexylene glycol	*Carassius auratus* (ASTM D1345)	LD_{50}	5000.0				62
F, Ethyltriglycol	*Carassius auratus* (ASTM D1345)	LD_{50}	5000.0				62
F, Diethanol amine	*Carassius auratus* (pH 7)	LD_{50}	5000.0				62
F, Triethanolamine, commercial	*Carassius auratus* (pH 7)	LD_{50}	5000.0				62
F, Phenol	*Carassius auratus* (S)	LC_{50}	60–200.0				146
F, Triethanolamine (85%)	*Carassius auratus* (pH 7)	LD_{50}	75,000.0				62
F, Hydroquinone	*Carassius auratus*	LC_{100}		0.287			84
F, Alkylbenzenesulfonate, linear, C = 10–15 (46.7% active)	*Carassius auratus*	LC_{50}		1.2			180
F, Alkylolefinsulfonate, C = 16–18	*Carassius auratus*	LC_{50}		1.9			180
F, *p*-Aminophenol	*Carassius auratus*	LC_{100}		2.0			84
F, Alkylbenzene sulfonate, linear, C = 10–15 (28.5% active)	*Carassius auratus*	LC_{50}		2.4			180
F, Alkylbenzene sulfonate (28.5% active)	*Carassius auratus*	LC_{50}		2.4			180
F, Algerite powder	*Carassius auratus*	LD_{100}		4.40			84
F, Nonylphenyl ethoxylate	*Carassius auratus*	LC_{50}		4.90			180
F, Alkylbenzene sulfonate, linear, C = 10–15 (15.4% active)	*Carassius auratus*	LC_{50}		4.9			180
F, *p*-Isopropyldiphenylamine	*Carassius auratus*	LC_{100}		5.7			84

F, Alkylethoxy sulphate, C = 12–15	*Carassius auratus*	LC_{50}		5.7			180
F, Alkylolefin sulfonate, C = 14–16	*Carassius auratus*	LC_{50}		5.7			180
F, 1,4-Diaminobenzene	*Carassius auratus*	LD_{100}		5.74			84
F, Catechol	*Carassius auratus*	LC_{100}		14.0			84
F, Pyrogallol	*Carassius auratus*	LC_{100}		18.0			84
F, *p*-Hydroxyphenylglycine	*Carassius auratus*	LC_{100}		20.0			84
F, Phenol	*Carassius auratus*	LD_{100}		28.9			84
F, Phenol	*Carassius auratus*	TL_m		44.5			183
F, Resorcinol	*Carassius auratus*	LC_{100}		57.40			84
F, Cuprous oxide	*Carassius* sp.	LC_{100}		60.0			799
F, 2,4-Diaminophenol, HCl	*Carassius auratus*	LC_{100}		80.0			84
F, Phloroglucinol	*Carassius auratus*	LD_{100}		630.0			8
F, Toxaphene	*Carassius auratus* (S, 25°C, pH 7.4)	LC_{50}	0.0082	0.0068	0.0056		801
F, Toxaphene	*Carassius auratus*	LC_{50}			0.014		122
F, DDT	*Carassius auratus*	LC_{50}			0.021		122
F, DDT	*Carassius auratus*	LC_{50}			0.027		802
F, Chlorodane	*Carassius auratus*	LC_{50}			0.082		785
F, Chlorodane	*Carassius auratus* (softW)	LC_{50}	0.166	0.087	0.082		785
F, Hexachlorobutadiene	*Carassius auratus* (17.5°C)	TL_m			0.09		217
F, Eulan	*Carassius auratus*	LC_0			0.100	1.0	2
F, Bromofenoxim	*Carassius auratus*	LC_{50}			0.13–0.24		804
F, Lindane	*Carassius auratus*	LC_{50}			0.131		122
F, Pentachlorophenol	*Carassius auratus*	LC_{50}	0.250		0.190		197
F, Pentachlorophenol	*Carassius auratus* (F, 25°C)	TL_m			0.22		205
F, Pentachlorophenol	*Carassius auratus*	LC_{50}	0.267		0.247		197
F, Heptachlor	*Carassius auratus* (S)	LC_{10}			0.320		805
F, Ethyl-*p*-nitrophenylthionobenzene phosphate	*Carassius auratus*	LC_{50}			0.45		806
F, Rotenone (5%)	*Carassius auratus* (S)	LC_{50}			0.497		28
F, Eulan	*Carassius auratus*	LC_{50}			0.500		2

Chemical	Species	Toxicity test	24 h	48 h	96 h	Chronic exposure	Ref.
F, Triphenyl phosphate	*Carassius auratus* (S, 25°C)	LC_{50}			0.70		807
F, Benfluralin (100% TG)	*Carassius auratus* (18°C, pH 7.1)	LC_{50}	0.88	0.85	0.81		404
F, Ionox 330	*Carassius auratus*	LC_0			1.20		345
F, Ammonia	*Carassius auratus*	TL_m	2.5		2.0		215
F, Azinphosmethyl	*Carassius auratus*, (F, 25°C)	TL_m			2.37		800
F, Parathion (25% EC)	*Carassius auratus* (25°C, pH 7.5)	LC_{50}	3.8	2.6	2.6	2.6 (72 h)	808
F, Parathion (TG 99%)	*Carassius auratus*	LC_{50}	2.7		2.7	2.7 (72 h)	808
F, Fenamiphos	*Carassius auratus*	LC_{50}			3.2		809
F, Baytex	*Carassius auratus*	LC_{50}			3.4		122
F, *p-t*-Butylbenzoic acid	*Carassius auratus* (pH 5)	LD_{50}	4.0		4.0		62
F, Azinphosmethyl	*Carassius auratus*	LC_{50}			4.3		211
F, Dobanic acid	*Carassius auratus* (pH 6)	LD_{50}	5.0		5.0		62
F, Dobanic acid	*Carassius auratus* (pH 7)	LD_{50}	7.0		5.0		62
F, Aniline	*Carassius auratus* (egg, larvae, hardW, F, 18–26°C, pH 7.86)	LC_{50}			7.60	4.60 (8 d)	810
F, Nonidet NP50	*Carassius auratus*	LD_{50}	8.0		7.0		62
F, Ionox 100	*Carassius auratus* (ASTM D1345)	LD_{50}	9.0		9.0		62
F, Aniline	*Carassius auratus* (egg, larvae, softW, F, 18–26°C, pH 7.86)	LC_{50}			9.30	5.50 (8 d)	810
F, Methyl parathion	*Carassius auratus* (fingerling, S, 18°C, pH 7.5)	LC_{50}			9.6		511
F, Acetic acid	*Carassius auratus*, (pH 7.3)	LC_0			10.0		220
F, Hydrogen sulfide	*Carassius auratus*	LC_{100}			10.0		215
F, Devrinol	*Carassius auratus*	LC_{50}			10.0		216
F, *n*-Amyl acetate	*Carassius auratus*	TL_m			10.0		239
F, Malathion	*Carassius auratus*	LC_{50}			10.7		122
F, Nonidet G2C/BX	*Carassius auratus*	LD_{50}	15.0		14.0		62
F, *o*-Chlorophenol	*Carassius auratus*	TL_m	15.0		12.0		50

F, Carbaryl	*Carassius auratus*	LC_{50}			13.2	122
F, Picloram	*Carassius auratus* (23.8°C)	LC_{50}	27.0–36.0	21.0–32.0	14.0–32.0	803
F, Ametryn	*Carassius auratus*	LC_{50}			14.1	811
F, *o*-Xylene	*Carassius auratus*	LC_{50}			16.9	202
F, *o*-Cresol	*Carassius auratus* (softW)	TL_m	49.1		19.0	50
F, Zectran	*Carassius auratus*	LC_{50}			19.1	122
F, Ethiofencarb	*Carassius auratus*	LC_{50}			20.0–40.0	813
F, Allyl chloride	*Carassius auratus* (softW)	TL_m	26.6	20.9	20.9	50
F, Toluene	*Carassius auratus*	LC_{50}			22.8	202
F, Toluene	*Carassius auratus* (S)	LC_{50}			22.8	814
F, Toluene	*Carassius auratus* (F, 17–19°C)	TL_m			23.0	218
F, Molinate	*Carassius auratus*	LC_{50}			30.0	815
F, Tris-treated fabric laundered and soaked on water	*Carassius auratus* (20°C)	NOAE			30.0	793
F, Allyl glycidyl ether	*Carassius auratus*	LD_{50}	78.0		30.0	62
F, *p*-*t*-Butylbenzoic acid	*Carassius auratus* (pH 7)	LD_{50}	33.0		33.0	62
F, Benzene	*Carassius auratus* (softW)	TL_m	34.4		34.4	154
F, *p*-Xylene	*Carassius auratus* (softW)	TL_m	36.8		36.8	50
F, Vinyl acetate	*Carassius auratus*	TL_m	42.3		42.3	50
F, Cyclohexane	*Carassius auratus*	TL_m	42.3		42.3	50
F, Phenylglycidyl ether	*Carassius auratus* (ASTM-D1345)	LD_{50}	69.0		43.0	62
F, Phenol	*Carassius auratus* (softW)	TL_m	49.9	49.1	44.5	50
F, Propoxur	*Carassius auratus* (S, 27°C)	LC_{50}			50.0	816
F, Chlorobenzene	*Carassius auratus*	TL_m	73.0		51.0	22
F, Chlorobenzene	*Carassius auratus* (S, 25°C, pH 7.5)	LC_{50}			51.62	817
F, Diphenamid	*Carassius auratus* (S, 18°C, pH 7.5)	LC_{50}			53.0	403b
F, Toluene	*Carassius auratus*	TL_m	57.7		57.7	50
F, Styrene	*Carassius auratus* (softW)	TL_m	64.7	64.7	64.7	50
F, Versatic 10	*Carassius auratus*	LC_{50}	80.0		80.0	62
F, Tris-(2-chloroethyl) phosphate	*Carassius auratus* (S)	LC_{50}			90.0	818
F, Ethylbenzene	*Carassius auratus* (softW)	TL_m	94.4	94.4	94.4	50

Chemical	Species	Toxicity test	24 h	48 h	96 h	Chronic exposure	Ref.
F, Acetic acid	*Carassius auratus* (pH 6.8)	LC_0			100.0		220
F, Sodium sulfite	*Carassius auratus*	LC_{100}			100.0		821
F, Ethanol amine	*Carassius auratus* (pH 10.1)	LD_{50}	190.0		170.0		62
F, Isoprene	*Carassius auratus*	TL_m	180.0	180.0	180.0		2
F, Benzoic acid	*Carassius auratus*	LC_{100}			200.0		84
F, Citric acid	*Carassius auratus* (hardW)	LC_0				625.0 (chr. exp.)	219
F, Citric acid	*Carassius auratus* (hardW)	LC_{100}				894.0 (chr. exp.)	219
F, Pivalic acid	*Carassius auratus* (pH 5)	LD_{50}	400.0		375.0		62
F, Asulam	*Carassius auratus* (S)	LC_{50}			5000.0		812
F, Copper oxychloride	*Carpiodes cyprinus*	LC_{100}				80.0 (h?)	819
F, Endosulfan	*Catostomus commersoni*	LC_{50}			0.003		85
F, Endothal	*Catostomus* sp.	LC_0				25.0 (8 d)	822
F, Biocallethrin (90% TG)	*Catostomus commersoni* (F, 12°C, pH 7.5)	LC_{50}			0.0124		597
F, Rotenone (5%)	*Catostomus catostomus* (S)	LC_{50}			0.057		28
F, Rotenone (5%)	*Catostomus commersoni* (S)	LC_{50}			0.068		28
F, Aldicarb	*Centropomus undecimalis* (embryo, S, 26–30°C)	LC_{50}				0.04 (36 h)	820
F, Aldicarb	*Centropomus undecimalis* (S, 26–30°C)	LC_{50}				0.04 (36 h)	820
F, Aldicarb	*Centropomus undecimalis* (juv., S, 26–30°C)	LC_{50}		0.10			820
F, Cupric sulfate	*Chaarogobius heptacanthus* (F, pH 8.2)	LC_{50}	6.5				823
F, Fenitrothion	*Channa punctatus* (F)	LC_{50}				1.5 (180 h)	165
F, Carbaryl	*Channa punctatus*	LC_{50}				2.0 (180 d)	165
F, Pentachlorophenol	*Channa gachua* (S)	LC_{50}	0.79	0.56	0.39	0.43 (72 h)	166
F, Methyl parathion	*Channa punctatus* (S, 26–29°C, pH 7.2)	LC_{50}	3.2	2.7	2.15	2.5 (72 h)	824
F, Methyl parathion	*Channa punctatus* (S, 29°C)	LC_{50}			3.0		825
F, Dimethoate	*Channa gachua*	TL_m	4.5		5.2		827
F, Dimethoate	*Channa punctatus*	ED_{50}			30.5		824
F, Dimethoate	*Channa punctatus*	LC_{50}	67.84	53.58	46.87		826
F, Abate	*Channa gachua*	LC_{50}			217.0		129

F, Phosalone	*Channa gachua* (S)	LC_{50}			0.081		827
F, Arsenic trioxide	*Chelon labrosus* (F, 11–13°C, pH 6.9–8.5)	LC_{50}	83.6	33.3	27.3	28.5 (72 h)	828
F, Endrin	*Chingatta* (S)	LC_{50}			0.0007		169
F, Endosulfan	*Chingatta* (S)	LC_{50}			0.011		170
F, Zolone (35% emulsified concentrate)	*Chingatta* (S)	LC_{50}			0.081		129
F, Leptophos	*Chingatta* (S)	LC_{50}			3.08–31.2		170
F, Dimethoate	*Chingatta* (S)	LC_{50}			4.48		129
F, Malathion	*Chingatta* (S)	LC_{50}			6.95–7.6		170
F, Fenitrothion	*Chingatta*(S)	LC_{50}			12.2		129
F, Parathion (Paramar 50)	*Cirrhinus mrigala* (26-30°C, pH 7.5)	LC_{50}		5.0			775
F, Bladex	*Cirrhinus mrigala*	LD_{50}			12.6		172
F, Bladex	*Cirrhinus mrigala*	LD_{100}			40.0		172
F, Endrin	*Clarius batrachus* (S)	LC_{50}		0.005			294, 295
F, Phosphorus, white	*Clupea harengus* (S, 9°C, pH 8)	LC_{50}				0.0025 (130 h)	832
F, 2,4-Dinitrophenol	*Clupea harengus*, embryo (S)	LC_{50}			5.50		662
F, Lindane	*Colisa fasciata* (S, 33°C)	TL_m				0.60 (12 h)	178
F, Lindane	*Colisa fasciata* (S, 18°C)	TL_m				0.87 (12 h)	178
F, DDT	*Colisa fasciata* (S)	LC_{50}	0.15	0.132	0.126		179
F, Lindane	*Colisa fasciata* (S, 33°C)	TL_m			0.41		178
F, Lindane	*Colisa fasciata* (S, 18°C)	TL_m			0.64		178
F, Arsenic trioxide	*Colisa fasciata* (S, juv., 29.8°C, pH 7.1)	LC_{50}	16.06	14.02	8.04	10.08 (72 h)	829
F, Endrin	*Coregonus lavaretus*, larvae (S)	LC_{50}				0.00006 (72 h)	162
F, Endrin	*Coregonus peled*, larvae (S)	LC_{50}				0.0006, 0.035 (72 h)	162
F, Endrin	*Coregonus lavaretus*, embryo (S)	LC_{50}				0.0034 (76 h)	162
F, Endrin	*Coregonus peled*, embryo (S)	LC_{50}				0.766 (72 h)	162
F, Arsenic trioxide	*Coregonus hoyi* (fry)	LC_{50}			17.0		830
F, Naphthalene	*Cottus* sp. (F, 4.6°C)	LC_{50}			1.40		833
F, Endothal	*Couesius plumbeus*	LC_0				25.0 (8 d)	822
C, Fluoride	Crab	NOAE			100.0		734
C, 1,2-Dichloroethane	*Crangon crangon* (S, 16°C)	LC_{50}		5.0			831

Chemical	Species	Toxicity test	24 h	48 h	96 h	Chronic exposure	Ref.
C, 1,2-Dichloroethane	*Crangon crangon* (S, 16°C)	LC_{50}		44.0			831
C, Toluene	*Crangon franciscorum* (S)	LC_{50}			3.7		455
F, Bromoform	*Crassostrea virginica* (S)	LD_{50}		1.0			181
F, Methomyl (TG)	*Salmo clarki* (S, 10°C, pH 7.5)	LC_{50}			6.8		513
F, Chloropyrifos	*Cymatogaster aggregata* (S)	TL_m			0.0035		467
F, Chloropyrifos	*Cymatogaster aggregata* (F)	TL_m			0.0037		467
F, Carbaryl	*Cymatogaster aggregata* (S, juv.)	TL_m	3.9				407
F, Endrin	*Cymatogaster aggregata* (IF)	TL_m			0.00012		186
F, DDT	*Cymatogaster aggregata* (IF)	TL_{50}			0.00026		186
F, Endrin	*Cymatogaster aggregata* (S)	TL_m			0.0008		186
F, Dieldrin (TG)	*Cymatogaster aggregata* (F)	TL_{50}			0.0015		186
F, Aldrin	*Cymatogaster aggregata* (IF)	TL_{50}			0.00226		186
F, Dieldrin (TG)	*Cymatogaster aggregata* (S)	TL_{50}			0.0037		186
F, Aldrin	*Cymatogaster aggregata* (S)	TL_{50}			0.0074		186
F, DDT	*Cymatogaster aggregata* (S)	TL_{50}			0.0076		186
F, Chlordecane (Kepone)	*Cyprinodon variegatus* (IF, 30°C)	MTD				0.000074 (141 d)	834
F, Mirex	*Cyprinodon variegatus* (F, 28–31°C)	LC_0				0.00012 (28 d)	428
F, Kepone	*Cyprinodon variegatus*	LC_{22}				0.00080 (28 d)	133
F, Arochlor, 1254	*Cyprinodon variegatus* (IF, 27–31°C)	LC_{50}				0.00093 (3 wk)	835
F, Kepone	*Cyprinodon variegatus* (IF, adult)	LC_{50}				0.0011 (28 d)	840
F, Tributyltin oxide	*Cyprinodon variegatus* (F, 29°C)	LC_{50}				0.001 (14 d)	765
F, Chlordecane (Kepone)	*Cyprinodon variegatus* (IF, 30°C)	LC_{50}				0.0013 (28 d)	840
F, Endosulfan	*Cyprinodon variegatus* (eggs to juvenile)	NOEL				0.0013 (28 d)	836
F, Tributyltin oxide	*Cyprinodon variegatus* (F, 29°C)	LC_{50}				0.0018 (7 d)	765
F, Kepone	*Cyprinodon variegatus*	LC_{80}				0.0019 (28 d)	133
F, Kepone	*Cyprinodon variegatus* (IF, juv.)	LC_{50}				0.0049 (36 d)	840
F, Chlordecane (Kepone)	*Cyprinodon variegatus* (embryo, IF, 30°C)	LC_{50}				0.0067 (36 d)	840
F, Aroclor 1254	*Cyprinodon variegatus* (F, 27–31°C)	LC_{100}				0.010 (3 wk)	835
F, Pentachlorophenol	*Cyprinodon variegatus* (embryo, F, 30°C)	NOAEL				0.047 (151 d)	841

F, Aldicarb	*Cyprinodon variegatus* (embryo, IF, 29°C)	NOAE				0.088 (28 d)	652
F, Pentachlorophenol	*Cyprinodon variegatus* (embryo, F, 30°C)	LC_{28}				0.195 (151 d)	841
F, Thalium	*Cyprinodon variegatus* (embryo, larvae)	NOAEL				4.3	837
F, Benzidine, dihydrochloride	*Cyprinodon variegatus* (F)	LC_{50}				11.5 (4 m)	843
F, 1,2-Dichloroethane	*Cyprinodon variegatus* (S, 28–31°C)	LC_{50}				130.0 (72 h)	842
F, Naphthalene	*Cyprinodon variegatus* (S)	LC_{50}	2.4				838
F, Endrin	*Cyprinodon variegatus* (F)	LC_{50}	0.0032				210
F, Decamethrin	*Cyprinodon macularius* (S)	LC_{50}		0.0006			188
F, Aldicarb	*Cyprinodon variegatus* (juv., S, 23–26°C)	LC_{50}		0.10			820
F, Aldicarb	*Cyprinodon variegatus* (juv., 23–26°C)	LC_{50}		0.1			820
F, Propoxur	*Cyprinodon variegatus* (S, 25°C)	NOEL		1.0			816
F, Endrin	*Cyprinodon variegatus*, adult (F)	LC_{50}			0.00036		187
F, Endrin	*Cyprinodon variegatus*, fry (F)	LC_{50}			0.00037		187
F, Endrin	*Cyprinodon variegatus*, juvenile (F)	LC_{50}			0.00034		187
F, Toxaphene	*Cyprinodon variegatus*	LC_{50}			0.0011		2
F, Toxaphene	*Cyprinodon variegatus* (F)	LC_{50}			0.0011		108
F, Heptachlor	*Cyprinodon variegatus* (F)	LC_{50}			0.00368		2
F, Heptachlor	*Cyprinodon variegatus* (S)	LC_{50}			0.0037		839
F, Chlordane	*Cyprinodon variegatus* (F)	LC_{50}			0.0125		190
F, Hexachlorobenzene	*Cyprinodon variegatus* (F, 32°C)	NOAE			0.0133		733
F, Chlordane	*Cyprinodon variegatus* (F)	LC_{50}			0.0245		369
F, Aldicarb	*Cyprinodon variegatus* (F, 27–29°C, pH 8.3–8.4)	LC_{50}			0.041		647
F, Aldicarb	*Cyprinodon variegatus* (28°C, pH 8.4)	LC_{50}			0.041		652
F, Malathion	*Cyprinodon variegatus* (F)	LC_{50}			0.051		189
F, Chlordecone (Kepone)	*Cyprinodon variegatus*	LC_{50}			0.0695		844
F, Kepone	*Cyprinodon variegatus* (F)	LC_{50}			0.0695		133
F, Lindane	*Cyprinodon variegatus* (F)	LC_{50}			0.1039		101
F, Chloropyriphos	*Cyprinodon variegatus* (F, 31.4°C)	LC_{50}			0.136		658
F, Isophorone	*Cyprinodon variegatus* (early lifestage) (F)	LC_{50}			0.140		845
F, Aldicarb	*Cyprinodon variegatus* (S, 27–29°C)	LC_{50}			0.168		647

Chemical	Species	Toxicity test	24 h	48 h	96 h	Chronic exposure	Ref.
F, Trifluralin	*Cyprinodon variegatus* (F)	LC_{50}			0.19		190
F, Pentachlorophenol	*Cyprinodon variegatus* (S, 6-wk-old fry)	LC_{50}			0.223		110
F, Chloropyriphos	*Cyprinodon variegatus* (juv., S, 25°C)	LC_{50}			0.270		654
F, Triphenyl phosphate	*Cyprinodon variegatus*	LC_{50}			0.3–0.56		849
F, Pentachlorophenol	*Cyprinodon variegatus* (S, 1-d-old fry)	LC_{50}			0.329		110
F, Pentachlorophenol	*Cyprinodon variegatus* (larvae, S, 30°C)	LC_{50}			0.329		698
F, Halowax 1014 (Hexa + penta + tetra)	*Cyprinodon variegatus*	LC_{50}			0.343		646
F, Carbofuran	*Cyprinodon variegatus* (F)	LC_{50}			0.386		850
F, Pentachlorophenol	*Cyprinodon variegatus* (F)	LC_{50}			0.442		190
F, Pentachlorophenol	*Cyprinodon variegatus* (juv., F, 29–31°C)	LC_{50}			0.442		846
F, Dowicide G	*Cyprinodon variegatus* (2-wk fry, F)	LC_{50}			0.516		191
F, 2,4,5-Trichlorophenol	*Cyprinodon variegatus* (juv., S, 25–31°C)	NOAE			1.0		851
F, Butylbenzyl phthalate	*Cyprinodon variegatus* (S, 20°C)	NOAE			1.0		655
F, 2,4,5-Trichlorophenol	*Cyprinodon variegatus* (S, 25–31°C)	LC_{50}			1.70		851
F, 1,3-Dichloropropane	*Cyprinodon variegatus* (S)	LC_{50}			1.770		651
F, 2,4,5-Trichlorophenol	*Cyprinodon variegatus* (juv., S, 25–31°C)	LC_{50}			1.9		851
F, Naphthalene (80%)	*Cyprinodon variegatus* (juv.)	LC_{50}			2.4		852
F, Butylbenzyl phthalate	*Cyprinodon variegatus* (S, 20°C)	LC_{50}			3.0		853
F, *p*-*t*-Butylphenyldiphenyl phosphate	*Cyprinodon variegatus*	LC_{50}			3.0		523
F, Butylbenzyl phthalate	*Cyprinodon variegatus* (S)	LC_{50}			3.0	1.0 (NOAE)	9
F, *p*-chlorphenol	*Cyprinodon variegatus* (S, 25–31°C)	NOAE			3.2		851
F, Selenium	*Cyprinodon variegatus* (S, 25–31°C)	LC_{50}	56.0		6.7		842
F, Selinium	*Cyprinodon variegatus* (S)	LC_{50}			6.71		854
F, *p*-Dichlorobenzene	*Cyprinodon variegatus* (S)	LC_{50}			7.4		660
F, Styrene	*Cyprinodon variegatus*	LC_{50}	9.1		9.1		842
F, 1,2-Dichlorobenzene	*Cyprinodon variegatus* (S, juv., 25–31°C)	LC_{50}	13.0		9.7		842
F, Chlorobenzene	*Cyprinodon variegatus* (S, juv., 25–31°C)	LC_{50}	20.0		10.0		842

F, Chlorobenzene	*Cyprinodon variegatus* (S)	LC_{50}			10.5		650
F, Epichlorhydrin	*Cyprinodon variegatus*	TL_m			11.8		259
F, Thallium	*Cyprinodon variegatus* (juv.)	LC_{50}	45.0		21.0		842
F, 1,2,4-Trichlorobenzene	*Cyprinodon variegatus* (S)	LC_{50}			21.4		650
F, *p*-nitrophenol	*Cyprinodon variegatus* (S)	LC_{50}			27.1		855
F, Tetrachlorethylene	*Cyprinodon variegatus* (juv.)	LC_{50}	52.0		29.0–52.0		842
F, Phthalic acid diethyl	*Cyprinodon variegatus* (S)	LC_{50}			29.6		853
F, Tetrachloroethylene	*Cyprinodon variegatus* (S)	LC_{50}			52.0		842
F, Nitrobenzene	*Cyprinodon variegatus* (S, juv., 25–31°C)	LC_{50}	120.0		59.0		842
F, Benzidine, DiHCl	*Cyprinodon variegatus* (F)	LC_{50}			60.0	40.0 (10 d)	843
F, Methylene chloride	*Cyprinodon variegatus* (S)	NOAE			130.0		842
F, 1,2-Dichloroethane	*Cyprinodon variegatus* (S, 28–31°C)	LC_{50}	230.0	230.0	130.0		842
F, Methyl chloride	*Cyprinodon variegatus* (S, juv., 25–31°C)	LC_{50}	300.0		170.0		842
F, Dicamba	*Cyprinodon variegatus*	LC_{50}			180.0		847
F, Toluene	*Cyprinodon variegatus*	LC_{50}			227.0–485.0		848
F, Ethylbenzene (80%)	*Cyprinodon variegatus* (juv., S, 25–31°C)	LC_{50}	300.0		280.0		856
F, Toluene	*Cyprinodon variegatus* (juv.)	LC_{50}	480.0		280.0		842
F, Methylene chloride	*Cyprinodon variegatus* (S)	LC_{50}	370.0		330.0		842
F, Isophorone	*Cyprinodon variegatus* (S, juv., 25–31°C)	LC_{50}	370.0		330.0		842
F, Octachloronaphthalene	*Cyprinodon variegatus* (S, 25–31°C)	LC_{50}			560.0		842
F, Fluoranthene	*Cyprinodon variegatus* (juv., S, 25–35°C)	LC_{50}			560.0		842
F, Fluoranthene	*Cyprinodon variegatus* (S, juv., 25–31°C)	LC_{50}	560.0		560.0		842
F, Trimethylol propane	*Cyprinodon variegatus*	LC_{50}		21,700.0	14,400.0		857
F, Pentaerythritol	*Cyprinodon variegatus*	LC_{20}			50,000.0		858
F, Endrin	*Cyprinus carpio*, juv., (S)	LC_{50}				0.0009 (72 h)	162
F, Rotenone	*Cyprinus carpio*	LC_{100}				0.010 (3 d)	557
F, Fenitrothion	*Cyprinus carpio*	NOAE				0.02 (4 wk)	81
F, Endrin	*Cyprinus carpio*, larvae (S)	LC_{50}				0.031 (72 h)	162
F, Noruron	*Cyprinus carpio* (S)	LC_0				0.2–5.2 (45 d)	860
F, Endrin	*Cyprinus carpio*, emb. (S)	LC_{50}				0.256, 0.049 (72 h)	162

Chemical	Species	Toxicity test	24 h	48 h	96 h	Chronic exposure	Ref.
F, Noruron	*Cyprinus carpio* (S)	LC_{42}				10.0 (8–35 d)	860
F, Furalaxyl	*Cyprinus carpio*	LC_{50}				38.4 (90 h)	861
F, Thiophanate methyl	*Cyprinus carpio*	TL_m				75.0	31
F, Flouride	*Cyprinus carpio*	LD_{50}				75–91.0 (480 h)	717
F, Hydrogen sulfide	*Cyprinus carpio*	LC_{100}	3.3				84
F, Linuron	*Cyprinus carpio*	LC_0	10.0				496
F, Linuron	*Cyprinus carpio*	LC_{50}	20.0–30.0				496
F, Linuron	*Cyprinus carpio*	LC_{100}	30.0				496
F, Noruron	*Cyprinus carpio* (S)	LC_{100}	50.0–100.0				860
F, Actellic	*Cyprinus carpio*	LC_{50}		0.005 ml/l			859
F, Endosulfan	*Cyprinus carpio* (fingerling)	LC_{50}		0.011			861
F, Ethyl-*p*-nitrophenylthiono-benzene phosphate	*Cyprinus carpio*	EC_{50}		0.350			862
F, Fenitrothion	*Cyprinus carpio* (S)	LC_{50}	3.31	2.55		2.3 (72 h)	164
F, Parathion	*Cyprinus carpio* (S, 20°C)	LC_0		3.0			863
F, Methyl parathion	*Cyprinus carpio* (S, 28–30°C, pH 7.3)	LC_{50}	4.0–6.0	3.0–4.0			679
F, Diazinon	*Cyprinus carpio* (S)	LC_{50}	3.18	3.14		3.11	164
F, Fenitrothion	*Cyprinus carpio*	LC_{50}		4.4			2
F, Parathion	*Cyprinus carpio* (S, 20°C)	LC_{100}		4.5			679
F, Linuron	*Cyprinus carpio*	LC_{50}		10.0			864
F, Quintozene	*Cyprinus carpio*	LC_{50}		10.0			492
F, Noruron	*Cyprinus carpio* (S)	LC_{75}		10.0		20.0 (15 d)	860
F, Methyl parathion	*Cyprinus carpio* (S, 28–30°C, pH 7.1–7.3)	LC_{50}	15.0				679
F, Parathion	*Cyprinus carpio* (S, 20°C)	LC_{50}		33.5			679
F, Bromacil (80%)	*Cyprinus carpio*	EC_{50}	164.0	164.0			865
F, Phosphamidon	*Cyprinus carpio* (S)	LC_{50}	177.7	169.3		163.4 (72 h)	164
F, Hydroxyethanediphosphonic acid	*Cyprinus carpio* (17.2°C)	LC_{50}		223.0			866
F, Hydroxyethanediphosphonic acid	*Cyprinus carpio* (17.2°C)	LC_{100}		240.0			866

F, 2,4,6-Trichlorophenyl-4′-nitro-phenyl ether	*Cyprinus carpio*	LC_{50}		290.0			163
F, Toxaphene	*Cyprinus carpio*	LC_{50}			0.004		122
F, Toxaphene	*Cyprinus carpio* (S, 18°C, pH 7)	LC_{50}			0.004		515
F, Aldrin	*Cyprinus carpio* (S)	LC_{50}			0.004		117
F, DDT	*Cyprinus carpio*	LC_{50}			0.010		122
F, Rotenone (5%)	*Cyprinus carpio* (S)	LC_{50}			0.050		28
F, Lindane	*Cyprinus carpio*	LC_{50}			0.090		122
F, Azinphosmethyl	*Cyprinus carpio*	LC_{50}			0.70		211
F, Propoxur	*Cyprinus carpio* (S, riverW)	NOEL			0.8		816
F, Edifenphos (dust)	*Cyprinus carpio*	LC_{50}			0.9		867
F, Baytex	*Cyprinus carpio*	LC_{50}			1.2		122
F, Edifenphos (emulsified concentrate)	*Cyprinus carpio*	LC_{50}			1.3		867
F, Cartap (97.7%)	*Cyprinus carpio* (juv., S, 18°C, pH 7.1)	LC_{50}			1.41		868
F, Malathion	*Cyprinus carpio* (S)	LC_{50}			1.900		2
F, Propoxur	*Cyprinus carpio* (S, riverW)	LC_{50}	4.61	4.06	2.98		816
F, Carbaryl	*Cyprinus carpio*	LC_{50}			5.3		122
F, Malathion	*Cyprinus carpio*	LC_{50}			6.6		122
F, Methyl parathion	*Cyprinus carpio* (fingerling, S, 18°C, pH 7.5)	LC_{50}			7.13		511
F, Ethiofencarb	*Cyprinus carpio*	LC_{50}			10.0–20.0		813
F, Carbaryl	*Cyprinus carpio*	LC_{50}	13.51	11.74	10.35		294
F, Propoxur	*Cyprinus carpio*	LC_{50}			10.7		816
F, Zectran	*Cyprinus carpio*	LC_{50}			13.4		122
F, Dimethyl parathion	*Cyprinus carpio* (S)	LC_{50}			14.8		116
F, Methyl parathion	*Cyprinus carpio* (S, 20°C, pH 7.2)	LC_{50}	27.6	21.2	14.8		772
F, 2,4,5-Trichlorophenoxyacetic acid	*Cyprinus carpio* (S)	LC_{50}			41.1		117
F, 2,4-Dichlorophenoxy acetic acid	*Cyprinus carpio* (S)	LC_{50}			96.5		64
F, Aroclor 1254	*Coregonus* sp. (22 d old)	LC_{50}				3.2 (5 d)	869
F, Permethrin	*Cyprinodon mascularis* (S)	LC_{50}		0.0050			188
F, Endosulfan	*Esox lucius* (fingerling)	LC_{100}	0.005				861

Chemical	Species	Toxicity test	24 h	48 h	96 h	Chronic exposure	Ref.
F, Parathion	*Esox lucius* (S, 18–21°C, pH 7.1–7.8)	LC_0	1.0				869
F, Parathion	*Esox lucius* (S, 18–21°C, pH 7.1–7.8)	LC_{100}	3.0				869
F, Neste A + oil	*Esox lucius* (free swimming, S)	LC_{50}		4.4			249
F, Neste A	*Esox lucius* (free swimming, S)	LC_{50}		5.2			249
F, Neste A	*Esox lucius* (1 mo old, S)	LC_{50}		10.0			249
F, Diquat	*Esox lucius*	LC_{50}		16.0			194
F, Neste A	*Esox lucius* (yolksac larvae, S)	LC_{50}		28.0			249
F, Neste A	*Esox lucius* (yolksac larvae, S)	LC_{50}		32.0			249
F, Neste A + oil	*Esox lucius* (yolksac-larvae, S)	LC_{50}		66.0			249
F, Biocallethrin (90% TG)	*Esox lucius* (fingerling, F, 12°C, pH 7.5)	LC_{50}			0.0033		597
F, Pyrethrum	*Esox lucius* (F)	LC_{50}			0.017		870
F, Rotenone (5%)	*Esox lucius* (S)	LC_{50}			0.033		28
F, Phorate, TG (100%)	*Esox lucius* (S, 15°C, pH 7.5)	LC_{50}			0.110		599
F, Ferbam	*Esox lucius*	NOEL			0.5–4.0		871
F, Methyl mercaptan	Fish	LC_{LO}				0.5 (120 h)	821
F, Diallate	Fish	NOAE				0.5 (3 m)	872
F, Methyl mercaptan	Fish	LC_{100}				0.9–1.75 (120 h)	821
F, Sulfur	Fish	LC_{100}				16.0 (1 h)	874
F, Picric acid	Fish	LD_0				30.0	8
F, *o*-Toluidine	Fish	TL_m				100.0 (h?)	8
F, Thiocyanate	Fish	LC_{100}				200.0 (h?)	873
F, α-Naphthoquinone	Fish	TL_m	0.30	0.60			8
F, Styphnic acid	Fish	EC_{50}			0.46		14
F, Trinitrotoluene	Fish	LC_{50}			1.6		14
F, Styphnic acid	Fish	LC_{50}			2.58		14
F, Thiosulfate	Fish	LC_{50}			2400.0		874
F, Trinitrotoluene	Fish (behavioral response)	EC_{50}			0.64		314
F, Dioxane	Freshwater fish	LC_{50}			6700.0		875
F, Naphthalene	*Fundulus heteroclitus*	LC_0				0.20 (15 d)	876

F, Cupric sulfate	*Fundulus heteroclitus* (1/2S, 24°C, larva)	LC_{50}			1.0–1.2 (7 d)	877
F, Naphthalene	*Fundulus heteroclitus* (S, 20°C, pH 7.6)	LC_{80}			2.0 (15 d)	876
F, Aroclor 1254	*Fundulus heteroclitus* (semiS, 24°C)	LC_{20}			10.00 (72 h)	878
F, Aroclor 1254	*Fundulus heteroclitus* (larvae, S, 24°C)	NOAE			10.0 (7 d)	878
F, Benzo (a)pyrene	*Fundulus* sp. (IPR)	NOEL			50.0 (10 min)	879
F, Nitrilotriacetic acid, Na, H_2O	*Fundulus heteroclitus* (S)	TL_{50}			5500.0 (168 h)	37
F, Endrin	*Fundulus similis* (F)	LC_{50}	0.00023			241
F, Permethrin, + (*cis*)	*Fundulus* sp.	NOAE		0.001		373
F, Permethrin, + (*trans*)	*Fundulus* sp.	NOAE		0.001		373
F, Permethrin (racemic)	*Fundulus* sp.	NOAE		0.001		373
F, Chloropyridyl phosphate	*Fundulus* sp. (juv.)	LC_{50}		0.0032		880
F, Chlordecane	*Fundulus similis* (juv., F, 31°C)	LC_{50}		0.004		417
F, Permethrin, + (*cis*)	*Fundulus* sp.	LC_{50}		0.013		373
F, Permethrin, + (*trans*)	*Fundulus* sp.	LC_{50}		0.017		373
F, Permethrin (racemic)	*Fundulus* sp.	LC_{50}		0.041		373
F, Kepone	*Fundulus* sp. (31°C)	LC_{50}	0.3	0.084		880
F, Tetramethrin (*cis*)	*Fundulus* sp.	LC_{50}		0.15		373
F, Tetramethrin (*trans*)	*Fundulus* sp.	LC_{50}		0.20		373
F, Dimethoate	*Fundulus* similis	LC_{50}	1.0	1.0		881
F, Permethrin (*trans*)	*Fundulus* sp.	NOAE		5.0		373
F, Permethrin (*cis*)	*Fundulus* sp.	NOAE		5.0		373
F, Tillam	*Fundulus* similis	TL_m		7.78		23
F, Eptam, technical	*Fundulus* similis	LC_{50}		10.0		25
F, Permethrin (*trans*)	*Fundulus* sp.	LC_{50}		10.0		373
F, Permethrin (*cis*)	*Fundulus* sp.	LC_{50}		10.0		373
F, 2-Methyl-4-chlorophenoxyacetic acid, amine	*Fundulus* sp. (S, 28°C)	LC_0		75.0		882
F, 2-Methyl-4-chlorophenoxyacetic acid, amine	*Fundulus* sp.	NOEL		75.0		883
F, DC Antifoam emulsion	*Fundulus heteroclitus* (S, 21°C)	LC_{50}		1000.0		884
F, Dimethylamine	*Fundulus* sp. (25°C)	LC_{50}	1000.0	1000.0		885

Chemical	Species	Toxicity test	24 h	48 h	96 h	Chronic exposure	Ref.
F, Endrin	*Fundulus heteroclitus* (S)	LC_{50}			0.0006		241, 115
F, DDT (*p,p′*)	*Fundulus majalis* (S)	LC_{50}			0.001		209
F, DDT (*p,p′*)	*Fundulus heteroclitus* (S)	LC_{50}			0.003		210
F, Aldrin	*Fundulus heteroclitus* (S)	LC_{50}			0.004		210
F, Dieldrin (100%)	*Fundulus majalis* (S)	LC_{50}			0.004		115
F, Chloropyrifos	*Fundulus* sp. (F, 30°C)	LC_{50}			0.0041		658
F, Chloropyrifos	*Fundulus heteroclitus* (F)	TL_{50}			0.0047		886
F, DDT (*p,p′*)	*Fundulus heteroclitus* (S)	LC_{50}			0.005		115
F, Dieldrin (100%)	*Fundulus heteroclitus* (S)	LC_{50}			0.005		115
F, Dieldrin	*Fundulus heteroclitus* (S)	LC_{50}			0.005		241
F, Aldrin	*Fundulus heteroclitus* (S)	LC_{50}			0.008		115
F, Chloropyrifos	*Fundulus heteroclitus* (S)	TL_{50}			0.0122		886
F, Aldrin	*Fundulus majalis* (S)	LC_{50}			0.017		115
F, Lindane	*Fundulus heteroclitus* (S)	LC_{50}			0.020		241
F, Aldrin	*Fundulus diaphanus* (S)	LC_{50}			0.021		117
F, Lindane	*Fundulus majalis* (S)	LC_{50}			0.028		115
F, Heptachlor	*Fundulus majalis* (S)	LC_{50}			0.032		115
F, Heptachlor	*Fundulus heteroclitus* (S)	LC_{50}			0.050		241, 115
F, Lindane	*Fundulus heteroclitus* (S)	LC_{50}			0.060		115
F, Malathion	*Fundulus heteroclitus* (S)	LC_{50}			0.070		241
F, Malathion	*Fundulus heteroclitus* (S)	LC_{50}			0.080		115
F, Mercuric chloride	*Fundulus heteroclitus* (S, 20°C, pH 8.0)	NOAEL			0.1		887
F, Malathion	*Fundulus diaphanus* (S)	LC_{50}			0.240		2
F, Malathion	*Fundulus majalis* (S)	LC_{50}			0.250		115
F, Pentachlorophenate, Na	*Fundulus similis* (F)	LC_{50}			0.306		109
F, Mercuric chloride	*Fundulus heteroclitus* (S, 20°C, pH 8.0)	LC_{50}			0.8		887
F, Mercuric chloride	*Fundulus heteroclitus* (S, seaW, 20°C, pH 8.0)	NOAE			0.86		887
F, Mercuric chloride	*Fundulus heteroclitus* (S, seaW, 20°C, pH 8.0)	LC_{50}			2.0		887

F, 2,2-Dichlorovinyldimethyl phosphate	*Fundulus majalis* (S)	LC_{50}	2.3		115
F, 2,2-Dichlorovinyldimethyl phospshate	*Fundulus heteroclitus* (S)	LC_{50}	2.68		115
F, Cupric sulfate	*Fundulus heteroclitus* (S, 20°C)	LC_{50}	3.1		823
F, $AlCl_3$ (salinity 0.66%)	*Fundulus heteroclitus* (S, 20°C, pH <4)	LC_{50}	3.6		888
F, $AlCl_3$ (salinity 0.79%)	*Fundulus heteroclitus* (S, 20°C, pH <4)	LC_{50}	3.6		888
F, 2,2-Dichlorovinyldimethyl phosphate	*Fundulus heteroclitis* (S)	LC_{50}	3.7		241
F, Dimethyl parathion	*Fundulus heteroclitus* (S)	LC_{50}	8.0		117
F, Dimethyl parathion	*Fundulus majalis* (S)	LC_{50}	13.8		117
F, Dimethyl parathion	*Fundulus diaphanus* (S)	LC_{50}	15.2		116
F, 2,4,5-Trichlorophenoxy-acetic acid	*Fundulus diaphanus* (S)	LC_{50}	17.4		117
F, 2,4-Dichlorophenoxyacetic acid	*Fundulus diaphanus* (S)	LC_{50}	26.7		64
F, $AlCl_3$ (salinity 1.7%)	*Fundulus heteroclitus* (S, 20°C, pH <4)	LC_{50}	27.5		888
F, $AlCl_3$ (salinity 1.88%)	*Fundulus heteroclitus* (S, 20°C, pH <4)	LC_{50}	31.5		888
F, Dimethyl parathion	*Fundulus heteroclitus* (S)	LC_{50}	58.0		117
F, Zinc chloride	*Fundulus heteroclitus* (S, 20°C, pH 8)	LC_{50}	60.0		889
F, Chromic acid, K_2	*Fundulus heteroclitus* (S, 19.5°C, pH 8.0)	NOAE	75.0	25.0 (168 h)	890
F, Chromic acid, K_2	*Fundulus heteroclitus* (S, 19.5°C, pH 8.0)	LC_{50}	91.0	44.0 (168 h)	890
F, Hexachlorobenzene	*Fundulus* grandis (seaW, F, 23°C, pH 6.6–7.9)	NOAE		125.0 (28 d)	891
F, Dimethylnitroso amine	*Fundulus heteroclitus* (S, 22°C)	LC_{50}	200.0		892
F, Nickel (2+)	*Fundulus heteroclitus* (S, 20°C, pH 7.8)	LC_0	250.0		680
F, Cobalt chloride	*Fundulus heteroclitus* (S, 20°C)	LC_{50}	275.0		888
F, Nickel (2+)	*Fundulus heteroclitus* (S, 20°C, pH 7.8)	LC_{50}	350.0		680
F, Cobalt carbonate	*Fundulus heteroclitus* (S, 20°C)	LC_{50}	1000.0		888
F, Nitrilotriacetic acid, Na, H_2O	*Fundulus heteroclitus* (S)	TL_{50}	5500.0		37
F, Phosphorus, white	*Gadus morrhua* (F, 10°C)	LC_{50}		0.0019 (125 h)	832
F, *o*-Xylene	*Gadus morrhua* (S, embryo, 4-6°C)	LC_{50}		0.016–0.035 (6 × 17 d)	892
F, *p*-Xylene	*Gadus morrhua* (embryo, 4–6°C)	LC_{95}		0.016–0.035 (6 h)	892
F, *o*-Xylene	*Gadus morrhua* (S, 4-6°C, egg)	NOAE		0.035 (7 h)	892

Chemical	Species	Toxicity test	24 h	48 h	96 h	Chronic exposure	Ref.
F, Venezuelan crude, water-soluble fraction	*Gadus morrhua*, embryo (S)	LC_{60}				10.0 (100 h)	893
F, 1,2-Dichloroethane	*Gadus morrhua* (S, renewal 6 h, 13–18°C)	LC_{50}		2.50			894
F, 1,2-Dichloroethane	*Gadus morrhua* (S, 16°C)	LC_{50}		10.0			831
F, Phosphorus, white	*Gadus morrhua* (F, 11°C)	LC_{50}			0.0065		832
F, *p*-Cresol	*Gadus morrhua* (24-h-old fertilized eggs, S, 5°C)	LC_{50}			5.0		895
F, *o*-Cresol	*Gadus morrhua* (24-h fertilized eggs, 5°C)	LC_{50}			12.0		895
F, *m*-Cresol	*Gadus morrhua* (24-h-old, fertilized eggs, S, 5°C)	LC_{50}			30.0		895
F, Heptachlor	*Gambusia affinis* (S)	LC_{50}				0.070 (30 h)	805
F, Parathion	*Gambusia affinis* (S)	LC_{50}				0.20 (72 h)	221
F, Tributyltin acetate	*Gambusia affinis* (aerated, S)	LC_{100}				0.2–1.0 (20 min–3 h)	765
F, Chloropyrifos	*Gambusia affinis* (S)	LC_{50}				0.26 (72 h)	2
F, Carbofuran	*Gambusia affinis*	LC_{50}				0.52 (72 h)	221
F, Propanil	*Gambusia affinis* (S)	LC_{50}				7.62 (72 h)	2
F, Furfural	*Gambusia affinis*	NOAE				10.0	8
F, *n*-Heptane	*Gambusia affinis*	TL_m				4924.0 (24–96 h)	243
F, *n*-Heptane	*Gambusia affinis*	NOAE				5600.0	8
F, *n*-Heptane	*Gambusia affinis*	LC_{100}				10,000.0	8
F, Acetamide	*Gambusia affinis*	TL_m				15,500.0–20,000.0 (72 h)	8
F, Decamethrin	*Gambusia affinis*	LC_{50}	0.001				896
F, Decamethrin	*Gambusia affinis*	LC_{90}	0.002				896
F, Toxaphene	*Gambusia affinis* (S)	LC_{50}	0.045				242
F, Methyl parathion	*Gambusia affinis* (23°C)	LC_0	0.10				897
F, Toxaphene	*Gambusia affinis* (S)	LC_{50}	0.860				515
F, Acetic acid	*Gambusia affinis* (RW)	TL_m	251.0				243
F, Pyridine	*Gambusia affinis*	TL_m	1350.0				243

F, Diethanolamnie	*Gambusia affinis* (riverW)	TL_m	1800.0			243
F, Cyclohexane	*Gambusia affinis* (turbid riverW)	TL_m	15,500.0			243
F, Diethylene glycol	*Gambusia affinis*	TL_m	32,000.0			243
F, Endrin	*Gambusia affinis* (susceptible strain)	LC_{50}		0.0006		245
F, Decamethrin	*Gambusia affinis* (S)	LC_{50}		0.001		188
F, Dieldrin	*Gambusia affinis* (susceptible strain)	LC_{50}		0.008		2
F, Toxaphene	*Gambusia affinis* (suscept.strain)	LC_{50}		0.012		245
F, DDT	*Gambusia affinis* (susceptible strain)	LC_{50}		0.019		245
F, Aldrin	*Gambusia affinis* (susceptible strain)	LC_{50}		0.036		245
F, Triton X100 emulsified concentrate	*Gambusia affinis* (susceptible strain) (S)	LC_{50}		0.05		898
P, DDT	*Gambusia affinis* (resistant strain)	LC_{50}		0.096		245
F, Permethrin	*Gambusia affinis* (S)	LC_{50}		0.0970		188
F, Parathion (Folidol E605)	*Gambusia affinis* (S, 30.5°C, pH 7)	LC_{50}		0.1		775
F, Folidol E605	*Gambusia affinis* (S, 30.5°C, pH 7)	LC_{100}		0.2		775
F, Triton X100 emulsified concentrate	*Gambusia affinis* (resistant strain) (S)	LC_{50}		0.2		898
F, Endrin	*Gambusia affinis* (resistant strain)	LC_{50}		0.314		245
F, Dieldrin	*Gambusia affinis* (resistant strain)	LC_{50}		0.434		245
F, Toxaphene	*Gambusia affinis* (resistant strain)	LC_{50}		0.459		245
F, Nitrofen	*Gambusia affinis*	LC_{50}		0.782		900
F, 2,4-D, butylthio ester	*Gambusia affinis* (susceptible strain)	LC_{50}		0.98		2
F, 2,4-D, ethylthio ester	*Gambusia affinis* (suscept. strain)	LC_{50}		1.34		244
F, 2,4-D, ethylthio ester	*Gambusia affinis* (resistant strain, S)	LC_{50}		1.58		899
F, 2,4-D, butylthio ester	*Gambusia affinis* (resistant strain, S)	LC_{50}		1.7		2
F, Aldrin	*Gambusia affinis* (resistant strain)	LC_{50}		2.735		245
F, Propanil	*Gambusia affinis* (S)	LC_{50}		8.45		901
F, Propanil	*Gambusia affinis* (S)	LC_{90}		12.14		901
F, *n*-Amyl acetate	*Gambusia affinis*	TL_m	65.0	65.0		22
F, Pyrene	*Gambusia affinis* (S, 24–27°C)	TL_m			0.0026	246
F, Permethrin	*Gambusia affinis* (S, 24°C, pH 8.4)	LC_{50}			0.015	902

Chemical	Species	Toxicity test	24 h	48 h	96 h	Chronic exposure	Ref.
F, Propanil	*Gambusia affinis* (S)	LC_{50}	11.3	11.0	9.46		242
F, Furfural	*Gambusia affinis*	TL_m			24.0		8
F, *m*-Cresol	*Gambusia affinis*	TL_m	24.0		24.0		22
F, Furfural	*Gambusia affinis*	TL_m	44.0		24.0		243
F, Carbaryl	*Gambusia affinis* (S)	LC_{50}	40.0	35.0	31.8		242
F, Phenol	*Gambusia affinis*	TL_m	22.70	22.20	56.0		22
F, Sodium arsenite	*Gambusia affinis* (S, 21–22°C)	LC_{50}	135.0	100.0	59.0		904
F, Thiocyanate	*Gambusia affinis* (16–23°C)	LC_{50}			114.0		873
F, Picloram	*Gambusia affinis* (S, 21-22°C)	LC_{50}			120.0		904
F, Carbon disulfide	*Gambusia affinis*	TL_m	162.0		135.0		22
F, Propylene oxide	*Gambusia affinis* (S, 23°C)	LC_{50}			141.0		905
F, Propene oxide	*Gambusia affinis* (S, 24°C)	TL_m			141.0		152
F, Naphthalene	*Gambusia affinis*	TL_m	220.0		150.0		22, 243
F, Benzoic acid	*Gambusia affinis*	LC_{50}	240.0	255.0	180.0		22
F, Ammonium acetate	*Gambusia affinis*	TL_m	238.0		238.0		22
F, Acetic acid	*Gambusia affinis*	TL_m	251.0		251.0		22
F, 4-Amino-*m*-tolouene	*Gambusia affinis*	TL_m	425.0	410.0	375.0		22
F, Benzene	*Gambusia affinis*	TL_m	395.0		395.0		154
F, Dicamba	*Gambusia affinis* (S, 12°C)	LC_{50}			465.0		902
F, Boron	*Gambusia affinis*	LC_{50}			980.0		491
F, Toluene	*Gambusia affinis* (turbid water)	TL_m	1340.0		1280.0		243
F, Sulfur	*Gambusia affinis*	LC_{50}			10,000.0		261
F, Acetone	*Gambusia affinis*	TL_m	13,000.0	13,000.0	13,000.0		243
F, Acetamide	*Gambusia affinis*	TL_m	26,300.0	26,300.0	13,300.0		22
F, α-Naphthylamine	*Gardon* (F)	TL_m				6.0	8
F, Toxaphene	*Gasterosteus aculeatus*	TL_m				0.0078	2
F, Endrin	*Gasterosteus aculeatus* (S)	TL_m			0.0005		211
F, Endrin	*Gasterosteus aculeatus* (S)	TL_m			0.0015		293
F, Azinphosmethyl	*Gasterosteus aculeatus*	TL_m			0.0048		211

F, Toxaphene	*Gasterosteus aculeatus* (S, 20°C, pH 7.4)	LC_{50}		0.012	0.0086		376
F, DDT	*Gasterosteus aculeatus* (S)	TL_m			0.0115		211
F, Dieldrin (TG)	*Gasterosteus aculeatus* (S)	TL_m			0.0131		211
F, Aldrin	*Gasterosteus aculeatus*	TL_m			0.0274		211
F, Lindane	*Gasterosteus aculeatus* (S)	TL_m			0.050		115
F, Malathion	*Gasterosteus aculeatus* (S)	LC_{50}			0.0769		211
F, Heptachlor	*Gasterosteus aculeatus* (S)	TL_m			0.1119		211
F, Carbaryl	*Gasterosteus aculeatus* (S, juv.)	TL_m	6.7		3.99		407
F, Acrylonitrile	*Gobius minutus* (seaW) 15°C	LC_{50}	20.0	15.0	14.0	1800.0 (1 h)	59
F, PDM5-silicone	*Gobiidae* sp.	NOEL			9.5		909
F, Endrin	*Gobio gobio* (S)	LC_{50}	0.0073			0.0015 (72 h)	162
F, Phenol	*Gobius minutus* (15°C)	LC_{50}				13.0 (12 h)	59
F, Phenol	*Gobius minutus* (15°C)	LC_{50}				15.0 (6 h)	59
F, Phenol	*Gobius minutus* (15°C)	LC_{50}				20.0 (3 h)	59
F, 4-Propylenebenzene sulfonate	*Gobius minutus* (softW, 15°C)	LC_{50}				25.0 (12 h)	59
F, 4-Propylenebenzene sulfonate	*Gobius minutus* (softW, 15°C)	LC_{50}				25.0 (3 h)	59
F, 4-Propylenebenzene sulfonate	*Gobius minutus* (softW, 15°C)	LC_{50}				25.0 (6 h)	59
F, 4-Propylenebenzene sulfonate	*Gobius minutus* (softW, 15°C)	LC_{50}				30.0 (1 h)	59
F, Phenol	*Gobius minutus* (15°C)	LC_{50}				85.0 (1 h)	59
F, 4-Propylenebenzene sulfonate	*Gobius minutus* (softW, 15°C)	LC_{50}				150.0 (27 min)	59
F, Phenol	*Gobius minutus* (15°C)	LC_{50}				320.0 (27 min)	59
F, 4-Propylenebenzene sulfonate	*Gobius minutus* (softW, 15°C)	LC_{50}				7000.0 (9 min)	59
F, 4-Propylenebenzene sulfonate	*Gobius minutus* (softW, 15°C)	LC_{50}				18000.0 (3 min)	59
F, Phenol	*Gobius minutus* (15°C)	LC_{50}				18000.0 (9 min)	59
F, Phenol	*Gobius minutus* (15°C)	LC_{50}	10.0	10.0	9.0	9.0 (72 h)	59
F, 4-Propylenebenzene sulfonate	*Gobius minutus* (softW, 15°C)	LC_{50}	25.0	25.0	25.0		59
F, 1,2-Dichloroethane	*Gobius minutus* (F, 15°C, pH 8)	LC_{50}			185.0		831
F, Phenol	*Notemigonus crysoleucas* (S)	LC_{50}	35–129.0				146
F, Laurylsulfate	*Jordanella floridae* (F, 25°C)	LC_{50}		10.0	8.10	6.90 (10 d)	155
F, Tri-*n*-butylamine	*Gudgeon*	TL_m				20.0–40.0 (h?)	8

Chemical	Species	Toxicity test	24 h	48 h	96 h	Chronic exposure	Ref.
F, *n*-Propanol	*Gudgeon*	TL_m	200.0–500.0				8
F, Parathion (Paramar M50)	*Heteropneustes fossilis* (S, 26°C)	LC_0			0.63		907
F, Parathion (Paramar M50)	*Heteropneustes fossilis* (Int, 26°C)	LC_{50}			26.0-29.0		907
F, Parathion (Paramar M50)	*Heteropneustes fossilis* (S, 26°C)	LC_{50}			32.0		907
F, Toxaphene	*Icatalurus punctatus* (18°C, pH 7.1)	LC_{50}			0.013		908
F, Toxaphene	*Ictalurus punctatus* (F)	MATC				0.000129 (240 d)	167
F, RU-11679 (aa pyrethroid)	*Ictalurus punctatus* (F)	TL_{50}				0.000194	138
F, RU-11679	*Ictalurus punctatus* (F)	LC_{50}				0.000305 (25 d)	138
F, RU-11679	*Ictalurus punctatus* (F)	LC_{50}				0.000305 (20 d)	138
F, RU-11679	*Ictalurus punctatus* (F)	LC_{50}				0.000305 (30 d)	138
F, RU-11679	*Ictalurus punctatus* (F)	LC_{50}				0.000410 (10 d)	138
F, RU-11679	*Ictalurus punctatus* (F)	LC_{50}				0.000410 (15 d)	138
F, RU-11679	*Ictalurus punctatus* (F)	LC_{50}				0.000750 (5 d)	138
F, Toxaphene	*Ictalurus punctatus* (F)	LC_{50}				0.015 (9 d)	167
F, Rotenone	*Ictalurus* sp.	LC_{100}				0.025 (3 d)	557
F, Aroclor 1242	*Ictalurus punctatus* (F)	LC_{50}				0.087 (30 d)	92
F, Aroclor 1248	*Ictalurus punctatus* (F)	LC_{50}				0.121 (15 d)	906
F, Aroclor 1254	*Ictalurus punctatus* (F)	LC_{50}				0.181 (25 d)	145
F, Aroclor 1260	*Ictalurus punctatus* (F)	LC_{50}				0.433 (30 d)	92
F, Methomyl	*Ictalurus punctatus* (S, 26°C, pH 7.0)	LC_{50}	0.92				910
F, Picloram K salt	*Ictalurus punctatus* (26.6°C)	LC_{50}	41.0				913
F, Picloram, triethanolamine salt	*Ictalurus punctatus* (26.6°C)	LC_{50}	70.5				913
F, Tordon 22K	Ictalurus melas	LC_{50}	91.0				803
F, Pyrethrum	*Ictalurus punctatus* (24°C)	LC_{50}		0.082			911
F, Ammonia	*Ictalurus* sp. (S)	LC_{50}		0.28			254
F, RU-11679	*Ictalurus punctatus* (S)	LC_{50}			0.000630		138
F, RU-11679	*Ictalurus punctatus* (F)	LC_{50}			0.000700		138
F, Endosulfan	*Ictalurus punctatus* (S, 18°C, pH 7.5)	LC_{50}			0.0015		600
F, Toxaphene	*Ictalurus melas* (S, 22.8°C, pH 7.9)	LC_{50}	0.012	0.0042	0.0018		783

F, Toxaphene	*Ictalurus melas* (S, 17.2°C, pH 7.9)	LC_{50}	0.015	0.043	0.0027	783
F, Dieldrin	*Ictalurus punctatus*	LC_{50}			0.0045	71
F, Toxaphene	*Ictalurus ameiurus*	LC_{50}			0.005	122
F, DDT	*Ictalurus melas*	LC_{50}			0.005	127
F, DDT	*Ictalurus ameiurus*	LC_{50}			0.005	122
F, Toxaphene	*Ictalurus* sp.	LC_{50}			0.013	122
F, Allethrin	*Ictalurus punctatus* (F)	LC_{50}			0.0146	138
F, Binapacryl (99%, TG)	*Ictalurus punctatus* (18°C, pH 7.1)	LC_{50}	0.062		0.015	914
F, DDT	*Ictalurus* sp.	LC_{50}			0.016	122
F, DDT	*Ictalurus punctatus*	LC_{50}			0.016	127
F, Toxaphene	*Ictalurus melas* (S, 11.6°C, pH 7.9)	LC_{50}	0.048	0.048	0.025	783
F, Bio-allethrin	*Ictalurus punctatus* (F, 12°C, pH 6.4–8.4)	LC_{50}			0.027	915
F, Allethrin (90% active)	*Ictalurus punctatus* (F)	LC_{50}			0.027	144
F, Allethrin (90% active)	*Ictalurus punctatus* (S)	LC_{50}			0.0301	138
F, Bio-allethrin	*Ictalurus punctatus* (S, 12°C, pH 6.4-8.4)	LC_{50}			0.0301	915
F, Lindane	*Ictalurus* sp.	LC_{50}			0.044	122
F, Lindane	*Ictalurus punctatus*	LC_{50}			0.044	127
F, Formaldehyde	*Ictalurus melas* (F)	LC_{50}			0.0621 ml/l	120
F, Lindane	*Ictalurus ameiurus*	LC_{50}			0.064	122
F, Lindane	*Ictalurus melas*	LC_{50}			0.064	127
F, Formaldehyde	*Ictalurus punctatus* (F)	LC_{50}			0.0658 ml/l	120
F, Biocallethrin, 90% TG	*Ictalurus punctatus* (F, 12°C, pH 7.5)	LC_{50}			0.096	597
F, Pyrethrins	*Ictalurus punctatus* (S)	LC_{50}			0.114	138
F, Dinoseb	*Ictalurus punctatus* (S)	LC_{50}			0.118	168
F, Aminocarb (17%)	*Ictalurus punctatus* (S, 22°C, pH 7.4)	LC_{50}			0.13	517
F, Bromofenoxim	*Ictalurus* sp.	LC_{50}			0.13–0.24	804
F, Aminocarb	*Ictalurus punctatus* (20°C, pH 7.5)	LC_{50}			0.130	609
F, Pyrethrins	*Ictalurus punctatus* (F)	LC_{50}			0.132	138
F, Rotenone (5%)	*Ictalurus punctatus* (S)	LC_{50}			0.164	28
F, Dimethrin	*Ictalurus punctatus* (F)	LC_{50}			0.165	168

Chemical	Species	Toxicity test	24 h	48 h	96 h	Chronic exposure	Ref.
F, Carbofuran, TG	*Ictalurus punctatus*	LC_{50}			0.21		912
F, Phorate	*Ictalurus punctatus* (12°C, pH 7.5)	LC_{50}			0.280		599
F, Methomyl, TG (liquid conc.)	*Ictalurus punctatus* (S, 22°C, pH 7.5)	LC_{50}			0.300		513
F, Bensulide	*Ictalurus punctatus* (S)	LC_{50}			0.379		168
F, Rotenone (5%)	*Ictalurus melas* (S)	LC_{50}			0.389		28
F, Trifluralin	*Ictalurus punctatus* (S)	LC_{50}			0.417		168
F, Chlordecone (Kepone)	*Ictalurus punctatus* (juv., F, 20–23°C)	LC_{50}	0.516	0.516	0.514		767
F, Methomyl, TG	*Ictalurus punctatus* (S, 22°C, pH 7.5)	LC_{50}			0.530		513
F, Captan	*Ictalurus punctatus* (fingerling, S)	LC_{50}			0.775		778
F, Tricresyl phosphate	*Ictalurus punctatus* (F, 12°C, pH 7.5)	LC_{50}			0.803		513
F, Furanace	*Ictalurus punctatus* (S)	LC_{50}			1.07		121
F, Dimethrin	*Ictalurus punctatus* (S)	LC_{50}			1.14		138,140
F, Propoxur	*Ictalurus punctatus* (S, 27°C)	LC_{50}			1.3		816
F, Baytex	*Ictalurus ameiurus*	LC_{50}			1.6		122
F, Baytex	*Ictalurus melas*	LC_{50}			1.62		127
F, Baytex	*Ictalurus punctatus*	LC_{50}			1.68		127
F, Baytex	*Ictalurus* sp.	LC_{50}			1.7		122
F, Tetradifon	*Ictalurus punctatus* (S, 18°C, pH 7.5)	LC_{50}			2.1		607
F, Pydraul 50E	*Ictalurus punctatus* (F)	LC_{50}			2.40		139
F, 4-Aminopyridine	*Ictalurus punctatus* (S, 12–22°C)	TL_m			2.43–5.8		916
F, Ciodrin	*Ictalurus punctatus*	LC_{50}			2.5		71
F, Parathion (98.7% active)	*Ictalurus punctatus* (fingerling, S, 18°C, pH 7.5)	LC_{50}			2.65		511
F, Pydraul 50E	*Ictalurus punctatus* (S)	LC_{50}	7.2		3.0		139
F, Azinphosmethyl	*Ictalurus punctatus*	LC_{50}			3.290		127
F, Azinphosmethyl	*Ictalurus melas*	LC_{50}			3.500		127
F, Propanil	*Ictalurus punctatus* (S)	LC_{50}			3.796		168
F, Carbofuran	*Ictalurus punctatus*	LC_{50}			4.1		912
F, Disulfoton	*Ictalurus punctatus* (18°C)	LC_{50}			4.700		941

F, Monocrotophos (38% active)	*Ictalurus punctatus* (S, 18°C, pH 7.1)	LC_{50}	32.0	4.93	917
F, Methyl parathion	*Ictalurus punctatus* (fingerling, 18°C, pH 7.5)	LC_{50}		5.24	511
F, Dimethyl parathion	*Ictalurus punctatus*	LC_{50}		5.71	127
F, Picloram	*Ictalurus punctatus* (S, 18°C, pH 7.5)	LC_{50}		6.3	411
F, Dimethyl parathion	*Ictalurus melas*	LC_{50}		6.64	127
F, Ethion	*Ictalurus punctatus*	LC_{50}		7.5	71
F, Malathion	*Ictalurus punctatus*	LC_{50}		8.970	127
F, Malathion	*Ictalurus* sp.	LC_{50}		9.0	122
F, Aminocarb (98% active)	*Ictalurus punctatus* (S, 22°C, pH 7.4)	LC_{50}		10.0	517
F, Aminocarb	*Ictalurus punctatus* (S, 20°C, pH 7.5)	LC_{50}		10.0	475
F, Imidan	*Ictalurus punctatus* (S)	LC_{50}		11.0	148
F, Phosmet	*Ictalurus punctatus* (S)	LC_{50}	13.0	11.0	918
F, Zectran	*Ictalurus punctatus*	LC_{50}		11.4	127
F, Zectran	*Ictalurus* sp.	LC_{50}		11.4	122
F, Malathion	*Ictalurus melas*	LC_{50}		12.900	127
F, Malathion	*Ictalurus ameiurus*	LC_{50}		12.9	122
F, Roundup	*Ictalurus punctatus* (22°C)	LC_{50}	13.0	13.0	562
F, Isopropylphenyldiphenyl phosphate	*Ictalurus punctatus* (F)	LC_{50}		15.0	849
F, Houghtosafe (phosphate esters mixture)	*Ictalurus punctatus* (F)	LC_{50}		15.0	139
F, Carbaryl	*Ictalurus punctatus*	NOEL		15.8	127
F, Zectran	*Ictalurus melas*	LC_{50}		16.7	127
F, Zectran	*Ictalurus ameiurus*	LC_{50}		16.7	122
F, Carbaryl	*Ictalurus melas*	NOEL		20.0	127
F, *p*-Chloraniline	*Ictalurus punctatus* (S)	LC_{50}		23.0	147
F, Isopropylphenyldiphenyl phosphate	*Ictalurus punctatus* (S, 20°C, pH 7.4)	LC_{50}		43.0	849
F, Houghtosafe	*Ictalurus punctatus* (S)	LC_{50}	130.0	43.0	139
F, Monuron	*Ictalurus nebulosus*	LC_{50}		57.0	789
F, Furalaxyl	*Ictalurus* sp.	LC_{50}		60.0	919

Chemical	Species	Toxicity test	24 h	48 h	96 h	Chronic exposure	Ref.
F, Tordon 22K	*Ictalurus melas*	LC_0			69.0		803
F, Phosphamidon	*Ictalurus punctatus*	LC_{50}			70.0		71
F, 2,6-Difluorobenzoic acld	*Ictalurus punctatus* (S)	LC_{50}			100.0		147
F, 4-Chlorophenylurea	*Ictalurus punctatus* (S)	LC_{50}			100.0		147
F, Altosid	*Ictalurus punctatus* (S)	TL_{50}			100.0		28
F, Pydraul 115E	*Ictalurus punctatus* (S)	LC_{50}	100.0		100.0		139
F, Glyphosate	*Ictalurus punctatus* (22°C)	LC_{50}	130.0		130.0		920
F, Endothal-di K	*Ictalurus punctatus* (fry, S, 12°C, pH 7.5)	LC_{50}			150.0		607
F, Ethylbenzene	*Ictalurus punctatus* (S, 22°C)	LC_{50}			210.0		921
F, Diethylene glycol dinitrate	*Ictalurus punctatus*	LC_{50}			278.3		1088
F, Diethylene glycol dinitrate	*Lepomis macrochirus*	LC_{50}			258.0		1088
F, Diflubenzuron	*Ictalurus punctatus* (S)	LC_{50}			370.0		147
F, Benzene	*Ictalurus punctatus* (S, 22°C)	LC_{50}			425.0		921
F, Asulam	*Ictalurus punctatus* (S)	LC_{50}			5000.0		812
F, Rotenone	*Ictiobus cyprinellus*	LC_{100}				0.0083 (3 d)	557
F, 2,6-Dichlorophenol	*Idus idus melanotus*	LC_0	2.0				223
F, 2,6-Dichlorophenol	*Idus idus melanotus*	LC_{50}	4.0				223
F, 2,6-Dichlorophenol	*Idus idus melanotus*	LC_{100}	5.0				223
F, Linuron	*Idus melanotus*	LC_0	8.5				496
F, Linuron	*Idus melanotus*	LC_{50}	12.0				496
F, Linuron	*Idus melanotus*	LC_{100}	18.0				496
F, Monolinuron	*Idus melanotus*	LC_{50}	65.0				496
F, Pentachlorophenol	*Idus idus melanotus*	LC_0		0.20			223
F, Pentachlorophenol	*Idus idus melanotus*	LC_{50}		0.30			223
F, Pentachlorophenol	*Idus idus melanotus*	LC_{100}		0.50			223
F, 2,4,5-Trichlorophenol	*Idus idus melanotus*	LC_0		1.0			223
F, *m*-Chlorophenol	*Idus idus melanotus*	LC_0		1.0			223
F, 2,4,5-Trichlorophenol	*Idus idus melanotus*	LC_{50}		1.3			223
F, 2,4,5-Trichlorophenol	*Idus idus melanotus*	LC_{100}		1.6			223

F, *p*-Chlorophenol	*Idus idus melanotus*	LC_0		2.0			223
F, *m*-Chlorophenol	*Idus idus melanotus*	LC_{50}		3.0			223
F, *p*-Chlorophenol	*Idus idus melanotus*	LC_{50}		3.5			223
F, *o*-Chlorophenol	*Idus idus melanotus*	LC_0		5.0			223
F, *p*-Chlorophenol	*Idus idus melanotus*	LC_{100}		5.5			223
F, *m*-Chlorophenol	*Idus idus melanotus*	LC_{100}		6.0			223
F, *o*-Chlorophenol	*Idus idus melanotus*	LC_{50}		8.5			223
F, *o*-Chlorophenol	*Idus idus melanotus*	LC_{100}		10.0			223
F, Alkylbenzene sulfonate (46.7% active)	*Idus idus*	LC_{50}		0.8–0.9	0.4–0.6		180
F, Alkylolefin sulfonate, C = 16–18	*Idus idus*	LC_{50}		1.0	0.9		180
F, Alkylbenzene sulfonate, 28.5% active	*Idus idus*	LC_{50}		1.3–1.7	1.2–1.3		180
F, Alkylbenzene sulfonate, 15.4% active	*Idus idus*	LC_{50}		2.1–2.9	1.9–2.9		180
F, Alkyl ethoxysulfate	*Idus idus*	LC_{50}		3.4–7.2	3.3–6.2		180
F, Alkylolefin sulfonate, C = 14–16	*Idus idus*	LC_{50}		3.7–6.8	3.4–4.9		180
F, Nonylphenol ethoxylate	*Idus idus melanotus*	LC_{50}		7.40–11.30	7.0–11.20		180
F, Endrin	*Jordanella floridae* (F)	MATC				0.00022 (110 d)	169
F, Malathion	*Jordanella floridae* (F)	MATC				0.0086 (110 d)	169
F, Sodium arsenite	*Jordanella floridae* (emb, F, 26°C, pH 7.2–8)	NOAEL				2.130 (31 d)	477
F, Sodium arsenite	*Jordanella floridae* (emb, F, 26°C, pH 7.2–8)	LOAEL				4.120 (31 d)	477
F, Endrin	*Jordanella floridae* (F)	LC_{50}			0.00085		169
F, Malathion	*Jordanella floridae* (F)	LC_{50}			0.349		169
F, Diazinon	*Jordanella floridae* (F)	LC_{50}			1.5–1.8		141
F, Pentachlorophenol	*Jordanella floridae* (F, 25°C)	LC_{50}		1.82	1.74	1.74 (10 d)	155
F, Tordon 22K	*Jordanella floridae* (F, 15°C)	LC_{50}			26.1	12.3	155
F, Phenol	*Jordanella floridae* (25°C, F)	LC_{50}		36.3	36.3	36.3 (10 d)	155
F, Aroclor 1254	*Lagodon rhomboides* (F, seaW)	LC_{50}				0.005 (12 d)	613
F, Acrylonitrile	*Lagodon rhomboides*	LD_0	20.0				922
F, Acrylonitrile	*Lagodon rhomboides* (S, 14–20°C)	LC_{50}	24.5				924

Chemical	Species	Toxicity test	24 h	48 h	96 h	Chronic exposure	Ref.
F, Acrylonitrile	*Lagodon rhomboides*	LD_{50}	24.5				922
F, Acrylonitrile	*Lagodon rhomboides*	LD_{100}	30.0				922
F, Endosulfan	*Lagodon rhomboides* (F)	LC_{50}			0.00003		85
F, Toxaphene	*Lagodon rhomboides* (F)	LC_{50}			0.0005		108
F, Toxaphene	*Lagodon rhomboides*	LC_{50}			0.00053		2
F, Heptachlor	*Lagodon rhomboides* (F)	LC_{50}			0.00377		2
F, Chlordane	*Lagodon rhomboides* (F)	LC_{50}			0.0064		369
F, Lindane	*Lagodon rhomboides* (F)	LC_{50}			0.0306		101
F, Pentachlorophenate, Na	*Lagodon rhomboides* (S, larvae)	LC_{50}			0.0380		110
F, Pentachlorophenate, Na	*Lagodon rhomboides* (larvae, semiS, 20°C)	LC_{50}			0.038		698
F, Pentachlorophenate, Na	*Lagodon rhomboides* (juv., F, 25°C)	LC_{50}			0.0532		698
F, Pentachlorophenate, Na	*Lagodon rhomboides* (F)	LC_{50}			0.0532		109
F, Dowicide	*Lagodon rhomboides* (48 h prolarvae, S)	LC_{50}			0.066		191
F, Aldicarb	*Lagodon rhomboides* (F, 23°C, pH 8.2)	LC_{50}			0.080		925
F, Aldicarb	*Lagodon rhomboides* (F, 22°C, pH 8.2)	LC_{50}			0.08		726
F, Hexachlorocyclohexane (mixed isomers)	*Lagodon rhomboides* (F, 22–26°C)	LC_{50}			0.0864		926
F, Lindane	*Lagodon rhomboides* (F)	LC_{50}			0.0864		101
F, Endrin	*Lebistes reticulatus* (S)	LC_{50}				0.00006, 0.00012 (72 h)	2
F, Pentachlorobenzene	*Lebistes reticulatus*	LC_{50}				0.178 (14 d)	207
F, 2,4,5-Trichlorotoluene	*Lebistes reticulatus*	LC_{50}				0.23	207
F, 1,2,4,5-Tetrachlorobenzene	*Lebistes reticulatus*	LC_{50}				0.3 (14 d)	207
F, α-Hexachlorocyclohexane	*Lebistes reticulatus*	LC_{10}				0.5 (35 d)	41
F, 1,2,3,5-Tetrachlorobenzene	*Lebistes reticulatus*	LC_{50}				0.8 (14 d)	207
F, Malathion	*Lebistes reticulatus*	LC_{50}				0.819 (1wk)	39
F, Pentachlorophenol	*Lebistes reticulatus* (pH 6)	LC_{50}				0.924 (21–38 min)	927
F, Pentachlorophenol	*Lebistes reticulatus* (pH 7.6)	LC_{50}				0.924 (72–93 min)	927

F, Binapacryl	*Lebistes reticulatus*	LD_{100}	1.0	2
F, Picloram	*Lebistes reticulatus*	NOEL	1.0 (6 m)	771
F, Ammonia	*Lebistes reticulatus* (fry, S)	LC_{50}	1.26 (72 h)	923
F, 2,4,5-Trichlorotoluene	*Lebistes reticulatus*	LC_{50}	1.7 (7 d)	207
F, 1,2,3-Trichlorobenzene	*Lebistes reticulatus*	LC_{50}	2.4 (14 d)	207
F, 1,2,4-Trichlorobenzene	*Lebistes reticulatus*	LC_{50}	2.4 (14 d)	207
F, *N*-Nitrosodiethylamine	*Lebistes reticulatus* (juv.)	LC_{26}	2.6-23.0 (56 d)	928
F, 1,3,5-Trichlorobenzene	*Lebistes reticulatus*	LC_{50}	3.3 (14 d)	207
F, *N*-Nitrosodiethylamine	*Lebistes reticulatus* (juv.)	LC_{82}	13.0–100.0 (74 wk)	928
F, Pentachloroethane	*Lebistes reticulatus*	LC_{50}	15.0 (7 d)	207
F, Perchloroethylene	*Lebistes reticulatus*	LC_{50}	18.0 (7 d)	207
F, *m*-Chlorotoluene	*Lebistes reticulatus*	LD_{50}	18.0 (7 d)	207
F, *o*-Xylene	*Lebistes reticulatus*	LC_{50}	35.0 (7 d)	207
F, *p*-Xylene	*Lebistes reticulatus*	LC_{50}	35.0 (7 d)	207
F, 1,1,2,2-Tetrachloroethane	*Lebistes reticulatus*	LC_{50}	37.0 (7 d)	207
F, *m*-Xylene	*Lebistes reticulatus*	LC_{50}	38.0 (14 d)	207
F, 1,2,3-Trichloropropane	*Lebistes reticulatus*	LC_{50}	42.0 (7 d)	207
F, Formaldehyde	*Lebistes reticulatus*	TL_m	50.0–200.0	8
F, Trichloroethylene	*Lebistes reticulatus*	LC_{50}	55.0 (7 d)	207
F, Carbon tetrachloride	*Lebistes reticulatus*	LC_{50}	67.0 (14 d)	207
F, Toluene	*Lebistes reticulatus*	LC_{50}	68.0 (14 d)	207
F, Ammonia	*Lebistes reticulatus* (fry, S)	LC_{50}	74.0 (72 h)	923
F, Cyclohexane	*Lebistes reticulatus*	LC_{50}	84.0 (7 d)	207
F, 1,1,2-Trichloroethane	*Lebistes reticulatus*	LC_{50}	94.0 (7 d)	207
F, *N*-Nitrosodiethylamine	*Lebistes reticulatus* (juv.)	LC_{26}	100.0 (28 d)	928
F, Ethylene dichloride	*Lebistes reticulatus*	LC_{50}	106.0 (7 d)	207
F, 1,2-Dichloroethane	*Lebistes reticulatus* (S, 22°C)	LC_{50}	106.0 (7 d)	929
F, 1,1,1-Trichloroethane	*Lebistes reticulatus*	LC_{50}	133.0 ppm	207
F, 3-Pentanol	*Lebistes reticulatus*	LC_{50}	989.0 (7 d)	207
F, Ethyl ether	*Lebistes reticulatus*	LC_{50}	2138.0 ppm (7 d)	207
F, *tert*-Butyl alcohol	*Lebistes reticulatus*	LC_{50}	3550.0 (7 d)	930

Chemical	Species	Toxicity test	24 h	48 h	96 h	Chronic exposure	Ref.
F, Isopropanol	*Lebistes reticulatus*	LC_{50}				7060.0 (7 d)	2
F, Methyl cellosolve	*Lebistes reticulatus*	LC_{50}				17,400.0 (7 d)	207
F, Ethylene glycol	*Lebistes reticulatus*	LC_{50}				49,300.0 (7 d)	207
F, Triethylene glycol	*Lebistes reticulatus*	LC_{50}				62,600.0	207
F, Endosulfan	*Lebistes reticulatus*	LC_{50}	0.005				396
F, Endosulfan	*Lebistes reticulatus*	LC_{100}	0.010				931
F, Pentachlorophenol	*Lebistes reticulatus* (pH 7.3)	LC_{50}	0.38				207
F, Potassium cyanide	*Lebistes reticulatus* (S, 20°C, pH 7.2–8.0)	LC_{10}	0.420				932
F, Isodrin/photoisodrin, toxicity ratio (TR)	*Lebistes reticulatus*	TR	0.45				44, 48, 49
F, 2,3,4,5-Tetrachlorophenol	*Lebistes reticulatus* (pH 7.3)	LC_{50}	0.77				207
F, Potassium cyanide	*Lebistes reticulatus* (S, 20°C, pH 7.2–8.0)	LC_{50}	0.800				932
F, Pentachlorophenol	*Lebistes reticulatus* (pH 9.0)	LC_{50}	0.924				927
F, 3,4,5-Trichlorophenol	*Lebistes reticulatus* (pH 7.3)	LC_{50}	1.1				207
F, 2,3,5,6-Tetrachlorophenol	*Lebistes reticulatus* (pH 7.3)	LC_{50}	1.37				207
F, 4-Chloro-*o*-cresol	*Lebistes reticulatus*	LC_{50}	1.45				933
F, 2,3,5-Trichlorophenol	*Lebistes reticulatus* (pH 7.3)	LC_{50}	1.6				207
F, 2,3,6-Trichlorophenol	*Lebistes reticulatus* (pH 7.3)	LC_{50}	5.1				207
F, Linuron	*Lebistes reticulatus*	LC_{0}	6.0				496
F, *m*-Chlorophenol	*Lebistes reticulatus* (pH 7.3)	LC_{50}	6.5				207
F, Linuron	*Lebistes reticulatus*	LC_{50}	8.0				496
F, *o*-Chlorophenol	*Lebistes reticulatus*	LC_{50}	11.0				207
F, Paraquat	*Lebistes reticulatus* (S)	LC_{50}	12.53				158
F, Phenol	*Lebistes reticulatus* (pH 7.3)	LC_{50}	30.0				207
F, Monolinuron	*Lebistes reticulatus*	LC_{50}	55.0				496
F, Ethybenzene (80%)	*Lebistes reticulatus* (S, juv., softW, 25°C, pH 7.5)	LC_{50}	97.0				797
F, Linuron	*Lebistes reticulatus*	LC_{100}	150.0				496
F, Actellic	*Lebistes reticulatus*	LC_{50}		0.004 ml/l			859

F, Triton X100 emulsified concentrate	*Lebistes reticulatus* (S, 25°C, pH 7.5)	LC_{50}	0.082	0.068			808
F, Lindane	*Lebistes reticulatus* (S, 24°C)	TL_m		0.80			2
F, Benomyl	*Lebistes reticulatus* (S, 24°C)	LC_{50}		3.4			474
F, Cabendazim	*Lebistes reticulatus* (S, 24°C, softW)	LC_0		8.0			406
F, Dichlobenil	*Lebistes reticulatus*	LD_{50}		18.0			2
F, Methyl-4-chlorophenoxyacetic acid	*Lebistes reticulatus*	LC_{50}		30.0–40.0			934
F, Dimethylamine	*Lebistes reticulatus* (S, 20°C)	LC_{50}	111.0	103.0		100.0 (72 h)	397
F, Zoalene	*Lebistes reticulatus* (S)	TL_m		200.0			222
F, Amprolium	*Lebistes reticulatus* (S, 24°C)	TL_m		270.0			222
F, Propylene glycol	*Lebistes reticulatus*	LC_{50}		>10,000.0			182
F, Aldrin	*Lebistes reticulatus* (S)	LC_{50}			0.02		117
F, Ethyl-*p*-nitrophenylthionobenzene phosphate	*Lebistes reticulatus*	LC_{50}			0.03		806
F, DDT	*Lebistes reticulatus*	LC_{50}			0.043		802
F, Chlorodane	*Lebistes reticulatus*	LC_{50}			0.190		785
F, Chlorodane	*Lebistes reticulatus* (softW)	LC_{50}	0.560	0.190	0.190		785
F, Hydrazine	*Lebistes reticulatus* (S, softW, 22–24°C)	LC_{50}	3.32	1.58	0.61	0.82 (72 h)	226
F, Cartap	*Lebistes reticulatus*	LC_{50}			0.710		516
F, Cartap (97.7%)	*Lebistes reticulatus* (juv., S, 18–22°C, pH 7.1)	LC_{50}			0.710		868
F, Aerozine-50	*Lebistes reticulatus* (softW, S, 22–24.5°C)	LC_{50}	5.08	2.86	1.17	1.95 (72 h)	226
F, Malathion	*Lebistes reticulatus* (S)	LC_{50}			1.200		2
F, α-Hexachlorocyclohexane	*Lebistes reticulatus* (semistatic, synthetic seaW, 24°C)	LC_{10}			1.31	0.5 (35 d)	393
F, α-Hexachlorocyclohexane	*Lebistes reticulatus*	EC_{50}		1.38	1.31		41
F, Lactonitrile	*Lebistes reticulatus*	TL_m			1.37		22
F, α-Hexachlorocyclohexane	*Lebistes reticulatus*	LC_{50}			1.4		41
F, Propoxur	*Lebistes reticulatus* (S, 27°C)	LC_{50}			1.4-2.2		816
F, Aerozine-50	*Lebistes reticulatus* (hardW, S, 22–24.5°C)	LC_{50}	12.3	4.37	2.25	2.72 (72 h)	226
F, Hydrazine	*Lebistes reticulatus* (S, hardW, 22–24°C)	LC_{50}	4.6	3.98	3.85	3.85 (72 h)	226

Chemical	Species	Toxicity test	24 h	48 h	96 h	Chronic exposure	Ref.
F, Dimethyl parathion	*Lebistes reticulatus* (S)	LC_{50}			6.2		117
F, Methyl parathion	*Lebistes reticulatus* (S)	LC_{50}	12.2	9.4	6.2		772
F, Furalaxyl	*Lebistes reticulatus*	LC_{80}			8.7		935
F, 1,1-Dimethylhydrazine	*Lebistes reticulatus* (S, hardW)	LC_{50}	78.4	29.9	10.1	17.2 (72 h)	226
F, Paraquat	*Lebistes reticulatus* (S)	LC_{50}			11.5		158
F, *o*-Cresol	*Lebistes reticulatus* (hardW)	LC_{50}	50.0		18.0		22
F, *o*-Chlorophenol	*Lebistes reticulatus*	TL_m	23.0		20.0		50
F, Furazolidine	*Lebistes reticulatus* (S)	TL_m			25.0		222
F, 1,1-Dimethylhydrazine	*Lebistes reticulatus* (S, softW)	LC_{50}	82.0	45.5	26.5	32.4 (72 h)	226
F, 2,4,5-Trichlorophenoxyacetic acid	*Lebistes reticulatus* (S)	LC_{50}			28.1		117
F, Vinyl acetate	*Lebistes reticulatus*	TL_m	31.1		31.1		50
F, Acrylonitrile	*Lebistes reticulatus* (softW)	TL_m			33.5		22
F, *p*-Xylene	*Lebistes reticulatus* (softW)	TL_m	34.7		34.7		50
F, Benzene	*Lebistes reticulatus* (softW)	TL_m	36.6		36.6		154
F, Phenol	*Lebistes reticulatus* (softW)	TL_m	49.9	49.9	39.2		50
F, Dimethyl amine	*Lebistes reticulatus* (S, 20°C)	NOAE			41.0		397
F, Chlorobenzene	*Lebistes reticulatus*	TL_m	45.0		45.0		22
F, Chlorobenzene	*Lebistes reticulatus* (S, 25°C, pH 7.5)	LC_{50}			45.53		817
F, Diquat	*Lebistes reticulatus* (S)	LC_{50}			50.2		158
F, Allyl chloride	*Lebistes reticulatus* (softW)	TL_m	57.7	53.5	51.8		50
F, Cyclohexane	*Lebistes reticulatus*	TL_m	57.7		57.7		50
F, Toluene	*Lebistes reticulatus*	TL_m	63.0		59.0		50
F, Toluene	*Lebistes reticulatus* (S)	LC_{50}			59.3		814
F, 2,4-Dichlorophenoxyacetic acid	*Lebistes reticulatus* (S)	LC_{50}			70.7		64
F, Styrene	*Lebistes reticulatus* (softW)	TL_m	74.8	74.8	74.8		50
F, Ethylbenzene	*Lebistes reticulatus* (softW)	TL_m	97.1	97.1	97.1		50
F, Isoprene	*Lebistes reticulatus*	TL_m	240.0	240.0	240.0		2
F, Benzonitrile	*Lebistes reticulatus* (softW)	TL_m			400.0		22
F, Adiponitrile	*Lebistes reticulatus* (softW)	TL_m			775.0		22

F, 2-Butoxyethanol	*Lebistes reticulatus* (semiS, 21–23°C)	LC_{50}				983.0 (7 d)	929
F, Acetonitrile	*Lebistes reticulatus* (softW)	TL_m			1650.0		22
F, Ethyleneurea	*Lebistes reticulatus*	LC_{50}			13,000.0		936
F, Endrin	*Leiostomus xantharus* (F)	LC_{50}	0.00045				210
F, Chlordecane	*Leiostomus xantharus* (juv., F, 22°C)	LC_{50}		0.17			938
F, Ametryn	*Leiostomus xantharus*	NOAE		1.0			936
F, Endosulfan	*Leiostomus xantharus* (F)	LC_{50}			0.00009		85
F, Heptachlor	*Leiostomus xantharus* (F)	LC_{50}			0.00085		65
F, Leptophos	*Leiostomus xantharus*	LC_{50}			0.0046		937
F, Aroclor 1254	*Leiostomus xantharus* (F, seaW)	LC_{50}				0.005 (18 d)	613
F, Kepone	*Leiostomus xantharus* (F)	LC_{50}			0.0066		133
F, Chlordecane	*Leiostomus xantharus* (F, 25°C)	LC_{50}			0.0066		418
F, Aldicarb	*Leiostomus xantharus* (S, 26°C)	LC_{50}			0.202		647
F, Hydrogen sulfide	*Lepomis macrochirus* (F, 20–22°C)	MATC				0.0004	143
F, Chlordane	*Lepomis macrochirus* (F)	MATC				0.00054	939
F, Lindane	*Lepomis macrochirus* (F, 27°C)	MATC				0.0091	145
F, Hydrogen sulfide	*Lepomis macrochirus* (egg, F, 22°C)	TL_m				0.019 (72 h)	136
F, Parathion	*Lepomis cyanellus* (S)	LC_{50}				0.020 (72 h)	221
F, Lindane	*Lepomis macrochirus* (F, 27°C)	TL_m				0.03 (2 yr)	145
F, Chloropyrifos	*Lepomis cyanellus* (S)	LC_{50}				0.04 (72 h)	221
F, Lindane, chlorinated, TG	*Lepomis macrochirus*	LC_{50}				0.057	83
F, Heptachlor	*Lepomis macrochirus* (tested in ponds)	LC_{90}				0.069 (171 d)	817
F, Aroclor 1242	*Lepomis macrochirus* (F)	LC_{50}				0.084 (30 d)	92
F, Chlorodane	*Lepomis macrochirus*	NOEL				0.100 (87 h)	785
F, Lindane, 1% chlorinated dust	*Lepomis macrochirus*	LC_{50}				0.138	83
F, Carbofuran	*Lepomis cyanellus*	LC_{50}				0.16 (72 h)	221
F, Pentachlorophenol	*Lepomis macrochirus* (F, 25°C)	TL_m				0.188 (406 h)	144
F, Chlorodane	*Lepomis macrochirus*	LC_{100}				0.200 (87 h)	785
F, Aroclor 1254	*Lepomis macrochirus* (F)	LC_{50}				0.239 (25 d)	145
F, Pentachlorophenol	*Lepomis macrochirus* (F, 25°C)	TL_m				0.251 (423 h)	144

Chemical	Species	Toxicity test	24 h	48 h	96 h	Chronic exposure	Ref.
F, Pentachlorophenol	*Lepomis macrochirus* (F, 25°C)	TL_m				0.303 (30 h)	144
F, Aroclor 1260	*Lepomis macrochirus* (F)	LC_{50}				0.4 (30 d)	92
F, Chlorodane	*Lepomis macrochirus*	LC_{100}				1.0 (5 h)	785
F, *o*-Nitrophenol	*Lepomis macrochirus*	NOAE				5.0 (h?)	942
F, Chlorodane	*Lepomis macrochirus*	LC_{100}				5.0 (2 h)	785
F, Atrazine	*Lepomis macrochirus* (F, 27°C)	TL_m				5.4–8.4 (2 yr)	940
F, Propanil	*Lepomis cyanellus* (S)	LC_{50}				5.85 (72 h)	221
F, Phenol	*Lepomis macrochirus*	TL_m				10.0 (25 h)	76
F, Monuron	*Lepomis cyanellus* (juv.)	LC_{100}				10.0 (8 d)	789
F, Phenol	*Lepomis macrochirus*	TL_m				15.0 (50 h)	76
F, Monuron	*Lepomis macrochirus* (juv.)	NOEL				20.0 (12 d)	789
F, Endothal	*Lepomis cyanellus*	LC_0				25.0 (8 d)	822
F, Endothal	*Lepomis macrochirus*	LC_0				25.0 (8 d)	822
F, Sodium arsenate	*Lepomis cyanellus* (juv., $^1/_2$S, 30°C, pH 8.3–8.46)	LC_{50}				30.0 (209 h)	915
F, Sodium arsenate	*Lepomis cyanellus* (juv., $^1/_2$S, 20°C, pH 8.3–8.46)	LC_{50}				30.0 (527 h)	915
F, Sodium arsenate	*Lepomis cyanellus* (juv., $^1/_2$S, 30°C, pH 8.3–8.46)	LC_{50}				60.0 (124 h)	915
F, Sodium arsenate	*Lepomis cyanellus* (juv., $^1/_2$S, 20°C, pH 8.3–8.46)	LC_{50}				60.0 (210 h)	915
F, Sodium arsenate	*Lepomis cyanellus* (juv., $^1/_2$S, 10°C, pH 8.3–8.46)	LC_{50}				60.0 (678 h)	915
F, Lindane, dechlorinated, dust	*Lepomis macrochirus*	LC_{50}				69.0	83
F, Lindane, dechlorinated, TG	*Lepomis macrochirus*	LC_{50}				82.065	83
F, Zinc-EDTA chelate	*Lepomis macrochirus* (S)	NOEL				320.0	135
F, Arsenic (5+)	*Lepomis cyanellus* (20°C)	LC_{50}				1000.0 (12 h)	941
F, 4-Nitrochlorobenzene 2-sulfonate, Na	*Lepomis macrochirus*	TL_m				6375.0 (25–100 h)	74
F, Phenolsulfonate, Na	*Lepomis macrochirus*	TL_m				19,616.0 (100 h)	74

F, Isodrin	*Lepomis macrochirus*	LC_{50}	0.012	2
F, Photoisodrin	*Lepomis macrochirus*	LC_{50}	0.025	44
F, Photodieldrin	*Lepomis macrochirus*	LC_{50}	0.030	44
F, Binapacryl (TG 99%)	*Lepomis macrochirus* (S, 18°C, pH 7.1)	LC_{50}	0.042	914
F, Dyfonate	*Lepomis macrochirus*	LC_{50}	0.045	116
F, Diazinon	*Lepomis macrochirus*	LC_{50}	0.052	2
F, Acrolein	*Lepomis macrochirus*	LC_{50}	0.08	230
F, Photoaldrin	*Lepomis macrochirus*	LC_{50}	0.09	44
F, Chlorodane	*Lepomis macrochirus*	NOEL	0.1	785
F, Aldicarb	*Lepomis macrochirus* (S, 24°C, pH 7.4)	LC_{50}	0.103	914
F, Malathion	*Lepomis macrochirus*	LC_{50}	0.120	130
F, Dieldrin	*Lepomis macrochirus*	LC_{50}	0.170	2
F, Hydrogen cyanide	*Lepomis humilis*	TL_m	0.18	290
F, Chlordene	*Lepomis macrochirus*	LC_{50}	0.218	44
F, Photochlordene	*Lepomis macrochirus*	LC_{50}	0.346	44
F, Isodrin/photoisodrin toxicity ratio (TR)	*Lepomis macrochirus*	TR	0.75	44, 258, 300
F, Carbaryl	*Lepomis macrochirus*	LC_{50}	3.4	116
F, 2,5-Dichlorobenzene sulfonate, Na	*Lepomis macrochirus*	TL_m	3.75	74
F, *p*-Dichlorobenzene	*Lepomis macrochirus* (S, 21–23°C, pH 6.5–7.9)	LC_{50}	4.5	852
F, Hydrogen sulfide	*Lepomis humilis*	LC_{100}	4.9–5.3	84
F, Ethylene dichloride	*Lepomis macrochirus*	NOEL	5.0	132
F, Dimethyl parathion	*Lepomis macrochirus*	LC_{50}	5.7	127
F, Glyphosate	*Lepomis macrochirus* (22°C)	LC_{50}	6.4	562
F, *o*-Chlorophenol	*Lepomis macrochirus*	TL_m	8.2	24
F, Zectran	*Lepomis macrochirus*	LC_{50}	11.2	116
F, Caproic acid	*Lepomis macrochirus*	TL_m	15.0–200.0	77
F, Phenol	*Lepomis macrochirus*	TL_m	22.7	124
F, Acrylonitrile	*Lepomis macrochirus* (softW)	TL_m	25.0	134
F, Dimethoate	*Lepomis macrochirus*	LC_{50}	28.0	1129

Chemical	Species	Toxicity test	24 h	48 h	96 h	Chronic exposure	Ref.
F, Monocrotophos (100%)	*Lepomis macrochirus* (S, 18°C, pH 7.1)	LC_{50}	32.0				917
F, Benzene	*Lepomis macrochirus*	LD_{100}	34.0			60.0 (2 h)	84
F, Ethylbenzene (80%)	*Lepomis macrochirus* (S, SW, 25°C, pH 7.5)	LC_{50}	35.0				797
F, Kepone	*Lepomis microlophus*	LC_{50}	0.62				880
F, *o*-Nitrophenol	*Lepomis macrochirus* (S)	LC_{50}	66.9				942
F, Picloram K salt	*Lepomis macrochirus* (26.6°C)	LC_{50}	69.0				913
F, Acetic acid	*Lepomis macrochirus*	TL_m	100.0				76
F, Picloram	*Lepomis cyanellus*	LC_{50}	150.0				706
F, 2-Methyl-4-chlorophenoxyacetic acid (amine)	*Lepomis macrochirus*	LC_{50}	164.0			10.0 (100 h)	934
F, Propionic acid	*Lepomis macrochirus*	TL_m	188.0				77
F, *n*-Butyric acid	*Lepomis macrochirus*	TL_m	200.0				77
F, Glutaric acid	*Lepomis macrochirus*	TL_m	330.0				77
F, Adipic acid	*Lepomis macrochirus*	TL_m	330.0				77
F, 2,2-Dichlorovinyldimethyl phosphate	*Lepomis macrochirus*	LC_{50}	1000.0				28
F, Adiponitrile	*Lepomis macrochirus* (softW)	TL_m	1250.0				134
F, 4-Nitrotoluene 2-sulfonate, Na	*Lepomis macrochirus*	TL_m	1440.0				74
F, Diethanolamine	*Lepomis macrochirus*	TL_m	2100.0				126
F, *p*-Chlorobenezenesulfonic acid	*Lepomis macrochirus*	TL_m	3219.0				74
F, *n*-Valeric acid, Na	*Lepomis macrochirus*	TL_m	5000.0				231
F, *n*-Butyric acid, Na	*Lepomis macrochirus*	TL_m	5000.0				231
F, Formic acid, Na	*Lepomis macrochirus*	TL_m	5000.0				231
F, Sodium propionate	*Lepomis macrochirus*	TL_m	5000.0				231
F, Phenol	*Lepomis macrochirus* (S)	LC_{50}	19.0–160.0				146
F, Trifluralin	*Lepomis macrochirus*	LC_{50}		0.019			116
F, Dichlone	*Lepomis macrochirus*	LC_{50}		0.070			230
F, Pyrethrum	*Lepomis macrochirus* (24°C)	LC_{50}		0.070			756
F, 2,4-Dichlorophenoxyacetic acid	*Lepomis macrochirus*	LC_{50}		0.9			64

F, Fenoprop, butoxyethyl ester	*Lepomis macrochirus*	LC_{50}		1.1		227
F, Tetraethyl lead	*Lepomis macrochirus*	LC_{50}	2.0	1.4		123
F, Fenitrothion	*Lepomis macrochirus*	LC_{50}		2.72		2
F, *n*-Butylmercaptan	*Lepomis macrochirus*	TL_m	7.4	5.5		8
F, Diuron	*Lepomis macrochirus*	LC_{50}		7.4		2
F, Chloroisopropyl-*N*-phenyl-carbamate	*Lepomis macrochirus*	LC_{50}		8.0		229
F, Propanil	*Lepomis cyanellus* (S)	LC_{50}		8.26		943
F, Propanil	*Lepomis cyanellus* (S)	LC_{90}		12.17		896
F, Trifene, Na	*Lepomis macrochirus*	LC_{50}		15.0		228
F, Fenoprop, isooctyl ester	*Lepomis macrochirus*	LC_{50}		16.0		227
F, Fenoprop, propylene glycol butyl ether ester	*Lepomis macrochirus*	LC_{50}		16.6		227
F, Ethylene bromide	*Lepomis macrochirus* (freshW)	LC_{50}		18.0		131
F, Diquat	*Lepomis macrochirus*	LC_{50}		19.0		128
F, Dichlobenil	*Lepomis macrochirus*	LC_{50}		20.0		2
F, Dicamba	*Lepomis macrochirus*	LC_{50}		20.0		228
F, Dichlobenil	*Lepomis macrochirus*	LC_{50}		20.0		98
F, Benzene	*Lepomis macrochirus*	LD_{50}	20.0	20.0		84
F, Chloroxuron	*Lepomis macrochirus*	LC_{50}		25.0		229
F, *o*-Nitrophenol	*Lepomis macrochirus*	TL_m	67.0	46.30		22, 124
F, 2-Nitrophenol	*Lepomis macrochirus* (S)	LC_{50}		46.3–51.6		790
F, *o*-Nitrophenol	*Lepomis macrochirus* (S)	LC_{100}	49.0	49.0		942
F, Bromacil (80%)	*Lepomis macrochirus*	EC_{50}	103.0	71.0		945
F, Fenoprop	*Lepomis macrochirus*	LC_{50}		83.0		227
F, 3-Amino-1,2,4-triaozle	*Lepomis macrochirus*	LC_{50}		100.0		116
F, Amitrol	*Lepomis macrochirus*	NOAE		100.0		43
F, 2,2-Dichloropropionic acid	*Lepomis macrochirus*	LC_{50}		115.0		116
F, Simazine	*Lepomis macrochirus*	LC_{50}		130.0		116
F, Dicamba	*Lepomis macrochirus*	LC_{50}		130.0		944
F, Arsenic (5+)	*Lepomis cyanellus* (20°C)	LC_{50}	175.0	150.0	350.0 (18 h)	941

Chemical	Species	Toxicity test	24 h	48 h	96 h	Chronic exposure	Ref.
F, Dacthal	*Lepomis macrochirus*	LC_{50}		700.0			229
F, Toxaphene	*Lepomis macrochirus*	LC_{50}			0.0004		116
F, Endrin	*Lepomis macrochirus*	LC_{50}			0.0006		232
F, Resmethrin	*Lepomis macrochirus* (F)	LC_{50}			0.000750		138
F, Endosulfan	*Lepomis macrochirus* (S, 18°C, pH 7.5)	LC_{50}			0.0012		600
F, Phorate	*Lepomis macrochirus* (12°C, pH 7.5)	LC_{50}			0.002		599
F, Resmethrin	*Lepomis macrochirus* (HW, S, 12°C, pH 7.8)	LC_{50}			0.002–0.003		915
F, Resmethrin	*Lepomis macrochirus* (softW)	LC_{50}			0.002–0.006		915
F, Resmethrin	*Lepomis macrochirus* (S)	LC_{50}			0.00262		138
F, Chloropyrifos	*Lepomis macrochirus*	LC_{50}			0.0026		71
F, DDT (chlorinated)	*Lepomis macrochirus*	LC_{50}			0.0034		83
F, DDT	*Lepomis humilis*	LC_{50}			0.005		122
F, DDT	*Lepomis microlophus*	LC_{50}			0.005		127
F, Azinphosmethyl	*Lepomis macrochirus*	LC_{50}			0.0052		252
F, Endrin	*Lepomis macrochirus*	LC_{50}			0.006		2
F, DDT	*Lepomis macrochirus*	LC_{50}			0.006–0.043		802
F, Resmethrin	*Lepomis macrochirus* (hardW, F, pH 7.9)	LC_{50}			0.006–0.009		915
F, DDT	*Lepomis macrochirus* (softW)	LC_{50}			0.0064		802
F, Dieldrin	*Lepomis gibbosus*	LC_{50}			0.0067		233
F, Dieldrin	*Lepomis macrochirus*	LC_{50}			0.008		2
F, DDT	*Lepomis macrochirus*	LC_{50}			0.008		122
F, DDT	*Lepomis macrochirus*	LC_{50}			0.008		127
F, Dieldrin	*Lepomis macrochirus*	LC_{50}			0.008		232
F, DDT	*Lepomis macrochirus* (hardW)	LC_{50}			0.0088		802
F, DDT (chlorinated emulsion)	*Lepomis macrochirus*	LC_{50}			0.009		83
F, Hydrogen sulfide	*Lepomis macrochirus* (F, 20–22°C)	TL_m			0.009–0.014		143
F, Toxaphene	*Lepomis humilis*	LC_{50}			0.013		122
F, Aldrin	*Lepomis macrochirus*	LC_{50}			0.013		232
F, Toxaphene	*Lepomis microlophus* (S, 18°C, pH 7.1)	LC_{50}			0.013		801

F, Aldrin	*Lepomis macrochirus*	LC_{50}	0.26		0.013	44
F, Hydrogen sulfide	*Lepomis macrochirus* (fry, F, 22°C)	TL_m			0.0131	136
F, Dichlorodiphenyltrichloroethane	*Lepomis macrochirus*	LC_{50}			0.016	116
F, Heptachlor	*Lepomis microlophus*	LC_{50}			0.017	234
F, Toxaphene	*Lepomis macrochirus*	LC_{50}			0.018	122
F, Heptachlor	*Lepomis macrochirus*	LC_{50}			0.019	232
F, Aldrin	*Lepomis gibbosus* (S)	LC_{50}			0.020	117
F, Tetraethyllead	*Lepomis macrochirus*	LC_{50}			0.02	125
F, Azinphosmethyl	*Lepomis macrochirus*	LC_{50}			0.02	211
F, Chlorodane	*Lepomis macrochirus*	LC_{50}			0.022	785
F, Chlordane	*Lepomis macrochirus*	LC_{50}			0.022	116
F, Leptophos	*Lepomis macrochirus* (S, 5°C, pH 7.5)	LC_{50}			0.022	678
F, Chlorodane	*Lepomis macrochirus* (softW)	LC_{50}	0.036	0.032	0.022	785
F, Dimethrin	*Lepomis macrochirus* (F)	LC_{50}			0.0223	138, 140
F, Biocallethrin, 90% TG	*Lepomis macrochirus* (F, 12°C, pH 7.5)	LC_{50}			0.0225	597
F, Parathion (100%)	*Lepomis macrochirus* (S, 18°C, pH 7.2–7.5)	LC_{50}			0.024	511
F, Kepone	*Lepomis macrochirus*	NOEL			0.024	880
F, Dioxathion	*Lepomis macrochirus*	LC_{50}			0.034	235
F, Dimethrin	*Lepomis macrochirus* (S)	LC_{50}			0.0375	138, 140
F, Pyrethrum	*Lepomis macrochirus* (S, 12°C)	LC_{50}			0.039–0.061	915
F, Binapacryl	*Lepomis macrochirus* (juv., 18°C, pH 7.1)	LC_{50}			0.04	914
F, Chlordane, TG	*Lepomis macrochirus*	LC_{50}			0.041	83
F, Binapacryl (TG 99%)	*Lepomis cyanellus* (S, 18°C, pH 7.1)	LC_{50}	0.053		0.043	914
F, Hydrogen sulfide	*Lepomis macrochirus* (F, 21–22°C)	TL_m			0.0448	136
F, Pyrethrum	*Lepomis macrochirus* (S, SW, pH 7.5,12°C)	LC_{50}			0.045	915
F, Pyrethrum	*Lepomis macrochirus* (S, very hardW, 12°C, pH 8.2)	LC_{50}			0.046	915
F, Hydrogen sulfide	*Lepomis macrochirus* (juv., F, 21–22°C)	TL_m			0.0478	136
F, Pyrethrin	*Lepomis macrochirus* (S)	LC_{50}			0.049	138
F, Azinphosmethyl	*Lepomis macrochirus*	LC_{50}			0.05	211
F, Chlorodane	*Lepomis macrochirus* (23.8°C)	LC_{50}	0.058	0.049	0.050	785

Chemical	Species	Toxicity test	24 h	48 h	96 h	Chronic exposure	Ref.
F, Kepone	*Lepomis macrochirus*	LC_{50}	0.257		0.051	0.024 (NOAE)	133
F, Kepone	*Lepomis macrochirus*	LC_{50}		0.180	0.051		880
F, Azinphosmethyl	*Lepomis macrochirus*	LC_{50}			0.052		127
F, Aldicarb	*Lepomis macrochirus* (S, 18°C, pH 7.4)	LC_{50}			0.052		914
F, Pyrethrum	*Lepomis macrochirus* (S, HW, pH 7.8, 12°C)	LC_{50}			0.054		915
F, Allethrin	*Lepomis macrochirus*	LC_{50}			0.056		71
F, Fensulfothion	*Lepomis macrochirus* (SW, 25°C, pH 8.2)	LC_{50}	0.20	0.11	0.056		946
F, Pyrethrum	*Lepomis macrochirus* (S, SW, pH 7.5, 22°C)	LC_{50}			0.058		915
F, Chlordane	*Lepomis macrochirus* (F)	LC_{50}			0.059		939
F, Pyrethrum	*Lepomis macrochirus* (S, SW, pH 7.5, 17°C)	LC_{50}			0.059		915
F, Dioxathion	*Lepomis cyanellus*	LC_{50}			0.061		235
F, Lindane	*Lepomis macrochirus*	LC_{50}			0.062		116
F, Pyrethrum	*Lepomis macrochirus* (S, very softW, pH 6.6, 12°C)	LC_{50}			0.062		915
F, Disulfoton	*Lepomis macrochirus*	LC_{50}			0.063		98
F, Parathion	*Lepomis macrochirus*	LC_{50}			0.0650		98
F, Kepone	*Lepomis macrochirus*	LC_{80}			0.065		880
F, Lindane	*Lepomis macrochirus*	LC_{50}			0.068		122
F, Lindane	*Lepomis macrochirus*	LC_{50}			0.068		127
F, Imidan	*Lepomis macrochirus* (S)	LC_{50}			0.070		148
F, Mevinphos	*Lepomis macrochirus*	LC_{50}			0.070		945
F, Fensulfothion	*Lepomis macrochirus* (HW, 25°C, pH 8.2)	LC_{50}	0.26	0.11	0.070		946
F, Aldicarb	*Lepomis macrochirus* (S, 18°C, pH 7.4)	LC_{50}			0.071		914
F, Pyrethrum	*Lepomis macrochirus* (25°C)	LC_{50}	0.080	0.074	0.074		870
F, Hydrogen cyanide	*Lepomis macrochirus* (juv., F)	LC_{50}			0.075–0.125		137
F, Pyrethrum	*Lepomis macrochirus* (F, 12°C, pH 7.8)	LC_{50}			0.080–0.135		915
F, Lindane	*Lepomis humilis*	LC_{50}			0.083		122
F, Hydrocyanic acid	*Lepomis macrochirus* (juv., F, 8.4°C, pH 7.9)	LC_{50}			0.0830		947

F, Lindane	*Lepomis microlophus*	LC_{50}			0.083		127
F, Hydrocyanic acid	*Lepomis macrochirus* (juv., F, 15°C, pH 7.9)	LC_{50}			0.0871		947
F, Malathion	*Lepomis macrochirus*	LC_{50}			0.100		122
F, Formaldehyde	*Lepomis macrochirus* (F)	LC_{50}			0.100 ml/l		120
F, Eulan	*Lepomis macrochirus*	LC_{50}			0.100		122
F, Aminocarb	*Lepomis macrochirus* (20°C, pH 7.5)	LC_{50}			0.100		475
F, Ethyl-*p*-nitrophenylthionobenzene phosphate	*Lepomis macrochirus*	LC_{50}			0.10		806
F, Ethyl-*p*-nitrophenylthionobenzene phosphate	*Lepomis macrochirus*	LC_{50}			0.100		70
F, Demeton	*Lepomis macrochirus*	LC_{50}			0.100		235
F, Aminocarb (17%)	*Lepomis macrochirus* (F, 20°C, pH 7.4)	LC_{50}			0.100		914
F, Aminocarb (17%)	*Lepomis macrochirus* (softW, 22°C, pH 7.4)	LC_{50}			0.10		610
F, Pyrethrin	*Lepomis macrochirus* (F)	LC_{50}			0.104		138
F, Malathion	*Lepomis macrochirus*	LC_{50}			0.110		236
F, Ethyl-*p*-nitrophenylthionobenzene phosphate	*Lepomis macrochirus* (S, 18°C, pH 7.5)	LC_{50}			0.110		602
F, Malathion	*Lepomis cyanellus*	LC_{50}			0.120		235
F, Diazinon (TG)	*Lepomis macrochirus*	LC_{50}			0.12		87
F, Hydrocyanic acid	*Lepomis macrochirus* (juv., F, 25°C, pH 7.9)	LC_{50}			0.120		947
F, Aroclor 1248	*Lepomis macrochirus*	LC_{50}			0.136	0.1 (25 d)	906
F, Anilazine	*Lepomis microlophus* (18°C)	LC_{50}			0.140		600
F, Kepone	*Lepomis humilis*	LC_{50}	0.62	0.27	0.14		133
F, Rotenone (5%)	*Lepomis cyanellus* (S)	LC_{50}			0.141		28
F, Rotenone (5%)	*Lepomis macrochirus* (S)	LC_{50}			0.141		28
F, Tricresyl phosphate	*Lepomis macrochirus* (F, 12°C, pH 7.5)	LC_{50}			0.150		946
F, Captan	*Lepomis macrochirus* (fingerling, S)	LC_{50}			0.150		778
F, Aldicarb	*Lepomis macrochirus* (S, 18°C, pH 7.4)	LC_{50}			0.160		914

Chemical	Species	Toxicity test	24 h	48 h	96 h	Chronic exposure	Ref.
F, Phosmet	*Lepomis macrochirus* (S, 20°C, pH 7.5)	LC_{50}			0.16–0.20		409
F, Malathion	*Lepomis humilis*	LC_{50}			0.170		122
F, Diazinon	*Lepomis macrochirus*	LC_{50}			0.17-0.53		73
F, Malathion	*Lepomis microlophus*	LC_{50}			0.170		127
F, Formaldehyde	*Lepomis cyanellus* (F)	LC_{50}			0.173 ml/l		120
F, Co-Ral	*Lepomis macrochirus*	LC_{50}			0.180		252
F, Naled	*Lepomis macrochirus*	LC_{50}			0.18		71
F, Heptachlor	*Lepomis macrochirus*	LC_{50}			0.190		2
F, Tetraethyllead	*Lepomis macrochirus*	TL_m	2.0	1.4	0.2		22
F, Ethion	*Lepomis macrochirus*	LC_{50}			0.220		71
F, Tri-Chloronate	*Lepomis macrochirus* (S, 24°C)	LC_{50}			0.22		952
F, 4,6-Dinitro-*o*-cresol	*Lepomis macrochirus* (S)	LC_{50}			0.230		662
F, Hydrogen cyanide	*Lepomis macrochirus* (swimup fry, F)	LC_{50}			0.232–0.365		137
F, Varbofuran, TG	*Lepomis macrochirus*	LC_{50}			0.24		912
F, Cyanogen bromide	*Lepomis macrochirus* (S, 23°C)	LC_{50}			0.24		224
F, Ciodrin	*Lepomis macrochirus*	LC_{50}			0.250		71
F, Hydrocyanic acid	*Lepomis macrochirus* (juv., F, 25°C, pH 7.9)	LC_{50}			0.276		947
F, Disulfoton	*Lepomis macrochirus* (24°C)	LC_{50}			0.300		410
F, Anilazine	*Lepomis macrochirus* (18°C)	LC_{50}			0.320		600
F, 4-Amino-3,5-xylenol	*Lepomis macrochirus* (S)	LC_{50}			0.32		949
F, Aramite	*Lepomis macrochirus* (S, 24°C, pH 7.5)	LC_{50}			0.350		609
F, Parathion (98.7% active)	*Lepomis macrochirus* (S, 18°C, pH 7.2–7.5)	LC_{50}			0.4		511
F, Parathion	*Lepomis cyanellus*	LC_{50}			0.425		235
F, Hydrazine	*Lepomis macrochirus* (F, 23–24°C, pH 7.2–8.4)	LC_0			0.43		951
F, Diazinon	*Lepomis macrochirus* (F)	LC_{50}			0.44–0.48		141
F, Malathion	*Lepomis gibbosus* (S)	LC_{50}			0.480		2
F, Tetrabromobisphenol A	*Lepomis macrochirus*	LC_{50}			0.51		953
F, Hydrogen cyanide	*Lepomis macrochirus* (egg, F)	LTC			0.535–0.693		137
F, Ferbam	*Lepomis macrochirus*	NOEL			0.5–4.0		871

F, Acetonecyanohydrin	*Lepomis macrochirus* (S, FW, 23°C)	LC_{50}			0.57		224
F, Actanilide	*Lepomis macrochirus* (S, 23°C, pH 7.6–7.9)	LC_{50}			0.57		954
F, 2,4-Dinitrophenol	*Lepomis macrochirus* (S)	LC_{50}			0.620		662
F, Chlorothion	*Lepomis macrochirus*	LC_{50}			0.700		235
F, Methomyl, TG (LC)	*Lepomis macrochirus* (S, 22°C, pH 7.5)	LC_{50}			0.710		513
F, Triton X100 emulsified concentrate	*Lepomis macrochirus* (S, 25°C, pH 7.4–8.2)	LC_{50}	0.83–2.8	0.71–1.5	0.71		955
F, Linear alkyl sulfonate	*Lepomis macrochirus* (S)	LC_{50}			0.72	0% biodeg	149
F, Dibutyl phthalate	*Lepomis macrochirus* (F)	LC_{50}			0.73		957
F, Benomyl	*Lepomis macrochirus* (S, 22°C, pH 7.5)	LC_{50}			0.85–1.2		475
F, 2,2-Dichlorovinyldimethyl phosphate	*Lepomis macrochirus*	LC_{50}			0.869		71
F, Strychnine	*Lepomis macrochirus* (S, 23°C, pH 7.9)	LC_{50}			0.87		956
F, Strychnine	*Lepomis macrochirus* (S, 23°C)	LC_{50}			0.87		224
F, Strychnine	*Lepomis macrochirus* (S, adult, 23°C, pH 7.6–7.9)	LC_{50}			0.87		954
F, Tetradifon	*Lepomis macrochirus* (S, 24°C, pH 7.5)	LC_{50}			0.880		607
F, Linear alkyl sulfonate	*Lepomis macrochirus* (S)	LC_{50}			0.89	36.7% biodeg	149
F, Lactonitrile	*Lepomis macrochirus*	TL_m	4.0		0.90		22
F, Parathion	*Lepomis cyanellus* (S, 18°C, pH 7.2–7.5)	LC_{50}			0.930		511
F, Hydrazine	*Lepomis macrochirus* (F, 15.5°C, pH 6.7–8.0)	LC_{50}	3.8		1.0		950
F, Methomyl, TG	*Lepomis macrochirus* (S, 22°C, pH 7.5)	LC_{50}			1.050		513
F, Hydrazine	*Lepomis macrochirus* (S, 23–24°C, pH 7.2–8.4)	LC_{50}			1.08		959
F, Linear alkyl sulfonate	*Lepomis macrochirus* (S)	LC_{50}			1.16	53.3% biodeg	149
F, *o*-Chloronitrobeneze	*Lepomis macrochirus* (FW, S, 23°C)	LC_{50}			1.2		224
F, *m*-Chloronitrobenzene	*Lepomis macrochirus* (S)	LC_{50}			1.2		948
F, Hydrazine	*Lepomis macrochirus* (F, 21.0°C, pH 6.7–8.0)	LC_{50}	1.7		1.2		950
F, Furfural	*Lepomis macrochirus*	TL_m	32.0		1.2		26
F, Amitraz	*Lepomis macrochirus*	LC_{50}			1.30		951
F, Baytex	*Lepomis macrochirus*	LC_{50}			1.38		237
F, Baytex	*Lepomis macrochirus*	LC_{50}			1.4		122

Chemical	Species	Toxicity test	24 h	48 h	96 h	Chronic exposure	Ref.
F, Benzethonium chloride	*Lepomis macrochirus*	LC_{50}			1.4		238
F, Monuron (in trichloroacetic acid)	*Lepomis macrochirus*	LC_{50}			1.5		789
F, DDT (dechlorinated emulsion)	*Lepomis macrochirus*	LC_{50}			1.519		83
F, Hydrazine	*Lepomis macrochirus* (F, 10°C, pH 6.7–8.0)	LC_{50}	7.7		1.6		950
F, Linear alkyl sulfonate	*Lepomis macrochirus* (S)	LC_{50}			1.64	76.0% biodeg	149
F, Butylbenzyl phthalate	*Lepomis macrochirus* (S)	LC_{50}			1.7	0.38 (NOAE)	9
F, Baytex	*Lepomis microlophus*	LC_{50}			1.88		237
F, Baytex	*Lepomis humilis*	LC_{50}			1.9		122
F, Methomyl	*Lepomis macrochirus* (23°C)	LC_{50}			2.0		910
F, *p*-Chloraniline	*Lepomis macrochirus* (S)	LC_{50}			2.0		147
F, Bromicide	*Lepomis macrochirus* (F, 21°C, pH 7.1)	LC_{50}			2.13		958
F, Pydraul 50E	*Lepomis macrochirus* (S)	LC_{50}	4.40		2.20		139
F, Carbofuran	*Lepomis macrochirus*	LC_{50}			2.3		912
F, Naphthalene (80%)	*Lepomis macrochirus* (21–23°C, pH 6.5–7.9)	LC_{50}	3.7		2.3		852
F, Igepal CO-520	*Lepomis macrochirus* (S)	LC_{50}			2.4		142
F, Furanace	*Lepomis cyanellus* (S)	LC_{50}			2.48		121
F, Pydraul 50E	*Lepomis macrochirus* (F)	LC_{50}			2.80		139
F, 4-Aminopyridine	*Lepomis macrochirus* (S, 12–22°C)	TL_m			2.82–7.56		916
F, Furanace	*Lepomis macrochirus* (S)	LC_{50}			3.00		121
F, Aminocarb	*Lepomis macrochirus* (S, 20°C, pH 7.5)	LC_{50}			3.1		475
F, Aminocarb (98%)	*Lepomis macrochirus* (320 mg/l $CaCO_3$, 22°C, pH 7.4)	LC_{50}			3.3–4.6		517
F, DDT (dechlorinated)	*Lepomis macrochirus*	LC_{50}			3.472		83
F, Crotanaldehyde	*Lepomis macrochirus* (S, 23°C, pH 7.9)	LC_{50}			3.50		954
F, Crotonaldehyde	*Lepomis macrochirus* (S, 23°C)	LC_{50}			3.5		224
F, Aminocarb (98%)	*Lepomis macrochirus* (SW, S, 12–22°C, pH 7.4)	LC_{50}			3.5–4.2		517
F, Dimethyl parathion	*Lepomis gibbosus* (S)	LC_{50}			3.6		116

F, Methyl parathion	*Lepomis gibbosus* (juv., S, 20°C, pH 7.2)	LC_{50}	4.9		3.6	772
F, Ametryn	*Lepomis macrochirus* (24°C)	LC_{50}			3.7	961
F, Chlorfos	*Lepomis macrochirus*	LC_{50}			3.8	237
F, Ametryn	*Lepomis macrochirus*	LC_{50}			4.1	960
F, *p*-Dichlorobenzene	*Lepomis macrochirus* (S)	LC_{50}			4.28	660
F, Methyl parathion	*Lepomis macrochirus* (fingerling, S, 17°C, pH 7.5)	LC_{50}			4.38	511
F, Phosphamidon	*Lepomis macrochirus*	LC_{50}			4.5	237
F, Altosid-SR10	*Lepomis macrochirus* (S)	TL_{50}			4.6	28
F, Propoxur	*Lepomis macrochirus* (S, 24°C, pH 7.5)	LC_{50}			4.8	606
F, Glyphosate (RoundUp)	*Lepomis macrochirus* (22°C)	LC_{50}			5.0	562
F, Dimethyl parathion	*Lepomis microlophus*	LC_{50}			5.17	237
F, Aminocarb (98% active)	*Lepomis macrochirus* (S, 22°C, pH 8.5)	LC_{50}			5.2	517
F, Tordon 22K	*Lepomis macrochirus* (18.3°C)	LC_{50}	8.2	7.3	5.4	803
F, Phenol	*Lepomis macrochirus*	TL_m	19.0	19.0	5.7	126
F, Dimethyl parathion	*Lepomis macrochirus*	LC_{50}			5.72	237
F, *N*-Nitrosodiphenylamine	*Lepomis macrochirus* (S, 21–23°C, pH 6.5–6.9)	LC_{50}	44.0		5.8	878
F, Diallate	*Lepomis macrochirus*	LC_{50}			5.9	914
F, 1,3-Dichloropropane	*Lepomis macrochirus*	LC_{50}			6.06	651
F, Igepal, CO-630	*Lepomis macrochirus* (F)	LC_{50}			6.3	142
F, Propoxur	*Lepomis macrochirus* (S, 22°C)	LC_{50}			6.6	816
F, Carbaryl	*Lepomis macrochirus*	NOEL			6.76	237
F, Carbaryl	*Lepomis macrochirus*	LC_{50}			6.8	122
F, 4-Dimethylamino-3,5-xylenol	*Lepomis macrochirus* (S)	LC_{50}			7.2	286
F, Tillam	*Lepomis macrochirus*	LC_{50}			7.40	23
F, Dimethyl sulfate	*Lepomis macrochirus* (S)	LC_{50}			7.5	224
F, Igepal CO-630	*Lepomis macrochirus* (F)	LC_{50}			7.9	142
F, Atrazine	*Lepomis macrochirus* (F, 18–20°C)	LC_{50}			8.0	535
F, *o*-Chlorophenol	*Lepomis macrochirus*	TL_m	12.0		8.0	50
F, *o*-Chlorophenol	*Lepomis macrochirus* (fingerlings)	TL_m			8.4	84
F, Benzyl alcohol	*Lepomis macrochirus* (S, freshW)	LC_{50}			10.0	224

Chemical	Species	Toxicity test	24 h	48 h	96 h	Chronic exposure	Ref.
F, *p*-Butylphenyldiphenyl phosphate	*Lepomis macrochirus*	LC_{50}			10.0–12.0		523
F, *m*-Cresol	*Lepomis macrochirus*	TL_m			10.0–13.6		183
F, Houghtosafe	*Lepomis macrochirus* (F)	LC_{50}			11.0		139
F, Methyl bromide	*Lepomis macrochirus* (S, adult, 23°C, pH 7.6–7.9)	LC_{50}			11.0		954
F, Isopropylphenyldiphenyl phosphate	*Lepomis macrochirus* (F, 20°C, pH 7.4)	LC_{50}			11.0		849
F, Carbaryl	*Lepomis humilis*	LC_{50}			11.2		122
F, Zectran	*Lepomis macrochirus*	LC_{50}			11.2		122
F, Zectran	*Lepomis macrochirus*	LC_{50}			11.2		237
F, Carbaryl	*Lepomis microlophus*	NOEL			11.2		237
F, Acrylonitrile	*Lepomis macrochirus* (softW)	TL_m			11.8		22
F, Isopropylphenyldiphenyl phosphate	*Lepomis macrochirus* (S, 20°C, pH 7.4)	LC_{50}			12.0		849
F, Houghtosafe	*Lepomis macrochirus* (S)	LC_{50}	32.0		12.0		139
F, Monocrotophos (100%)	*Lepomis macrochirus* (18°C, pH 7.1)	LC_{50}			12.1		404
F, Tetrachloroethylene	*Lepomis macrochirus* (juv., S, 23°C, pH 6.5–7.9)	LC_{50}			13.0		878
F, Aminocarb	*Lepomis macrochirus* (S, 22°C, pH 6.5)	LC_{50}			14.0		962
F, Zectran	*Lepomis humilis*	LC_{50}			16.7		122
F, Zectran	*Lepomis macrochirus*	LC_{50}			16.7		237
F, Vinyl acetate	*Lepomis macrochirus*	TL_m	18.0		18.0		50
F, 2,4,5-Trichloroacetic acid	*Lepomis gibbosus* (S)	LC_{50}			20.0		117
F, *o*-Cresol	*Lepomis macrochirus* (softW)	TL_m	22.2		20.8		50
F, *p*-Xylene	*Lepomis macrochirus* (softW)	TL_m	24.0		20.9		50
F, Picloram	*Lepomis macrochirus* (17.2°C)	LC_{50}	26.5	22.5	21.0		803
F, Benzene	*Lepomis macrochirus* (softW)	TL_m	22.5		22.5		154
F, Picloram	*Lepomis macrochirus* (S, 32°C, pH 7.5)	LC_{50}			23.0		411
F, Phenol	*Lepomis macrochirus* (softW)	TL_m	25.8	23.9	23.9		50

F, Toluene	*Lepomis macrochirus* (S)	LC_{50}			24.0	814
F, Chlorobenzene	*Lepomis macrochirus*	TL_m	24.0		24.0	22
F, Styrene	*Lepomis macrochirus* (softW)	TL_m	25.1	25.1	25.1	50
F, Fluorescent whitening agent, 4A	*Lepomis macrochirus* (S)	LC_{50}			26.0	208
F, Eptam, technical	*Lepomis macrochirus*	LC_{50}			27.0	25
F, *o*-Dichlorobenzene	*Lepomis macrochirus* (S)	LC_{50}			27.0	224
F, Actanilide	*Lepomis macrochirus* (S, 23°C, pH 7.6–7.9)	LC_{50}			27.0	954
F, Molinate	*Lepomis macrochirus*	LC_{50}			29.0	963
F, Fluorescent, 3A	*Lepomis macrochirus* (S)	LC_{50}			32.0	208
F, 2-Ethylhexyldiphenyl phosphate	*Lepomis macrochirus*	LC_{50}			32.0	523
F, *n*-Butyl amine	*Lepomis macrochirus* (S)	LC_{50}			32.0	224
F, Ethylbenzene	*Lepomis macrochirus* (softW)	TL_m	35.1	32.0	32.0	50
F, Monuron (80% active)	*Lepomis macrochirus*	LC_{50}			33.0	789
F, Cyclohexane	*Lepomis macrochirus*	TL_m	43.0		34.0	50
F, Diquat	*Lepomis macrochirus*	LC_{50}			35.0	194
F, Epichlorhydrin	*Lepomis macrochirus* (semiS, 23°C, pH 7.9)	LC_{50}			35.0	954
F, Epichlorhydrin	*Lepomis macrochirus* (S, 23°C)	LC_{50}			35.0	224
F, Epichlorhydrin	*Lepomis macrochirus* (S, adult, 23°C, pH 7.6–7.9)	LC_{50}			35.0	954
F, Brucine	*Lepomis macrochirus* (S)	LC_{50}			36.0	224
F, Picloram	*Lepomis cyanellus* (10°C)	LC_0			39.0	803
F, Monuron, 25% pellet	*Lepomis macrochirus*	LC_{50}			40.0	789
F, Atrazine, wet powder	*Lepomis macrochirus* (S, 22°C, pH 7.1)	LC_{50}	48.0		42.0	404
F, Allyl chloride	*Lepomis macrochirus* (softW)	TL_m	59.3	42.3	42.3	50
F, Isoprene	*Lepomis macrochirus*	TL_m	42.5	42.5	42.5	2
F, Trichloroethylene	*Lepomis macrochirus* (S)	LC_{50}			44.7	965
F, Methylcellosolve acetate	*Lepomis macrochirus* (S, 23°C)	LC_{50}			45.0	224
F, Dicamba	*Lepomis macrochirus* (S)	LC_{50}			50.0	402
F, Acetaldehyde	*Lepomis humilis*	TL_m			53.0	84
F, Saniticizer 148 + 6% TPP	*Lepomis macrochirus* (S, 22°C, pH 7.6)	LC_{50}			72.0	964
F, Acetic acid	*Lepomis macrochirus*	TL_m			75.0	51

Chemical	Species	Toxicity test	24 h	48 h	96 h	Chronic exposure	Ref.
F, Acetic acid	*Lepomis macrochirus*	TL_m			75.0		51
F, Benzonitrile	*Lepomis macrochirus* (softW)	TL_m			78.0		134
F, Tetramethyllead	*Lepomis macrochirus* (S, 23°C)	LC_{50}			84.0		224
F, Tetramethyllead	*Lepomis macrochirus* (adult, 23°C, pH 7.6–7.9)	LC_{50}			84.0		954
F, Ethylbenzene	*Lepomis macrochirus* (S, 17°C)	LC_{50}			88.0		921
F, Glycol diacetate	*Lepomis macrochirus* (S, 23°C)	LC_{50}			90.0		224
F, Picloram	*Lepomis cyanellus* (10°C)	LC_{50}	91.0		91.0		803
F, 2,4-Dichlorophenoxyacetic acid	*Lepomis gibbosus* (S)	LC_{50}			94.6		64
F, 2,6-Difluorobenzoic acid	*Lepomis macrochirus* (S)	LC_{50}			100.0		147
F, Diethylthiophosphate	*Lepomis macrochirus*	LC_{50}			100.0		87
F, Actanilide	*Lepomis macrochirus* (S, 23°C, pH 7.6–7.9)	LC_{50}			100.0		954
F, 4-Chlorophenylurea	*Lepomis macrochirus* (S)	LC_{50}			100.0		147
F, *n*-Butyl acetate	*Lepomis macrochirus* (S)	LC_{50}			100.0		224
F, Benzene	*Lepomis macrochirus* (S, 22°C)	LC_{50}			100.0		921
F, EDTA	*Lepomis macrochirus* (S)	NOEL			100.0		135
F, Pydraul 115E	*Lepomis macrochirus* (S)	LC_{50}	100.0		100.0		139
F, Carbon tetrachloride	*Lepomis macrochirus* (S, FW, 23°C)	LC_{50}			125.0		224
F, Propionaldehyde	*Lepomis macrochirus* (S, 23°C)	LC_{50}			130.0		224
F, Glyphosate	*Lepomis macrochirus* (22°C)	LC_{50}	150.0		140.0		562
F, Diquat dibromide	*Lepomis macrochirus* (25°C, pH 8.1)	LC_{50}	410.0	210.0	140.0		463
F, Triclopyr	*Lepomis macrochirus*	LC_{50}			148.0		461
F, Ethylbenzene (80% active)	*Lepomis macrochirus* (juv., S) Solubility?	LC_{50}	169.0		150.0		856
F, EDTA	*Lepomis macrochirus* (S)	LC_{50}			159.0		135
F, Aquathol K-di-K	*Lepomis macrochirus*	LC_{50}			160.0		238
F, Thallium acetate	*Lepomis macrochirus* (S, 23°C)	LC_{50}			170.0		224
F, Thallium acetate	*Lepomis macrochirus* (S, adult, 23°C, pH 7.6–7.9)	LC_{50}			170.0		954
F, Endothal-diNa	*Lepomis macrochirus*	LC_{50}	450.0	320.0	180.0		968
F, Propylene oxide	*Lepomis macrochirus* (S, 23°C)	LC_{50}			215.0		905

F, Propene oxide	*Lepomis macrochirus* (S, 24°C)	TL_m		215.0	152
F, 1,1-Dichloroethylene	*Lepomis macrochirus* (S)	LC_{50}		220.0	224
F, 1,1-Dichloroethylene	*Lepomis macrochirus* (S, adult, 23°C, pH 7.6–7.9)	LC_{50}		220.0	954
F, Ammonium picrate	*Lepomis macrochirus* (S, 23°C, freshW)	LC_{50}		220.0	224
F, Isophorone	*Lepomis macrochirus* (S)	LC_{50}		224.0	539
F, EDTA, tetraammonium salt	*Lepomis macrochirus* (S)	NOEL		240.0	135
F, Toluene	*Lepomis macrochirus*	TL_m	240.0	240.0	50
F, 2,2-Diphenolic acid	*Lepomis macrochirus*	LC_{50}		290.0	238
F, Triphenyl phosphate	*Lepomis macrochirus* (S, 23°C, pH 7.6–7.9)	LC_{50}		290.0	954
F, Triphenyl phosphate	*Lepomis macrochirus* (S, 23°C, freshW)	LC_{50}		290.0	224
F, 1,2-Dichloropropane	*Lepomis macrochirus* (S)	LC_{50}		300.0	538
F, 1,2-Dichloropropane	*Lepomis macrochirus* (S)	LC_{50}		320.0	224
F, EDTA, Cu chelate	*Lepomis macrochirus*	NOAE		320.0	135
F, Endothal-diK	*Lepomis macrochirus* (S, 22°C, pH 7.5)	LC_{50}		343.0	607
F, Diacetone alcohol	*Lepomis macrochirus* (S)	LC_{50}		420.0	224
F, 1,2-Dichloroethane	*Lepomis macrochirus* (S, 21–23°C, pH 6.5–7.9)	LC_{50}	600.0	430.0	852
F, EDTA, tetraNa salt	*Lepomis macrochirus*	NOEL		456.0	135
F, Fluorescent whitening agent, 2A	*Lepomis macrochirus* (S)	LC_{50}		474.0	208
F, EDTA, tetraNa salt	*Lepomis macrochirus* (S)	LC_{50}		486.0	135
F, 1,1-Dichloroethane	*Lepomis macrochirus* (S)	LC_{50}		550.0	2
F, Methyl chloride	*Lepomis macrochirus* (S, adult, 23°C, pH 7.6–7.9)	LC_{50}		550.0	954
F, Versenol 120	*Lepomis macrochirus* (S)	NOEL		560.0	135
F, Bis(2-chloroethyl)ether	*Lepomis macrochirus* (juv., S, 21–23°C, pH 6.5–7.9)	LC_{50}		600.0	852
F, Bentazone (TG)	*Lepomis macrochirus*	LC_{50}		616.0	967
F, *n*-Amyl acetate	*Lepomis macrochirus* (S, freshW)	LC_{50}		650.0	224
F, Diflubenzuron	*Lepomis macrochirus* (S)	LC_{50}		660.0	147
F, EDTA Zn chelate	*Lepomis macrochirus* (S)	LC_{50}		685.0	135

Chemical	Species	Toxicity test	24 h	48 h	96 h	Chronic exposure	Ref.
F, EDTA, tetraammonium salt	*Lepomis macrochirus* (S)	LC_{50}			705.0		135
F, Adiponitrile	*Lepomis macrochirus* (softW)	TL_m			720.0		22
F, DTPA, pentaNa	*Lepomis macrochirus*	NOEL			750.0		135
F, Versenol 120	*Lepomis macrochirus* (S)	LC_{50}			808.0		135
F, EDTA, tetraNa salt	*Lepomis macrochirus*	NOEL			870.0		135
F, Igepal CO-880	*Lepomis macrochirus* (S)	LC_{50}			1000.0		142
F, Fluorescent whitening agent, 1A	*Lepomis macrochirus* (S)	LC_{50}			1000.0		208
F, EDTA, dihydrate-Ca chelate, diNa	*Lepomis macrochirus* (S)	NOEL			1000.0		135
F, EDTA, tetraNa salt	*Lepomis macrochirus* (S)	LC_{50}			1030.0		135
F, Cyclohexanol	*Lepomis macrochirus* (S, freshW, 23°C)	LC_{50}			1100.0		224
F, Diethylene triamine pentaacetic acid	*Lepomis macrochirus* (S)	LC_{50}			1115.0		135
F, Butyldigol	*Lepomis macrochirus* (S)	LC_{50}			1300.0		2
F, EDTA, diammonium salt	*Lepomis macrochirus* (S)	NOEL			1350.0		135
F, *m*-Nitrobenzene sulfonate, Na	*Lepomis macrochirus*	TL_m	1350.0				629
F, EDTA, magnesium chelate	*Lepomis macrochirus* (S)	NOEL			1350.0		135
F, Butylcellosolve	*Lepomis macrochirus* (S)	LC_{50}			1490.0		224
F, Fire-trol 931	*Lepomis macrochirus* (S)	LC_{50}			1500.0		151
F, Fire-trol 100	*Lepomis macrochirus* (S)	LC_{50}			1500.0		151
F, Polypropylene glycol	*Lepomis macrochirus* (S, 23°C, pH 7.6–7.9)	LC_{50}			1700.0		954
F, Polypropylene glycol	*Lepomis macrochirus* (S, 23°C)	LC_{50}			1700.0		224
F, Versene	*Lepomis macrochirus* (S)	NOEL			1800.0		135
F, Acetonitrile	*Lepomis macrochirus* (softW)	TL_m			1850.0		134
F, EDTA, copper chelate	*Lepomis macrochirus* (S)	LC_{50}			2340.0		135
F, EDTA, diammonium salt	*Lepomis macrochirus* (S)	LC_{50}			2340.0		135
F, EDTA, dihydrate, calcium chelate	*Lepomis macrochirus* (S)	LC_{50}			2340.0		135
F, EDTA, Mg chelate	*Lepomis macrochirus* (S)	LC_{50}			2520.0		135
F, 2-Butoxyethanol	*Lepomis macrochirus* (S)	LC_{50}	2950.0		2950.0		969
F, Asulam	*Lepomis macrochirus* (S)	LC_{50}			3000.0		812

F, Versena	*Lepomis macrochirus* (S)	LC_{50}	3092.0		135
F, Versenol AG	*Lepomis macrochirus* (S)	NOEL	5600.0		135
F, Ethiofencarb	*Lepomis macrochirus*	LC_{50}	6700.0		849
F, Isodecyldiphenyl phosphate	*Lepomis macrochirus* (S, 23°C)	LC_{50}	6700.0		239
F, Diisopropyl ether	*Lepomis macrochirus*	LC_{50}	7000.0		224
F, Tricresyl phosphate	*Lepomis macrochirus* (S)	LC_{50}	7000.0		224
F, Tricresyl phosphate	*Lepomis macrochirus* (S, 23°C, pH 7.9)	LC_{50}	7000.0		954
F, Diethylene glycol, monomethyl ether	*Lepomis macrochirus* (S)	LC_{50}	7500.0		224
F ,Versenol AG	*Lepomis macrochirus* (S)	LC_{50}	8100.0		135
F, Acetone	*Lepomis macrochirus*	LC_{50}	8300.0		182
F, Cellosolve	*Lepomis macrochirus* (S, 23°C)	LC_{50}	10,000.0		224
F, 1,4-Dioxane	*Lepomis macrochirus*	LC_{50}	10,000.0		224
F, Diethyl ether	*Lepomis macrochirus* (S, adult, 23°C, pH 7.6–7.9)	LC_{50}	10,000.0		954
F, Cellosolve	*Lepomis macrochirus* (S, adult, 23°C, pH 7.6–7.9	LC_{50}	10,000.0		954
F, Hexylene glycol	*Lepomis macrochirus* (S, 23°C)	LC_{50}	10,000.0		224
F, Ethyl digol	*Lepomis macrochirus* (S)	LC_{50}	10,000.0		224
F, *p*-Dioxane	*Lepomis macrochirus* (S, adult, 23C, pH 7.6–7.9)	LC_{50}	10,000.0		954
F, Triethylene glycol	*Lepomis macrochirus* (S, 23°C, freshW)	LC_{50}	10,000.0		224
F, Ethylene glycol monomethyl ether	*Lepomis macrochirus* (S, adult, 23°C, pH 7.6–7.9	LC_{50}	10,000.0		954
F, Methyl cellosolve	*Lepomis macrochirus* (S, 23°C)	LC_{50}	10,000.0		224
F, Ethyl ether	*Lepomis macrochirus* (S, 23°C)	LC_{50}	10,000.0		224
F, Polydimethylsiloxane	*Lepomis macrochirus* (S, 24°C)	TL_m	10,000.0		153
F, Endrin	*Leucaspius delineatus* (S)	LC_{50}		0.002, 0.013, 0.021 (72 h)	162
F, Triethyl phosphate	*Leuciscus idus*	NOAE		1926.0 (h?)	493
F, Triethyl phosphate	*Leuciscus idus*	LC_{50}		2140.0 (h?)	493

Chemical	Species	Toxicity test	24 h	48 h	96 h	Chronic exposure	Ref.
F, Triethyl phosphate	*Leuciscus idus*	LC_{100}				2300.0 (h?)	493
F, Endrin	*Leuciscus idus* (S)	LC_0		0.002			493
F, Hexachlorobenzene	*Leuciscus idus* (S)	LC_0		0.003			493
F, Endrin	*Leuciscus idus* (S)	LC_{50}		0.003			493
F, Endrin	*Leuciscus idus* (S)	LC_{100}		0.004			493
F, Hexachlorobenzene	*Leuciscus idus* (S)	LC_{50}		0.007			493
F, Dieldrin	*Leuciscus idus* (adult, S)	LC_0		0.01			493
F, Lindane	*Leuciscus idus* (S)	LC_{50}		0.01			493
F, Dieldrin	*Leuciscus idus* (adult, S)	LC_{50}		0.019			493
F, Hexachlorobenzene	*Leuciscus idus* (S)	LC_{100}		0.02			493
F, Lindane	*Leuciscus idus* (S)	LC_{100}		0.02			493
F, Dieldrin	*Leuciscus idus* (adult, S)	LC_{100}		0.023			493
F, Pentachlorophenol	*Leuciscus idus* (S)	LC_0		0.08			493
F, Heptachlor	*Leuciscus idus* (S)	LC_{50}		0.09			493
F, Pentachlorophenol	*Leuciscus idus* (S)	LC_{50}		0.1			493
F, Pentachlorophenol	*Leuciscus idus* (S)	LC_{100}		0.12			493
F, Aldrin	*Leuciscus idus* (adult, S)	LC_0		0.17			493
F, Aldrin	*Leuciscus idus* (adult, S)	LC_{50}		0.18			493
F, Aldrin	*Leuciscus idus* (adult, S)	LC_{100}		0.19			493
F, Isobenzan	*Leuciscus idus* (S)	LC_0		0.27			493
F, Isobenzan	*Leuciscus idus* (S)	LC_{50}		0.35			493
F, Isobenzan	*Leuciscus idus* (S)	LC_{100}		0.45			493
F, Pentachlorobenzene	*Leuciscus idus* (S)	LC_0		0.6			493
F, 1,2,4-Trichlorobenzene	*Leuciscus idus* (S)	LC_0		0.6			493
F, 1,2,4-Trichlorobenzene	*Leuciscus idus* (S)	LC_{50}		0.7			493

F, 1,2,4-Trichlorobenzene	*Leuciscus idus* (S)	LC_{100}	0.8	493
F, 2,4,5-Trichlorobenzene	*Leuciscus idus* (S)	LC_0	0.9	493
F, Pentachlorobenzene	*Leuciscus idus* (S)	LC_{50}	0.9	493
F, 2,4,5-Trichlorobenzene	*Leuciscus idus* (S)	LC_{50}	1.0	493
F, 2,4,5-Trichlorobenzene	*Leuciscus idus* (S)	LC_{100}	1.1	493
F, 1,2-Dichloroethane	*Leuciscus idus* (adult, S)	LC_0	1.3	493
F, Pentachlorobenzene	*Leuciscus idus* (S)	LC_{100}	1.3	493
F, 1,2-Dichloroethane	*Leuciscus idus* (adult, S)	LC_{50}	1.8	493
F, 1,2-Dichloroethane	*Leuciscus idus* (adult, S)	LC_{100}	2.4	493
F, Hexachloro-1,3-butadiene	*Leuciscus idus* (S)	LC_{50}	3.0	493
F, Carbon tetrachloride	*Leuciscus idus* (S)	LC_0	5.0	493
F, Epichlorhydrin	*Leuciscus idus melanotus* (S, pH 7.5)	LC_{50}	12.0	970
F, Carbon tetrachloride	*Leuciscus idus* (S)	LC_{50}	13.01	493
F, Butylbenzyl phthalate	*Leuciscus idus* (adult, S)	LC_0	14.0	493
F, Butylbenzyl phthalate	*Leuciscus idus* (adult, S)	LC_{50}	17.0	493
F, Butylbenzyl phthalate	*Leuciscus idus* (adult, S)	LC_{100}	21.0	493
F, Carbon tetrachloride	*Leuciscus idus* (S)	LC_{100}	33.0	493
F, Epichlorhydrin	*Leuciscus idus melanotus* (S, pH 7.5)	LC_{100}	35.0	970
F, Disodium arsenate	*Leuciscus idus* (juv., S, 19–21°C, pH 7–8)	LC_0	43.0	970
F, Ethybenzene (80%)	*Leuciscus idus* (juv., S, aeration, 19–21°C, pH 7–8)	LC_{50}	44.0	856
F, Isopropylbenzene	*Leuciscus idus* (juv., S, 19–21°C, pH 7–8)	LC_{50}	47.0–207.0	970
F, Nitrobenzene	*Leuciscus idus* (juv., S, 19–21°C, pH 7–8)	NOAE	48.0	970
F, Chloroform	*Leuciscus idus* (adult, S)	LC_0	51.0	493
F, Isopropylbenzene	*Leuciscus idus* (juv., S, 19–21°C, pH 7–8)	LC_{100}	52.0–345.0	970
F, Disodium arsenate	*Leuciscus idus* (S, juv.,12–21°C, pH 7–8)	LC_{50}	70.0	970
F, Nitrobenzene	*Leuciscus idus* (juv., S, 19–21°C, pH 7–8)	LC_{100}	72.0	970
F, Tetrachloroethylene	*Leuciscus idus* (S)	LC_0	81.0	493
F, Chloroform	*Leuciscus idus* (adult, S)	LC_{50}	92.0	493
F, Maleic anhydride	*Leuciscus idus* (S)	LC_0	100.0	493
F, Disodium arsenate	*Leuciscus idus* (S, juv., 12–21°C, pH 7–8)	LC_{50}	103.0	970

Chemical	Species	Toxicity test	24 h	48 h	96 h	Chronic exposure	Ref.
F, Maleic anhydride	*Leuciscus idus* (S)	LC_{50}		115.0			493
F, Maleic anhydride	*Leuciscus idus* (S)	LC_{100}		125.0			493
F, Nitrobenzene	*Leuciscus idus* (juv., S, 19–21°C, pH 7–8)	LC_{100}		130.0			970
F, Tetrachloroethylene	*Leuciscus idus* (S)	LC_{50}		130.0			493
F, Chloroform	*Leuciscus idus* (adult, S)	LC_{100}		151.0			493
F, Disodium arsenate	*Leuciscus idus* (juv., S, 19–21°C, pH 7–8)	LC_{100}		192.0			970
F, Tetrachloroethylene	*Leuciscus idus* (S)	LC_{100}		201.0			493
F, Hydroxyethanediphosphonic acid	*Leuciscus idus melanotus*	LC_{50}		207.0			505
F, Hydroxyethanediphosphonic acid	*Leuciscus idus melanotus*	LC_{100}		240.0			505
F, Methyl-*t*-butyl ether	*Leuciscus idus melanotus*	LC_0		1000.0			671
F, *n*-Butyl alcohol	*Leuciscus idus*	LC_{50}		1200.0			970
F, *e*-Caprolactam	*Leuciscus idus* (adult, S)	LC_0		1300.0			493
F, 2-Butoxyethanol	*Leuciscus idus* (juv., S, 19–21°C, pH 7–8)	NOAE		1350.0			971
F, *e*-Caprolactam	*Leuciscus idus* (adult, S)	LC_{50}		1450.0			493
F, 2-Butoxyethanol	*Leuciscus idus* (juv., S, 19–21°C, pH 7–8)	LC_{50}		1575.0			971
F, *e*-Caprolactam	*Leuciscus idus* (adult, S)	LC_{100}		1600.0			493
F, 2-Butoxyethanol	*Leuciscus idus* (juv., S, 19–21°C, pH 7–8)	LC_{100}		1650.0			971
F, Methyl-*t*-butyl ether	*Leuciscus idus melanotus*	LC_{100}		2000.0			671
F, Triethyl phosphate	*Leuciscus idus* (S)	LC_{50}		2140.0			493
F, Triethyl phosphate	*Leuciscus idus* (S)	LC_{100}		2300.0			493
F, *sec*-Butyl alcohol	*Leuciscus idus*	LC_{50}		3520.0			509
F, *n*-Propanol	*Leuciscus idus melanotus* (S, 20°C, pH 7.8)	LC_0		4000.0			970
F, *n*-Propanol	*Leuciscus idus melanotus* (S, 20°C, pH 7.8)	LC_{50}		4560.0			970
F, Triethanolamine	*Leuciscus idus*	LC_0		10,000.0			505
F, Potassium cyanide	*Leuciscus idus* (S, 20°C)	LC_{10}			0.070		400
F, Potassium cyanide	*Leuciscus idus* (S, 20°C)	LC_{50}			0.209		400
F, Ethiofencarb	*Leuciscus idus*	LC_{50}			8.0–10.0		813
F, Triethyl phosphate	*Leuciscus idus*	NOAE			1926.0		669
F, Triethyl phosphate	*Leuciscus idus*	LC_{50}			2140.0		669

F, Triethyl phosphate	*Leuciscus idus*	LC_{100}			2300.0		669
F, Pentaerythritol	*Leuciscus carpio*	LC_0			5000.0		493
F, Chloroform	*Limanda* sp.	LC_{50}			28.0		629
F, Chloroform	*Limanda* sp.	LC_{50}			28.0		629
F, Arsenic trioxide	*Limanda limanda* (F, 11–13°C, pH 6.9–8.5)	LC_{50}		43.9	28.5	31.6 (72 h)	828
F, Ethylene dichloride	*Limanda limanda*	NOAE			60.0		113
F, Ethylene dichloride	*Limanda limanda* (F)	LC_{50}			115.0		570
F, Phenol	*Mollienesia latipinna*	TL_m				22.0 (50 h)	76
F, Phenol	*Mollienesia latipinna*	TL_m				63.0 (25 h)	76
F, Cartap, HCl (95%)	*Aplocheilus latipes* (15°C)	LC_{50}		1.40			972
F, Cartap, HCl (95%)	*Aplocheilus latipes* (25°C)	LC_{50}		0.80			972
F, Cartap, HCl (95%)	*Aplocheilus latipes* (35°C)	LC_{50}		0.41			972
F, Cartap (granules)	*Aplocheilus latipes* (no soil added to water, S, 15°C)	LC_{100}				3.0 (11 d)	972
F, Cartap (granules)	*Aplocheilus latipes* (no soil added to water, S, 25°C)	LC_{100}				3.0 (60 d)	972
F, Cartap (granules)	*Aplocheilus latipes* (soil added to water, S, 25°C)	LC_{100}			3.0		972
F, Cartap (granules)	*Aplocheilus latipes* (soil added to water, S, 35°C)	LC_{100}		3.0	3.0		972
F, Chloropyriphos	*Menidia beryllina* (embryo, F, 25–28°C)	LC_0				0.00048 (28 d)	973
F, South Louisiana Crude	*Menidia beryllina* (23–27°C)	LC_{30}				5.0 (30 d)	975
F, Triphenyl phosphate	*Menidia beryllina* (freshW and softW mix, S, 20°C)	LC_{30}				100.0 (72 h)	954
F, Triphenyl phosphate	*Menidia beryllina* (FW and SW mix, S, 20°C)	LC_{90}	320.0				954
F, Hydro#2 fuel oil (water-soluble-fraction)	*Menidia beryllina* (S, 18–22°C)	LC_{50}		5.2			656
F, Kuwait Crude (water-soluble-fraction)	*Menidia beryllina* (S, 18–22°C)	LC_{50}		6.6			656
F, Kuwait Crude	*Menidia beryllina* (S, 18–22°C)	LC_{50}		38.0			976
F, Hydro#2 fuel oil	*Menidia beryllina* (S, 18–22°C)	LC_{50}		40.0			656

Chemical	Species	Toxicity test	24 h	48 h	96 h	Chronic exposure	Ref.
F, Triphenyl phosphate	*Menidia beryllina* (freshW and softW mix, S, 20°C)	LC_{60}		180.0			954
F, Endrin	*Menidia menidia* (S)	LC_{50}			0.00005		115
F, DDT (*p,p′*)	*Menidia menidia* (S)	LC_{50}			0.0004		209
F, Chloropyriphos	*Menidia beryllina* (F, 27.5°C)	LC_{50}			0.0017		658
F, Heptachlor	*Menidia menidia* (S)	LC_{50}			0.003		115
F, Dieldrin (100%)	*Menidia menidia* (S)	LC_{50}			0.005		115
F, Lindane	*Menidia menidia* (S)	LC_{50}			0.009		115
F, Aldrin	*Menidia menidia* (S)	LC_{50}			0.013		115
F, Chlordecane (Kepone)	*Menidia menidia* (F, 17–18°C)	LC_{50}			0.0288		782
F, Malathion	*Menidia menidia* (S)	LC_{50}			0.125		115
F, Cyanogen bromide	*Menidia beryllina* (S, 23°C)	LC_{50}			0.47		224
F, Acetonecyanohydrin	*Menidia beryllina* (S, syn. seaW, 23°C)	LC_{50}			0.50		224
F, *o*-Chloronitrobenezne	*Menidia beryllina* (seaW, S, 23°C)	LC_{50}			0.55		224
F, Strychnine	*Menidia menidia* (S, 23°C)	LC_{50}			0.95		224
F, 2,2-Dichlorovinyldimethyl phosphate	*Menidia menidia* (S)	LC_{50}			1.25		115
F, Crotonaldehyde	*Menidia peninsulae* (S, 20°C, syn. seaW)	LC_{50}			1.30		954
F, Crotonaldehyde	*Menidia beryllina* (S, 23°C)	LC_{50}			1.3		224
F, Dimethyl parathion	*Menidia menidia* (S)	LC_{50}			5.7		117
F, *o*-Dichlorobenzene	*Menidia beryllina* (S)	LC_{50}			7.3		224
F, *o*-Dichlorobenzene	*Menidia beryllina* (20°C, pH 7.9, S)	LC_{50}			7.30		974
F, Tetramethyllead	*Menidia beryllina* (S, softW, 23°C)	LC_{50}			13.5		224
F, Dimethyl sulfate	*Menidia beryllina* (S)	LC_{50}			15.0		224
F, Benzyl alcohol	*Menidia beryllina* (S, seaW)	LC_{50}			15.0		224
F, Epichlorhydrin	*Menidia beryllina* (S, 23°C, mild aeration)	LC_{50}			18.0		224
F, Epichlorhydrin	*Menidia beryllina*	LC_{50}			18.0		954

F, Brucine	*Menidia beryllina* (S)	LC_{50}	20.0	224
F, *n*-Butylamine	*Menidia beryllina* (S)	LC_{50}	24.0	224
F, Thallium acetate	*Menidia beryllina* (S, softW, 23°C)	LD_{50}	31.0	224
F, Methylcellosolve acetate	*Menidia beryllina* (S, 23°C)	LC_{50}	40.0	224
F, Ammonium picrate	*Menidia beryllina* (S, 23°C, softW)	LC_{50}	66.0	224
F, Glycol diacetate	*Menidia beryllina* (S, 23°C, mild aeration)	LC_{50}	78.0	224
F, Triphenyl phosphate	*Menidia beryllina* (S, 23°C, softW)	LC_{50}	95.0	224
F, Triphenyl phosphate	*Menidia beryllina* (FW and SW mix, S, 20°C)	LC_{50}	95.0	954
F, Propionaldehyde	*Menidia beryllina* (S, 23°C, softW)	LC_{50}	100.0	224
F, Acetanilide	*Menidia beryllina* (S, 23°C, softW)	LC_{50}	115.0	224
F, Acetanilide	*Menidia beryllina* (seaW, 23°C, S)	LC_{50}	115.0	224
F, Carbon tetrachloride	*Menidia beryllina* (S, softW, 23°C)	LC_{50}	150.0	224
F, *n*-Amyl acetate	*Menidia beryllina*	LC_{50}	180.0	224
F, *n*-Butyl acetate	*Menidia beryllina* (S)	LC_{50}	185.0	224
F, 1,2-Dichloropropane	*Menidia beryllina* (S)	LC_{50}	240.0	224
F, 1,1-Dichloroethylene	*Menidia beryllina* (S)	LC_{50}	250.0	224
F, Diacetone alcohol	*Menidia beryllina* (S)	LC_{50}	420.0	224
F, 1,2-Dichloroethane	*Menidia beryllina* (S, 20°C, pH 7.7)	LC_{50}	480.0	954
F, 1,1-Dichloroethane	*Menidia beryllina* (S)	LC_{50}	480.0	224
F, Polypropylene glycol	*Menidia beryllina* (S, 23°C)	LC_{50}	650.0	224
F, Polypropylene glycol	*Menidia beryllina* (S, 20°C)	LC_{50}	650.0	954
F, Cyclohexanol	*Menidia beryllina* (S, seaW, 23°C)	LC_{50}	720.0	224
F, Butyl cellosolve	*Menidia beryllina* (seaW, S)	LC_{50}	1250.0	224
F, Isodecyldiphenyl phosphate	*Menidia beryllina* (S, 23°C)	LC_{50}	1400.0	224
F, Ethiofencarb	*Menidia beryllina*	LC_{50}	1400.0	849
F, Butyl digol	*Menidia beryllina* (S)	LC_{50}	2000.0	224
F, Diisopropyl ether	*Menidia beryllina*	LC_{50}	6600.0	224
F, Dioxane	*Menidia beryllina* (syn. seaW, S, 20°C, pH 7.9)	LC_{50}	6700.0	954
F, 1,4-Dioxane	*Menidia beryllina*	LC_{50}	6700.0	224
F, Tricresyl phosphate	*Menidia beryllina* (S, synthetic seaW)	LC_{50}	8700.0	224

Chemical	Species	Toxicity test	24 h	48 h	96 h	Chronic exposure	Ref.
F, Tricresyl phosphate	*Menidia beryllina* (S, 20°C, pH 7.9)	LC_{50}			8700.0		954
F, Hexylene glycol	*Menidia menidia* (S, 23°C)	LC_0			10,000.0		224
F, Cellosolve	*Menidia beryllina* (S, 23°C)	LC_{50}			10,000.0		224
F, Ethyl digol	*Menidia beryllina* (S)	LC_{50}			10,000.0		224
F, 2-Ethoxyethanol	*Menidia beryllina* (S, 20°C)	LC_{50}			10,000.0		954
F, Triethylene glycol	*Menidia beryllina* (S, 23°C, seaW)	LC_{50}			10,000.0		224
F, Ethyl ether	*Menidia beryllina* (S, 23°C)	LC_{50}			10,000.0		224
F, Methyl cellosolve	*Menidia beryllina* (S, 23°C)	LC_{50}			10,000.0		224
F, Endrin	*Micrometrus minimus* (IF)	TL_m			0.00013		186
F, Endrin	*Micrometrus minimus* (S)	TL_m			0.0006		186
F, Aldrin	*Micrometrus minimus* (IF)	TL_{50}			0.00203		186
F, Dieldrin (TG)	*Micrometrus minimus* (F)	TL_{50}			0.00244		186
F, DDT	*Micrometrus minimus* (S)	TL_{50}			0.0046		186
F, Dieldrin (TG)	*Micrometrus minimus* (S)	TL_{50}			0.005		186
F, Aldrin	*Micrometrus minimus* (S)	TL_{50}			0.018		186
F, Endosulfan	*Micropterus salmoides*	LC_{50}				0.010 (1 h)	396
F, Hexachlorobenzene	*Micropterus salmoides* (23°C, pH 6.65–7.9)	NOAE				0.010 (15 d)	974
F, Endosulfan	*Micropterus salmoides*	LC_{95}				0.025 (3 h)	396
F, Chlorodane	*Micropterus* sp.	NOEL				0.100 (30 h)	785
F, Chlorodane	*Micropterus* sp.	LC_{100}				0.200 (30 h)	785
F, Endothal	*Micropterus dolomieui* (fry)	LC_0				2.05 (8 d)	822
F, Aniline	*Micropterus salmoides* (eggs and larvae, hardW, F, 8–25°C)	LC_{50}				7.1 (7.5 d)	810
F, Quinoline	*Micropterus salmoides* (F, 20–23°C, pH 7.4–8.1)	LC_{50}				7.42 (7 d)	595
F, Aniline	*Micropterus salmoides* (softW, F, 18–20°C, pH 7.6)	LC_{50}				11.8 (7.5 d)	810
F, Aniline	*Micropterus salmoides* (eggs and larvae, HW, F, 18–25°C	LC_{50}				29.9 (3.5 d)	810

F, Aniline	*Micropterus salmoides* (softW, F, 18–20°C, pH 7.86)	LC_{50}			32.7 (3.5 d)	810
F, Endrin	*Micropterus salmoides* (S, 19°C)	TL_m	0.00027			187
F, Dichlone	*Micropterus salmoides*	LC_{50}	0.120			228
F, Monuron	*Micropterus salmoides* (20°C)	LC_{10}	230.0			789
F, Rotenone	*Micropterus salmoides*	LC_{50}		0.0016–0.012		557
F, Toxaphene	*Micropterus salmoides* (S, 18°C, pH 7.4)	LC_{50}		0.002		515
F, DDT	*Micropterus salmoides*	LC_{50}		0.002		127
F, Azinphosmethyl	*Micropterus salmoides*	LC_{50}		0.005		127
F, Phorate, TG (91%)	*Micropterus salmoides* (S, 15°C, pH 7.5)	LC_{50}		0.005		599
F, Biocallethrin, 90% TG	*Micropterus dolomieui* (fingerling, F, 12°C, pH 7.5)	LC_{50}		0.0077		597
F, Biocallethrin, 90% TG	*Micropterus salmoides* (F, 12°C, pH 7.5)	LC_{50}		0.012		597
F, Lindane	*Micropterus salmoides*	LC_{50}		0.032		127
F, Dioxathion	*Micropterus salmoides*	LC_{50}		0.036		235
F, Disulfoton	*Micropterus salmoides* (18°C)	LC_{50}		0.060		410
F, Rotenone (5%)	*Micropterus dolomieui* (S)	LC_{50}		0.079		28
F, Mevinphos	*Micropterus salmoides*	LC_{50}		0.110		945
F, Formaldehyde	*Micropterus dolomieui* (F)	LC_{50}		0.136 ml/l		120
F, Rotenone (5%)	*Micropterus salmoides* (S)	LC_{50}		0.142		28
F, Formaldehyde	*Micropterus salmoides* (F)	LC_{50}		0.143 ml/l		120
F, Phosmet	*Micropterus dolomieui* (S, 20°C, pH 7.5)	LC_{50}		0.150		409
F, Ethion	*Micropterus salmoides*	LC_{50}		0.150		71
F, Imidan	*Micropterus dolomieui* (S)	LC_{50}		0.150		148
F, Parathion	*Micropterus salmoides*	LC_{50}		0.190		235
F, Malathion	*Micropterus salmoides*	LC_{50}		0.285		127
F, Ferbam	*Micropterus salmoides*	NOEL		0.5-4.0		871
F, Parathion (100%)	*Micropterus salmoides* (fingerling, S, 18°C, pH. 7.5)	LC_{50}		0.620		511
F, Methomyl, TG (liquid concentrate, 24%)	*Micropterus salmoides* (S, 22°C, pH 7.5)	LC_{50}		0.760		513

Chemical	Species	Toxicity test	24 h	48 h	96 h	Chronic exposure	Ref.
F, Ciodrin	*Micropterus salmoides*	LC_{50}			1.1		71
F, Methomyl, TG	*Micropterus salmoides* (S, 22°C, pH 7.5)	LC_{50}			1.250		513
F, Baytex	*Micropterus salmoides*	LC_{50}			1.54		127
F, 1,3-Dichloropropane	*Micropterus salmoides* (S, 18°C, pH 7.5)	LC_{50}			3.65		758
F, Methyl parathion	*Micropterus salmoides* (fingerling, S, 18°C, pH 7.5)	LC_{50}			5.22		511
F, Dimethyl parathion	*Micropterus salmoides*	LC_{50}			5.22		127
F, Carbaryl	*Micropterus* sp.	LC_{50}			6.4		122
F, Carbaryl	*Micropterus salmoides*	NOEL			6.4		127
F, Quinoline	*Micropterus salmoides* (larvae F, 20–23°C, pH 7.4–8.1)	LC_{50}			7.50		977
F, Diquat	*Micropterus salmoides*	LC_{50}			7.8		238
F, Zectran	*Micropterus salmoides*	LC_{50}			14.7		127
F, Aquathol K-diNa	*Micropterus salmoides*	LC_{50}			120.0		238
F, Aquathol K-diNa	*Micropterus salmoides*	LC_{50}			200.0		238
F, Endothal-diNa	*Micropterus salmoides* (25°C, pH 8.1)	LC_{50}	560.0	320.0	200.0		968
F, Fire-trol 931	*Micropterus salmoides* (S)	LC_{50}			1500.0		151
F, Fire-trol 100	*Micropterus salmoides* (S)	LC_{50}			1500.0		151
F, Isodrin	Minnow	LC_{50}	0.006				2
F, Photoheptachlor	Minnow	LC_{50}	0.008				44
F, Photoisodrin	Minnow	LC_{50}	0.010				44
F, Photodieldrin	Minnow	LC_{50}	0.010				44
F, Dieldrin	Minnow	LC_{50}	0.024				44
F, Hydrogen sulfide	Minnow	LC_{100}	5.0–6.0				84
F, Toxaphene	Minnow	LC_{50}			0.014		122
F, DDT	Minnow	LC_{50}			0.019		122
F, Lindane	Minnow	LC_{50}			0.087		122
F, Baytex	Minnow	LC_{50}			2.4		122
F, Malathion	Minnow	LC_{50}			8.7		122

F, Zectran	Minnow	LC_{50}			17.0		122
F, Heptachlor	Minnow (F)	LC_{50}	0.013				65
F, Hydrogen sulfide	Minnows, shiners	LC_{100}				1.0 (h?)	84
F, Benzene	Minnows	LD_{100}				5.0–7.0 (6 h)	84
F, 2-Chlorotoluene 5-sulfonate, Na	*Mollienesia latipinna*	TL_m	115.2	66.1 (50 h)			74
F, Bromine	*Morone saxatilis* (egg)	LC_{50}	697.4				978
F, Tetraethyl lead	*Morone laborax* (6 mm larvae)	LC_0		0.010			40
F, Tetramethyl lead	*Morone laborax* (6 mm larvae)	LC_0		0.045			40
F, Tetraethyl lead	*Morone laborax* (6 mm larvae)	LC_{50}		0.065			40
F, Tetramethyl lead	*Morone laborax* (6 mm larvae)	LC_{50}		0.10			40
F, Tetraethyl lead	*Morone laborax* (6 mm larvae)	LC_{100}		0.130			40
F, Tetramethyl lead	*Morone laborax* (6 mm larvae)	LC_{100}		0.25			40
F, Tillam	*Mugil curema*	TL_m		6.25			23
F, Chloropyriphos	*Morone saxatilis* (juv., F, 13°C)	LC_{50}			0.00058		979
F, *p*-Xylene	*Morone saxatilis* (S, 16°C)	LC_{50}			0.002		455
F, Toxaphene	*Morone saxatilis* (S, 17°C)	LC_{50}			0.0044		515
F, *m*-Xylene	*Morone saxatilis* (S, 16°C)	LC_{50}			0.0092		455
F, Aldrin	*Morone saxatilis* (S)	LC_{50}			0.010		117
F, *o*-Xylene	*Morone saxatilis* (S, juv.,16°C)	LC_{50}			0.011		455
F, Ethyl-*p*-nitrophenylthiono, benzene phosphate	*Morone saxatilis* (F)	LC_{50}			0.06		806
F, Aluminum nitrate	*Morone saxatilis* (S, pH 7.1)	LC_{100}	0.37–0.52		0.37–0.52		980
F, Aluminum nitrate	*Morone saxatilis* (S, pH 6–6.55)	LC_{99}			0.480–1.30		980
F, Aluminum nitrate	*Morone saxatilis* (S, pH 6–6.55)	LC_{95}			1.0–2.9		980
F, Malathion	*Morone americana* (S)	LC_{50}			1.100		2
F, Aluminum nitrate	*Morone saxatilis* (S, pH 6.33–6.35)	LC_{98}			1.5–3.2		980
F, *p*-Xylene	*Morone saxatilis*	LC_{50}			2.0		61
F, Toluene	*Morone saxatilis*	LC_{50}			6.30		455
F, Toluene	*Morone saxatilis* (S)	LC_{50}			6.3		455
F, *m*-Xylene	*Morone saxatilis*	LC_{50}			9.2		61
F, Arsenic ion	*Morone saxatilis*	LC_{50}			10.3		1086

Chemical	Species	Toxicity test	24 h	48 h	96 h	Chronic exposure	Ref.
F, Molybdenum ion	*Morone saxatilis*	LC_{50}			79.8		1086
F, Strontium ion	*Morone saxatilis*	LC_{50}			92.8		1086
F, Lithium ion	*Morone saxatilis*	LC_{50}			105.0		1086
F, Benzene	*Morone saxatilis*	LC_{50}			10.9		61
F, *o*-Xylene	*Morone saxatilis*	LC_{50}			11.0		202
F, Dimethyl parathion	*Morone americana* (S)	LC_{50}			14.0		116
F, Methyl parathion	*Morone saxatilis* (juv., S, 20°C, pH 7.2)	LC_{50}	16.8	14.2	14.0		772
F, Bromine	*Morone saxatilis* (egg, larvae, $^1/_2$S, 20°C)	LC_{50}			30.8		978
F, 2,4,5-Trichlorophenoxyacetic acid	*Morone americana* (S)	LC_{50}			16.4		117
F, Bromine	*Morone saxatilis* (egg, larvae, $^1/_2$S, 20°C)	LC_{50}			30.8		978
F, Molybdenum ion	*Morone saxatilis*	LC_{50}			79.8		1086
F, Strontium ion	*Morone saxatilis*	LC_{50}			92.8		1086
F, Lithium ion	*Morone saxatilis*	LC_{50}			105.0		1086
F, 2,4-Dichlorophenoxyacetic acid	*Morone americana* (S)	LC_{50}			40.0		64
F, Abate	*Mugil carinatus* (S)	LC_{50}				0.023 (72 h)	639
F, Abate	*Mugil cephalus* (S)	LC_{50}				0.600 (72 h)	639
F, Endrin	*Mugil cephalus* (F)	LC_{50}	0.0026				241
F, Monuron	*Mugil cephalus* (S)	LC_{50}		16.3			789
F, Sodium fluoride	*Mugil curema* (S, 21°C)	NOAE		100.0			981
F, Endrin	*Mugil cephalus* (S)	LC_{50}			0.0003		115
F, Endosulfan	*Mugil cephalus* (F)	LC_{50}			0.00038		247
F, DDT (*p,p′*)	*Mugil cephalus* (S)	LC_{50}			0.0009		209
F, Chloropyriphos	*Mugil cephalus* (F, 24.8°C)	LC_{50}			0.0054		658
F, Dieldrin (100%)	*Mugil cephalus* (S)	LC_{50}			0.023		115
F, Lindane	*Mugil cephalus* (S)	LC_{50}			0.066		115
F, Aldrin	*Mugil cephalus* (S)	LC_{50}			0.100		115
F, Pentachlorophenol	*Mugil cephalus* (juv., F, 24.7°C)	LC_{50}			0.112		698
F, Pentachlorophenate,Na	*Mugil cephalus* (F)	LC_{50}			0.1121		109
F, Heptachlor	*Mugil cephalus* (S)	LC_{50}			0.194		115

F, 2,2-Dichlorovinyl-dimethyl-phosphate	*Mugil cephalus* (S)	LC_{50}			0.2		115
F, Halowax 1014	*Mugil cephalus* (juv.)	LC_{50}			0.263		646
F, Malathion	*Mugil cephalus* (S)	LC_{50}			0.550		115
F, Dimethyl parathion	*Mugil cephalus* (S)	LC_{50}			5.2		117
F, Mirex	*Mugil cephalus* (adult, F, 23–26°C)	LC_{0}			10.0		982
F, Mirex	*Mugil cephalus* (larvae, S, 22°C)	LC_{55}			10.0		982
F, Fluoride	*Mugill cephalus*	NOAE				5.5 (113 d)	717
F, Chlordecane	*Mugill curema* (F, 31°C)	LC_{50}		0.055			983
F, Fluoride	*Mugill cephalus*	NOAE			100.0		717
F, Fluoride	*Mugil* sp.	NOAE				5.50 (113 d)	734
O, Phenanthrene	*Neanthes arenaceodentata* (S)	TL_{m}			0.60		248
O, Benzo (a)pyrene	*Neanthes arenaceodentata* (SeaW)	TL_{m}			1.0		248
O, Dibenzanthracene	*Neanthes arenaceodentata* (S)	TL_{m}			1.0		2
O, Chrysene	*Neanthes arenaceodentata* (22°C)	TL_{m}			1.0		248
O, 2,6-Dimethyl naphthalene	*Neanthes arenaceodentata* (S)	TL_{m}			2.6		248
C, Triethyl phosphate	*Nitroca spinipes*	LC_{50}			950.0		666
F, Toxaphene	*Notemigonus crysoleucas* (S, 11.6°C)	LC_{50}	0.0125				515
F, Toxaphene	*Notemigonus crysoleucas* (S, 17°C, pH 7.9)	LC_{50}	0.027	0.007	0.005	0.0062 (72 h)	515
F, Toxaphene	*Notemigonus crysoleucas* (S, 22.8°C)	LC_{50}	0.0134	0.0066	0.006		515
F, DDT	*Notopterus notopterus* (S)	LC_{50}	0.084	0.062	0.043		179
F, 1,3-Dichloropropane	*Notropis atherinoides*	LC_{0}				1.0 (72 h)	651
F, Pentabromochlorocyclohexane	*Notropis atherinoides*	LC_{100}				3.0 (72 h)	984
F, 1,3-Dichloropropane	*Notropis atherinoides*	LC_{100}				10.0 (72 h)	758
F, RU-11679	*Oncorhynchus kisutch* (F)	LC_{50}				0.000064 (35 d)	138
F, RU-11679	*Oncorhynchus kisutch* (F)	LC_{50}				0.000089 (25 d)	138
F, RU-11679	*Oncorhynchus kisutch* (F)	LC_{50}				0.000093 (20 d)	138
F, RU-11679	*Oncorhynchus kisutch* (F)	LC_{50}				0.000069 (30 d)	138
F, Endrin	*Oncorhynchus mykiss* (larvae, S)	LC_{50}				0.00006, 0.00012 (72 h)	162
F, RU-11679	*Oncorhynchus kisutch* (F)	LC_{50}				0.0000293	138
F, RU-11679	*Oncorhynchus kisutch* (F)	LC_{50}				0.000160 (10 d)	138

Chemical	Species	Toxicity test	24 h	48 h	96 h	Chronic exposure	Ref.
F, RU-11679	*Oncorhynchus kisutch* (F)	LC_{50}				0.000109 (15 d)	138
F, RU-11679	*Oncorhynchus kisutch* (F)	LC_{50}				0.000176 (5 d)	138
F, DDT	*Oncorhynchus mykiss*	LC_{50}				0.00026 (15 d)	71
F, DDT	*Oncorhynchus mykiss* (F)	LC_{50}				0.0023 (5 d)	580
F, Toxaphene	*Oncorhynchus mykiss* (yearling)	LC_{50}				0.0036 (14 d)	515
F, Endrin	*Oncorhynchus mykiss* (embryo, S)	LC_{50}				0.0077, 0.0006 (72 h)	162
F, *o*-Xylene	*Oncorhynchus mykiss*	LOAC				0.01	193
F, Aroclor 1254	*Oncorhynchus mykiss* (F)	LC_{50}				0.027 (25 d)	145
F, Aroclor 1260	*Oncorhynchus mykiss* (F)	LC_{50}				0.051 (30 d)	92
F, Hydrogen cyanide	*Oncorhynchus mykiss*	LC_{100}				0.1–0.15	84
F, Acrolein	*Oncorhynchus mykiss*	LOAC				0.1	193
F, Boron	*Oncorhynchus kisutch* (alevin, 11°C)	LC_{100}				0.113 (283 h)	986
F, Chlorodane	*Oncorhynchus mykiss*	LC_{100}				0.1 (5 h)	785
F, *N*-Nitrosodiethylamine	*Oncorhynchus mykiss* (fry, F, 8°C)	LC_{37}				0.170	987
F, *N*-Nitrosodiethylamine	*Oncorhynchus mykiss* (eyed eggs, F, 8°C)	LC_{8}				0.170 mg/egg	987
F, *N*-Nitrosodiethylamine	*Oncorhynchus mykiss* (8°C, 11 min post exp.)	Liver CARC				0.170 mg/egg	987
F, Captan	*Oncorhynchus mykiss*	LC_{0}				0.18 (72 h)	989
F, Tris	*Oncorhynchus mykiss* (sac-fry, S)	LC_{50}				0.240 (69 h)	626
F, Arsenic trioxide	*Oncorhynchus kisutch* (juv., F, 13.8°C)	LC_{0}				0.300 (6 min)	990
F, Diflubenzuron	*Oncorhynchus mykiss* (eyed eggs and fingerlings, F)	NOAE				0.30 (30 d)	403b
F, Isopropyl xanthate, Na	*Oncorhynchus mykiss*	LC_{100}				0.3 (3 d exp.)	279
F, Trifluoromethyl-4-nitrophenol	*Oncorhynchus kisutch* (S, green eggs, 12°C)	TL_{m}				0.57 (192 h)	161
F, Arsenic trioxide	*Oncorhynchus mykiss* (SW, F, 14–16°C, pH 6.9–7.30)	LC_{0}				0.961 (28 d)	478
F, Hydrogen sulfide	*Oncorhynchus tshawytscha*	LC_{100}				1.0 (h?)	84
F, 2,2-Dichloropropionic acid	*Oncorhynchus mykiss*	LOAC				1.0	193
F, 2,4-D, dimethylamine salt	*Oncorhynchus mykiss*	LOAC				1.0	193
F, Arsenic acid, diNa, $7H_2O$	*Oncorhynchus mykiss* (in diet, 15.1°C, pH 7.6)	NOAEL				1.0–127.0 mg/kg 8 wk)	988

F, Arsenic trioxide	*Oncorhynchus mykiss* (juv., in diet 8 wk, 15°C, pH 7.6)	NOAEL	1.0–180.0 mg/kg	988
F, Amylxanthate phosphate	*Oncorhynchus mykiss* (F)	LC_{100}	1.0 (28 d)	279
F, Chlorodane	*Oncorhynchus mykiss*	LC_{100}	1.0 (3 h)	785
F, Magniflox 512 C	*Oncorhynchus mykiss* (F, 15–17°C)	TL_m	1.10 (14 d)	278
F, Atrazine	*Oncorhynchus mykiss* (egg, larv., F, 12.5°C, pH 7.8)	LC_{50}	1.11 (23 d)	810
F, Trifluoromethyl-4-nitrophenol	*Oncorhynchus kisutch* (S, green eggs,12°C)	TL_m	1.1 (192 h)	161
F, Disodium sulfide (as sulfur)	*Oncorhynchus mykiss*	LC_{50}	3.2 (10 min)	874
F, Trifluoromethyl-4-nitrophenol	*Oncorhynchus kisutch* (S, green egg, 12°C)	TL_m	3.65 (192 h)	161
F, Phenol	*Oncorhynchus mykiss*	TL_m	4.0 (18 wk)	183
F, Flouride	*Oncorhynchus mykiss*	LC_{95}	4.0–8.5 (504 h)	717
F, *p*-Cresol	*Oncorhynchus mykiss*	LC_{100}	5.0	985
F, 3-Bromo-4-nitrophenol	*Oncorhynchus mykiss*	LD_{10}	5.0	84
F, *o*-Nitrophenol	*Oncorhynchus mykiss*	NOAE	5.0 (h?)	942
F, Chlorodane	*Oncorhynchus mykiss*	LC_{100}	5.0 (1 h)	785
F, Phenol	*Oncorhynchus mykiss*	LC_{100}	5.0 (3 h)	215
F, Fluoride	*Oncorhynchus mykiss*	LC_{50}	5.9 (10 d)	717
F, Diquat	*Oncorhynchus mykiss*	LOAC	10.0	193
F, Aquathol-K	*Oncorhynchus mykiss*	LOAC	10.0	193
F, Hydrogen sulfide	*Oncorhynchus mykiss*	LC_{100}	10.0 (15 min)	84
F, Rhodamine WT	*Oncorhynchus kisutch* (juv., S, 22°C)	NOAE	10.0 (17.5 h)	992
F, Triphenyl phosphate	*Oncorhynchus mykiss*	LC_{50}	10.0 (6 min)	991
F, 2-Bromo-4-nitrophenol	*Oncorhynchus mykiss*	LD_{10}	13.0	84
F, 2,5-Dichloro-4-nitrophenol	*Oncorhynchus mykiss*	LD_{10}	13.0	84
F, 3,4,6-Chloro-2-nitrophenol	*Oncorhynchus mykiss*	LD_{10}	17.0	84
F, *n*-Butyric acid	*Oncorhynchus mykiss*	TL_m	20.0–40.0	8
F, Fire-trol 100	*Oncorhynchus mykiss* (F)	LC_{50}	43.0 (20 d)	151
F, Fluoride	*Oncorhynchus mykiss* (fry)	LC_{50}	61.0–85.0 (825 h)	717
F, Fire-trol 100	*Oncorhynchus kisutch* (F)	LC_{50}	73.0 (20 d)	151
F, *dl*-Lactic acid	*Oncorhynchus mykiss*	TL_m	100.0 (18 h)	2
F, *p*-Cresol	*Oncorhynchus kisutch* (HW, F, 16.5–19.5°C)	NOAE	10.0 (5 wk)	993

Chemical	Species	Toxicity test	24 h	48 h	96 h	Chronic exposure	Ref.
F, Arsenic acid, diNa, $7H_20$	*Oncorhynchus mykiss* (in diet, 15.1°C, pH 7.6, 8 wk)	$LD_{3\text{-}11}$				137.0–1053.0 mg/kg (8 wk)	988
F, Hydrazine	*Oncorhynchus mykiss*	LC_{100}				146.0 (22 min)	84
F, Arsenic trioxide	*Oncorhynchus mykiss* (juv., in diet 8 wk, 15°C, pH. 7.6)	$LC_{3\text{-}9}$				180.0–1477.0 mg/kg (8 wk)	98
F, Acetylene	*Oncorhynchus mykiss*	TL_m				200.0 (33 h)	8
F, Rhodamine WT	*Oncorhynchus kisutch* (juv., S, 22°C)	NOAE				375.0 (3.2 h)	992
F, *N*-Nitrosodiethyl amine	*Oncorhynchus mykiss* (in diet, fed 52 wk)	CARC				800.0 mg/kg (52 wk) (72 wk)	994
F, Diflubenzuron	*Oncorhynchus kisutch*	LC_0				1000.0 (15 min)	175
F, Diflubenzuron	*Oncorhynchus mykiss* (juv.)	LC_0				1000.0 (15 min)	177
F, Aroclor 1260	*Oncorhynchus mykiss*, oral	LD_{50}				1500.0 mg/kg	92
F, *N*-Dimethylnitrosoamine	*Oncorhynchus mykiss* (IP)	LD_{50}				1770.0 mg/kg	280
F, *N*-Nitrosodiethylamine	*Oncorhynchus mykiss* (IPR)	LD_{50}				1770.0 mg/kg (10 d)	994
F, Boron	*Oncorhynchus kisutch* (underyearling, 8°C)	LC_{50}				12,200.0 (283 h)	986
F, Endosulfan	*Oncorhynchus mykiss* (S, 12.7°C)	LC_{50}	0.0032				995
F, Permethrin, (*cis* + *trans*)	*Oncorhynchus mykiss*	LC_{50}	0.008	0.006			896
F, Fenpropanate (Formul. G)	*Oncorhynchus mykiss* (S)	LC_{50}	0.0086				266
F, Endosulfan	*Oncorhynchus mykiss* (fry)	LC_{100}	0.010				861
F, Cypermethrin	*Oncorhynchus mykiss* (S)	LC_{50}	0.011				266
F, Fenvalerate, formulated product	*Oncorhynchus mykiss* (S)	LC_{50}	0.021				266
F, Fenvalerate, emulsified concentrate	*Oncorhynchus mykiss* (S, 5–15°C)	LC_{50}	0.021				996
F, Tri-*n*-butyltin oxide	*Oncorhynchus mykiss*	EC_{50}	0.0308				292
F, Chlorpyrifos	*Oncorhynchus mykiss* (S)	LC_{50}	0.053				420
F, Cypermethrin,TG	*Oncorhynchus mykiss* (S)	LC_{50}	0.055				266
F, Pyrethrin	*Oncorhynchus mykiss*	LC_{50}	0.056				265
F, Pyrethrum	*Oncorhynchus mykiss*	LC_{50}	0.056				996
F, Permethrin (formulated product)	*Oncorhynchus mykiss* (S)	LC_{50}	0.061				266

F, Ammonia	*Oncorhynchus mykiss* (85-d-old fry, S)	LC_{50}	0.068		997
F, Fenvalerate, technical grade	*Oncorhynchus mykiss* (S)	LC_{50}	0.076		266
F, Fenvalerate, technical grade	*Oncorhynchus mykiss* (S, 5–15°C)	LC_{50}	0.076		996
F, Ammonia	*Oncorhynchus mykiss* (adult, S)	LC_{50}	0.097		997
F, Malathion	*Oncorhynchus mykiss*	LC_{50}	0.100		116
F, Dyfonate	*Oncorhynchus mykiss*	LC_{50}	0.110		116
F, Permethrin (technical grade)	*Oncorhynchus mykiss* (S)	LC_{50}	0.135		266
F, Permethrin	*Oncorhynchus mykiss* (S, 10°C, pH 7.5)	LC_{50}	0.135		902
F, Diazinon	*Oncorhynchus mykiss*	LC_{50}	0.38		141
F, Carbofuran	*Oncorhynchus kisutch*	LC_{50}	0.530		912
F, Hydrogen sulfide	*Oncorhynchus mykiss*	LC_{100}	0.86		84
F, Aldicarb	*Oncorhynchus mykiss* (S, 12°C, pH 7.4)	LC_{50}	1.0		404
F, Chlorobenzene	*Oncorhynchus mykiss*	LD_{50}	1.8 ml/kg		998
F, *o*-Cresol	*Oncorhynchus mykiss* (embryo)	TL_m	2.0		183
F, Kelevan	*Oncorhynchus mykiss* (S, 10°C, pH 6)	LC_{50}	2.0	1.0 (72 h)	966
F, Picloram, isocotyl ester	*Oncorhynchus mykiss* (18.3°C)	LC_{50}	2.2		771
F, Dimethyl parathion	*Oncorhynchus mykiss*	LC_{50}	2.7		116
F, Carbaryl	*Oncorhynchus mykiss*	LC_{50}	3.5		116
F, Picloram, isocotyl ester	*Oncorhynchus mykiss* (12.7°C)	LC_{50}	3.56		771
F, Ammonia	*Oncorhynchus mykiss* (fertilized eggs, S)	LC_{50}	3.58		997
F, Ammonia	*Oncorhynchus mykiss* (alevins, S)	LC_{50}	3.58		997
F, Trifluoromethyl-4-nitrophenol	*Oncorhynchus mykiss* (S)	LC_{50}	3.85		276
F, Trifluoromethyl-4-nitrophenol, commercial	*Oncorhynchus mykiss* (green eggs, 6 h old, hardW)	LC_{50}	4.0		275
F, *p*-Cresol	*Oncorhynchus mykiss* (embryo)	LC_{50}	4.0		183
F, Phosphamidon	*Oncorhynchus mykiss*	LC_{50}	5.0		116
F, Anthracene	*Oncorhynchus mykiss*	NOAE	5.0		8
F, Ethylene dichloride	*Oncorhynchus mykiss*	NOEL	5.0		132
F, Phenol	*Oncorhynchus mykiss* (embryo)	TL_m	5.0		183
F, Phenol	*Oncorhynchus mykiss* (S)	LC_{50}	5.6–11.3		146
F, 3,4-Xylenol	*Oncorhynchus mykiss* (embryo)	TL_m	7.0		184

Chemical	Species	Toxicity test	24 h	48 h	96 h	Chronic exposure	Ref.
F, *m*-Cresol	*Oncorhynchus mykiss* (embryo)	TL_m	7.0				183
F, Formaldehyde (37% active)	*Oncorhynchus mykiss* (S)	LC_{50}	7.2				120
F, Zectran	*Oncorhynchus mykiss*	LC_{50}	10.2				116
F, Trifluoromethylnitrophenol (96% active)	*Oncorhynchus mykiss* (sac fry, 27 d old, hardW)	LC_{50}	10.8				274
F, Diquat	*Oncorhynchus mykiss*	LC_{50}		11.2			194
F, Trifluoromethylnitrophenol, commercial	*Oncorhynchus mykiss* (32 d old, hardW)	LC_{50}	11.3				275
F, Trifluoromethylnitrophenol (96% active)	*Oncorhynchus mykiss* (swim-up, 46 d old, hardW)	LC_{50}	11.3				274
F, Monocrotophos (38% active)	*Oncorhynchus mykiss* (S, 13°C, pH 7.1)	LC_{50}	12.0				404
F, Trifluoromethylnitrophenol (96% active)	*Oncorhynchus mykiss* (fingerlings)	LC_{50}	14.2				274
F, Trifluoromethylnitrophenol commercial	*Oncorhynchus mykiss* (21 d old, hardW)	LC_{50}	14.3				275
F, Trifluoromethylnitrophenol (96% active)	*Oncorhynchus mykiss* (fry)	LC_{50}	19.2				274
F, Trifluoromethylnitrophenol (96% active)	*Oncorhynchus mykiss* (green egg, 40 d old, hardW)	LC_{50}	19.6				274
F, Dimethoate	*Oncorhynchus mykiss*	LC_{50}	20.0				116
F, 2,4-Xylenol	*Oncorhynchus mykiss* (embryo)	TL_m	28.0				184
F, Trifluoromethylnitrophenol (96% active)	*Oncorhynchus mykiss* (eyed egg, 16 d old, HW)	LC_{50}	40.0				274
F, Trifluoromethylnitriophenol, commercial grade	*Oncorhynchus mykiss* (10 d old, HW)	LC_{50}	43.0				275
F, Ethylbenzene	*Oncorhynchus kisutch*, (young, artificial seawater, 8°C)	LC_{100}	50.0				173
F, Ethybenzene	*Oncorhynchus kisutch* (S, 8°C, pH 8.1)	LC_{100}	50.0				999
F, Toluene	*Oncorhynchus kisutch* (young, softW, 8°C)	LC_{90}	50.0				173
F, 3,5-Xylenol	*Oncorhynchus mykiss* (embryo)	TL_m	50.0				183

F, Benzene	*Oncorhynchus kisutch* (young, seaW, 8°C)	LC_{100}	100.0				173
F, Toluene	*Oncorhynchus kisutch* (young, softW, 8°C, S)	LC_{90}	100.0				173
F, *o*-Xylene	*Oncorhynchus kisutch* (softW, 8°C)	LD_{100}	100.0				173
F, Picloram	*Oncorhynchus mykiss*	LC_{50}	150.0–230.0				1000
F, Dicamba	*Oncorhynchus kisutch* (juv.)	LC_{50}	151.0				944
F, Formaldehyde (37% active)	*Oncorhynchus mykiss* (S)	LC_{50}	214.0				120
F, Formaldehyde (37% active)	*Oncorhynchus mykiss* (S)	LC_{50}	249.0				120
F, Acetone	*Oncorhynchus mykiss* (fingerling, F)	LC_{50}	6100.0				267
F, Ethanol	*Oncorhynchus mykiss* (fingerling, F)	LC_{50}	11,200.0				267
F, Propylene glycol	*Oncorhynchus mykiss* (fingerling, S, 10°C)	NOAE	50,000.0				267
F, Decamethrin	*Oncorhynchus mykiss* (S)	LC_{50}			0.0005		188
F, Decamethrin	*Oncorhynchus mykiss*	LC_{50}		0.0007	0.0005		896
F, Decamethrin	*Oncorhynchus mykiss*	LC_{90}		0.00113	0.0009		896
F, Actellic	*Oncorhynchus mykiss*	LC_{50}			0.001 ml/l		859
F, Permethrin	*Oncorhynchus mykiss* (S)	LC_{50}			0.0060		188
F, Permethrin (*cis*)	*Oncorhynchus mykiss*	LC_{50}		0.010	0.007		896
F, Permethrin (*cis* + *trans*)	*Oncorhynchus mykiss*	LC_{90}		0.017	0.010		896
F, Trifluralin	*Oncorhynchus mykiss*	LC_{50}			0.011		116
F, Permethrin (*cis*)	*Oncorhynchus mykiss*	LC_{90}		0.015	0.014		896
F, Kepone	*Oncorhynchus mykiss*	LC_{100}			0.100		880
F, Fenbutatin oxide	*Oncorhynchus mykiss*	LC_{50}			0.27		809
F, Dactinol	*Oncorhynchus mykiss* (softW, F)	LC_{50}		0.58	0.47		145
F, Benomyl	*Oncorhynchus mykiss* (15°C)	LC_{50}			0.48		474
F, Lindane	*Oncorhynchus mykiss* (S, 12°C)	TL_m			1.05		277
F, 2,4-Dichlorophenoxyacetic acid	*Oncorhynchus mykiss*	LC_{50}			1.1		64
F, Fenitrothion	*Oncorhynchus mykiss*	LC_{50}			1.28		20
F, Baytex	*Oncorhynchus kisutch*	LC_{50}				1.3	122
F, Carbendazim	*Oncorhynchus mykiss* (S, 15°C)	LC_{50}			1.8		1002
F, 4-Chloro-*o*-cresol	*Oncorhynchus mykiss*	LD_{50}			2.12		1003

Chemical	Species	Toxicity test	24 h	48 h	96 h	Chronic exposure	Ref.
F, Diuron	*Oncorhynchus mykiss*	LC_{50}		4.3			116
F, Dactinol	*Oncorhynchus mykiss* (hardW, F)	LC_{50}	7.3	5.8			145
F, Simazine	*Oncorhynchus kisutch*	LC_{50}		6.6			230
F, Diuron	*Oncorhynchus kisutch*	LC_{50}		16.0			230
F, Diquat	*Oncorhynchus mykiss*	LC_{50}		20.0			116
F, Dichlobenil	*Oncorhynchus mykiss*	LC_{50}		22.0			116
F, Diquat	*Oncorhynchus tschawytscha*	LC_{50}		28.5			230
F, Paraquat	*Oncorhynchus mykiss*	LC_{50}		62.0			116
F, Bromacil (80%)	*Oncorhynchus mykiss*	EC_{50}	102.0	75.0		38.0 (72 h)	945
F, Simazine	*Oncorhynchus mykiss*	LC_{50}		85.0			116
F, Ethyl xanthate, K	*Oncorhynchus mykiss* (S)	LC_{100}		100.0		0.5 (2d exp. F)	279
F, Monuron	*Oncorhynchus kisutch*	LC_{50}		110.0			230
F, Monuron	*Oncorhynchus kisutch*	LC_{50}		110.0			943
F, Zoalene	*Oncorhynchus mykiss* (S)	TL_m		200.0			222
F, Eythylene diamine	*Oncorhynchus mykiss* (yearlings, S)	LC_{50}		230.0			159
F, Amitrol	*Oncorhynchus kisutch*	LC_{50}		325.0			230
F, Boron	*Oncorhynchus mykiss*	LC_{50}		339.0			491
F, 2,2-Dichloropropionic acid	*Oncorhynchus kisutch*	LC_{50}		340.0			230
F, Amprolium	*Oncorhynchus mykiss* (S, 15°C)	TL_m		1550.0			222
F, *n*-Propanol	*Oncorhynchus mykiss* (15°C, pH 7–8)	LC_0		2000.0			1001
F, *n*-Propanol	*Oncorhynchus mykiss* (15°C, pH 7–8)	LC_{50}		3200.0			1001
F, RU-11679	*Oncorhynchus kisutch* (F)	LC_{50}			0.000151		138
F, RU-11679	*Oncorhynchus mykiss* (F)	LC_{50}			0.000100		138
F, RU-11679	*Oncorhynchus mykiss* (S)	LC_{50}			0.000110		138
F, Resmethrin	*Oncorhynchus kisutch* (F)	LC_{50}			0.000277		138
F, Resmethrin	*Oncorhynchus mykiss* (F)	LC_{50}			0.000275		138
F, Resmethrin	*Oncorhynchus mykiss* (F, 12°C)	LC_{50}			0.0002-0.0003		918
F, Endosulfan	*Oncorhynchus mykiss*	LC_{50}			0.0003		85
F, Resmethrin	*Oncorhynchus kisutch* (F, pH 7.9, 12°C)	LC_{50}			0.0003		918

F, Resmethrin	*Oncorhynchus mykiss* (S, 12°C)	LC_{50}			0.0003–0.0006		918
F, Resmethrin	*Oncorhynchus mykiss* (S)	LC_{50}			0.00045		138
F, Endrin	*Oncorhynchus kisutch*	LC_{50}			0.0005		252
F, RU-11679	*Oncorhynchus kisutch* (S)	LC_{50}			0.000635		138
F, Endrin	*Oncorhynchus mykiss*	LC_{50}			0.0006		252
F, DDT	*Oncorhynchus mykiss* (fry)	LC_{10}	0.008	0.0019	0.0008		145
F, Bio-allethrin	*Oncorhynchus mykiss* (S, 12°C, pH 6.4–8.4)	LC_{50}			0.00097		918
F, Xylene	*Oncorhynchus kisutch* (S, 8°C, juv.)	LC_0			0.001		1004
F, Endrin	*Oncorhynchus tschawytscha*	LC_{50}			0.0012		252
F, Endosulfan	*Oncorhynchus mykiss* (13°C, pH 7.5)	LC_{50}			0.0014		600
F, Rotenone	*Oncorhynchus mykiss*	LC_{50}			0.0016–0.012		557
F, Toxaphene	*Oncorhynchus mykiss* (S, 18.3°C)	LC_{50}	0.005	0.0028	0.0018		801
F, DDT	*Oncorhynchus mykiss* (fry)	LC_{50}	0.103	0.0058	0.0024	0.0001 (3 m)	145
F, Toxaphene	*Oncorhynchus tschawytscha* (S, 20°C, pH 7.1)	LC_{50}	0.0079	0.0033	0.0025		801
F, Bio-allethrin, 90% TG	*Oncorhynchus kisutch* (F, 12°C, pH 7.5)	LC_{50}			0.0026		597
F, Toxaphene	*Oncorhynchus mykiss* (S, 12.8°C)	LC_{50}	0.0076	0.0044	0.0027		801
F, DDT	*Oncorhynchus kisutch*	LC_{50}			0.004		122
F, DDT	*Oncorhynchus kisutch*	LC_{50}			0.004		127
F, Toxaphene	*Oncorhynchus mykiss* (S, 7.2°C)	LC_{50}	0.016	0.0084	0.0054		801
F, Toxaphene	*Oncorhynchus mykiss*	LC_{50}			0.0055		130
F, Dieldrin	*Oncorhynchus tschawytscha*	LC_{50}			0.006		252
F, DDT	*Oncorhynchus mykiss*	LC_{50}			0.007		127
F, DDT	*Oncorhynchus mykiss*	LC_{50}			0.007		122
F, Endrin	*Oncorhynchus mykiss*	LC_{50}			0.007		116
F, Chlorodane	*Oncorhynchus mykiss* (12.8°C)	LC_{50}	0.022	0.010	0.0078		785
F, Toxaphene	*Oncorhynchus kisutch*	LC_{50}			0.008		122
F, Toxaphene	*Oncorhynchus mykiss*	LC_{50}			0.008		116
F, Aldrin	*Oncorhynchus tschawytscha*	LC_{50}			0.008		252
F, Allethrin	*Oncorhynchus kisutch* (F)	LC_{50}			0.0094		138, 140
F, Bio-allethrin	*Oncorhynchus kisutch* (F, 12°C, pH 6.4–8.4)	LC_{50}			0.0094		1004

Chemical	Species	Toxicity test	24 h	48 h	96 h	Chronic exposure	Ref.
F, Toxaphene	*Oncorhynchus kisutch* (S, 20°C, pH 7.1)	LC_{50}	0.013	0.0105	0.0094		252
F, DDT	*Oncorhynchus mykiss*	LC_{50}			0.0096		130
F, Allethrin	*Oncorhynchus mykiss* (F)	LC_{50}			0.0097		138, 140
F, Xylene	*Oncorhynchus kisutch* (S, 8°C, juv.)	LC_{30}			0.010		918
F, Dieldrin	*Oncorhynchus mykiss*	LC_{50}			0.010		801
F, Aldrin	*Oncorhynchus mykiss*	LC_{50}			0.010		130
F, Triphenyltin hydroxide	*Oncorhynchus mykiss* (fry, F)	LC_{10}	0.055	0.019	0.01		145
F, Toxaphene	*Oncorhynchus mykiss*	LC_{50}			0.011		122
F, Dieldrin	*Oncorhynchus kisutch*	LC_{50}			0.011		252
F, Chloropyrifos	*Oncorhynchus mykiss*	LC_{50}			0.011		71
F, Lindane	*Oncorhynchus mykiss*	LC_{50}			0.013		130
F, Phorate	*Oncorhynchus mykiss* (S, 12°C, pH 7.5)	LC_{50}			0.013		599
F, Azinphosmethyl	*Oncorhynchus mykiss*	LC_{50}			0.014		127
F, Kepone	*Oncorhynchus mykiss*	NOEL			0.014		880
F, Triphenyltin hydroxide	*Oncorhynchus mykiss* (fry, F)	LC_{50}	0.078	0.03	0.015	0.0004 (3 m)	145
F, Azinphosmethyl	*Oncorhynchus kisutch*	LC_{50}			0.017		127
F, Heptachlor	*Oncorhynchus tschawytscha*	LC_{50}			0.017		253
F, Bio-allethrin	*Oncorhynchus mykiss* (S, 12°C, pH 6.4–8.4)	LC_{50}			0.0175		918
F, Allethrin	*Oncorhynchus mykiss* (S)	LC_{50}			0.0175		138, 140
F, Aldrin	*Oncorhynchus mykiss*	LC_{50}			0.0177		252
F, Hydrocyanic acid	*Oncorhynchus mykiss* (juv., F, 6°C, pH 8.0)	LC_{0}			0.018		1005
F, DDT	*Oncorhynchus mykiss*	LC_{50}			0.018		116
F, Pyrethrum	*Oncorhynchus kisutch* (F, pH 7.8)	LC_{50}			0.018–0.030		918
F, Niclosamide	*Oncorhynchus mykiss* (F, 12°C)	LC_{50}			0.018		1006
F, Lindane	*Oncorhynchus mykiss* (20°C, pH 8.1)	LC_{10}	0.026	0.02	0.018		145
F, Lindane	*Oncorhynchus mykiss* (fry, F)	LC_{10}	0.026	0.02	0.018		145
F, Dieldrin	*Oncorhynchus mykiss*	LC_{50}			0.019		116
F, Heptachlor	*Oncorhynchus mykiss*	LC_{50}			0.019		252
F, Azinphosethyl	*Oncorhynchus mykiss*	LC_{50}			0.019		71

F, Pyrethroid	*Oncorhynchus mykiss*	LC_{50}			0.019		71
F, Kepone	*Oncorhynchus mykiss*	LC_{50}			0.020		1007
F, Leptophos	*Oncorhynchus mykiss* (S, 12°C, pH 7.5)	LC_{50}			0.020		678
F, Kepone	*Oncorhynchus mykiss*	LC_{50}	0.066	0.038	0.02		133
F, Allethrin	*Oncorhynchus kisutch* (S)	LC_{50}			0.022		138, 140
F, Chlordane	*Oncorhynchus mykiss*	LC_{50}			0.022		116
F, Lindane	*Oncorhynchus mykiss* (20°C, pH 8.1)	LC_{50}	0.037	0.023	0.022	0.01 (3 m)	145
F, Lindane	*Oncorhynchus mykiss* (fry, F)	LC_{50}	0.037	0.023	0.022	0.02 (3 m)	145
F, Bio-allethrin	*Oncorhynchus kisutch* (S, 12°C, pH 6.4–8.4)	LC_{50}			0.0222		918
F, Pyrethrins	*Oncorhynchus mykiss* (F)	LC_{50}			0.0225		138
F, Pyrethrins	*Oncorhynchus kisutch* (F)	LC_{50}			0.023		138
F, Pyrethrins	*Oncorhynchus mykiss* (S)	LC_{50}			0.0246		138
F, Pyrethrum	*Oncorhynchus mykiss* (12°C)	LC_{50}			0.025		870
F, Lindane	*Oncorhynchus mykiss*	LC_{50}			0.027		122
F, Lindane	*Oncorhynchus mykiss*	LC_{50}			0.027		127
F, Hydrocyanic acid	*Oncorhynchus mykiss* (juv., F, 6°C, pH 8.0)	LC_{50}			0.028		1005
F, Hydrocyanic acid	*Oncorhynchus mykiss* (juv., F, 12°C, pH 8.1)	LC_0			0.032		1008
F, Lindane	*Oncorhynchus mykiss* (F, 20–25°C)	TL_m	0.051		0.032		198
F, Pyrethrum	*Oncorhynchus kisutch* (S, 12°C)	LC_{50}			0.033–0.046		918
F, Aroclor 1248	*Oncorhynchus mykiss* (F)	LC_{50}			0.034 (5 d)	0.0034 (25 d)	906
F, Leptophos	*Oncorhynchus mykiss* (S, 12°C, pH 7.5)	LC_{50}			0.035		678
F, Aldrin	*Oncorhynchus mykiss*	LC_{50}			0.036		116
F, Kepone	*Oncorhynchus mykiss*	LC_{50}	0.066		0.036		880
F, Kepone	*Oncorhynchus mykiss*	LC_{50}	0.156		0.036	0.04 (NOAE)	133
F, Rotenone (5%)	*Oncorhynchus tschawytscha* (S)	LC_{50}			0.0369		28
F, Hydrocyanic acid	*Oncorhynchus mykiss* (juv., F, 6°C, pH 8.0)	LC_{100}			0.037		1005
F, Pentachlorophenate, Na	*Oncorhynchus kisutch* (S)	LC_{50}			0.037–0.096		176
F, Kepone	*Oncorhynchus mykiss*	LC_{80}			0.037		880
F, Aminocarb (17% active)	*Oncorhynchus mykiss* (yolk-sac fry)	LC_{50}			0.038		404
F, Pyrethrin	*Oncorhynchus kisutch* (S)	LC_{50}			0.039		138

Chemical	Species	Toxicity test	24 h	48 h	96 h	Chronic exposure	Ref.
F, Pyrethrum	*Oncorhynchus kisutch* (12°C)	LC_{50}			0.039		870
F, Lindane	*Oncorhynchus kisutch*	LC_{50}			0.041		122
F, Lindane	*Oncorhynchus kisutch*	LC_{50}			0.041		127
F, Hydrocyanic acid	*Oncorhynchus mykiss* (juv., F, 12°C, pH 8.1)	LC_{50}			0.042		1008
F, Aldrin	*Oncorhynchus kisutch*	LC_{50}			0.046		252
F, Rotenone (5%)	*Oncorhynchus mykiss* (S)	LC_{50}			0.046		28
F, Sodium cyanide	*Oncorhynchus mykiss* (juv., S, 12°C, pH 8.0)	LC_{50}			0.0463–0.0748		1010
F, Pentachlorophenate, Na	*Oncorhynchus mykiss* (S)	LC_{50}			0.048–0.10		176
F, Pentachlorophenate, Na	*Oncorhynchus nerka* (S)	LC_{50}			0.050–0.130		176
F, Binapacryl (TG, 99%)	*Oncorhynchus mykiss* (S, 13°C, pH 7.1)	LC_{50}	0.050		0.050		404
F, Hydrocyanic acid	*Oncorhynchus mykiss* (juv., F, 11–13°C, pH 7.1–7.6)	LC_{50}			0.052		1009
F, Hydrocyanic acid	*Oncorhynchus mykiss* (juv., F, 12°C, pH 8.1)	LC_{100}			0.053		1008
F, Aminocarb (17%)	*Oncorhynchus mykiss* (hardW, F, 10°C, pH 7.4)	LC_{50}			0.054		404
F, Pyrethrum	*Oncorhynchus mykiss* (12°C)	LC_{50}	0.056		0.054		870
F, Ciodrin	*Oncorhynchus mykiss*	LC_{50}			0.055		71
F, Bromofenoxim	*Oncorhynchus mykiss*	LC_{50}			0.056-0.1		804
F, Captan	*Oncorhynchus kisutch* (fingerling, F)	LC_{50}			0.0565		778
F, Hydrogen cyanide	*Oncorhynchus mykiss* (F)	LC_{50}			0.057		137
F, Hydrocyanic acid	*Oncorhynchus mykiss* (juv., F, 10°C, pH 8.8, DO = 8.8 mg/l)	LC_{50}			0.0572		947
F, Heptachlor	*Oncorhynchus kisutch*	LC_{50}			0.059		252
F, Hydrocyanic acid	*Oncorhynchus mykiss* (juv., F, 18°C, pH 7.9)	LC_0			0.060		1008
F, Lindane	*Oncorhynchus mykiss*	LC_{50}			0.060		116
F, Rotenone (5%)	*Oncorhynchus kisutch* (S)	LC_{50}			0.062		28
F, Aroclor 1242	*Oncorhynchus mykiss* (F)	LC_{50}			0.067	0.012 (25 d)	92
F, Hydrocyanic acid	*Oncorhynchus mykiss* (juv., F, 18°C, pH 7.9)	LC_{50}			0.068		1008
F, Pentachlorophenate, Na	*Oncorhynchus tschawytscha* (F)	LC_{50}			0.078		171
F, Potassium cyanide	*Oncorhynchus mykiss* (20°C)	LC_{10}			0.082		400
F, Hydrocyanic acid	*Oncorhynchus mykiss* (juv., F, 18°C, pH 7.9)	LC_{100}			0.087		1008

F, Potassium cyanide	*Oncorhynchus mykiss* (20°C)	LC_{50}			0.097		400
F, Xylene	*Oncorhynchus kisutch* (S, 8°C, juv.)	LC_{100}			0.10		1004
F, Malathion	*Oncorhynchus kisutch*	LC_{50}			0.100		122
F, Malathion	*Oncorhynchus kisutch*	LC_{50}			0.101		127
F, Captan	*Oncorhynchus mykiss* (fingerling, S)	LC_{50}			0.102		778
F, Fenamiphos	*Oncorhynchus mykiss*	LC_{50}			0.11		809
F, Formaldehyde	*Oncorhynchus mykiss* (F)	LC_{50}			0.118 ml/l		120
F, Eulan	*Oncorhynchus mykiss*	LC_{50}			0.127		2
F, Aminocarb (17%)	*Oncorhynchus mykiss* (eggs, F, pH 7.4, hardW)	LC_{50}			0.130		404
F, Naled	*Oncorhynchus mykiss*	LC_{50}			0.132		71
F, Anilzaine	*Oncorhynchus mykiss* (13°C)	LC_{50}			0.140		600
F, Aminocarb (17%)	*Oncorhynchus mykiss* (softW, S, 17°C, pH 7.4)	LC_{50}			0.14		610
F, Tri-chloronate	*Oncorhynchus mykiss* (S, 12°C)	LC_{50}			0.14		952
F, Resmethrin (SBP1382)	*Oncorhynchus kisutch* (S)	LC_{50}			0.150		138
F, Heptachlor	*Oncorhynchus mykiss*	LC_{50}			0.150		116
F, 2,4,5-T (emulsion)	*Oncorhynchus mykiss*	LC_{50}			0.15		268
F, Imidan	*Oncorhynchus tschawytscha* (S)	LC_{50}			0.150		148
F, Phosmet	*Oncorhynchus tschawytscha* (S, 10°C, pH 7.5)	LC_{50}	0.180		0.150		409
F, Abate	*Oncorhynchus mykiss*	LC_{50}			0.158		71
F, Malathion	*Oncorhynchus mykiss*	LC_{50}			0.170		122
F, Malathion	*Oncorhynchus mykiss*	LC_{50}			0.170		130
F, Malathion	*Oncorhynchus mykiss*	LC_{50}			0.170		127
F, Benomyl, TG (99%)	*Oncorhynchus mykiss* (S, 12°C, pH 7.5)	LC_{50}			0.17		609
F, Ethyl-*p*-nitrophenylthiono benzene phosphate	*Oncorhynchus mykiss* (S, 13°C, pH 7.5)	LC_{50}			0.210		602
F, Isopimaric acid	*Oncorhynchus kisutch* (juv.)	LC_{50}			0.22		174
F, *o*-Chlorbenzylidene malonitrile (OCBM)	*Oncorhynchus mykiss*	LC_{50}	0.45	0.42	0.22	1.28 (12 h)	1015
F, Pentachlorophenol	*Oncorhynchus mykiss* (F, 15°C)	LC_{50}		0.25	0.23	0.23 (10 d)	155
F, Tricresyl phosphate	*Oncorhynchus mykiss* (F, 12°C, pH 7.5)	LC_{50}			0.260		946
F, Carbofuran, TG	*Oncorhynchus mykiss*	LC_{50}			0.28		912

Chemical	Species	Toxicity test	24 h	48 h	96 h	Chronic exposure	Ref.
F, Triphenyl phosphate	*Oncorhynchus mykiss* (S)	LC_{50}			0.30		285
F, Triphenyl phosphate	*Oncorhynchus mykiss* (sac-fry, fingerlings, S)	LC_{50}			0.30		1011
F, Phosmet	*Oncorhynchus mykiss* (S, 10°C, pH 7.5)	LC_{50}			0.300		409
F, Isopimarol	*Oncorhynchus mykiss* (S, pH 7.0)	TL_m			0.3		296
F, Benomyl, technical grade (50%)	*Oncorhynchus mykiss* (S, 12°C, pH 7.5)	LC_{50}			0.31		609
F, Endothal-diNa	*Oncorhynchus mykiss* (fry, S, 13°C, pH 7.5)	LC_{50}			0.31		607
F, Pimaric acid	*Oncorhynchus kisutch* (juv.)	LC_{50}			0.32		174
F, 3,4,5,6-Chloroguaicol	*Oncorhynchus mykiss* (juv.)	LC_{50}			0.32		269
F, Palustric resin acid	*Oncorhynchus mykiss* (juv., S)	LC_{50}			0.32		174
F, Aramite	*Oncorhynchus mykiss* (S, 13°C, pH 7.5)	LC_{50}			0.320		961
F, Carbendazim	*Oncorhynchus mykiss*	LC_{50}			0.36		1012
F, Aminocarb (19.5%)	*Oncorhynchus mykiss* (juv., S, 8–9°C)	LC_{50}			0.36		1013
F, Triphenyl phosphate	*Oncorhynchus mykiss* (S, fry, 12°C, pH 7.5)	LC_{50}			0.36		1014
F, Triphenyl phosphate	*Oncorhynchus mykiss*	LC_{50}			0.40		849
F, Tetrabromo-bisphenol A	*Oncorhynchus mykiss*	LC_{50}			0.40		953
F, Isopimaric acid	*Oncorhynchus mykiss* (S, pH 7.0)	TL_m			0.4		270
F, Abietic acid	*Oncorhynchus kisutch* (juv.)	LC_{50}			0.41		174
F, Ammonia	*Oncorhynchus kisutch* (F)	LC_{50}			0.45		1016
F, Frescon, emulsified concentrate	*Oncorhynchus mykiss* (S)	LC_{50}	0.74	0.62	0.52		145
F, Chlorpyrifos	*Oncorhynchus mykiss* (S)	LC_{50}	0.0071		0.550		420
F, Ethion	*Oncorhynchus mykiss*	LC_{50}			0.560		71
F, Imidan	*Oncorhynchus mykiss* (fingerling, S)	LC_{50}			0.560		148
F, Aldicarb	*Oncorhynchus mykiss* (S, 18°C, pH 7.4)	LC_{50}			0.560		404
F, Phosmet	*Oncorhynchus mykiss* (fingerling)	LC_{50}	0.760		0.560		918
F, Trifluoromethyl-4-nitrophenol	*Oncorhynchus kisutch* (S, green eggs, 12°C)	TL_m			0.57		161
F, Dichlorodehydroabietic acid	*Oncorhynchus mykiss* (juv., S)	LC_{50}			0.6		269
F, Houghtosafe	*Oncorhynchus mykiss* (F)	LC_{50}			0.65		139

F, Isopropylphenyldiphenyl phosphate	*Oncorhynchus mykiss* (F, 17°C, pH 7.4)	LC_{50}		0.65		849
F, Dehydroabietic acid	*Oncorhynchus mykiss* (F)	TL_m	1.35-1.99	0.65–0.92		198
F, Aldicarb	*Oncorhynchus mykiss* (S, 18°C, pH 7.4)	LC_{50}		0.660		404
F, Pydraul 50E	*Oncorhynchus mykiss* (F)	LC_{50}		0.67		139
F, Abietic acid	*Oncorhynchus mykiss*	TL_m		0.70		270
F, Pydraul 50E	*Oncorhynchus mykiss* (S)	LC_{50}	1.3	0.72		139
F, Dehydroabietic acid	*Oncorhynchus kisutch*	LC_{50}		0.75		174
F, 3,4,5-Chloroguaiacol	*Oncorhynchus mykiss* (juv., S)	LC_{50}		0.75		2
F, Parathion (100%)	*Oncorhynchus mykiss* (fingerling, S, 12°C?)	LC_{50}		0.75		511
F, Carbaryl	*Oncorhynchus kisutch*	LC_{50}		0.76		122
F, Carbaryl	*Oncorhynchus kisutch*	NOEL			0.764	127
F, Dehydroabietol	*Oncorhynchus mykiss* (S)	LC_{50}		0.8		174
F, Pimaric acid	*Oncorhynchus mykiss* (S, pH 7.0)	TL_m		0.80		270
F, δ-4′-dehydrojuvabione	*Oncorhynchus mykiss* (renewal)	TL_m		0.8		281
F, Baytex	*Oncorhynchus mykiss*	LC_{50}		0.9		174
F, Baytex	*Oncorhynchus mykiss*	LC_{50}		0.930		127
F, 2,4,5-Trichlorophenoxyacetic acid	*Oncorhynchus mykiss*	LC_{50}		0.98		268
F, Furanace	*Oncorhynchus mykiss* (S)	LC_{50}		1.0		121
F, Formaldehyde (37% active)	*Oncorhynchus mykiss* (green egg, S)	LC_{50}		1.02		120
F, Dehydroabietic acid	*Oncorhynchus mykiss* (S)	LC_{50}		1.03–1.74		176
F, Dehydroabietic acid	*Oncorhynchus mykiss* (S)	TL_m		1.1		270
F, *p*-Dichlorobenzene	*Oncorhynchus mykiss* (F)	LC_{50}		1.12		660
F, Dehydroabietic acid	*Oncorhynchus nerka* (S)	LC_{50}		1.18–1.38		176
F, Naphthalene	*Oncorhynchus gorbuscha* (F, 8°C, juv.)	LC_{50}		1.20		1019
F, Methomyl (liquid concentrate)	*Oncorhynchus mykiss* (S, 12–17°C, pH 7.5)	LC_{50}		1.20		513
F, Trifluoromethyl-4-nitrophenol	*Oncorhynchus kisutch* (S, green egg, 12°C)	TL_m		1.2		161
F, Naphthalene	*Oncorhynchus gorbuscha* (S, 12°C, juv.)	LC_{50}		1.24		1018
F, Naphthalene	*Oncorhynchus gorbuscha* (S, 12°C)	TL_m		1.24		88
F, Molinate	*Oncorhynchus mykiss*	LC_{50}		1.30		815

Chemical	Species	Toxicity test	24 h	48 h	96 h	Chronic exposure	Ref.
F, Baytex	*Oncorhynchus kisutch*	LC_{50}			1.32		127
F, Diazinon (technical grade)	*Oncorhynchus mykiss*	LC_{50}			1.35		87
F, Naphthalene	*Oncorhynchus gorbuscha* (S, 4°C)	TL_m			1.37		88
F, Dehydroabietic acid	*Oncorhynchus kisutch* (S)	LC_{50}			1.38–1.76		176
F, Tris (2,3-dibromopropyl) phosphate	*Oncorhynchus mykiss* (fingerling, S)	LC_{50}			1.450		1017
F, Co-Ral	*Oncorhynchus mykiss*	LC_{50}			1.500		252
F, Kelevan	*Oncorhynchus mykiss* (juv., 14°C, softW)	LC_{50}			1.5		966
F, Juvabione	*Oncorhynchus mykiss* (8°C)	TL_m			1.5		281
F, Frescon, paste	*Oncorhynchus mykiss* (S)	LC_{50}	8.2	2.5	1.5		145
F, Methomyl, technical grade	*Oncorhynchus mykiss* (S, 12–17°C, pH 7.5)	LC_{50}			1.60		513
F, Zectran	*Oncorhynchus kisutch*	LC_{50}			1.7		122
F, Isopropylphenyldiphenyl phosphate	*Oncorhynchus mykiss* (S, 10°C, pH 7.4)	LC_{50}			1.7		849
F, Houghtosafe	*Oncorhynchus mykiss* (S)	LC_{50}	4.2		1.7		139
F, Zectran	*Oncorhynchus kisutch*	LC_{50}			1.73		127
F, Dehydrojuvabione	*Oncorhynchus mykiss* (Renewal)	TL_m			1.8		281
F, Naphthalene	*Oncorhynchus gorbuscha* (S, 8°C)	TL_m			1.84		88
F, Disulfoton	*Oncorhynchus mykiss* (13°C)	LC_{50}			1.850		410
F, *p*-Butylphenyldiphenyl phosphate	*Oncorhynchus mykiss*	LC_{50}			2.0		1020
F, Juvabiol	*Oncorhynchus mykiss* (8°C)	TL_m			2.0		281
F, Methyl parathion	*Oncorhynchus mykiss* (S, 12°C, pH 7.5, fry)	LC_{98}			2.1–2.8		1014
F, Kelevan	*Oncorhynchus mykiss* (juv., 14°C, hardW)	LC_{50}			2.2		966
F, Trifluoromethyl-nitrophenol	*Oncorhynchus kisutch* (S, fingerling, 12°C)	TL_m			2.2		161
F, 9,10-Dichlorostearic acid	*Oncorhynchus mykiss* (juv., S)	LC_{50}			2.5		174
F, Carbaryl	*Oncorhynchus keeta* (S, juv.)	TL_m			2.5		211
F, *o*-Chlorobenzaldehyde	*Oncorhynchus mykiss*	LC_{50}	3.6	2.8	2.5	5.2 (12 h)	1015
F, Trifluoromethyl-4-nitrophenol	*Oncorhynchus kisutch* (S, fingerling, 12°C)	TL_m			2.7		161
F, Dimethyl parathion	*Oncorhynchus mykiss*	LC_{50}			2.75		127

F, Methyl parathion	*Oncorhynchus mykiss* (S, 12°C, pH 7.5, fry)	LC_{50}			2.8		1014
F, 2,5-Xylenol	*Oncorhynchus mykiss* (S)	LC_{50}			3.2-5.6		279
F, Butylbenzyl phthalate	*Oncorhynchus mykiss* (S)	LC_{50}			3.3	0.36 (NOAE)	9
F, Carbofuran	*Oncorhynchus mykiss*	LC_{50}			4.0		912
F, Propoxur	*Oncorhynchus mykiss* (S)	LC_{50}			4.0		816
F, Carbaryl	*Oncorhynchus mykiss*	LC_{50}			4.3		122
F, Carbaryl	*Oncorhynchus mykiss*	NOEL			4.34		127
F, Altosid-SR10	*Oncorhynchus mykiss* (S)	TL_{50}			4.39		177
F, Ro-neet (technical grade)	*Oncorhynchus mykiss*	LC_{50}			4.5-5.6		24
F, Lauryl sulfate	*Oncorhynchus mykiss* (F, 15°C)	LC_{50}		5.95	4.62	2.85 (10 d)	155
F, Chlorfos	*Oncorhynchus mykiss* (S)	LC_{50}			4.85		283
F, Tetrachloroethylene	*Oncorhynchus mykiss* (11.6°C, pH 7.1)	LC_{50}			5.0		1021
F, Aminocarb (98% active)	*Oncorhynchus mykiss* (S, 17°C, pH 9.5)	LC_{50}			5.0		517
F, Tributyl phosphate	*Oncorhynchus mykiss* (S)	LC_{50}			5.0–9.0		2
F, Dimethyl parathion	*Oncorhynchus kisutch*	LC_{50}			5.30		127
F, Methyl parathion	*Oncorhynchus kisutch* (fingerling, S, 12°C, pH 7.5)	LC_{50}			5.3		511
F, Trifluoromethyl-4-nitrophenol	*Oncorhynchus mykiss* (F)	LC_{50}			6.1		157
F, Toluene	*Oncorhynchus gorbuscha* (4°C, S)	TL_m			6.41		88
F, RoundUp	*Oncorhynchus mykiss* (12°C)	LC_{50}	8.3		7.0		562
F, Saniticizer-148	*Oncorhynchus mykiss*	LC_{50}			6.0–7.0		523
F, Azinphosmethyl	*Oncorhynchus mykiss* (fingerlings, S)	LC_{50}			7.10		283
F, Tillam	*Oncorhynchus mykiss*	LC_{50}			7.40		23
F, Toluene	*Oncorhynchus gorbuscha* (8°C, S)	TL_m			7.63		88
F, Diallate	*Oncorhynchus mykiss*	LC_{50}			7.9		914
F, Magniflox 905V	*Oncorhynchus mykiss* (S, 14–16°C)	TL_m			8.0		278
F, Toluene	*Oncorhynchus gorbuscha* (12°C, S)	TL_m			8.09		88
F, Propoxur	*Oncorhynchus mykiss* (S, 13°C, pH 7.5)	LC_{50}			8.2		606
F, Dimethoate	*Oncorhynchus mykiss*	LC_{50}	20.0		8.5		881
F, 2,4,5-Trichlorophenoxyacetic acid	*Oncorhynchus mykiss* (acetone sol.)	LC_{50}			8.7		268
F, Magniflox 512C	*Oncorhynchus mykiss* (S, 14–16°C)	TL_m			8.70		278

Chemical	Species	Toxicity test	24 h	48 h	96 h	Chronic exposure	Ref.
F, Ametryn	*Oncorhynchus mykiss*	LC_{50}			8.8		811
F, Triethyllead chloride	*Oncorhynchus mykiss*	LC_{50}			9.0 (Pb)		282
F, Benzene	*Oncorhynchus mykiss* (S, 12°C)	LC_{50}			9.2		921
F, Benzene	*Oncorhynchus kisutch* (young, seaW, 8°C)	LC_0			10.0		173
F, Dow Chemical Z14	*Oncorhynchus mykiss* (S)	LC_{50}			10.0–100.0		279
F, Dow Chemical Z4	*Oncorhynchus mykiss* (S)	LC_{50}			10.0–50.0		279
F, Toluene	*Oncorhynchus kisutch* (young, SW, 8°C)	LC_0	10.0		10.0		173
F, Ethylbenzene	*Oncorhynchus kisutch* (Y, artificial seaW, 8°C)	LC_0	10.0	10.0	10.0		173
F, *o*-Xylene	*Oncorhynchus kisutch* (SW, 8°C)	LD_0	10.0	10.0	10.0		173
F, Dow Chemical Z3	*Oncorhynchus mykiss* (S)	LC_{50}			10.0–100.0		279
F, Zectran	*Oncorhynchus mykiss*	LC_{50}			10.2		127
F, Sodium arsenite	*Oncorhynchus mykiss* (F, pH 6.4–8.3)	LC_{50}			10.8		1022
F, Quinoline	*Oncorhynchus mykiss* (larvae, F, 14°C, pH 7.4–8.1)	LC_{50}			11.0		977
F, Phenol	*Oncorhynchus mykiss* (15°C, F)	LC_{50}		11.6	11.6	11.6 (10 d)	155
F, Aminocarb (98%)	*Oncorhynchus mykiss* (320 mg/l $CaCO_3$, S, 7–12°C, pH 7.4)	LC_{50}			12.0–25.0		517
F, Picloram	*Oncorhynchus mykiss* (12°C, pH 7.5)	LC_{50}			12.5		411
F, *o*-Xylene	*Oncorhynchus mykiss* (S)	LC_{50}			13.5		284
F, Aminocarb	*Oncorhynchus mykiss* (12°C, pH 7.5)	LC_{50}			13.5		475
F, *p*-Chloraniline	*Oncorhynchus mykiss* (S)	LC_{50}			14.0		147
F, Ethylbenzene	*Oncorhynchus mykiss* (S, 12°C)	LC_{50}			14.0		921
F, Co-Ral	*Oncorhynchus kisutch*	LC_{50}			15.00		1024
F, Ethylhexyldiphenyl phosphate	*Oncorhynchus mykiss*	LC_{50}			15.0		507
F, Aminocarb (98% active)	*Oncorhynchus mykiss* (S, 17°C, pH 8.5)	LC_{50}			17.0		517
F, Dimethylamine	*Oncorhynchus mykiss* (SW, 20 mg $CaCO_3$, 15°C)	LC_{50}			17.0		503
F, Amylxanthate, K (DowChem Z6)	*Oncorhynchus mykiss* (S)	LC_{50}			18.0		279

F, Aminocarb (98% active)	*Oncorhynchus mykiss* (softW, S, 7–12°C, pH 7.4)	LC_{50}			18.0–27.0		517
F, Eptam, technical grade	*Oncorhynchus mykiss*	LC_{50}			19.0		25
F, Zectran	*Oncorhynchus mykiss* (fingerling, S)	LC_{50}			20.0		283
F, Aminocarb	*Oncorhynchus mykiss* (S, 8.5–9.5°C)	LC_{50}			21.3		1013
F, Aminocarb (98% active)	*Oncorhynchus mykiss* (S, 17°C, pH 7.5)	LC_{50}			22.0		517
F, Atrazine, wet powder	*Oncorhynchus mykiss* (S, 12°C, pH 7.3)	LC_{50}			24.0		404
F, Picloram	*Oncorhynchus mykiss* (12.7°C)	LC_{50}	34.0	25.0	24.0		803
F, Saniticizer, 148 + 6% TPP	*Oncorhynchus mykiss* (S, 12°C, pH 7.6)	LC_{50}			26.0		964
F, Tordon 22K	*Oncorhynchus mykiss* (F, 15°C)	LC_{50}		31.0	26.0	22.2 (10 d)	155
F, Dicamba	*Oncorhynchus mykiss* (S)	LC_{10}			28.0		402
F, Versatic 10	*Oncorhynchus mykiss* (S)	LC_{50}			28.0–32.0		286
F, Cyanamid C 325	*Oncorhynchus mykiss* (S)	LC_{50}			29.0–37.0		279
F, Aminocarb (oil suspension)	*Oncorhynchus mykiss* (juv., 7.5–8.5°C)	LC_{50}			29.1		1013
F, Furazolidine	*Oncorhynchus mykiss* (S)	TL_m			30.0		222
F, Trimethyllead chloride	*Oncorhynchus mykiss*	LC_{50}			32.0 (Pb)		282
F, 2-Ethyl-1-hexanol	*Oncorhynchus mykiss* (S)	LC_{50}			32.0–37.0		286
F, Isopropylxanthate, K (Dow Chemical Z9)	*Oncorhynchus mykiss* (S)	LC_{50}			32.0–320.0		279
F, Methomyl	*Oncorhynchus mykiss* (eyed eggs, S, pH 7.5)	LC_{50}			32.0		513
F, Paraquat	*Oncorhynchus mykiss*	LC_{50}	109.0	62.0	32.0		1023
F, Amylxanthate, K (Cyanamid C350)	*Oncorhynchus mykiss*	LC_{50}			32.0–56.0		279
F, Furalaxyl	*Oncorhynchus mykiss*	LC_{50}			32.5		861
F, Fenethcarb	*Oncorhynchus mykiss*	LC_{50}			35.0		130
F, Dinopropylamine	*Oncorhynchus mykiss* (20 mg $CaCO_3$, SW, 15°C)	LC_{50}			37.0		503
F, Isopropylthionocarbamate	*Oncorhynchus mykiss* (S)	LC_{50}			45.0–48.0		287
F, Pydraul 115E	*Oncorhynchus mykiss* (S)	LC_{50}	100.0		45.0		139
F, Di-2(ethylhexyl)phthlalic acid	*Oncorhynchus mykiss* (S)	LC_{50}			48.0–54.0		286
F, Benzene	*Oncorhynchus kisutch* (young, seaW, 8°C)	LC_{50}	50.0		50.0		173

Chemical	Species	Toxicity test	24 h	48 h	96 h	Chronic exposure	Ref.
F, Toluene	*Oncorhynchus kisutch* (young, SW, 8°C)	LC_{100}		50.0	50.0		173
F, Cyanamid C303	*Oncorhynchus mykiss* (S)	LC_{50}			52.0		279
F, Pyrethrin	*Oncorhynchus mykiss* (S)	LC_{50}			52.2		283
F, Benzethonium chloride	*Oncorhynchus kisutch*	LC_{50}			53.0		251
F, Cyanamid C 317	*Oncorhynchus mykiss* (S)	LC_{50}			56.0–100.0		279
F, Ethylxanthate, Na salt	*Oncorhynchus mykiss* (S)	LC_{100}			56.0	1.0 (8 d exp F)	279
F, Formaldehyde (37%)	*Oncorhynchus mykiss* (fingerling, S)	LC_{50}			61.9		120
F, 4-Chlorophenylurea	*Oncorhynchus mykiss* (S)	LC_{50}			72.0		147
F, Formaldehyde (37%)	*Oncorhynchus mykiss* (fingerling, S)	LC_{50}			73.5		120
F, Altosid	*Oncorhynchus kisutch*	LC_{50}			86.0		177
F, Formaldehyde (37%)	*Oncorhynchus mykiss* (sac larvae, S)	LC_{50}			89.5		120
F, Fire-trol	*Oncorhynchus kisutch* (S)	LC_{50}			90.0–1500.0		151
F, Formaldehyde (37%)	*Oncorhynchus mykiss* (fingerling, S)	LC_{50}			94.5		120
F, Formaldehyde (37%)	*Oncorhynchus mykiss* (sac larvae, S)	LC_{50}			96.0		120
F, *n*-Heptane	*Oncorhynchus kisutch* (young, seaW)	LC_0			100.0		173
F, *n*-Hexane	*Oncorhynchus kisutch* (8°C)	LC_0			100.0		173
F, *n*-Pentane	*Oncorhynchus kisutch* (seaW, 8°C)	LC_0			100.0		173
F, Hexane	*Oncorhynchus kisutch*	LC_0			100.0		1026
F, Cyclohexene	*Oncorhynchus kisutch* (seaW, 8°C)	LC_0			100.0		173
F, Cyclohexane	*Oncorhynchus kisutch* (seaW, 8°C)	LC_0			100.0		173
F, Cyclopentane	*Oncorhynchus kisutch* (seaW, 8°C)	LC_0			100.0		173
F, Fire-trol 100	*Oncorhynchus mykiss* (F)	LC_{50}			100.0		151
F, 2,6-Difluorobenzoic acid	*Oncorhynchus mykiss* (S)	LC_{50}			100.0		147
F, Isopropyl xanthate, (Dow Chemical Z11)	*Oncorhynchus mykiss* (S)	LC_{50}			100.0–180.0		279
F, Endothal-diK	*Oncorhynchus kisutch* (fry, S, 13°C, pH 7.5)	LC_{50}			100.0		607
F, Toluene	*Oncorhynchus kisutch* (young, SW, 8°C, S)	LC_{100}		100.0	100.0		173
F, *sec*-Butylxanthate	*Oncorhynchus mykiss* (S)	LC_{50}			100.0–166.0		279

F, Formaldehyde (37%)	*Oncorhynchus mykiss* (fingerling, S)	LC_{50}			106.0	120
F, Altosid-SR10	*Oncorhynchus mykiss* (juv.)	LC_{50}			106.0	177
F, Altosid-SR10	*Oncorhynchus mykiss* (aerated)	TL_{50}			106.0	28
F, Prudhoe crude oil	*Oncorhynchus gorbuscha* (S, 11.5°C)	LC_{50}			110.0	893
F, Formaldehyde (37%)	*Oncorhynchus mykiss* (sac-larvae, S)	LC_{50}			112.0	120
F, Triclopyr	*Oncorhynchus mykiss*	LC_{50}			117.0	461
F, Dimethyl amine	*Oncorhynchus mykiss* (hardW, 320 mg $CaCO_3$, 15°C)	LC_{50}			118.0	503
F, Formaldehyde (37%)	*Oncorhynchus mykiss* (fingerling, S)	LC_{50}			123.0	120
F, Formaldehyde (37%)	*Oncorhynchus mykiss* (fingerling, S)	LC_{50}			134.0	120
F, Dicamba	*Oncorhynchus mykiss*	LC_{50}		35.0	135.0	1000
F, Aquathol K Di-Na	*Oncorhynchus tschawytscha*	LC_{50}			136.0	1025
F, Glyphosate	*Oncorhynchus mykiss* (12°C)	LC_{50}	140.0		140.0	569
F, Formaldehyde (37%)	*Oncorhynchus mykiss* (fingerling, S)	LC_{50}			145.0	120
F, Diflubenzuron	*Oncorhynchus kisutch*	LC_{50}			150.0	177
F, Diflubenzuron	*Oncorhynchus mykiss* (juv.)	LC_{50}			150.0	177
F, Fire-trol 100	*Oncorhynchus mykiss* (S)	LC_{50}			150.0–100.0	151
F, Isopropylxanthate, Na	*Oncorhynchus mykiss* (S)	LC_{100}			180.0	279
F, Bentazone, TG	*Oncorhynchus mykiss*	LC_{50}			190.0	967
F, Disopropyl amine	*Oncorhynchus mykiss*	LC_{50}			196.0	503
F, Diisopropylamine	*Oncorhynchus mykiss* (320 mg $CaCO_3$, HW, 15°C)	LC_{50}			196.0	503
F, Formaldehyde (37%)	*Oncorhynchus mykiss* (eyed egg, S)	LC_{50}			198.0	120
F, Isopropylxanthate, (Cyanamid C343)	*Oncorhynchus mykiss* (S)	LC_{50}			217.0	279
F, Diflubenzuron	*Oncorhynchus mykiss* (S)	LC_{50}			250.0	147
F, Fire-trol	*Oncorhynchus kisutch* (F)	LC_{50}			280.0	151
F, Diethylene glycol dinitrate	*Oncorhynchus mykiss*	LC_{50}			284.1	1088
F, Formaldehyde (37%)	*Oncorhynchus mykiss* (eyed egg, S)	LC_{50}			289.0	120
F, *sec*-Butyl xanthate	*Oncorhynchus mykiss* (S)	LC_{50}			320.0	451
F, Formaldehyde (37%)	*Oncorhynchus mykiss* (eyed egg, S)	LC_{50}			338.0	120
F, *n*-Pentanol	*Oncorhynchus mykiss* (S, 10°C)	TL_m			370.0–490.0	1029

Chemical	Species	Toxicity test	24 h	48 h	96 h	Chronic exposure	Ref.
F, Formaldehyde (37%)	*Oncorhynchus mykiss* (eyed egg, S)	LC_{50}			435.0		120
F, Formaldehyde (37%)	*Oncorhynchus mykiss* (S)	LC_{50}			440.0		120
F, Formaldehyde (37%)	*Oncorhynchus mykiss* (green egg, S)	LC_{50}			565.0		120
F, Fire-trol 1931	*Oncorhynchus kisutch* (S)	LC_{50}			580.0–1000.0		151
F, Formaldehyde (37%)	*Oncorhynchus mykiss* (S)	LC_{50}			610.0		120
F, Formaldehyde (37%)	*Oncorhynchus mykiss* (S)	LC_{50}			618.0		120
F, Formaldehyde (37%)	*Oncorhynchus mykiss* (green egg, S)	LC_{50}			631.0		120
F, Formaldehyde (37%)	*Oncorhynchus mykiss* (green egg, S)	LC_{50}			700.0		120
F, Fire-trol 931	*Oncorhynchus mykiss* (S)	LC_{50}			700.0–1000.0		151
F, Polypropylene glycol	*Oncorhynchus mykiss* (juv., S, 9.5–10.5°C, pH 7.9–8.2)	LC_0			1000.0		894
F, Asulam	*Oncorhynchus mykiss* (S)	LC_{50}			5000.0		812
F, Alaska crude oil	*Oncorhynchus kisutch* (S, 10°C)	LC_{50}			7500.0		1028
F, Imidan	*Oncorhynchus mykiss* (eyed eggs, S)	LC_{50}			10,000.0		148
F, Imidan	*Oncorhynchus mykiss* (yolk-sac fry, S)	LC_{50}			10,000.0		148
F, Polypropylene glycol	*Oncorhynchus mykiss* (juv., S, 9.5–10.5°C, pH 7.9–8.2)	LC_{50}			10,000.0		894
F, Polydimethyl siloxane	*Oncorhynchus mykiss* (S, 13°C)	TL_m			10,000.0		153
F, Polypropylene glycol	*Oncorhynchus mykiss* (S, 10°C)	TL_m			10,000.0		1029
F, Methyl cellosolve	*Oncorhynchus mykiss* (12°C)	LC_0			12,610.0		264
F, Methyl cellosolve	*Oncorhynchus mykiss* (fingerlings, 12°C)	LC_{50}			15,520.0		264
O, 1,2-Dichloroethane	*Ophiura texturata* (S, renewal 6 h, 13–18°C)	LC_{50}		23.0			572
F, Endrin	*Ophiocephalus punctatus*	LC_{50}			0.033		250
F, Phenol	*Ophiocephalus punctatus* (S)	LC_{50}		46.0			254
F, Benzoic acid	Orangespotted fish	LC_{100}				550.0–570.0 (1 h)	84
C, DDT	*Orconectes nais* (S)	LC_{50}			0.1		580
F, *n*-Propanol	*Oryzias latipes* (24°C, pH 8.2–8.4)	LC_0		4400.0			1001
F, *n*-Propanol	*Oryzias latipes* (S, 24°C, pH 8.2–8.4)	LC_{50}		5900.0			1001
F, Triphenyl phosphate	*Oryzias latipes* (S, 25°C)	LC_{50}			1.20		818

F, Nitrilotriacetic acid	*Oryzias latipes*	NOEL			100.0		372
F, Tris (2-chloroethyl)phosphate	*Oryzias latipes* (S, 25°C)	LC_{50}			210.0		818
C, Cupric acetate	*Palaemonetes vulgaris*	LC_{50}			37.0		93
F, Carbaryl	*Parophrys vetulus* (S, juv.)	TL_m	4.1				407
C, DDT	*Pegurus longicarpus*	LC_{50}			0.006		1030
C, Dieldrin	*Pegurus longicarpus*	LC_{50}			0.018		1030
C, Heptachlor	*Pegurus longicarpus* (F)	LC_{50}			0.055		1030
C, Aldicarb	*Penaeus duorarum* (F, 23°C, pH 8.3)	LC_{50}			0.012		925
C, Chloroform	*Penaeus duorarum*	LC_{50}			81.5		629
F, Potassium cyanide	*Perca fluviatilis* (F, 11–12°C, pH 7.5–7.9)	LC_0				0.087 (6 d)	1027
F, Potassium cyanide	*Perca fluviatilis* (F, 11–12°C, pH 7.5–7.9)	LC_{100}				0.125 (6 d)	1027
F, Potassium cyanide	*Perca fluviatilis* (F, 11–12°C, pH 7.5–7.9)	LC_{50}				0.1 (6 d)	1027
F, Phenol	*Perca fluviatilis*	LC_{100}				9.0 (1 h)	183
F, 2-Methyl-4-chlorophenoxy acetic acid (MCPA)	*Perca fluviatilis*	LC_{50}		200.0–215.0			934
F, Resmethrin	*Perca flavescens* (F, 12°C)	LC_{50}			0.00044–0.0006		1915
F, Resmethrin	*Perca flavescens* (F)	LC_{50}			0.000513		138
F, Resmethrin	*Perca flavescens* (S, 12°C)	LC_{50}			0.002–0.0028		915
F, Resmethrin	*Perca flavescens* (S)	LC_{50}			0.00236		138
F, Allethrin	*Perca flavescens* (S)	LC_{50}			0.0078		138
F, DDT	*Perca flavescens*	LC_{50}			0.009		127
F, DDT	*Perca fluviatilis*	LC_{50}			0.009		122
F, Allethrin (90% active)	*Perca flavescens* (F)	LC_{50}			0.0099		140
F, Toxaphene	*Perca flavescens* (S, 18°C, pH 7.1)	LC_{50}			0.012		515
F, Toxaphene	*Perca fluviatilis*	LC_{50}			0.012		122
F, Azinphosmethyl	*Perca flavescens*	LC_{50}			0.013		127
F, Dimethrin	*Perca flavescens* (S)	LC_{50}			0.028		138, 140
F, Pyrethrum	*Perca flavescens* (F, 12°C, pH 7.8)	LC_{50}			0.036–0.054		915
F, Pyrethrin	*Perca flavescens* (F)	LC_{50}			0.0445		138
F, Pyrethrum	*Perca flavescens* (S, 12°C, pH 7.8)	LC_{50}			0.050		915
F, Pyrethrin	*Perca flavescens* (S)	LC_{50}			0.0501		138

Chemical	Species	Toxicity test	24 h	48 h	96 h	Chronic exposure	Ref.
F, Lindane	*Perca flavescens*	LC_{50}			0.068		127
F, Lindane	*Perca fluviatilis*	LC_{50}			0.068		122
F, Rotenone (5% active)	*Perca flavescens* (S)	LC_{50}			0.070		28
F, Hydrogen cyanide	*Perca flavescens* (juv., F)	LC_{50}			0.076–0.108		137
F, Hydrocyanic acid	*Perca flavescens* (juv., F, 15°C, pH 7.82)	LC_{50}			0.094		947
F, Hydrocyanic acid	*Perca flavescens* (juv., F, 15°C, pH 7.82)	LC_{50}			0.102		947
F, Malathion	*Perca fluviatilis*	LC_{50}			0.260		122
F, Malathion	*Perca flavescens*	LC_{50}			0.263		127
F, Hydrogen cyanide	*Perca flavescens* (eggs, F)	LC_{50}			0.276–0.389		137
F, Hydrogen cyanide	*Perca flavescens* (swim-up fry, F)	LC_{50}			0.295–0.395		137
F, Tricresyl phosphate	*Perca flavescens* (F, 12°C, pH 7.5)	LC_{50}			0.502		946
F, Carbaryl	*Perca flavescens*	NOEL			0.745		127
F, Carbaryl	*Perca fluviatilis*	LC_{50}			0.75		122
F, Baytex	*Perca flavescens*	LC_{50}			1.65		127
F, Zectran	*Perca flavescens*	LC_{50}			2.48		127
F, Zectran	*Perca fluviatilis*	LC_{50}			2.5		122
F, Dimethyl parathion	*Perca flavescens*	LC_{50}			3.06		127
F, Ethylene glycol, dinitrate	*Perca fluviatilis*	LOEL			5.0		8
F, Chlorodane	*Petromyzon marinus*	LC_{100}				1.0 (14 h)	785
F, 2,5-Dichloro-4-nitrophenol	*Petromyzon marinus* (larvae)	LD_{100}				3.0	84
F, 3-Bromo-4-nitrophenol	*Petromyzon marinus* (larvae)	LD_{100}				3.0	84
F, 2-Bromo-4-nitrophenol	*Petromyzon marinus* (larvae)	LD_{100}				5.0	84
F, 3,4,6-Chloro-2-nitrophenol	*Petromyzon marinus* (larvae)	LD_{100}				5.0	84
F, Niclosamide	*Petromyzon marinus* (S, 11–18°C, pH 8.2)	LC_{50}	0.049				1031
F, Trifluoromethyl-4-nitrophenol	*Petromyzon marinus* (S)	LC_{50}	0.78				276
F, Trifluoromethyl-4-nitrophenol	*Petromyzon marinus* (larvae, S)	LC_{50}			0.35–1.3		157
F, Trifluoromethyl-4-nitrophenol	*Petromyzon marinus* (free-swim. larvae, F)	LC_{50}			1.48		157
F, Trifluoromethyl-4-nitrophenol	*Petromyzon marinus* (burrowed larvae, F)	LC_{50}			1.68		157
F, Boric acid	*Phoxinus phoxinus* (20°C)	LC_{100}				19,000.0 (6 h)	1032

F, Ammonium salt of saturated carboxylic acid, a dispersant	*Phoxinus phoxinus*	LD_{50}	300.0				70
F, Azinphosmethyl	*Phoxinus phoxinus*	LC_{50}			0.24		211
F, Ethyl-*p*-nitrophenylthionobenzene phosphate	*Phoxinus phoxinus*	LC_{50}			0.25		806
F, Carbaryl	*Phoxinus phoxinus*	LC_{50}			14.6		122
F, Acrylonitrile	*Phoxinus phoxinus*	TL_m		37.4	24.0		1036
F, Perchloroethylene	*Pimephales promelas* (F)	LC_{50}			18.4		256
F, Perchloroethylene	*Pimephales promelas* (S)	LC_{50}			21.4		256, 257
F, Toxaphene	*Pimephales promelas* (F)	MATC				0.000025 (245 d)	167
F, Endosulfan	*Pimephales promelas* (F, 25°C)	TL_m				0.000086	201
F, Endosulfan	*Pimephales promelas* (F)	MATC				0.0002	201
F, Trifluralin	*Pimephales promelas* (F, 25°C)	MATC				0.0019	140
F, Aroclor 1260	*Pimephales promelas* (F)	LC_{50}				0.0033 (30 d)	1037
F, Hydrogen sulfide	*Pimephales promelas* (F, 6–24°C)	MATC				0.0037	143
F, Toxaphene	*Pimephales promelas* (F)	LC_{50}				0.0048 (10 d)	167
F, Heptachlor	*Pimephales promelas* (F)	LC_{50}				0.007 (10 d)	1033
F, Lindane	*Pimephales promelas* (F, 25°C)	MATC				0.0091	145
F, Acrolein	*Pimephales promelas* (F)	MATC				0.0114	261
F, Aldicarb	*Pimephales promelas* (embryo, hardW, F, 22–24°C, pH 7.8–8.0)	LC_0				0.0186 (30 d)	1034
F, Aldicarb	*Pimephales promelas* (juv., F, 22–24°C, pH 7.8–8.0)	LC_{10}				0.020–0.078 (30 d)	1034
F, Captan	*Pimephales promelas*	LC_{24}				0.0395 (45 wk)	778
F, Dinoseb	*Pimephales promelas* (juv. fry, F, 24–26°C pH 7.5)	LC_{78}				0.0485 (22 d)	1035
F, Saniticizer 154	*Pimephales promelas* (in pond with sediment)	LC_{18}				0.062 (28 d)	528
F, Captan	*Pimephales promelas*	LC_{100}				0.0635 (45 wk)	778
F, Lindane	*Pimephales promelas* (F, 25°C)	TL_m				0.069 (1 yr)	145
F, Hydroquinone	*Pimephales promelas*	LC_{50}				0.1–0.18	20

Chemical	Species	Toxicity test	24 h	48 h	96 h	Chronic exposure	Ref.
F, Trifluralin	*Pimephales promelas* (F, 25°C)	TL_m				0.115	140
F, Butylbenzyl phthalate	*Pimephales promelas* (larvae)	MATC				0.14–0.36	9
F, Aldicarb	*Pimephales promelas* (juv., F, 22–24°C, pH 7.8–8.0)	LC_{42}				0.156 (30 d)	1034
F, Carbaryl	*Pimephales promelas*	NOEL				0.210 (6 min)	1038
F, 2,4-D, butoxyethyl ester	*Pimephales promelas*	NOEL				0.300 (10 min)	237
F, Aldicarb	*Pimephales promelas* (juv., F, 22–24°C, pH 7.8–8.0)	LC_{20}				0.340 (30 d)	1034
F, Saniticizer 154	*Pimephales promelas* (in pond without sediment)	LC_{91}				0.826 (28 d)	525
F, 1-Phenyl-3-pyrazolidone	*Pimephales promelas*	LC_{50}				1.0–10.0	20
F, 1,3-Dichloropropane	*Pimephales promelas*	LC_0				1.0 (72 h)	1033
F, Alachlor	*Pimephales promelas* (embryo, juv., F,) 24–26°C, pH 7.5)	NOAE				1.10 (64 d)	1035
F, Sodium arsenite	*Pimephales promelas* (F, 23–26°C, pH 7.2–8.1)	NOAE				2.13 (29 d)	477
F, Butylbenzyl phthalate	*Pimephales promelas* (F)	LC_{50}				2.25 (14 d)	9
F, *o*-Dichlorobenzene	*Pimephales promelas*	LC_0				3.0 (72 h)	259
F, Hydroxylamine sulfate	*Pimephales promelas*	LC_{50}				7.20	20
F, Pentachlorophenol	*Pimephales promelas* (S)	LC_{50}				8.0 (1 h)	196
F, Atrazine	*Pimephales promelas* (F)	NOAE				8.0 (8 d)	542
F, Hydroquinone sulfonate, Na	*Pimephales promelas*	LC_{50}				10.0 (h?)	20
F, *t*-Butylamine borane	*Pimephales promelas*	LC_{50}				10.0–18.0	20
F, *o*-Dichlorobenzene	*Pimephales promelas*	LC_{100}				10.0 (72 h)	259
F, Atrazine	*Pimephales promelas* (F, 25°C)	TL_m				11.0–20.0 (1 yr)	940
F, 3,4-Xylenol	*Pimephales promelas* (S, lakeW, 18–22°C, 1 h)	LC_{50}				20.0	196
F, *o*-Xylene	*Pimephales promelas* (S, lakeW, 18–22°C, 1 h)	LC_{50}				46.0	202
F, Furfural	*Pimephales promelas* (S)	LC_{50}				50.0 (1 h)	196
F, *n*-Valeric acid	*Pimephales promelas* (lakeW, S, 18–22°C, 1 h)	LC_{50}				100.0	196
F, Thiourea	*Pimephales promelas*	LC_{50}				100.0	20

F, Pyrrolidone	*Pimephales promelas*	LC_{50}		100.0	20
F, Nitrilotriacectic acid	*Pimephales promelas*	LC_{50}		100.0	20
F, Diethylene glycol	*Pimephales promelas*	LC_{50}		100.0	20
F, β-Phenylethyleneamine sulfate	*Pimephales promelas*	LC_{50}		100.0	20
F, Diethanolamine	*Pimephales promelas*	LC_{50}		100.0	20
F, Citrazinic acid	*Pimephales promelas*	LC_{50}		100.0	20
F, Vaniline	*Pimephales promelas* (lakeW, S, 18–22°C, 1 h)	LC_{50}		173.0	196
F, Vaniline	*Pimephales promelas* (riverW, S, 18–22°C, 1 h)	LC_{50}		173.0	196
F, EDTA, ammonium ferric salt	*Pimephales promelas*	LC_{50}		190.0	20
F, Diethylene triamine pentaacetic acid	*Pimephales promelas*	LC_{50}		300.0	135
F, 1,3-Diamino-2-propanol tetra-acetic acid	*Pimephales promelas*	LC_{50}		300.0	20
F, EDTA, triNa	*Pimephales promelas*	LC_{50}		300.0	20
F, Benzyl alcohol	*Pimephales promelas* (S)	LC_{50}		480.0 (72 h)	196
F, Ethylene dichloride	*Pimephales promelas*	LC_{50}		500.0 ppm	196
F, Ethylene diamine	*Pimephales promelas*	LC_{50}		1000.0	20
F, *n*-Butanol	*Pimephales promelas* (lake water S)	LC_{50}		1940.0 (1 h)	196
F, *n*-Butanol	*Pimephales promelas* (river water S)	LC_{50}		1950.0 (1 h)	196
F, Aroclor 1248	*Pimephales promelas*	LC_{50}		4700.0 (1 min)	906
F, Toxaphene	*Pimephales promelas* (S, 24°C, pH 7.4)	LC_{50}	0.0057		801
F, Toxaphene	*Pimephales promelas* (S, 10°C, pH 7.4)	LC_{50}	0.020		801
F, Acrolein	*Pimephales promelas* (F)	TL_m	0.084		201
F, Isodrin/photoisodrin toxicity ratio (TR)	*Pimephales promelas*	TR	0.3		44, 258, 300
F, Pentachlorophenate, Na	*Pimephales promelas* (S)	TL_m	0.330		195
F, Heptachlor/photoheptachlor toxicity ratio	*Pimephales promelas*	TR	1.23		44, 300
F, Heptachlor/photoheptachlor txicity ratio	*Pimephales promelas*	TR	1.63		258

Chemical	Species	Toxicity test	24 h	48 h	96 h	Chronic exposure	Ref.
F, Ethybenzene (80%)	*Pimephales promelas* (juv., hardW, 25°C, pH 7.8)	LC_{50}	42.0				797
F, Ethybenzene (80%)	*Pimephales promelas* (juv., softW, 25°C, pH 7.8)	LC_{50}	49.0				797
F, Pentachlorophenate, Na	*Pimephales promelas* (F, 15°C)	TL_m		0.21			199
F, Pentachlorophenate, Na	*Pimephales promelas* (F, 25°C)	TL_m		0.37			144
F, Linear alkyl sulfonate homolog C14	*Pimephales promelas* (S)	LC_{50}	0.60	0.40			859
F, Linear alkyl sulfonate homolog C13	*Pimephales promelas* (S)	LC_{50}	1.70	0.40			859
F, Hydrogen sulfide	*Pimephales promelas*	TL_m		1.38			84
F, 2,4-D, butoxyethyl ester	*Pimephales promelas* (eggs)	LC_{100}		1.5			237
F, Linear alkyl sulfonate commercial	*Pimephales promelas* (S)	LC_{50}	1.9	1.7			859
F, Linear alkyl sulfonate homolog C12	*Pimephales promelas* (S)	LC_{50}	4.70	4.70			859
F, Dialkyltetralinindane sulfonate, C = 14	*Pimephales promelas*	LC_{50}	8.1	5.3			859
F, Alachlor	*Pimephales promelas* (juv., F, 24–26°C, pH 7.5)	LC_{50}	9.9	6.3–7.0			1039
F, Linear alkyl sulfonate homolog C11	*Pimephales promelas* (S)	LC_{50}	17.0	16.0			859
F, Dialkyltetralinindane sulfonate, C = 12	*Pimephales promelas*	LC_{50}	24.8	21.5			859
F, Linear alkyl sulfonate homolog C10	*Pimephales promelas* (S)	LC_{50}	48.0	43.0			859
F, Linear alkyl sulfonate Model Int. C11	*Pimephales promelas* (S)	LC_{50}	85.9	76.6			859
F, Dialkyltetralinindane linear sulfonate, C = 10	*Pimephales promelas*	LC_{50}	87.0	86.1			859
F, *n*-Propanol	*Pimephales promelas* (20°C, S)	LC_0		2600.0			1001
F, *n*-Propanol	*Pimephales promelas* (S, 20°C)	LC_{50}		5000.0			1001

F, Linear alkyl sulfonate Model Int. C4	*Pimephales promelas* (S)	LC_{50}	10,000.0	10,000.0			859
F, Linear alkyl sulfonate Model Int. C5	*Pimephales promelas* (S)	LC_{50}	10,000.0	10,000.0			859
F, Heptachlor	*Pimephales promelas* (F)	MATC			0.00086		201
F, Endrin	*Pimephales promelas*	LC_{50}			0.001		232
F, Endosulfan	*Pimephales promelas* (S, 18°C, pH 7.5)	LC_{50}			0.0015		600
F, Sulfur	*Pimephales promelas*	LC_0			0.003		874
F, Hexachlorocyclopentadiene	*Pimephales promelas* (F)	LC_{50}			0.007		1042
F, Heptachlor	*Pimephales promelas* (F)	TL_m			0.00702		201
F, Hexachlorocyclopentadiene	*Pimephales promelas* (juv., F, 23–27°C, pH 7.2–7.7)	LC_{50}			0.007	0.0065 (30 d)	1040
F, Hexachlorocyclopentadiene	*Pimephales promelas* (larva, F)	LC_{50}			0.007	0.0067 (30 d)	203
F, Hydrogen sulfide	*Pimephales promelas* (F, 6–24°C)	TL_m			0.0071–0.55		143
F, Toxaphene	*Pimephales promelas* (S, 25°C, pH 7.1)	LC_{50}	0.013	0.0075	0.0075		801
F, *p*-Chloro-*m*-cresol	*Pimephales promelas* (S)	TL_m			0.01		82
F, Dieldrin	*Pimephales promelas*	LC_{50}			0.016		232
F, DDT	*Pimephales promelas* (S)	LC_{50}			0.019		127
F, Aldrin	*Pimephales promelas*	LC_{50}			0.028		232
F, 4-Chloro-*m*-cresol	*Pimephales promelas* (S)	LC_{50}			0.03		1041
F, Sulfur	*Pimephales promelas* (20°C)	LC_{50}			0.032		874
F, Chlordane	*Pimephales promelas* (F)	LC_{50}			0.0369		939
F, Bio-allethrin, 90% TG	*Pimephales promelas* (F, 12°C, pH 7.5)	LC_{50}			0.048		597
F, Chlorodane	*Pimephales promelas*	LC_{50}			0.052		785
F, Chlorodane	*Pimephales promelas* (SW)	LC_{50}	0.079	0.069	0.052		785
F, Allethrin	*Pimephales promelas* (F)	LC_{50}			0.053		138
F, Heptachlor	*Pimephales promelas*	LC_{50}			0.056		232
F, Chlorodane	*Pimephales promelas* (hardW)	LC_{50}	0.098	0.069	0.069		785
F, Aminocarb	*Pimephales promelas* (20°C, pH 7.5)	LC_{50}			0.075		475
F, Heptachlor	*Pimephales promelas* (hardW, S)	LC_{50}			0.078		805
F, Aminocarb (17% active)	*Pimephales promelas* (S, 22°C, pH 7.4)	LC_{50}			0.08		610

Chemical	Species	Toxicity test	24 h	48 h	96 h	Chronic exposure	Ref.
F, Allethrin	*Pimephales promelas* (S)	LC_{50}			0.08		138
F, Hydrogen cyanide	*Pimephales promelas* (swim-up fry, F)	LC_{50}			0.082–0.122		137
F, Hydrogen cyanide	*Pimephales promelas* (juv., F)	LC_{50}			0.082–0.137		137
F, Heptachloronorbornene	*Pimephales promelas* (F)	LC_{50}			0.0856		204
F, Hexachloronorbornene	*Pimephales promelas* (larva, F)	LC_{50}			0.0856	0.0601 (30 d)	203
F, Lindane	*Pimephales promelas*	LC_{50}			0.087		127
F, Dinoseb (technical)	*Pimephales promelas* (12°C, softwater)	LC_{50}			0.088		1087
F, Azinphosmethyl	*Pimephales promelas*	LC_{50}			0.093 (95 h)		252
F, Hydrocyanic acid	*Pimephales promelas* (juv., F, 20°C, pH 7.9)	LC_{50}			0.0991		947
F, Benfluralin	*Pimephales promelas* (18°C, pH 7.1)	LC_{50}	1.0		0.1		404
F, Hydrocyanic acid	*Pimephales promelas* (larva, F, 25°C, pH 8.0)	LC_{100}			0.1071		1043
F, Hydrogen cyanide	*Pimephales promelas* (eggs, F)	LC_{50}			0.121–0.352		137
F, Heptachlor	*Pimephales promelas* (SW, S)	LC_{50}			0.130		805
F, Rotenone (5%)	*Pimephales promelas* (S)	LC_{50}			0.142		28
F, 2,3-Dichloro-1,4-naphthaquinone	*Pimephales promelas*	TL_m	0.24		0.15		8
F, Hydrogen cyanide	*Pimephales promelas* (wild juv., F)	LC_{50}			0.157–0.191		137
F, Hexachloronorbornadiene	*Pimephales promelas* (larva, F)	LC_{50}			0.188	0.123 (30 d)	203
F, Pentachlorophenol	*Pimephales promelas* (14 wk old)	LC_{50}	0.200		0.190		197
F, Hydrocyanic acid	*Pimephales promelas* (egg, F, 15°C, pH 7.9)	LC_{50}			0.196		947
F, Pentachlorophenol	*Pimephales promelas* (4 wk old)	LC_{50}	0.222		0.198		197
F, Pentachlorophenol	*Pimephales promelas* (large)	LC_{50}	0.213		0.203		197
F, Pentachlorophenate, Na	*Pimephales promelas* (F, 25°C)	TL_m			0.21		205
F, Pentachlorophenate, Na	*Pimephales promelas* (F, 15°C)	TL_m			0.21		144
F, Pentachlorophenol	*Pimephales promelas* (11 wk old)	LC_{50}	0.232		0.222		197
F, Pentachlorophenol	*Pimephales promelas* (small)	LC_{50}	0.240		0.227		197
F, Pentachlorophenol	*Pimephales promelas* (7 wk old)	LC_{50}	0.245		0.230		197
F, Hydrocyanic acid	*Pimephales promelas* (egg, F, 20°C, pH 7.9)	LC_{50}			0.273		947
F, Pentachlorophenate, Na	*Pimephales promelas* (F, 25°C)	TL_m			0.285		144
F, Ronnel	*Pimephales promelas*	LC_{50}			0.305		260

F, Pentachlorophenate, Na	*Pimephales promelas* (F, 25°C)	TL_m			0.34		144
F, Hydrocyanic acid	*Pimephales promelas* (egg, F, 15°C, pH 7.9)	LC_{50}			0.352		947
F, Bromicide	*Pimephales promelas* (adult, F, 21°C, pH 6.9–7.1)	LC_{50}			0.39		958
F, Cupric acetate	*Pimephales promelas*	LC_{50}	0.48	0.42	0.39		93
F, Quinoline	*Pimephales promelas* (juv., S, 20°C, pH 7.8)	LC_{50}			0.44		595
F, Trinitrotoluene	*Pimephales promelas* (F, behavioral) response, 24°C)	EC_{50}			0.46		206
F, Triton X100	*Pimephales promelas* (F, 24°C, pH 7.94)	LC_{50}		0.5	0.50		472
F, Pentachlorophenol	*Pimephales promelas* (S, 18–22°C)	LC_{50}	0.60	0.6	0.6	0.6 (72 h)	196
F, Di-*n*-butyl-*t*-phthalate	*Pimephales promelas*	LC_{50}			0.61		1080
F, Dinoseb	*Pimephales promelas* (juv., pH 7.5)	LC_{50}	0.8–0.9	0.7–0.8	0.6–0.7		1035
F, Triphenyl phosphate	*Pimephales promelas*	LC_{50}			0.66		849
F, Furanace	*Pimephales promelas* (S)	LC_{50}			0.82		121
F, Lactonitrile	*Pimephales promelas*	TL_m			0.9		22
F, Di-*n*-butyl-*iso*-phthalate	*Pimephales promelas*	LC_{50}			0.90		1080
F, Triethyl phosphate	*Pimephales promelas*	LC_{50}			1.0		555
F, 4-Chloro-*m*-cresol	*Pimephales promelas* (S)	LC_{80}			1.0		1041
F, Chloranil	*Pimephales promelas* (S)	TL_m			1.0		82
F, 2,4,6-Trichlorophenol	*Pimephales promelas* (S)	TL_m			1.0–0.10		82
F, 2,4,6-Trichloroaniline	*Pimephales promelas* (S)	TL_m			1.0		82
F, Di-*n*-butyl-*o*-phthalate	*Pimephales promelas*	LC_{50}			1.1		1080
F, Methyl parathion, TG (10 wk old)	*Pimephales promelas* (S, 24–26°C, pH 7.4–7.8)	LC_{50}			1.22		1044
F, Pydraul 50E	*Pimephales promelas* (S)	LC_{50}	2.5		1.3		139
F, Aldicarb	*Pimephales promelas* (juv., HW, F, 21–23°C, pH 7.8–8)	LC_{50}			1.370		1034
F, Bromicide	*Pimephales promelas* (juv., 21°C, pH 6.9–7.1)	LC_{50}			1.37		958
F, Parathion	*Pimephales promelas*	LC_{50}			1.410		260
F, Benzethonium chloride	*Pimephales promelas*	LC_{50}			1.6		238
F, Triton X100	*Pimephales promelas* (S, 22°C, pH 7.94)	LC_{50}		2.02	1.60		472

Chemical	Species	Toxicity test	24 h	48 h	96 h	Chronic exposure	Ref.
F, Methomyl, (liquid concentrate)	*Pimephales promelas* (S, 17–22°C, pH 7.5)	LC_{50}			1.80		513
F, Azinphosmethyl	*Pimephales promelas* (F, 25°C)	TL_m			1.9		800
F, Benomyl	*Pimephales promelas* (S, 22°C, pH 7.5)	LC_{50}			1.9–2.2		475
F, 4,6-Dinitro-*o*-cresol	*Pimephales promelas* (F)	LC_{50}			2.030		1045
F, Pydraul 50E	*Pimephales promelas* (F)	LC_{50}			2.1		139
F, Butylbenzyl phthalate	*Pimephales promelas* (S)	LC_{50}			2.1–5.3	1.0–2.2 (NOAE)	9
F, RoundUp	*Pimephales promelas*	LC_{50}	2.4		2.3		562
F, Butylbenzyl phthalate	*Pimephales promelas* (F)	LC_{50}			2.32		9
F, Baytex	*Pimephales promelas*	LC_{50}			2.44		127
F, Trinitrotoluene	*Pimephales promelas* (F, 24°C)	TL_m			2.58		206
F, Chlorothion	*Pimephales promelas*	LC_{50}			2.8		235
F, Methomyl, TG	*Pimephales promelas* (S, 17–22°C, pH 7.5)	LC_{50}			2.8		513
F, Demeton	*Pimephales promelas*	LC_{50}			3.2		235
F, *p*-Butylphenyldiphenyl phosphate	*Pimephales promelas*	LC_{50}			3.4		1020
F, Santicizer 154	*Pimephales promelas*	LC_{50}			3.40		525
F, Methyl parathion, capsules (10 wk old)	*Pimephales promelas* (S, 24–26°C, pH 7.4–7.8)	LC_{50}			3.47		1044
F, Disulfoton	*Pimephales promelas*	LC_{50}			3.7		235
F, Diazinon	*Pimephales promelas*	LC_{50}			3.7–10.0		73
F, *p*-Dichlorobenzene	*Pimephales promelas* (F)	LC_{50}			4.0		660
F, 1,3-Dichloropropane	*Pimephales promelas* (S, 18°C, pH 7.5)	LC_{50}			4.1		758
F, Aluminum nitrate	*Pimephales promelas* (S, 22°C, pH 7.4)	LC_{50}			4.25		1047
F, Disulfoton	*Pimephales promelas* (18°C)	LC_{50}			4.30		410
F, Aluminum sulfate	*Pimephales promelas* (S, 22°C, pH 7.4)	LC_{50}			4.40		1047
F, Phenol	*Pimephales promelas* (20–25°C, F)	LC_{50}	8.21		5.02		198
F, Sodium arsenite	*Pimephales promelas* (F, 23–26°C, pH 7.2–8.1)	LC_0			5.06		477
F, 2,4-D, butoxyethyl ester	*Pimephales promelas*	LC_{50}			5.6		237
F, Butylphenyl ether	*Pimephales promelas* (larvae, <24 h old)	LC_{50}			>5.88		1079
F, Hydrazine	*Pimephales promelas* (SW, F, 20°C, pH 6.95)	LC_{50}			5.98		499

F, Benzyl chloride	*Pimephales promelas*	LC_{50}	11.6	7.3	6.0		93
F, Methyl parathion, capsules	*Pimephales promelas* (F, 24–26°C, pH 7.4–7.8)	LC_{50}			6.91		1044
F, Phosmet	*Pimephales promelas* (S, 20°C, pH 7.5)	LC_{50}			7.3		409
F, Imidan	*Pimephales promelas* (S)	LC_{50}			7.3		148
F, 4-Chloro-*m*-cresol	*Pimephales promelas* (F)	LC_{50}	13.3	11.4	7.56		1046
F, Methyl parathion, capsules (5 wk old)	*Pimephales promelas* (S, 24–26°C, pH 7.4–7.8)	LC_{50}			8.17		1044
F, Ammonia	*Pimephales promelas* (hardW)	TL_m			8.2		22
F, Aminocarb (98% active)	*Pimephales promelas* (S, 22°C, pH 7.4)	LC_{50}			8.5		517
F, Aminocarb	*Pimephales promelas* (S, 20°C, pH 7.5)	LC_{50}			8.5		475
F, Dimethyl parathion	*Pimephales promelas*	LC_{50}			8.9		127
F, Malathion	*Pimephales promelas*	LC_{50}			9.00		237
F, Carbaryl	*Pimephales promelas*	NOEL			9.0		1038
F, 1,1,1-Trichloroethane	*Pimephales promelas* (F)	EC_{10}	10.5	10.0	9.0	9.0 (72 h)	256
F, Dioxathion	*Pimephales promelas*	LC_{50}			9.3		235
F, Diazinon (technical grade)	*Pimephales promelas*	LC_{50}			10.3		87
F, Benzophenone	*Pimephales promelas* (larvae, <24 h old)	LC_{50}			10.89		1079
F, *o*-Chlorophenol	*Pimephales promelas*	TL_m	22.0		11.0		50
F, 1,1,1-Trichloroethane	*Pimephales promelas* (F)	EC_{50}	12.1	11.5	11.1	11.1 (72 h)	256
F, *p*-Chloraniline	*Pimephales promelas* (S)	LC_{50}			12.0		147
F, Dioxathion	*Pimephales promelas* (F)	LC_{50}	24.0	14.0	12.0		808
F, *o*-Cresol	*Pimephales promelas* (hardW)	TL_m	18.0		13.4		50
F, Trichloroethylene	*Pimephales promelas* (F)	EC_{10}	15.20	16.90	13.7	15.5 (72 h)	207
F, 1,1,1-Trichloroethane	*Pimephales promelas* (F)	EC_{90}	14.1	13.2	13.8	13.8 (72 h)	256
F, Diquat	*Pimephales promelas*	LC_{50}			14.0		238
F, Ethylhexyldiphenyl phosphate	*Pimephales promelas*	LC_{50}			14.0		523
F, Indene	*Pimephales promelas* (S, lakeW, 22°C)	LC_{50}	39.0	14.0	14.0		196
F, 3,4-Xylenol	*Pimephales promelas* (S, lakeW, 18–22°C)	LC_{50}	20.0	15.0	14.0	14.0 (72 h)	196
F, Diquat-dibromide	*Pimephales promelas* (28°C, pH 7.1)	LC_{50}	56.0	23.0	14.0		463
F, Sodium arsenite	*Pimephales promelas* (F, 23–26°C, pH 7.2–8.1)	LC_{50}	18.9	15.9	14.1	14.7 (72 h)	477
F, Acrylonitrile	*Pimephales promelas* (hardW)	TL_m			14.3		22

Chemical	Species	Toxicity test	24 h	48 h	96 h	Chronic exposure	Ref.
F, Atrazine	*Pimephales promelas* (renewal F, 18–20°C)	LC_{50}			15.0		535
F, Benzene	*Pimephales promelas* (larvae, <24 h old)	LC_{50}			15.59		1079
F, 2,4-Dinitrophenol	*Pimephales promelas* (juv., F)	LC_{50}			16.7		662
F, Isopropylphenyldiphenyl phosphate	*Pimephales promelas* (17°C, pH 7.4)	LC_{50}			17.0		849
F, Zectran	*Pimephales promelas*	LC_{50}			17.0		127
F, Houghtosafe	*Pimephales promelas* (F)	LC_{50}			17.0		139
F, Toluene	*Pimephales promelas* (larvae, <24 h old)	LC_{50}			17.03		1079
F, Trichloroethylene	*Pimephales promelas* (F)	LC_{10}	34.7	27.7	17.4	20.9 (72 h)	207
F, Santicizer-148	*Pimephales promelas*	LC_{50}			18.0		523
F, Co-Ral	*Pimephales promelas*	LC_{50}			18.0		252
F, Acrylonitrile	*Pimephales promelas* (SW)	TL_m			18.1		22
F, Tetrachloroethylene	*Pimephales promelas* (F, 12°C, pH 8)	LC_{50}			18.4		1048
F, Vinyl acetate	*Pimephales promelas*	TL_m	39.0		19.0		50
F, *p*-Cresol	*Pimephales promelas* (S, lakeW, 22°C)	LC_{50}	26.0	21.0	19.0	30.0 (1 h)	196
F, Allyl chloride	*Pimephales promelas* (SW)	TL_m	24.0	24.0	19.8		50
F, Pentabromochlorocyclohexane	*Pimephales promelas* (S)	NOAE			20.0		984
F, Tetrachloroethylene	*Pimephales promelas* (F, 12°C, pH 8)	LC_{50}			21.4		1048
F, Tetrachloroethylene	*Pimephales promelas* (F, lakeW)	LC_{50}			21.4		1048
F, Trichloroethylene	*Pimephales promelas* (F)	EC_{50}	23.0	22.70	21.9	22.2 (72 h)	207
F, Allyl chloride	*Pimephales promelas* (hardW)	TL_m	25.9	24.0	24.0		50
F, Phenol	*Pimephales promelas* (25°C, F)	TL_m		28.0	24.0		199
F, Sodium arsenite	*Pimephales promelas* (F, 23–26°C, pH 7.2–8.1)	LC_{100}			25.0		477
F, Propoxur	*Pimephales promelas* (S, 18°C, pH 7.5)	LC_{50}			25.0		606
F, *p*-Xylene	*Pimephales promelas* (SW)	TL_m	28.8		26.7		50
F, *p*-Xylene	*Pimephales promelas* (hardW)	TL_m	28.8		28.8		50
F, Chlorobenzene	*Pimephales promelas*	TL_m	39.0		29.0		22
F, Tordon 22K	*Pimephales promelas* (10°C)	LC_{50}	52.0	32.0	29.0		803
F, Leptophos	*Pimephales promelas* (S, 5°C, pH 7.5)	LC_{50}			30.0		678

F, 1,1,1-Trichloroethane	*Pimephales promelas* (F)	LC_{10}			30.8	34.1 (72 h)	256
F, Aniline	*Pimephales promelas* (S, 19–21°C, pH 6.5–8.5)	LC_{50}			32.0		408
F, Benzene	*Pimephales promelas* (hardW)	TL_m	24.4		32.0		154
F, Cyclohexane	*Pimephales promelas*	TL_m	43.0		32.0		50
F, Furfural	*Pimephales promelas* (S)	LC_{50}	48.0	37.0	32.0	32.0 (72 h)	196
F, Phenol	*Pimephales promelas* (hardW)	TL_m	38.6	38.6	32.0		50
F, Phenol	*Pimephales promelas* (18–22°C, S, LW)	LC_{50}	50.0	50.0	32.0	50.0 (1 h)	196
F, Benzene	*Pimephales promelas* (SW)	TL_m	35.5		33.5		154
F, *p*-Dichlorobenzene	*Pimephales promelas* (S, 21–23°C, pH 7.9)	LC_{50}			33.7		696
F, *p*-Dichlorobenzene	*Pimephales promelas*	LC_{50}	35.4	35.4	33.7		93
F, Toluene	*Pimephales promelas*	TL_m	56.0		34.0		50
F, Phenol	*Pimephales promelas* (SW)	TL_m	40.6	40.6	34.3		50
F, Trichloroethylene	*Pimephales promelas* (F)	EC_{90}	36.20	30.60	34.9	31.8 (72 h)	207
F, Isopropylphenyldiphenyl phosphate	*Pimephales promelas* (S, 17°C, pH 7.4)	LC_{50}			35.0		849
F, Houghtosafe	*Pimephales promelas* (S)	LC_{50}	90.0		35.0		139
F, Benzoyl chloride	*Pimephales promelas*	LC_{50}	43.0	35.0	35.0		93
F, Phenol	*Pimephales promelas* (15°C, F)	TL_m		41.0	36.0		199
F, Toluene	*Pimephales promelas* (S)	LC_{50}			38.1		814
F, Trichloroethylene	*Pimephales promelas* (F)	LC_{50}			40.7		1048
F, Trichloroethylene	*Pimephales promelas* (F)	LC_{50}			40.7		207
F, Trichloroethylene	*Pimephales promelas* (F)	LC_{50}	52.4	53.3	40.7	39.0 (72 h)	207
F, *o*-Xylene	*Pimephales promelas* (S, LW, 18–22°C)	LC_{50}	42.0	42.0	42.0	42.0 (72 h)	202
F, Ethylbenzene	*Pimephales promelas* (HW)	TL_m	42.3	42.3	42.3		50
F, Nitrobenzene	*Pimephales promelas* (larvae, <24 h old)	LC_{50}			44.10		1079
F, Styrene	*Pimephales promelas* (SW)	TL_m	56.7	53.6	46.4		50
F, Diphenamid	*Pimephales promelas* (S, 18°C, pH 7.5)	LC_{50}			48.0		403b
F, Ethylbenzene	*Pimephales promelas* (SW)	TL_m	48.8	48.5	48.5		50
F, Monocrotophos (100%)	*Pimephales promelas* (18°C, pH 7.1)	LC_{50}			50.0		404
F, 1,1,1-Trichloroethane	*Pimephales promelas* (F)	LC_{50}			52.80		256
F, 1,1,1-Trichloroethane	*Pimephales promelas* (F)	LC_{50}			52.8		207

Chemical	Species	Toxicity test	24 h	48 h	96 h	Chronic exposure	Ref.
F, 1,1,1-Trichloroethane	*Pimephales promelas* (F)	LC_{50}			52.8	55.4 (72 h)	256
F, Resorcinol	*Pimephales promelas*	LC_{50}	88.60	72.60	53.4		93
F, *o*-Dichlorobenzene	*Pimephales promelas*	LC_{50}	105.0	76.0	57.0		93
F, Styrene	*Pimephales promelas* (hardW)	TL_m	62.8	62.8	59.3		50
F, Trichloroethylene	*Pimephales promelas* (S)	LC_{50}			66.8		1048
F, Trichloroethylene	*Pimephales promelas* (S)	LC_{50}			66.8		207
F, Aniline	*Pimephales promelas* (larvae, <24 h old)	LC_{50}			68.63		1079
F, 2,6-Difluorobenzoic acid	*Pimephales promelas* (S)	LC_{50}			69.0		147
F, Sulfamic acid	*Pimephales promelas* (S, 21–23°C, pH 7.2–7.9)	LC_{50}			70.3		704
F, Isoprene	*Pimephales promelas*	TL_m	87.0		74.0		2
F, *n*-Valeric acid	*Pimephales promelas* (LW, S, 18–22°C)	LC_{50}	100.0	77.0	77.0	77.0 (72 h)	196
F, Benzonitrile	*Pimephales promelas* (hardW)	TL_m			78.0	41.0	207
F, Acetic acid	*Pimephales promelas* (recon. W, S, 18–22°C	LC_{50}	106.0	106.0	79.0	175.0 (1 h)	196
F, Ethylene oxide	*Pimephales promelas* (S, 22°C, pH 7.0)	LC_{50}			84.0		390
F, Caproic acid	*Pimephales promelas* (S, LakeW, 22°C)	LC_{50}	88.0	88.0	88.0	140.0 (1 h)	196
F, Acetic acid	*Pimephales promelas* (lakeW, S, 18–22°C)	LC_{50}	122.0	92.0	88.0	315.0 (1 h)	196
F, 2-Chloroethanol	*Pimephales promelas* (S, 22°C, pH 7.0)	LC_{50}			90.0		390
F, 1,1,1-Trichloroethane	*Pimephales promelas* (F)	LC_{90}			90.8	88.9 (72 h)	256
F, Cyclohexane	*Pimephales promelas* (S, LW, 22°C)	LC_{50}	93.0	93.0	93.0	93.0 (72 h)	196
F, Trichloroethylene	*Pimephales promelas* (F)	LC_{90}	79.1	102.6	95.0	72.6 (72 h)	207
F, Glyphosate	*Pimephales promelas* (22°C)	LC_{50}	97.0		97.0		562
F, Adipic acid	*Pimephales promelas* (S, lakeW, 18–22)	LC_{50}	172.0	114.0	97.0	300.0 (1 h)	196
F, Phosphamidon	*Pimephales promelas*	LC_{50}			100.0		71
F, 4-Chlorophenyl urea	*Pimephales promelas* (S)	LC_{50}			100.0		147
F, 1,1,1-Trichloroethane	*Pimephales promelas* (S)	LC_{50}			105.0		256
F, 1,1,1-Trichloroethane	*Pimephales promelas* (S)	LC_{50}			105.0		207
F, Chlorfos	*Pimephales promelas*	LC_{50}			109.0		237
F, Aquathol K-diNa	*Pimephales notatus*	LC_{50}			110.0		1025

F, Ethyl-*p*-nitrophenylthiono-benzene phosphate	*Pimephales promelas*	LC_{50}			110.0		260
F, Camphor	*Pimephales promelas* (S, lakeW, 22°C)	LC_{50}	112.0	111.0	110.0	145.0 (1 h)	196
F, 1,2-Dichloroethane	*Pimephales promelas* (F, 25°C, pH 6.7–7.6)	LC_{50}			116.0		549
F, Vaniline	*Pimephales promelas* (RW, S, 18–22°C)	LC_{50}	125.0	116.0	116.0	116.0 (72 h)	196
F, Cyclohexane	*Pimephales promelas* (S, RCW, 22°C)	LC_{50}	117.0	117.0	117.0	117.0 (72 h)	196
F, Vaniline	*Pimephales promelas* (LW, S, 18–22°C)	LC_{50}	131.0	123.0	121.0	121.0 (72 h)	196
F, α-*w*-Butylene di[9-0-4-hydroxy-butoxycarbonyl-benzoate]	*Pimephales promelas*	LC_{50}			121.0		1080
F, Diquat-dibromide	*Pimephales promelas* (25°C, pH 8.2)	LC_{50}	260.0	222.0	130.0		463
F, Benzonitrile	*Pimephales promelas* (softW)	TL_m			135.0		22
F, 1,2-Dichloropropane	*Pimephales promelas* (F)	LC_{50}			139.0		458
F, Toluene 2,6-diisocyanate	*Pimephales promelas*	LC_{50}	195.0	172.0	164.0		93
F, Methylene chloride	*Pimephales promelas* (F, 12°C, pH 8, lakeW)	LC_{50}	268.0		193.0		1048
F, 2,2-Dichloropropionic acid	*Pimephales promelas*	LC_{50}			290.0		238
F, Methylene chloride	*Pimephales promelas* (12°C, pH 8)	LC_{50}			310.0		1048
F, Aquathol-K di-K	*Pimephales promelas*	LC_{50}			320.0		238
F, Ammonium fluoride	*Pimephales promelas*	LC_{50}	438.0	417.0	364.0		93
F, Diflubenzuron	*Pimephales promelas* (S)	LC_{50}			430.0		147
F, Daconate (technical formulation of MSMA)	*Pimephales promelas* (12°C, softW)	LC_{50}			448.0		1087
F, Benzyl alcohol	*Pimephales promelas* (S, lakeW)	LC_{50}	770.0	770.0	460.0	770.0 (1 h)	196
F, Diethylene glycol dinitrate	*Pimephales promelas*	LC_{50}			491.4		1088
F, Thioacetamide	*Pimephales promelas* (S, 21–23°C, pH 7.2–7.9)	LC_0			600.0		704
F, Endothal-diNa	*Pimephales promelas* (25°C, pH 8.5)	LC_{50}	680.0	660.0	610.0		968
F, *N*-Nitrosodiethylamine	*Pimephales promelas* (S, 22–23°C, pH 8.25)	LC_{50}			775.0		567
F, Adiponitrile	*Pimephales promelas* (HW)	TL_m			820.0		22
F, 2-Butanone oxime	*Pimephales promelas* (F, 24–26°C, pH 7.5)	LC_{50}			844.0		1049
F, *N*-Nitrosodiethylamine	*Pimephales promelas* (S, 22–23°C, pH 8.25)	LC_{50}			940.0		567
F, Acetonitrile	*Pimephales promelas* (softW)	TL_m			1000.0		22
F, Acetonitrile	*Pimephales promelas* (hardW)	TL_m			1020.0		22

Chemical	Species	Toxicity test	24 h	48 h	96 h	Chronic exposure	Ref.
F, Cyclohexanol	*Pimephales promelas* (S, LakeW, 18-22°C)	LC_{50}	1033.0	1033.0		1033.0 (72 h)	196
F, Monosodium methanearsonate, (MSMA), TG	*Pimephales promelas* (12°C, softW)	LC_{50}			1210.0		1087
F, Adiponitrile	*Pimephales promelas* (softW)	TL_m			1250.0		22
F, Fire-trol 100	*Pimephales promelas* (S)	LC_{50}			1500.0		151
F, Fire-trol 931	*Pimephales promelas* (S)	LC_{50}			1500.0		151
F, *n*-Butanol	*Pimephales promelas* (riverW, S)	LC_{50}	1950.0	1950.0	1910.0	1950.0 (72 h)	196
F, *sec*-Butyl alcohol	*Pimephales promelas* (18–22°C)	LC_{50}			1940.0	1940.0 (1 h)	1050
F, *n*-Butanol	*Pimephales promelas* (lakeW, S)	LC_{50}	1940.0	1940.0	1940.0	1940.0 (72 h)	196
F, *n*-Butyl alcohol	*Pimephales promelas* (S, 18–22°C)	LC_{50}			1950.0		1050
F, Trichloroacetic acid	*Pimephales promelas*	LC_{50}			2000.0		73
F, Tetrahydrofuran	*Pimephales promelas* (30d old, 24–26°C, pH 7.5)	LC_{50}			2160.0		1051
F, 1,2-Ethanediol	*Pimephales promelas* (S, 22°C, pH 7.0)	LC_{50}			10,000.0		390
F, Isopropanol	*Pimephales promelas* (S, lakeW, 22°C)	LC_{50}	11,160.0	11,130.0	11,130.0		196
F, Ethanol	*Pimephales promelas* (S, lakeW, 18–22°C)	LC_{50}	18,000.0	13,480.0	13,480.0	13,480.0 (72 h)	196
F, Hydrogen cyanide	Pinperch (salt water)	TL_m	0.05				261
F, Hydrogen cyanide	Pinperch	TL_m	0.069				22
F, Lactonitrile	Pinperch (24 h seawater)	TL_m	0.215				261
F, Acrylonitrile	Pinperch (seawater)	TL_m	24.5				261
F, Acetaldehyde	Pinperch	TL_m	70.0				261
F, 1,3-Butadiene	Pinperch	TL_m	71.5				22
F, 1,1-Dichloroethane	Pinperch	TL_m	160.0				51
F, Acrylonitrile	Pinperch (soft water)	TL_m			24.5		22
F, Cupric sulfate	*Platichthys flesus* (S, 15°C)	LC_{50}		1.0–3.3			443
F, 1,2-Dichloroethane	*Pleuronectes platessa* (S, renewal 6 h, 13–18°C)	LC_{50}		9.0			572
F, Tetramethyllead	*Pleuronectes platessa*	LC_{50}			0.05		114
F, Tetraethyllead	*Pleuronectes platessa*	LC_{50}			0.23		114

F, Triethyllead chloride	*Pleuronectes platessa*	LC_{50}		1.7 (Pb)		114
F, Swedish EDC tar	*Pleuronectes platessa*	LC_{50}	9.0			572
F, Trimethyllead chloride	*Pleuronectes platessa*	LC_{50}		24.60 (Pb)		114
F, Diethyllead chloride	*Pleuronectes platessa*	LC_{50}		75.0		114
F, Dimethyllead chloride	*Pleuronectes platessa*	LC_{50}		300.0		114
F, Cobalt chloride	*Pleuronectes platessa* (S, 15°C, pH 7.8)	LC_{50}		454.0–680.0		777
F, DC-X2-3168	*Pleuronectes platessa* (F, 10–14°C, pH 7–8)	LC_{50}		3110.0		1054
F, DC-silicone fluid	*Pleuronectes platessa* (S, 14°C, pH 8)	NOEL		10,000.0		1054
F, Dieldrin	*Poecilia latipinna*	LC_{50}			0.003 (19 wk)	262
F, α,α′-Dichloro-*m*-xylene	*Poecilia reticulata*	LC_{50}			0.12 (14 d)	2
F, Hexachlorobenzene	*Poecilia reticulata*	LD_{50}			0.32 (14 d)	207
F, Hexachlorobutadiene	*Poecilia reticulata*	LC_{50}			0.4 ppm (14 d)	207
F, *p*-Dichlorobenzene	*Poecilia reticulata*	LC_{50}			4.0 (14 d)	207
F, 2,4-Dichlorotoluene	*Poecilia reticulata*	LC_{50}			4.6 (14 d)	207
F, 3,4-Dichlorotoluene	*Poecilia reticulata*	LC_{50}			5.0 (7 d)	207
F, *o*-Dichlorobenzene	*Poecilia reticulata*	LC_{50}			5.9 (14 d)	207
F, *m*-Dichlorobenzene	*Poecilia reticulata*	LC_{50}			7.4 (14 d)	207
F, 1,5-Dichloropentane	*Poecilia reticulata*	LC_{50}			11.0 (7 d)	84
F, 2,4-Dichloroaniline	*Poecilia reticulata*	LC_{50}			11.7 (14 d)	207
F, Disopropyl amine	*Poecilia reticulata* (semistatic)	NOEL			32.0 (4 wk)	480
F, *trans*-1,4-Dichlorobutene	*Poecilia reticulata*	LC_{50}			40.0 (7 d)	207
F, Benzene	*Poecilia reticulata*	LC_{50}			63.0 (14 d)	207
F, 1,3-Dichloropropane	*Poecilia reticulata*	LC_{50}			84.0 (7 d)	207
F, *n*-Butyl chloride	*Poecilia reticulata*	LC_{50}			97.0 (7 d)	207
F, 2,3-Dichloro-1-propene	*Poecilia reticulata*	LC_{50}			100.0 (7 d)	207
F, 1,2-Dichloropropane	*Poecilia reticulata*	LC_{50}			116.0 (7 d)	207
F, 1,1-Dichloroethane	*Poecilia reticulata*	LC_{50}			202.0 (7 d)	207
F, Butyl cellosolve	*Poecilia reticulata*	LC_{50}			983.0 (7 d)	207
F, *t*-Butanol	*Poecilia reticulata*	LC_{50}			3550.0 (7 d)	207
F, Acetone	*Poecilia reticulata*	LC_{50}			7032.0 (14 d)	207

Chemical	Species	Toxicity test	24 h	48 h	96 h	Chronic exposure	Ref.
F, Ethanol	*Poecilia reticulata*	LC_{50}				11,050.0	207
F, Cellosolve	*Poecilia reticulata*	LC_{50}				16,400.0 (7 d)	207
F, Diethylene glycol	*Poecilia reticulata*	LC_{50}				61,072.0	207
F, 2,5-Dichlorophenol	*Poecilia reticulata*	LC_{50}	2.7				207
F, 2,4-Dichlorophenol	*Poecilia reticulata*	LC_{50}	4.2				207
F, Tricresyl phosphate	*Poecilia reticulata*	NOAE			1.0	1.0 (4 wk)	480
F, Tricresyl phosphate	*Poecilia reticulata*	LC_{50}			4.0		480
F, Sodium arsenite	*Poecilia latipinna* (S, 23–25°C)	LC_{50}		15.0	12.5		1053
F, Diisopropyl amine	*Poecilia reticulata* (semistatic)	NOEL			56.0		480
F, Sodium arsenate	*Poecilia latipinna* (S, 23–25°C)	LC_{50}		70.0	64.0		480
F, Nitrilotriacetic acid	*Poecilia reticulata*	NOEL			100.0		372
F, Dimethylamine	*Poecilia reticulata* (semistatic)	LC_{50}			210.0		936
F, 2,2-Dichloropropionic acid	*Poecilia reticulata* (S)	LC_{50}			223.0		158
F, Nitrilotriacetic acid	*Poecilia reticulata* (S, 23°C)	LC_{50}			560.0–1000.0		372
F, Diisopropyl amine	*Poecilia reticulata* (semistatic)	LC_{100}			1700.0		480
F, Monocrotophos	*Puntius conchonius* (pH 7.5)	LC_{50}			0.160		1052
F, New blitane	*Rasbora heteromorpha* (F)	LC_{10}	8.20				145
F, Dactinol	*Rasbora heteromorpha* (F)	LC_{50}	9.5				145
F, Diallate	*Rasbora heteromorpha*	LC_{50}	12.0				1055
F, Polyram	*Rasbora heteromorpha* (F)	LC_{50}	1000.0				145
F, NRDC 107 (pyrethroids)	*Rasbora heteromorpha* (F)	LC_{10}	0.018	0.015			145
F, Triphenyltin hydroxide	*Rasbora heteromorpha* (F)	LC_{10}	0.038	0.024			145
F, Triphenyltin hydroxide	*Rasbora heteromorpha* (F)	LC_{50}	0.062	0.042			145
F, Frescon, paste	*Rasbora heteromorpha* (F)	LC_{10}	0.078	0.094			145
F, Captan (89% active)	*Rasbora heteromorpha* (F)	LC_{10}	0.23	0.14			145
F, Frescon, emulsified concentrate	*Rasbora heteromorpha* (F)	LC_{10}	0.2	0.17			145
F, Gusathion	*Rasbora heteromorpha* (F)	LC_{10}	0.27	0.22			145
F, Cutrine	*Rasbora heteromorpha* (F)	LC_{10}	0.7	0.26			145
F, Aroclor 1232	*Rasbora heteromorpha* (F)	LC_{10}	0.52	0.27			145

F, Aroclor 1242	*Rasbora heteromorpha* (F)	LC_{10}	0.63	0.275			145
F, Busan 77	*Rasbora heteromorpha* (F)	LC_{10}	0.47	0.32			145
F, Roccal	*Rasbora heteromorpha* (F)	LC_{10}	1.85	0.59			145
F, 2,4,6-Trichlorophenyl-4′-nitrophenyl ether	*Rasbora heteromorpha*	LC_{10}	1.4	0.66			145
F, Balan	*Rasbora heteromorpha*	LC_{10}	1.0	0.95			145
F, Chlormephos	*Rasbora heteromorpha* (F)	LC_{10}	3.5	2.8			145
F, Alkyl ethoxysulfate, C = 12–15	*Rasbora heteromorpha*	LC_{50}		3.9			180
F, Cyanox	*Rasbora heteromorpha* (F)	LC_{10}	20.0	6.7 (40% active)			145
F, Cyanox	*Rasbora heteromorpha* (F)	LC_{50}	36.0	14.0 (40% active)			145
F, Propoxur	*Rasbora heteromorpha* (softW)	LC_{50}		15.0			816
F, Polyram	*Rasbora heteromorpha*	LC_{50}	32.0	17.0			28
F, Propoxur	*Rasbora heteromorpha* (hardW)	LC_{50}	33.0	28.0			816
F, Epichlorhydrin	*Rasbora heteromorpha* (F, 20°C, pH 7.2)	LC_{50}		36.0			1055
F, Epichlorhydrin	*Rasbora heteromorpha* (static and flow-through)	LC_{50}		36.0			96
F, Dosanex	*Rasbora heteromorpha* (F)	LC_{10}	105.0	37.0			145
F, Picloram K Salt	*Rasbora heteromorpha* (20°C)	LC_{50}	66.0	44.0			1055
F, Diuron	*Rasbora heteromorpha* (F)	LC_{10}	110.0	150.0			145
F, Diuron	*Rasbora heteromorpha* (F)	LC_{50}	200.0	190.0			145
F, Thion	*Rasbora heteromorpha* (F)	LC_{10}	0.047	0.024	0.013		145
F, NRDC 107 (pyrethroids)	*Rasbora heteromorpha* (F)	LC_{50}	0.025	0.025	0.014	0.01 (3 m)	145
F, Thion	*Rasbora heteromorpha* (F)	LC_{50}	0.09	0.045	0.022	0.001 (3 m)	145
F, Dihydroheptachlor	*Rasbora heteromorpha* (F)	LC_{10}	0.05	0.036	0.031		145
F, Dinoterpacetate	*Rasbora heteromorpha* (F)	LC_{10}	0.045	0.038	0.031		145
F, Busan 72	*Rasbora heteromorpha* (F)	LC_{10}	0.08	0.044	0.031		145
F, Busan 74	*Rasbora heteromorpha* (F)	LC_{10}	0.12	0.052	0.035		145
F, Busan 72	*Rasbora heteromorpha* (F)	LC_{50}	0.13	0.075	0.036	0.006 (3 m)	145
F, Dinoterpacetate	*Rasbora heteromorpha* (F)	LC_{50}	0.068	0.051	0.039	0.03 (3 m)	145
F, Dihydroheptachlor	*Rasbora heteromorpha* (F)	LC_{50}	0.071	0.056	0.044	0.04 (3 m)	145

Chemical	Species	Toxicity test	24 h	48 h	96 h	Chronic exposure	Ref.
F, Busan 74	*Rasbora heteromorpha* (F)	LC_{50}	0.21	0.084	0.045	0.001 (3 m)	145
F, Frescon, paste	*Rasbora heteromorpha* (F)	LC_{50}	0.135	0.115	0.08		145
F, Frescon, emulsified concentrate	*Rasbora heteromorpha* (F)	LC_{50}	0.33	0.28	0.1		145
F, Gusathion	*Rasbora heteromorpha* (F)	LC_{50}	0.45	0.34	0.17	0.1 (3 m)	145
F, Busan 77	*Rasbora heteromorpha* (F)	LC_{50}	0.66	0.39	0.17	0.01 (3 m)	145
F, Cutrine	*Rasbora heteromorpha* (F)	LC_{50}	1.2	0.35	0.24	0.01 (3 m)	145
F, Busan 76	*Rasbora heteromorpha* (F)	LC_{10}	0.31	0.29	0.29		145
F, Melprex 65	*Rasbora heteromorpha* (F)	LC_{10}	1.3	0.43	0.29		145
F, Captan (89%)	*Rasbora heteromorpha* (F)	LC_{50}	0.46	0.33	0.3	0.2 (3 m)	145
F, Busan 76	*Rasbora heteromorpha* (F)	LC_{50}	0.47	0.35	0.31	0.26 (3 m)	145
F, Aroclor 1232	*Rasbora heteromorpha* (F)	LC_{50}	0.9	0.56	0.32	0.03 (m)	145
F, Busan 25	*Rasbora heteromorpha* (F)	LC_{10}	0.6	0.43	0.34		145
F, Busan 70	*Rasbora heteromorpha* (F)	LC_{10}	0.48	0.37	0.36		145
F, Aroclor 1242	*Rasbora heteromorpha* (F)	LC_{50}	0.96	0.6	0.37	0.05 (3 m)	145
F, Busan 25	*Rasbora heteromorpha* (F)	LC_{50}	1.0	0.57	0.42	.07 (3 m)	145
F, Busan 70	*Rasbora heteromorpha* (F)	LC_{50}	0.76	0.47	0.43	.3 (3 m)	145
F, Alkylolefin sulfonate, C = 16–18	*Rasbora heteromorpha*	LC_{50}		0.9	0.5		180
F, Aroclor 1254	*Rasbora heteromorpha* (F)	LC_{10}	1.6	0.82	0.56		145
F, Chandor[trifluralin (24%) + linuron (12%)]	*Rasbora heteromorpha* (F)	LC_{10}	0.87	0.58	0.58		145
F, Chandor[trifluralin (24%) + linuron (12%)]	*Rasbora heteromorpha* (F)	LC_{50}	1.1	0.74	0.6	0.3 (3 m)	145
F, Melprex 65	*Rasbora heteromorpha* (F)	LC_{50}	1.7	0.82	0.60	0.10 (3 m)	145
F, Roccal	*Rasbora heteromorpha* (F)	LC_{50}	2.45	1.10	0.62	0.04 (3 m)	145
F, Alkylbenzene sulfonate linear, C = 10–15 (46.7% active material)	*Rasbora heteromorpha*	LC_{50}		0.9	0.70		180
F, 2,4,6-Trichlorophenyl-4′-nitro-phenyl ether	*Rasbora heteromorpha*	LC_{50}	2.3	1.3	0.77	0.08 (3 m)	145
F, Bidisin	*Rasbora heteromorpha*	LC_{10}	1.2	1.3	0.8		145

F, Aroclor 1221	*Rasbora heteromorpha* (F)	LC_{10}	1.1	1.05	0.98		145
F, Nalfloc-N206	*Rasbora heteromorpha* (F)	LC_{10}	1.35	1.30	1.0		145
F, Aroclor 1221	*Rasbora heteromorpha* (F)	LC_{50}	1.3	1.15	1.05	0.5 (3 m)	145
F, Aroclor 1254	*Rasbora heteromorpha* (F)	LC_{50}	6.2	1.45	1.1	0.1 (3 m)	145
F, Bidisin	*Rasbora heteromorpha*	LC_{50}	1.85	1.7	1.1	0.6 (3 m)	145
F, Balan	*Rasbora heteromorpha*	LC_{50}	1.4	1.3	1.2	1.0 (3 m)	145
F, Nalfloc-N206	*Rasbora heteromorpha* (F)	LC_{50}	2.50	1.90	1.40	0.5 (3 m)	145
F, *N*-Dodecyldi (aminoethyl)glycine	*Rasbora heteromorpha* (F)	LC_{10}	6.8	3.7	1.95		145
F, Gamlenoil spill remover	*Rasbora heteromorpha* (F)	LC_{10}	6.0	4.0	2.2		145
F, Ekatin	*Rasbora heteromorpha* (F)	LC_{10}	3.3	2.2	2.3		145
F, Imugan	*Rasbora heteromorpha* (F)	LC_{10}	5.0	3.0	2.4		145
F, Dearcide 706	*Rasbora heteromorpha*	LC_{10}	3.8	2.5	2.5		145
F, Chlormephos	*Rasbora heteromorpha* (F)	LC_{50}	4.8	3.5	2.5	1.5 (3 m)	145
F, 5,5′-Dichloro-2,2′-dihydroxy-diphenyl methane	*Rasbora heteromorpha* (F)	LC_{10}	4.4	3.8	2.7		145
F, Ekatin	*Rasbora heteromorpha* (F)	LC_{50}	4.7	3.7	3.2	1.2 (3 m)	145
F, *N*-Dodecyldi (aminoethyl)glycine	*Rasbora heteromorpha* (F)	LC_{50}	8.7	5.9	3.2	1.0 (3 m)	145
F, Dichlobenil	*Rasbora heteromorpha* (F)	LC_{10}	4.3	3.4	3.3		145
F, Dearcide 706	*Rasbora heteromorpha*	LC_{50}	5.6	4.6	3.3	1.4 (3 m)	145
F, Alkylolefin sulfonate, C = 14–16	*Rasbora heteromorpha*	LC_{50}		4.8	3.3		180
F, 5,5′-Dichloro-2,2′-dihydroxy-diphenyl methane	*Rasbora heteromorpha* (F)	LC_{50}	5.4	4.8	3.6	3.4 (3 m)	145
F, Imugan	*Rasbora heteromorpha* (F)	LC_{50}	7.2	5.2	3.7	1.0	145
F, Alkyl ethoxysulfate, C = 12–14	*Rasbora heteromorpha*	LC_{50}			3.9		180
F, Gamlenoil spill remover	*Rasbora heteromorpha* (F)	LC_{50}	7.5	6.4	4.0	1.0 (3 m)	145
F, Dichlobenil	*Rasbora heteromorpha* (F)	LC_{50}	6.2	4.7	4.2	4.0 (3 m)	145
F, Alkylbenzenesulfonate linear, C = 10–15 (28.5% active material)	*Rasbora heteromorpha*	LC_{50}		5.1	4.60		180
F, Alkylbenzenesulfonate linear, C = 10–15 (15.4% active material)	*Rasbora heteromorpha*	LC_{50}		7.6	6.10		180

Chemical	Species	Toxicity test	24 h	48 h	96 h	Chronic exposure	Ref.
F, Paraquat	*Rasbora trilineata* (S)	LC_{50}			6.99		158
F, Nonylphenol ethoxylate	*Rasbora heteromorpha*	LC_{50}		11.30	8.60		180
F, Cyanatrine	*Rasbora heteromorpha* (F)	LC_{10}	15.0	9.0	9.0		145
F, New blitane	*Rasbora heteromorpha* (F)	LC_{50}	16.50	19.0	9.60		145
F, Copper oxychloride	*Rasbora heteromorpha* (F, 20°C, pH 8.1)	LC_{50}	16.5	19.0	9.6		1056
F, Warfarin	*Rasbora heteromorpha* (S)	LC_{50}	17.0	14.0	12.0		145
F, Cyanatrine	*Rasbora heteromorpha* (F)	LC_{50}	35.0	18.0	15.0	5.0 (3 m)	145
F, Diquat	*Rasbora trilineata* (S)	LC_{50}			29.9		2
F, Giv-Gard DXN	*Rasbora heteromorpha* (F)	LC_{10}	74.0	36.0	34.0		145
F, Dosanex	*Rasbora heteromorpha* (F)	LC_{50}	200.0	54.0	40.0	20.0 (3 m)	145
F, Giv-Gard DXN	*Rasbora heteromorpha* (F)	LC_{50}	92.0	54.0	44.0	30.0 (3 m)	145
F, Chloramine-T	*Rasbora heteromorpha* (F, 20°C, pH 7.9)	LC_{50}	120.0	90.0	84.0		1056
F, Acrylamide	*Rasbora heteromorpha* (F)	LC_{10}	390.0	220.0	103.0		145
F, Carbetamex	*Rasbora heteromorpha* (F)	LC_{10}	170.0	150.0	125.0		145
F, Metribuzin	*Rasbora heteromorpha* (F)	LC_{10}	105.0	130.0	130.0		145
F, Acrylamide	*Rasbora heteromorpha* (F)	LC_{50}	460.0	250.0	130.0	10.0 (3 m, extrapolated)	145
F, 2,2-Dichloropropionic acid	*Rasbora trilineata* (S)	LC_{50}			135.0		2
F, Metribuzin	*Rasbora heteromorpha*	LC_{50}	145.0	140.0	140.0	100.0 (3 m)	145
F, Zineb	*Rasbora heteromorpha* (F)	LC_{10}	380.0	320.0	150.0		145
F, Carbetamex	*Rasbora heteromorpha* (F)	LC_{50}	220.0	190.0	165.0	100.0 (3 m)	145
F, Benzthiazuron	*Rasbora heteromorpha* (F)	LC_{10}	850.0	700.0	200.0		145
F, Zineb	*Rasbora heteromorpha* (F)	LC_{50}	560.0	400.0	250.0	100.0 (3 m)	145
F, Azodrin	*Rasbora heteromorpha* (F)	LC_{10}	580.0	580.0	280.0		145
F, Konsin	*Rasbora heteromorpha* (F)	LC_{10}	470.0	370.0	300.0		145
F, Benzthiazuron	*Rasbora heteromorpha* (F)	LC_{50}	1300.0	920.0	400.0	100.0 (3 m)	145
F, Azodrin	*Rasbora heteromorpha* (F)	LC_{50}	750.0	730.0	450.0	150.0 (3 m)	145
F, Presco 5	*Rasbora heteromorpha* (F)	LC_{10}	600.0	550.0	460.0		145
F, Mecoprop	*Rasbora heteromorpha* (F, 20°C, pH 8.1)	LC_{50}		630.0	560.0	700.0 (25 h)	1056

F, Presco 5	*Rasbora heteromorpha* (F)	LC_{50}	700.0	630.0	560.0	500.0 (3 m)	145
F, Konsin	*Rasbora heteromorpha* (F)	LC_{50}	1000.0	820.0	600.0	100.0 (3 m)	145
F, Imsol A (90% isopropanol)	*Rasbora heteromorpha* (F)	LC_{10}	6000.0	3700.0	1500.0		145
F, Imsol A	*Rasbora heteromorpha* (F)	LC_{50}	7100.0	4900.0	4200.0	2000.0 (3 m)	145
F, Nitrilotriacetic acid, Na, H_2O	*Roccus saxatilis* (S, juv.)	TL_{100}				3000.0 (168 h)	37
F, Nitrilotriacetic acid, Na, H_2O	*Roccus saxatilis* (S, juv.)	TL_{50}				5500.0 (168 h)	37
F, Nitrilotriacetic acid, Na, H_2O	*Roccus saxatilis* (S, juv.)	TL_0				10,000.0 (168 h)	37
F, Malathion	*Roccus saxatilis* (S)	LC_{50}			0.039		2
F, Aldrin	*Roccus americanus* (S)	LC_{50}			0.042		117
F, Dimethyl parathion	*Roccus saxatilis* (S)	LC_{50}			14.0		116
F, 2,4,5-Trichlorophenoxyacetic acid	*Roccus saxatilis* (S)	LC_{50}			14.6		117
F, 2,4-Dichlorophenoxyacetic acid	*Roccus saxatilis* (S)	LC_{50}			70.1		64
F, Nitrilotriacetic acid, Na, H_2O	*Roccus saxatilis* (S, juv.)	TL_{50}			5500.0		37
F, *n*-Heptane	*Rutilus rutilus*	LC_{100}				30.0 (1–4 h)	84
F, *p*-Cresol	*Rutilus rutilus*	LC_{50}	1.07				183
F, Phenol	*Rutilus rutilus*	TL_m	15.0				183
F, *o*-Cresol	*Rutilus rutilus*	TL_m	16.0				183
F, 3,4-Xylenol	*Rutilus rutilus*	TL_m	16.0				184
F, *m*-Cresol	*Rutilus rutilus*	TL_m	23.0				183
F, 2-Methyl-4-chlorophenoxyacetic acid	*Rutilus rutilus*	LC_{50}		200.0–215.0			934
F, Chlordane	*Saccobranchus fossilis* (S)	LC_{50}			0.42		288
F, Ekalux	*Saccobranchus fossilis* (S)	LC_{50}			1.55		288
F, Ekatin	*Saccobranchus fossilis* (S)	LC_{50}			11.0		288
F, Fenitrothion	*Saccobranchus fossilis* (S)	LC_{50}			12.5		288
F, Permethrin	*Salmo salar* (S, juv.)	LTC				0.009	289
F, Dinocap	*Salmo salar* (juv., S)	LC_{100}				0.02	97
F, Aluminum nitrate	*Salmo salar* (S, pH 4.9–5.06)	LC_0				0.038 (20 d)	1058
F, Dinoseb	*Salmo salar* (S)	LC_{100}				0.07	97
F, Aluminum nitrate	*Salmo salar* (S, pH 4.9–5.06)	LC_{50}				0.137 (38 h)	1058
F, Aluminum nitrate	*Salmo salar* (S, pH 4.9–5.06)	LC_{50}				0.177 (32 h)	1058

Chemical	Species	Toxicity test	24 h	48 h	96 h	Chronic exposure	Ref.
F, 4,6-Dinitro-*o*-cresol	*Salmo salar* (juv., S)	LC_{100}				0.2	97
F, Ammonia	*Salmo clarki* (fry, F)	LC_{50}				0.56 (36 d)	1016
F, 2,4-Dinitrophenol	*Salmo salar* (juv., S)	LC_{100}				0.7	97
F, Fluoride	*Salmo trutta* (fry)	NOAE				0.9 (240 h?)	717
F, Hydrogen sulfide	*Salmo clarki*	LC_{100}				1.0 (h?)	84
F, Propoxur	*Salmo aguabonita* (S)	NOEL				1.0 (168 h)	816
F, Trifluoromethyl-4-nitrophenol	*Salmo trutta* (S, green egg, 12°C)	TL_m				1.39 (192 h)	161
F, 3-Bromo-4-nitrophenol	*Salmo trutta*	LD_{10}				5.0	84
F, *o*-Nitrophenol	*Salmo trutta*	NOAE				5.0 (h?)	942
F, Trifluoromethyl-4-nitrophenol	*Salmo trutta* (S, green egg, 12°C)	TL_m				5.0 (192 h)	161
F, 2,5-Dichloronitrophenol	*Salmo trutta*	LD_{10}				7.0	84
F, Thiophanate methyl	*Salmo irideus*	TL_m				8.8	31
F, 2-Bromo-4-nitrophenol	*Salmo trutta*	LD_{10}				11.0	84
F, Benzene	*Salmo trutta* (yearling)	LC_{50}				12.0 (1 h)	159
F, 3,4,6-Chloro-2-nitrophenol	*Salmo trutta*	LD_{10}				15.0	84
F, Fluoride	*Salmo trutta* (fry)	LC_{50}				15.0 (240 h)	717
F, Lauryl sulfate	*Salmo trutta* (median survival time)	MST				18.0 (45 h)	160
F, Polychlorinated dibenzofurans (generic)	*Salmo salar* (juv., in diet, F, 10°C)	LD_{50}				18.4 (90–150 d)	1057
F, 4,6-Dinitro-*o*-*sec*-amylphenol	*Salmo salar* (S)	LC_{100}				30.0	97
F, Lauryl sulfate	*Salmo trutta*	MST				32.0 (32 h)	160
F, Lauryl sulfate	*Salmo trutta*	MST				56.0 (6.5 h)	160
F, Lauryl sulfate	*Salmo trutta*	MST				100.0 (2.15 h)	160
F, Lauryl sulfate	*Salmo trutta*	MST				120.0 (0.86 h)	160
F, Lauryl sulfate	*Salmo trutta*	MST				150.0 (0.26 h)	160
F, Lauryl sulfate	*Salmo trutta*	MST				180.0 (0.15 h)	160
F, Picloram	*Salmo clarki* (fry)	NOEL				290.0 (h?)	1059
F, Lauryl sulfate	*Salmo trutta*	MST				320.0 (0.08 h)	160
F, *dl*-Lactic acid	Salmo irideus	TL_m				400.0 (h?)	2

F, *n*-Butyric acid	Salmo irideus	TL_m				400.0	8
F, Lauryl sulfate	*Salmo trutta*	MST				560.0 (0.07 h)	160
F, Lauryl sulfate	*Salmo trutta*	MST				1000.0 (0.07 h)	160
F, Hydrocyanic acid	*Salmo salar*, smolts (F, 11°C, pH 8, 3.5 mg/l DO)	LC_{50}	0.024				1060
F, Acrolein	*Salmo trutta*	LC_{50}	0.046				1065
F, Hydrocyanic acid	*Salmo salar*, smolts (F, 11°C, pH 8, 10.3 mg/l DO)	LC_{50}	0.073				1060
F, 4-Chloro-*m*-cresol	*Salmo trutta* (S, 5°C)	LC_{50}	1.3				852
F, Propoxur	*Salmo salar* (S, 8°C, pH 6.8)	LC_{50}	6.0				1062
F, Linuron	*Oncorhynchus mykiss*	LC_{100}	10.0				496
F, 4-Chlorophenoxyacetic acid	*Salmo trutta*	LD_{50}	147.0				1066
F, Permethrin	*Salmo salar* (S)	LTC		0.0088			119
F, Pyrethrins	*Salmo salar* (S)	LTC		0.032			119
F, *N*-Tritylmorpholine	*Salmo trutta* (S)	LC_{50}		0.083			30
F, Benzene sulfonyl chloride	*Salmo trutta* (yearling) (S)	LC_{50}		3.0			159
F, Diquat-dibromide	*Salmo trutta* (fingerling)	LC_{25}		17.5			1064
F, Gensol, No.1	*Salmo trutta* (yearlings, S)	LC_{50}		50.0			128
F, Gensol, No.2	*Salmo trutta* (yearlings, S)	LC_{50}		80.0			128
F, Paraquat	*Salmo trutta* (S)	LC_{50}		82.0			158
F, 2-Methy-4-chlorophenoxyacetic acid	*Salmo trutta* (11°C)	LC_{50}		147.0			614
F, Acrylamide	*Salmo trutta* (yearlings, S)	LC_{50}		400.0			159
F, Diquat	*Salmo trutta* (yearlings, S)	LC_{50}		570.0			128
F, Permethrin (*cis*)	*Salmo salar*	LC_{50}			0.0013		623
F, Decamethrin	*Salmo salar* (10°C)	LC_{50}			0.00197		623
F, DDT	*Salmo trutta*	LC_{50}			0.002		127
F, Lindane	*Salmo trutta*	LC_{50}			0.002		127
F, DDT	*Salmo trutta*	LC_{50}			0.002		122
F, Lindane	*Salmo trutta*	LC_{50}			0.002		122
F, Toxaphene	*Salmo trutta*	LC_{50}			0.003		122

Chemical	Species	Toxicity test	24 h	48 h	96 h	Chronic exposure	Ref.
F, Azinphosmethyl	*Salmo trutta*	LC_{50}			0.004		127
F, Leptophos	*Salmo clarki* (S, 10°C, pH 7.5)	LC_{50}			0.0053		678
F, Phorate	*Salmo clarki* (S, 12°C, pH 7.5)	LC_{50}			0.006		599
F, Permethrin (*cis* + *trans*)	*Salmo salar*	LC_{50}			0.0088		623
F, Rotenone (5%)	*Salmo salar* (S)	LC_{50}			0.0215		28
F, Thallium acetate	*Salmo salar*	LD_{50}			0.03		53
F, Dinoseb	*Salmo clarki* (S)	TL_m			0.041–1.35		2
F, Captan	*Salmo clarki* (fingerling, F)	LC_{50}			0.0485		778
F, Aluminum nitrate	*Salmo salar* (S, pH 4.9–5.06)	LC_{50}			0.075		1058
F, Cartap (98%)	*Salmo trutta* (juv., S, 18°C, pH 7.1)	LC_{50}			0.08		868
F, Cartap	*Salmo trutta*	LC_{50}			0.080		1063
F, Alkylbenzene sulfonate (46.7% active)	*Salmo trutta*	LC_{50}		0.2–0.4	0.1–0.5		180
F, Ethyl-*p*-nitrophenylthiono-benzene phosphate	*Salmo clarki* (S, 13°C, pH 7.5)	LC_{50}			0.160		602
F, Formaldehyde	*Salmo salar* (F)	LC_{50}			0.173 ml/l		120
F, Malathion	*Salmo trutta*	LC_{50}			0.200		127
F, Malathion	*Salmo trutta*	LC_{50}			0.200		122
F, Ammonia	*Salmo clarki* (fry, F)	LC_{50}			0.5–0.8		923
F, Alkylolefin sulfonate, C = 16–18	*Salmo trutta*	LC_{50}		0.3–0.6	0.5		180
F, 2,4-Dinitrophenol	*Salmo salar* (juv.)	LC_{100}			0.70		624
F, Ethion	*Salmo clarki*	LC_{50}			0.720		71
F, Alkylbenzene sulfonate (15.4%)	*Salmo trutta*	LC_{50}		2.0–5.3	0.9–4.6		180
F, Nonylphenol ethoxylate	*Salmo trutta*	LC_{50}		2.70	1.0		180
F, Alkyl ethoxysulfate, AES	*Salmo trutta*	LC_{50}		1.4–2.6	1.0–2.5		180
F, Methomyl, technical grade	*Salmo salar* (S, 12°C, pH 7.5)	LC_{50}			1.12		513
F, Aroclor 1221	*Salmo clarki*	LC_{50}			1.2		92
F, Baytex	*Salmo trutta*	LC_{50}			1.3		122
F, Baytex	*Salmo trutta*	LC_{50}			1.33		127

F, Methomyl (liquid concentrate)	*Salmo salar* (S, 12°C, pH 7.5)	LC_{50}			1.40		513
F, ABS (28.5%)	*Salmo trutta*	LC_{50}		0.7-2.3	1.4		180
F, Furanace	*Salmo salar* (S)	LC_{50}			1.41		121
F, Triton X100	*Salmo trutta* (F,12°C, pH 7.81)	LC_{50}	2.15	1.93	1.51	1.13 (144 h)	472
F, Trifluoromethyl-4-nitrophenol	*Salmo trutta* (S, green egg, 12°C)	TL_m			1.52		161
F, Methyl parathion	*Salmo clarki* (fingerling, S, 12°C, pH 7.5)	LC_{50}			1.85		511
F, Carbaryl	*Salmo trutta*	NOEL			1.95		127
F, Aroclor 1232	*Salmo clarki*	LC_{50}			2.5		92
F, Alkylolefin sulfonate, C = 14–16	*Salmo trutta*	LC_{50}		2.5–5.0	2.5–5.0		180
F, Picloram	*Salmo clarki* (S, 10°C)	TL_m			3.45–8.6		185
F, Methyl parathion	*Salmo trutta* (fingerling, S, 12°C, pH 7.5)	LC_{50}			4.70		511
F, Dimethyl parathion	*Salmo trutta*	LC_{50}			4.74		127
F, Picloram	*Salmo clarki* (12°C, pH 7.5)	LC_{50}			4.8		411
F, Picloram	*Salmo clarki* (S, 12°C, pH 7.5)	LC_{50}			4.8		411
F, Monocrotophos (38%)	*Oncorhynchus mykiss* (juv., S, 13°C, pH 7.1)	LC_{50}			5.2		917
F, Aroclor 1242	*Salmo clarki* (S)	LC_{50}			5.4		92
F, Aroclor 1248	*Salmo clarki*	LC_{50}			5.7		92
F, Aminocarb	*Salmo salar* (12°C, pH 7.5)	LC_{50}			7.6		475
F, Zectran	*Salmo trutta*	LC_{50}			8.1		122
F, Zectran	*Salmo trutta*	LC_{50}			8.1		127
F, Aminocarb	*Salmo salar* (juv., S, 10°C, pH 7.5)	LC_{50}			8.7		460
F, Zectran	*Salmo gairdnerii*	LC_{50}			10.2		122
F, Aminocarb	*Salmo trutta* (12°C, pH 7.5)	LC_{50}			15.0		475
F, Tordon 22K	*Salmo trutta* (10°C)	LC_{50}			22.0		803
F, Aminocarb	*Salmo clarki* (S, 10°C, pH 7.5)	LC_{50}			31.0		475
F, Aroclor 1254	*Salmo clarki*	LC_{50}			42.0		92
F, Tordon 22K	*Salmo trutta* (10°C)	LC_{50}	52.0		52.0		803
F, Aroclor 1260	*Salmo clarki* (S)	LC_{50}			61.0		92
F, Diquat	*Salmo trutta* (F)	LC_{50}			300.0		128
F, Dioxane	Saltwater fish	LC_{50}			10,000.0		875

Chemical	Species	Toxicity test	24 h	48 h	96 h	Chronic exposure	Ref.
F, Diazinon	*Salvelinus fontinalis* (F)	MATC				0.0032 (274 d)	156
F, Hydrogen sulfide	*Salvelinus fontinalis* (F, 8–12.5°C)	MATC				0.0055	143
F, Lindane	*Salvelinus fontinalis* (F, 9–16°C)	MATC				0.0088	145
F, Lindane	*Salvelinus fontinalis* (F, 9–16°C)	TL_m				0.026 (1.5 yr)	145
F, Pentachlorophenate, Na	*Salvelinus fontinalis* (F, 15°C)	TL_m				0.118 (219 h)	144
F, Pentachlorophenate, Na	*Salvelinus fontinalis* (F, 15°C)	TL_m				0.118 (336 h)	144
F, Atrazine	*Salvelinus fontinalis*	NOAE				0.72 (44 wk)	542
F, Trifluoromethyl-4-nitrophenol	*Salvelinus namaycush* (S, green egg, 12°C)	TL_m				1.4 (192 h)	161
F, *o*-Nitrophenol	*Salvelinus fontinalis*	NOAE				5.0 (h?)	944
F, Triton X100	*Salvelinus fontinalis* (S, 12°C)	LC_{50}	2.25				472
F, Propoxur	*Salvelinus fontinalis* (S, 8°C, pH 6.8)	LC_{50}	11.2				1062
F, Chlordane	*Salvelinus fontinalis* (F)	LOEL			0.00032		939
F, Diuron	*Salvelinus namaycush* (S, 10°C, pH 7.5)	LC_{50}			0.0027		410
F, Permethrin	*Salvelinus fontinalis* (S, 12°C, pH 7.5)	LC_{50}			0.003		521
F, Bio-allethrin, 90% TG	*Salvelinus namaycush* (F, 12°C, pH 7.5)	LC_{50}			0.016		597
F, Hydrogen sulfide	*Salvelinus fontinalis* (F, 8–12.5°C)	TL_m			0.0216–0.0308		143
F, Rotenone (5%)	*Salvelinus namaycush* (S)	LC_{50}			0.0269		28
F, Dinoseb	*Salvelinus namaycush* (S)	TL_m			0.032–1.40		185
F, Niclosamide	*Salvelinus fontinalis* (F, 12°C)	LC_{50}			0.034		1031
F, Pyrethrum	*Salvelinus namaycush* (12°C)	LC_{50}			0.037		870
F, Lindane	*Salvelinus fontinalis* (F)	LC_{50}			0.0443		529
F, Rotenone (5%)	*Salvelinus fontinalis* (S)	LC_{50}			0.0443		28
F, Chlorodane	*Salvelinus fontinalis*	LC_{50}			0.047		714
F, Chlordane	*Salvelinus fontinalis* (F)	LC_{50}			0.047		939
F, Hydrogen cyanide	*Salvelinus fontinalis* (juv., F)	LC_{50}			0.053-0.143		137
F, Hydrocyanic acid	*Salvelinus fontinalis* (swim-up fry, 10°C, pH 7.8, DO = 3.9 mg/l)	LC_{50}			0.0558		947
F, Hydrogen cyanide	*Salvelinus fontinalis* (swim-up fry, F)	LC_{50}			0.056–0.106		137

F, Hydrocyanic acid	*Salvelinus fontinalis* (swim-up fry, 10°C, pH 7.8, DO = 6 mg/l)	LC_{50}		0.0886	947
F, Formaldehyde	*Salvelinus namaycush* (F)	LC_{50}		0.100ml/l	120
F, Hydrocyanic acid	*Salvelinus fontinalis* (swim-up fry, 10°C, pH 7.8, DO = 8 mg/l)	LC_{50}		0.100	947
F, Hydrocyanic acid	*Salvelinus fontinalis* (sac fry, 10°C, pH 7.8, DO = 3.50 mg/l)	LC_{50}		0.108	947
F, Hydrogen cyanide	*Salvelinus fontinalis* (sac fry, F)	LC_{50}		0.108–0.518	137
F, Pentachlorophenate, Na	*Salvelinus fontinalis* (F, 15°C)	TL_m		0.135	144
F, Hydrogen cyanide	*Salvelinus fontinalis* (eggs, F)	LC_{50}		0.212–0.242	137
F, Hydrocyanic acid	*Salvelinus fontinalis* (F, 7.1°C, pH 7.8)	LC_{50}		0.212	947
F, Hydrocyanic acid	*Salvelinus fontinalis* (sac fry, 10°C, pH 7.8, DO = 6.03 mg/l)	LC_{50}		0.350	947
F, Diazinon	*Salvelinus fontinalis* (F)	LC_{50}		0.45–1.05	141
F, Hydrocyanic acid	*Salvelinus fontinalis* (sac fry, 10°C, pH 7.8, DO = 7.96 mg/l)	LC_{50}		0.518	947
F, Captan	*Salvelinus namaycush* (fingerling, S)	LC_{50}		0.752	778
F, Ferbam	*Salvelinus fontinalis* (fingerling)	LC_{100}		1.0–2.0	871
F, Trifluoromethyl-4-nitrophenol	*Salvelinus namaycush* (S, fingerling, 12°C)	TL_m		1.4	161
F, Methomyl, TG	*Salvelinus fontinalis* (S, 12°C, pH 7.5)	LC_{50}		1.500	513
F, Picloram	*Salvelinus namaycush* (S, 10°C)	TL_m		1.55–4.95	185
F, Triton X100	*Salvelinus fontinalis* (F, 12°C)	LC_{50}	2.57	1.76	472
F, Carbaryl	*Salvelinus fontinalis*	LC_{50}		2.0	122
F, Trifluoromethyl-4-nitrophenol	*Salvelinus namaycush* (S, green egg, 12°C)	TL_m		2.1	161
F, Methomyl, TG (liquid concentrate)	*Salvelinus fontinalis* (S, 12°C, pH 7.5)	LC_{50}		2.200	513
F, Picloram	*Salvelinus namaycush* (S)	LC_{50}		4.3	411
F, Trifluoromethyl-4-nitrophenol	*Salvelinus fontinalis* (F)	LC_{50}		5.95	157
F, Atrazine	*Salvelinus fontinalis* (F, 12–14°C)	LC_{50}		6.3	535
F, Aminocarb	*Salvelinus fontinalis* (12°C, pH 7.5)	LC_{50}		16.0	475
F, Picloram	*Salvelinus fontinalis* (10°C)	LC_0		69.0	803

Chemical	Species	Toxicity test	24 h	48 h	96 h	Chronic exposure	Ref.
F, Picloram	*Salvelinus fontinalis* (10°C)	LC_{50}	91.0		91.0		803
F, Bladex	*Sarotherodon mossambicus*	LD_{50}			24.5		172
F, Bladex	*Sarotherodon mossambicus*	LD_{100}			64.0		172
F, Phenol	*Semotilus atromaculatus* (riverW)	LD_0				10.0	21
F, Phenol	*Semotilus atromaculatus* (riverW)	LD_{100}				20.0	21
F, Ethylamine	*Semotilus atromaculatus* (riverW)	LD_0				30.0	21
F, Ethylamine	*Semotilus atromaculatus* (riverW)	LD_{100}				50.0	21
F, Isobutylamine	*Semotilus atromaculatus* (riverW)	LD_0	20.0				21
F, Ethylenediamine	*Semotilus atromaculatus*	LC_{50}	30.0–60.0				20
F, *n*-Amylamine	*Semotilus atromaculatus*	LD_0	30				21
F, Tri-*n*-propylamine	*Semotilus atromaculatus* (riverW)	LD_0	30.0				21
F, *n*-Amylamine	*Semotilus atromaculatus* (riverW)	LD_0	30.0				21
F, Isopropylamine	*Semotilus atromaculatus*	LC_{50}	40.0–80.0				84
F, Diisopropylamine	*Semotilus atromaculatus*	LD_0	40.0				21
F, Triethylamine	*Semotilus atromaculatus* (riverW)	LD_0	50.0				21
F, *n*-Amyl acetate	*Semotilus atromaculatus* (riverW)	LD_0	50.0				21
F, *n*-Amylamine	*Semotilus atromaculatus*	LD_{100}	50.0				21
F, *n*-Amylamine	*Semotilus atromaculatus* (riverW)	LD_{100}	50.0				21
F, Diisopropylamine	*Semotilus atromaculatus*	LD_{100}	60.0				21
F, Isobutylamine	*Semotilus atromaculatus* (riverW)	LD_{100}	60.0				21
F, Diethylamine	*Semotilus atromaculatus* (riverW)	LD_0	70.0				21
F, Tri-*n*-propylamine	*Semotilus atromaculatus* (riverW)	LD_{100}	70.0				21
F, Triethylamine	*Semotilus atromaculatus* (riverW)	LD_{100}	80.0				21
F, Acetic acid	*Semotilus atromaculatus* (riverW)	LD_0	100.0				21
F, Diethylamine	*Semotilus atromaculatus* (riverW)	LD_{100}	100.0				21
F, Acetic acid	*Semotilus atromaculatus* (riverW)	LDo	100.0				21
F, *n*-Amyl acetate	*Semotilus atromaculatus* (riverW)	LD_{100}	120.0				21
F, *n*-Propanol	*Semotilus atromaculatus* (riverW)	LD_0	200.0				8
F, Acetic acid	*Semotilus atromaculatus* (riverW)	LD_{100}	200.0				21

F, Acetic acid	*Semotilus atromaculatus* (riverW)	LD_{100}	200.0				21
F, *n*-Pentanol	*Semotilus atromaculatus* (riverW)	LD_0	350.0				21
F, *n*-Propanol	*Semotilus atromaculatus* (riverW)	LD_{100}	500.0				8
F, *n*-Pentanol	*Semotilus atromaculatus* (riverW)	LD_{100}	500.0				21
F, Isopropanol	*Semotilus atromaculatus* (riverW)	LD_0	900.0				21
F, *n*-Butanol	*Semotilus atromaculatus*	LD_0	1000.0				21
F, Isopropanol	*Semotilus atromaculatus* (riverW)	LD_{100}	1100.0				21
F, *n*-Butanol	*Semotilus atromaculatus*	LD_{100}	1400.0				21
F, *t*-Butanol	*Semotilus atromaculatus* (riverW)	LD_0	3000.0				21
F, *t*-Butanol	*Semotilus atromaculatus* (riverW)	LD_{100}	6000.0				21
F, Ethanol	*Semotilus atromaculatus*	LC_{50}	7000.0				182
F, Ethanol	*Semotilus atromaculatus* (riverW)	LD_0	7000.0				21
F, Ethanol	*Semotilus atromaculatus* (riverW)	LD_{100}	9000.0				21
F, *n*-Propanol	*Semotilus atromaculatus* (S, 15–21°C, pH 8.3)	LC_0				200.0 (h?)	1067
F, *n*-Propanol	*Semotilus atromaculatus* (S, 15–21°C, pH 8.3)	LC_{100}				500.0 (h?)	1067
F, Diethyl phthalate	*Cyprinodon variegatus* (S, juv., 25–31°C)	LC_{50}	69.0		30.0		842
F, Rotenone (5%)	*Sitzostedion vitreum* (S)	LC_{50}			0.0165		28
F, Tributyltin oxide	*Solea solea* (semiS, larvae)	LC_{50}			0.0021		765
F, Tributyltin oxide	*Solea solea* (semistatic)	LC_{50}			0.036		765
F, Endrin	*Sphaeroides maculatus* (S)	LC_{50}			0.0031		115
F, Dieldrin (100%)	*Sphaeroides maculatus* (S)	LC_{50}			0.034		115
F, Lindane	*Sphaeroides maculatus* (S)	LC_{50}			0.035		115
F, Aldrin	*Sphaeroides maculatus*	LC_{50}			0.036		115
F, DDT (*p,p'*)	*Sphaeroides maculatus* (S)	LC_{50}			0.089		209
F, Heptachlor	*Sphaeroides maculatus* (S)	LC_{50}			0.188		115
F, 2,2-Dichlorovinyldimethyl phosphate	*Sphaeroides maculatus* (S)	LC_{50}			2.25		115
F, Malathion	*Sphaeroides maculatus* (S)	LC_{50}			3.250		115
F, Dimethyl parathion	*Sphaeroides maculatus* (S)	LC_{50}			75.8		60
F, *dl*-Lactic acid	*Squalis leuciscus*	TL_m				1000.0 (h?)	2
F, Nitrilotriacetic acid, Na, H_2O	*Stenotomus chrysops* (S)	TL_{50}				3150.0 (168 h)	37

Chemical	Species	Toxicity test	24 h	48 h	96 h	Chronic exposure	Ref.
F, Pydraul 50E	*Stenotomus chrysops* (S)	LC_{50}	1.5		0.56		139
F, Houghtosafe	*Stenotomus chrysops* (S)	LC_{50}	3.1		0.7		139
F, Pydraul 50E	*Stenotomus chrysops* (F)	LC_{50}			1.5		139
F, Nitrilotriacetic acid, Na, H_2O	*Stenotomus chrysops* (S)	TL_{50}			3150.0		37
F, Endrin	*Stizostedion lucioperca* (S)	LC_{50}	0.0075			0.0048 (72 h)	162
F, Ethyl-*p*-nitrophenylthionobenzene phosphate	*Stizostedion vitreum* (S, 18°C, pH 7.5)	LC_{50}			0.35		602
F, 1,3-Dichloropropane	*Stizostedion vitreum* (S, 18°C, pH 7.5)	LC_{50}			1.08		758
F, Diquat	*Stizostedion vitreum*	LC_{50}			2.1		194
F, Endrin	*Fundulis majalis* (S)	LC_{50}			0.0003		115
F, Endrin	*Thalassoma bifasciatum* (S)	LC_{50}			0.0001		115
F, Heptachlor	*Thalassoma bifasciatum* (S)	LC_{50}			0.0008		115
F, Dieldrin (100%)	*Thalassoma bifasciatum* (S)	LC_{50}			0.006		115
F, DDT (*p,p′*)	*Thalassoma bifasciatum* (S)	LC_{50}			0.007		209
F, Aldrin	*Thalassoma bifasciatum*	LC_{50}			0.012		115
F, Lindane	*Thalassoma bifasciatum* (S)	LC_{50}			0.014		115
F, Malathion	*Thalassoma bifasciatum* (S)	LC_{50}			0.027		115
F, 2,2-Dichlorovinyldimethyl phosphate	*Thalassoma bifasciatum* (S)	LC_{50}			1.44		115
F, Dimethyl parathion	*Thalassoma bifasciatum* (S)	LC_{50}			12.3		117
F, Fluoride	*Therapan jarbua*	NOAE			100.0		717
F, Mercuric chloride	*Therapon jarbua* (juv., S, seaW)	LC_{90}				0.018 (18 h)	1068
F, Mercuric chloride	*Therapon jarbua* (juv., S, seaW)	LC_{100}				0.100 (18 h)	1068
F, Cupric sulfate	*Therapon jarbua* (F, pH 8.2)	LC_{50}	4.0				823
F, Arsenic trioxide	*Therapon jarbua* (seaW, juv., S)	LC_0		1.4			1068
F, Mercuric chloride	*Therapon jarbua* (adult, seaW)	LC_{50}	0.071	0.071	0.06	0.07 (72 h)	1068
F, Arsenic trioxide	*Therapon jarbua* (seaW, juv., S)	LC_{10}			1.03		1068
F, Arsenic trioxide	*Therapon jarbua* (seaW, juv., S)	LC_{50}	5.5	4.08	3.38	3.68 (72 h)	1068
F, Arsenic trioxide	*Therapon jarbua* (seaW, juv., S)	LC_{100}			6.2		1068

F, Tributyltin oxide	*Tilapia rendalli*	EC_{50}	0.0532				292
F, Decamethrin	*Tilapia mossambica* (S)	LC_{50}		0.0008			188
F, Permethrin	*Tilapia mossambica* (S)	LC_{50}		0.0440			188
F, Nitrofen (TOK, 50% N_2)	*Tilapia mossambica*	LC_{50}		1.02			900
F, Rotenone	*Tilapia sparmanii*	LC_{50}			0.0016–0.012		557
F, Niclosamide	*Tilapia leucosticta* (S, 25°C, 9-d-old larvae)	LC_{50}			1.5		1069
F, Niclosamide	*Tilapia leucosticta* (S, 25°C, 3-d-old eggs)	LC_{50}			8.4–11.4		1069
F, Monocrotophos	*Tilapia mossambica* (27°C, pH 5.7–6.7)	LC_{100}			17.38–18.62		1070
F, Sulfur (H_2SO_3)	*Tinca tinca*	LC_{100}				1.0 (2 h)	874
F, Hydrogen sulfide	*Tinca tinca*	LC_{100}				100.0 (3 h)	84
F, Sodium sulfate	*Tinca tinca* (S)	NOEL				500.0 (4 wk)	791
F, 2,4-Xylenol	*Tinca tinca*	TL_m	13.0				184
F, *o*-Cresol	*Tinca tinca*	TL_m	15.0				183
F, *p*-Cresol	*Tinca tinca*	LC_{50}	16.0				183
F, Phenol	*Tinca tinca*	TL_m	17.0				183
F, 3,4-Xylenol	*Tinca tinca*	TL_m	18.0				184
F, *m*-Cresol	*Tinca tinca*	TL_m	21.0				183
F, 3,5-Xylenol	*Tinca tinca*	TL_m	52.0				183
F, Cupric sulfate	*Trachinotus carolinus* (S, juv., 20–25°C)	LC_{50}			2.0		1070
F, Endrin	*Tinca tinca* (juv., S)	LC_{50}	0.0026			0.0091, 0.016 (72 h)	162
F, Malathion	*Umbra pygmaeo* (S, 16°C)	TL_m			0.24	0.14 (14 d) (F)	192
F, Diethylfumarate	*Umbra pygmaeo* (F)	TL_m			4.2		192
F, Diethylfumarate	*Umbra pygmaeo* (S)	TL_m			8.5		192
F, *p*-Nitrophenol	Vairon (F, distilledW)	TL_m				4.0 (6 h)	8
F, *m*-Nitrophenol	Vairon (F, distilledW)	TL_m				9.0 (6 h)	8
F, *o*-Nitrophenol	Vairon (F, distilledW)	TL_m				14.0 (6 h)	8
F, Nitrobenzene	Vairon (F, distilledW)	TL_m				20.0 (6 h)	8
F, *m*-Nitrophenol	Vairon (F, hardW)	TL_m				20.0 (6 h)	8
F, *p*-Nitrophenol	Vairon (F, hardW)	TL_m				30.0 (6 h)	8
F, Nitrobenzene	Vairon (F, hardW)	TL_m				90.0 (6 h)	8

Chemical	Species	Toxicity test	24 h	48 h	96 h	Chronic exposure	Ref.
F, *o*-Nitrophenol	Vairon (F, hardW)	TL_m				125.0 (6 h)	8
F, Linuron	*Xiphorus helleri*	LC_0	1.5				496
F, Linuron	*Xiphorus helleri*	LC_{50}	4.6				496
F, Monolinuron	*Xiphorus helleri*	LC_0	12.0				496
F, Linuron	*Xiphorus helleri*	LC_{100}	15.0				496
F, Monolinuron	*Xiphorus helleri*	LC_{50}	20.0				496
F, Monolinuron	*Xiphorus helleri*	LC_{100}	30.0				496
F, Acetanilide	*Lepomis macrochirus* (freshW, 23°C, S)	LC_{50}			100.0		224
F, Fenpropanate (TG)	*Onchorhynchus mykiss* (S)	LC_{50}	0.0767				266
F, Baytex	*Perca flavescens*	LC_{50}			1.7		122
F, Acetophenone	*Pimephales promelas* (S, lakeW, 18–22°C)	LC_{50}	200.0	103.0	155.0	200.0 (1 h)	196
I, Parathion	*Acroneuria lycorias*	LC_{50}				0.000013 (30 d)	91
I, Malathion	*Acroneuria lycorias*	NOEL				0.00017 (30 d)	91
I, Diazinon	*Acroneuria lycorias*	NOEL				0.00083 (30 d)	91
I, Carbaryl	*Acroneuria lycorias*	NOEL				0.0013 (30 d)	91
I, Azinphosmethyl	*Acroneuria lycorias*	LC_{50}				0.0015	91
I, Carbaryl	*Acroneuria lycorias*	LC_{50}				0.0022 (30 d)	91
I, Endrin	*Acroneuria pacifica*	LC_{50}			0.00032	0.00003	70
I, Malathion	*Acroneuria lycorias*	LC_{50}			0.001	0.0003 (30 d)	91
I, Diazinon	*Acroneuria lycorias*	LC_{50}			0.0017	0.00125 (30 d)	91
I, Parathion	*Acroneuria pacifica*	LC_{50}			0.0030	0.00044 (30 d)	297
I, Disulfoton	*Acroneuria pacifica*	LC_{50}			0.0082	0.0014 (30 d)	297
I, Chlorfos	*Acroneuria pacifica*	LC_{50}			0.0165	0.0087 (30 d)	297
I, Dieldrin	*Acroneuria pacifica*	LC_{50}			0.024	0.0002 (30 d)	298
I, Photo-aldrin	*Aedes aegypti* (larvae 3rd instar)	LC_{50}	0.0005				44
I, Photoheptachlor	*Aedes aegypti* (larvae 3rd instar)	LC_{50}	0.002				44
I, Photodieldrin	*Aedes aegypti* (larvae 3rd instar)	LC_{50}	0.003				44
I, Aldrin	*Aedes* sp. (late 3rd instar larvae)	LC_{50}	0.003				44
I, Heptachlor	*Aedes aegypti* (late 3rd instar larvae)	LC_{50}	0.005				2

I, Dieldrin	*Aedes aegypti* (3rd instar)	LC_{50}	0.006			2
I, Isodrin (aldrin isomer)	*Aedes aegypti* (late 3rd instar larvae)	LC_{50}	0.019			44
I, Photoisodrin	*Aedes aegypti* (larvae 3rd instar)	LC_{50}	0.019			44
I, Chlordene	*Aedes aegypti* (larvae)	LC_{50}	0.130			44
I, Photochlordene	*Aedes aegypti* (larvae 3rd ins)	LC_{50}	0.150			44
I, Isodrin/photoisodrin toxicity ratio	*Aedes aegypti* (larvae)	TR	0.30			44, 48, 49
I, Microencapsulated permethrin	*Aedes aegypti* (3rd instar), 1 + 1 h exp.	LC_{50}			0.1799	1081
I, Microencapsulated permethrin	*Aedes aegypti* (3rd instar), 2 h exp.	LC_{50}			0.2498	1081
I, Carbaryl	*Aedes aegypti* (3rd instar), 1 + 1 h exp.	LC_{50}			3.040	1081
I, Carbaryl	*Aedes aegypti* (3rd instar), 2 h exp.	LC_{50}			3.470	1081
I, Permethrin, technical	*Aedes aegypti* (3rd instar), 1 + 1 h exp.	LC_{50}			0.00203	1081
I, Permethrin, technical	*Aedes aegypti* (3rd instar), 2 h exp.	LC_{50}			0.00232	1081
I, Carbofuran	*Aedes aegypti* (3rd instar), 1 + 1 h exp.	LC_{50}			1.590	1081
I, Carbofuran	*Aedes aegypti* (3rd instar), 2 h exp.	LC_{50}			2.130	1081
I, Heptachlor/photoheptachlor toxicity ratio	*Aedes aegypti* (larvae)	TR	2.5			258
I, Heptachlor/photoheptachlor toxicity ratio	*Aedes aegypti* (larvae)	TR	2.80			44, 49
I, Nitrilotriacetic acid	*Aedes aegypti*	NOEL		100.0		372
I, Abate	Agrion	LC_{95}			0.04 (1 h)	574
I, Chlorpyrifos	Agrion	LC_{95}			0.2 (1 h)	574
I, Fenvalerate	*Atherix variegata* (F, 15°C, pH 7.8)	LC_{50}			0.00003 (28 d)	575
I, Premethrin	*Baetis* sp.	LC_{100}			0.001 (1 h)	576
I, Chloropyrifos	*Baetis rhodani*	LC_{95}			0.01–0.02 (1 h)	574
I, Temephos	*Baetis* sp.	LC_{91}	0.1			587
I, Abate	*Baetis rhodani*	LC_{95}			0.001 (1 h)	574
I, Malathion	*Boyeria vinosa*	NOEL			0.00165 (30 d)	91
I, Malathion	*Boyeria vinosa*	LC_{50}			0.0023 (30 d)	91
I, Premethrin	*Brachycentrus* sp.	LC_{100}			0.001 (1 h)	576
I, Abate	*Brachycentrus subnubilis*	LC_{95}			0.1 (1 h)	574

Chemical	Species	Toxicity test	24 h	48 h	96 h	Chronic exposure	Ref.
I, Chlorpyrifos	*Brachycentrus subnubilis*	LC_{95}				0.2–0.5 (1 h)	574
I, Temephos	*Brachycentrus* sp.	LC_{100}	1.0				587
I, Dichlobenil	*Callibaetis* sp.	LC_{50}			10.3		98
I, Diquat	*Callibaetis* sp.	LC_{50}			16.4		98
I, Baytex	*Chaoborus* sp. (larvae)	LC_{50}		0.008			90
I, Lindane	*Chaoborus* sp. (larvae)	LC_{50}		0.008			299
I, Carbaryl	*Chaoborus* sp. (larvae)	LC_{50}		0.296			90
I, Dieldrin	*Chironomus riparius* (4th instar larvae)	LC_{50}	0.0005				299
I, Malathion	*Chironomus riparius* (4th instar larvae)	LC_{50}	0.0019				299
I, Parathion	*Chironomus riparius* (4th instar larvae)	LC_{50}	0.0025				299
I, Lindane	*Chironomus riparius* (4th instar larvae)	LC_{50}	0.0036				299
I, DDT	*Chironomus riparius* (4th instar larvae)	LC_{50}	0.0047				299
I, Malaoxon	*Chironomus riparius* (4th instar larvae)	LC_{50}	0.0054				299
I, Aldrin	*Chironomus riparius* (4th instar larvae)	LC_{50}	0.008				299
I, Allethrin	*Chironomus riparius* (4th instar larvae)	LC_{50}	0.049				299
I, Landrin	*Chironomus riparius* (4th instar larvae)	LC_{50}	0.0514				299
I, Baygon	*Chironomus riparius* (4th instar larvae)	LC_{50}	0.0644				299
I, Carbaryl	*Chironomus riparius* (4th instar larvae)	LC_{50}	0.1045				299
I, Aminocarb	*Chironomus riparius*	LC_{50}	0.377				299
I, Chloropyrifos	*Claassenia sabulosa*	LC_{50}			0.00057		70
I, Dieldrin	*Claassenia sabulosa*	LC_{50}			0.00058		70
I, Endrin	*Claassenia sabulosa*	LC_{50}			0.00076		70
I, Toxaphene	*Claassenia sabulosa*	LC_{50}			0.0013		70
I, Parathion	*Claassenia sabulosa*	LC_{50}			0.0015		70
I, Heptachlor	*Claassenia sabulosa*	LC_{50}			0.0028		70
I, Malathion	*Claassenia sabulosa*	LC_{50}			0.0028		70
I, DDT	*Claassenia sabulosa*	LC_{50}			0.0035		70
I, Carbaryl	*Claassenia sabulosa*	NOEL			0.0056		70
I, Chlorfos	*Claassenia sabulosa*	LC_{50}			0.022		70

I, Carbaryl	*Cloeon* sp. (larvae)	LC_{50}		0.48			90
I, Baytex	*Cloeon* sp. (larvae)	LC_{50}		0.012			90
I, Lindane	*Cloeon* sp. (larvae)	LC_{50}		0.092			299
I, Propionic acid	*Culex* sp. (larvae)	TL_m	1000.0	1000.0			77
I, Propionate, Na	*Culex* sp.	TL_m		2320.0			231
I, Diquat	*Enallagma* sp.	LC_{50}			100.0		98
I, Dichlobenil	*Enallagma* sp.	LC_{50}			20.7		98
I, Furfural	*Epeorus assimilis*	PertL				300.0	8
I, Azinphosmethyl	*Ephemerella subvaria*	NOAE				0.0025	91
I, Azinphosmethyl	*Ephemerella subvaria*	LC_{50}				0.0045	91
I, Acrolein	*Ephemerella walkeri*	LOAC				0.1	193
I, Diquat	*Ephemerella walkeri*	LOAC				1.0	193
I, 2,4-D, dimethylamine, salt	*Ephemerella walkeri*	LOAC				10.0	193
I, 2,2-Diphenyl amine	*Ephemerella walkeri*	LOAC				10.0	193
I, *o*-Xylene	*Ephemerella walkeri*	LOAC				10.0	2
I, Parathion	*Ephemerella subvaria*	LC_{50}			0.00016	0.000056 (30 d)	91
I, Hydrogen sulfide	*Ephemerella* sp.	TL_m			0.316		46
I, Fenvalerate	*Ephemerella* sp. (F, 15.6°C, pH 7.8)	LC_{50}			0.00013		575
I, Co-Ral	*Hexagenia* sp.	LC_{50}	0.430				1024
I, Malathion	*Hydropsyche bettoni*	NOEL				0.00024 (30 d)	91
I, Malathion	*Hydropsyche bettoni*	LC_{50}				0.00034 (30 d)	91
I, Carbaryl	*Hydropsyche bettoni*	NOEL				0.0018 (30 d)	91
I, Azinphosmethyl	*Hydropsyche bettoni*	NOAE				0.00494	91
I, Azinphosmethyl	*Hydropsyche bettoni*	LC_{50}				0.0074 (30 d)	91
I, Abate	*Hydropsyche pellucidula*	LC_{95}				0.5 (1 h)	574
I, Chloropyrifos	*Hydropsyche pellucidula*	LC_{95}				0.5 (1 h)	574
I, Co-Ral	*Hydropsyche* sp.	LC_{50}	0.005				1024
I, Parathion	*Hydropsyche bettoni*	LC_{50}				0.00045 (30 d)	91
I, Diazinon	*Hydropsyche bettoni*	NOEL				0.00179 (30 d)	302
I, Carbaryl	*Hydropsyche bettoni*	LC_{50}				0.0027 (30 d)	91

Chemical	Species	Toxicity test	24 h	48 h	96 h	Chronic exposure	Ref.
I, Diazinon	*Hydropsyche bettoni*	LC_{50}				0.00354 (30 d)	302
I, Temephos	*Hydropsyche* sp.	LC_{62}	1.0				587
I, Toxaphene	*Ischnura verticalis*	LC_{50}		0.086			515
I, Endosulfan	*Ischnura* sp.	LC_{50}			0.0718		85
I, Dichlobenil	*Limnephilus* sp.	LC_{50}			13.0		98
I, Diquat	*Limnephilus*	LC_{50}			33.0		98
I, Aldrin	*Musca domestica* (3-d-old female)	LD_{50}				0.0014	44
I, Photoheptachlor	*Musca domestica* (3-d-old female)	LD_{50}				0.0056 mg/fly	44
I, Photo-aldrin	*Musca domestica* (3-d-old female)	LD_{50}				0.0069 mg/fly	44
I, Photodieldrin	*Musca domestica* (3-d-old female)	LD_{50}				0.0083 mg/fly	44
I, Dieldrin	*Musca domestica* (3-d-old Musca)	LD_{50}				0.0098 mg/fly	44
I, Heptachlor	*Musca domestica* (3-d-old female)	LD_{50}				0.011 mg/fly	44
I, Photoisodrin	*Musca domestica* (3-d-old female)	LD_{50}				0.113 mg/fly	44
I, Photochlordene	*Musca domestica* (3-d-old female)	LD_{50}				0.179 mg/fly	44
I, Phenthoate	*Musca domestica*	LD_{50}				5.0 mg/kg	301
I, Isodrin	*Musca domestica* (3-d-old female)	LD_{50}				54 µg/fly	44
I, Chlordene	*Musca domestica* (3-d-old female)	LD_{50}				158 µg/fly	44
I, Isodrin/photoisodrin toxicity ratio	*Musca aegyptica*	TR	0.41				44, 48, 49
I, Heptachlor/photoheptachlor toxicity ratio	*Musca domestica*	TR	1.90				44, 49
I, Heptachlor/photoheptachlor toxicity ratio	*Musca domestica*	TR	1.96				258
I, Aroclor 1242	*Odonata* (naiad) (S)	LC_{50}				0.8 (7 d)	92
I, Aroclor 1254	*Odonata* (naiad) (S)	LC_{50}				1.0 (7 d)	92
I, Aroclor 1254	*Odonata* (naiad) (F)	LC_{50}			0.20		92
I, Aroclor 1242	*Odonata* (naiad) (F)	LC_{50}			0.40		92
I, Malathion	*Ophiogomphus rupinsulensis*	NOEL				0.00028 (30 d)	91

I, Malathion	*Ophiogomphus rupinsulensis*	LC_{50}			0.00052 (30 d)	91
I, Diazinon	*Ophiogomphus rupinsulensis*	NOEL			0.00129 (30 d)	91
I, Azinphosmethyl	*Ophiogomphus rupinsulensis*	NOAE			0.00173 (30 d)	91
I, Diazinon	*Ophiogomphus rupinsulensis*	LC_{50}			0.0022 (30 d)	91
I, Parathion	*Ophiogomphus rupinsulensis*	LC_{50}		0.00325	0.00022 (30 d)	91
I, Azinphosmethyl	*Ophiogomphus rupinsulensis*	LC_{50}		0.012	0.0022 (30 d)	91
I, Aldrin	*Pteronarcys californica*	LC_{50}		0.0013		298
I, Aldrin	*Pteronarcys californica*	LC_{50}		0.18	0.0025 (30 d)	298
I, Dieldrin	*Pteronarcella badia*	LC_{50}		0.0005		70
I, Endrin	*Pteronarcella badia*	LC_{50}		0.00054		70
I, Heptachlor	*Pteronarcella badia*	LC_{50}		0.0009		2
I, Malathion	*Pteronarcella badia*	LC_{50}		0.0011		70
I, DDT	*Pteronarcella badia*	LC_{50}		0.0019		70
I, Toxaphene	*Pteronarcella badia*	LC_{50}		0.003		70
I, Chlorfos	*Pteronarcella badia*	LC_{50}		0.011		70
I, Fenvalerate	*Pteronarcys dorsata* (F, 15.6°C, pH 7.8)	EC_{50}			0.00013 (28 d)	575
I, Diazinon	*Pteronarcys dorsata*	NOEL			0.0032 (30 d)	91
I, Diazinon	*Pteronarcys dorsata*	LC_{50}			0.0046 (30 d)	91
I, Malathion	*Pteronarcys dorsata*	NOEL			0.0094 (30 d)	91
I, Malathion	*Pteronarcys dorsata*	LC_{50}			0.0111 (30 d)	91
I, Carbaryl	*Pteronarcys dorsata*	NOEL			0.0115 (30 d)	91
I, Carbaryl	*Pteronarcys dorsata*	LC_{50}			0.023 (30 d)	91
I, Toxaphene	*Pteronarcys* sp.	LC_{50}	0.0023			263
I, Endosulfan	*Pteronarcys* sp.	LC_{50}	0.0023			263
I, Lindane	*Pteronarcys californica*	LC_{50}	0.0045			263
I, Parathion	*Pteronarcys* sp.	LC_{50}	0.0054			263
I, Diuron	*Pteronarcys* sp.	LC_{50}	1.2			263
I, 2,2-Dichlorovinyldimethyl-phosphate	*Pteronarcys californica*	LC_{50}		0.00010		70
I, Endrin	*Pteronarcys lacustris*	LC_{50}		0.00025		70
I, Chloropyrifos	*Pteronarcys badia*	LC_{50}		0.00038		70

Chemical	Species	Toxicity test	24 h	48 h	96 h	Chronic exposure	Ref.
I, Dieldrin	*Pteronarcys californica*	LC_{50}			0.0005		70
I, Pyrethrins	*Pteronarcys californica*	LC_{50}			0.001		70
I, Heptachlor	*Pteronarcys californica*	LC_{50}			0.0011		2
I, Azinphosmethyl	*Pteronarcys californica*	LC_{50}			0.0015		70
I, Carbaryl	*Pteronarcys badia*	NOEL			0.0017		70
I, Allethrin	*Pteronarcys californica*	LC_{50}			0.0021		70
I, Endosulfan	*Pteronarcys californica*	LC_{50}			0.0023		70
I, Endrin	*Pteronarcys californica*	LC_{50}			0.0024	0.0012 (30 d)	298
I, Ethion	*Pteronarcys californica*	LC_{50}			0.0028		70
I, Parathion	*Pteronarcys dorsata*	LC_{50}			0.0030	0.00090 (30 d)	91
I, Lindane	*Pteronarcys californica*	LC_{50}			0.0045		70
I, Baytex	*Pteronarcys californica*	LC_{50}			0.0045		70
I, Carbaryl	*Pteronarcys californica*	NOEL			0.0048		70
I, Disulfoton	*Pteronarcys californica*	LC_{50}			0.005		70
I, Disulfoton	*Pteronarcys californica* (S, 15°C, pH 7.5)	LC_{50}			0.005		410
I, DDT	*Pteronarcys californica*	LC_{50}			0.007		70
I, Naled	*Pteronarcys californica*	LC_{50}			0.008		70
I, Zectran	*Pteronarcys californica*	LC_{50}			0.01		70
I, Abate	*Pteronarcys californica*	LC_{50}			0.010		70
I, Malathion	*Pteronarcys californica*	LC_{50}			0.010		70
I, Chloropyrifos	*Pteronarcys californica*	LC_{50}			0.010		235
I, Baygon	*Pteronarcys californica*	LC_{50}			0.013		70
I, Chlordane	*Pteronarcys californica*	LC_{50}			0.015		70
I, Azinphosmethyl	*Pteronarcys dorsata*	LC_{50}			0.021	0.0049 (30 d)	91
I, Disulfoton	*Pteronarcys californica*	LC_{50}			0.024	0.0019 (30 d)	297
I, Diazinon	*Pteronarcys californica*	LC_{50}			0.025		70
I, Parathion	*Pteronarcys californica* (S, 12°C, FW)	LC_{50}			0.032		744
I, Chlorfos	*Pteronarcys californica*	LC_{50}			0.035		70
I, Parathion	*Pteronarcys californica*	LC_{50}			0.036	0.0022 (30 d)	297

I, Dieldrin	*Pteronarcys californica*	LC_{50}			0.039	0.002 (30 d)	298
I, Difolitan	*Pteronarcys californica*	LC_{50}			0.04		70
I, Chlorfos	*Pteronarcys californica*	LC_{50}			0.069	0.0098 (30 d)	297
I, Phosphamidon	*Pteronarcys californica*	LC_{50}			0.150		70
I, Rotenone	*Pteronarcys californica*	LC_{50}			0.38		70
I, DDD	*Pteronarcys californica*	LC_{50}			0.380		70
I, Diuron	*Pteronarcys californica*	LC_{50}			1.2		66
I, Diuron	*Pteronarcys californica* (S, 15C, pH 7.2-7.5)	LC_{50}			1.2		410
I, 2,4-D-butoxyethyl ester	*Pteronarcys californica*	LC_{50}			1.6		43
I, DEF	*Pteronarcys californica*	LC_{50}			2.10		70
I, Triflurlin	*Pteronarcys californica*	LC_{50}			3.0		70
I, Dichlobenil	*Pteronarcys californica*	LC_{50}			7.0		70
I, Dexon	*Pteronarcys californica*	LC_{50}			24.0		70
I, Picloram	*Pteronarcys californica*	LC_{50}			48.0		70
I, Trifene, Na	*Pteronarcys californica*	LC_{50}			55.0		70
I, Paraquat	*Pteronarcys californica*	NOAE			100.0		70
I, 2,2-Dichloropropionic acid	*Pteronarcys californica*	NOEL			100.0		70
I, Trichloronate	*Pteronarcys* (16°C, pH 7.2–7.5)	LC_{50}			0.0001		952
I, Aldrin	*Pteronarcys* sp.	LC_{50}		0.043			263
I, Chloropyrifos	*Simulium ornatum*	LC_{95}				0.05–0.1 (1 h)	574
I, Abate	*Simulium ornatum*	LC_{95}				0.2 (1 h)	574
I, Temephos	*Simulium* sp.	LC_{60}	0.1				587
I, Temephos	*Simulium* sp. (early instar)	LC_{50}	0.2				587
I, Temephos	*Simulium* sp. (late instar)	LC_{90}	0.2				587
I, Dichlobenil	*Tendipedidae*	LC_{50}			7.8		98
I, Diquat	*Tendipedidae*	LC_{50}			100.0		98
Mo, Malathion	*Anadonta cygnea*	NOAE	1.0				39
Mo, Carbaryl	*Clinocardium nuttalli* (adult)	TL_m	7.3				407
Mo, Carbaryl	*Clinocardium nuttalli* (juv.)	TL_m			3.85		1071
Mo, Phygon	*Crassostrea virginica* (larvae, S)	TL_m				0.041 (14 d)	64

Chemical	Species	Toxicity test	24 h	48 h	96 h	Chronic exposure	Ref.
Mo, Fenoprop	*Crassostrea virginica* (larvae, S)	TL_m				0.710 (14 d)	64
Mo, 2,4-Dichlorophenoxyacetic acid	*Crassostrea virginica* (larvae, S)	TL_m				0.740 (14 d)	64
Mo, Tri-*o*-cresyl phosphate	*Crassostrea virginica* (larvae, S)	TL_m				1.0	64
Mo, Co-Ral	*Crassostrea virginica* (larvae, S)	TL_m				1.0 (14 d)	64
Mo, Carbaryl	*Crassostrea virginica* (larvae)	TL_m				3.0 (14 d)	64
Mo, Disulfoton	*Crassostrea virginica* (larvae, S)	TL_m				3.67 (14 d)	2
Mo, Phygon	*Crassostrea virginica* (egg, S)	TL_m		0.014			64
Mo, Co-Ral	*Crassostrea virginica* (egg, S)	TL_m		0.110			64
Mo, Tri-*o*-cresyl phosphate	*Crassostrea virginica* (egg, S)	TL_m		0.6			64
Mo, Azinphosmethyl	*Crassostrea virginica* (S)	TL_m		0.620			64
Mo, Endrin	*Crassostrea virginica* (S)	TL_m		0.790			64
Mo, Chlorfos	*Crassostrea virginica* (larvae, S)	TL_m		1.0			365
Mo, Carbaryl	*Crassostrea gigas* (larvae)	EC_{50}		2.2			407
Mo, Carbaryl	*Crassostrea virginica* (eggs)	TL_m		3.0			64
Mo, Disulfoton	*Crassostrea virginica* (eggs, S)	TL_m		5.86			2
Mo, Fenoprop	*Crassostrea virginica* (egg, S)	TL_m		5.9 ppm			64
Mo, 2,4-Dichlorophenoxyacetic acid	*Crassostrea virginica* (egg, S)	TL_m		8.0			64
Mo, Phenol	*Crassostrea virginica* (egg, S)	TL_m		58.25			365, 64
Mo, Ammonium salt of saturated carboxylic acid, a dispersant	*Crassostrea gigas*	LD_{50}		1000.0			2
Mo, Toxaphene	*Crassostrea virginica*	EC_{50}			0.016		2
Mo, Tillam	*Crassostrea virginica*	EC_{50}			1.0		23
Mo, *m*-Cresol	*Glossosiphonia complanata*	PL				1.1	8
Mo, Acetic acid	*Lymnaea ovata*	TT				12.0	8
Mo, Ammonium salt of saturated carboxylic acid, a dispersant	*Littoridina littorea*	LD_{50}		1000.0			2
Mo, Ammonium chloride	*Lymnaea* sp. (eggs)	TL_m	241.0	173.0	70.0	73.0 (72 h)	77
Mo, *n*-Butyric acid	*Lymnaea ovata*	LC_{100}				50.0	77
Mo, Baytex	*Lymnaea stagnalis*	LC_{50}		6.4			90

Mo, Lindane	*Lymnaea stagnalis*	LC_{50}		7.3			367
Mo, Carbaryl	*Lymnaea stagnalis*	LC_{50}		21.0			90
Mo, 2-Chlorotoluene-5-sulfonate, Na	*Lymnaea* sp.	TL_m	30.0 (25 h)				74
Mo, Anthraquinone"α"sulfonic acid	*Lymnaea* sp. (eggs)	TL_m	186.0		186.0		77
Mo, 2,5-Dichlorobenzene sulfonate, Na	*Lymnaea* sp.	TL_m	4981.0	4513.0	3144.0	3984.0 (72 h)	77
Mo, *p*-Chlorobenzene sulfonic acid	*Lymnaea* sp.	TL_m	8600.0	7633.0	5053.0	6343.0 (72 h)	77
Mo, *p*-Phenol sulfonate, Na	*Lymnaea* sp.	TL_m	10,700.0	9122.0	8828.0	8828.0	77
Mo, Cyanatrine	*Lymnaea peregra* (eggs)	NOAE				0.2	489
Mo, Cyanatrine	*Lymnaea peregra* (adult)	LC_0				10.0 (8 d)	489
Mo, Cyanatrine	*Lymnaea peregra* (adult)	LC_{60}		20.0			489
Mo, 3,3,4-Trichloro carbanilide	*Mercenaria mercenaria* (larvae, S)	TL_m				0.037 (12 d)	64
Mo, Roccal	*Mercenaria mercenaria* (larvae)	TL_m				0.14 (12 d)	64
Mo, Dowicide G	*Mercenaria mercenaria* (larvae, S)	TL_m				0.250 (12 d)	2
Mo, Toxaphene	*Mercenaria mercenaria* (larvae, S)	TL_m				0.25 (12 d)	2
Mo, Dowicide A	*Mercenaria mercenaria* (larvae, S)	TL_m				0.75 (12 d)	64
Mo, Azinphosmethyl	*Mercenaria mercenaria* (larvae, S)	TL_m				0.860 (12 d)	64
Mo, Disulfoton	*Mercenaria mercenaria* (larvae, S)	TL_m				1.39 (12 d)	64
Mo, Phygon	*Mercenaria mercenaria* (larvae, S)	TL_m				1.75 (12 d)	64
Mo, Carbaryl	*Mercenaria mercenaria* (larvae)	TL_m				2.5 (14 d)	64
Mo, Carbaryl	*Mercenaria mercenaria* (eggs)	TL_m				3.82 (14 d)	64
Mo, Phenol	*Mercenaria mercenaria* (larvae, S)	TL_m				55.0 (12 d)	365, 64
Mo, Phygon	*Mercenaria mercenaria* (egg, S)	TL_m		0.014			64
Mo, 3,3,4-Trichlorocarbanilide	*Mercenaria mercenaria* (egg, S)	TL_m		0.032			64
Mo, Roccal	*Mercenaria mercenaria* (egg)	TL_m		0.19			64
Mo, Dowicide G	*Mercenaria mercenaria* (eggs, S)	TL_m		0.250			2
Mo, Azinphosmethyl	*Mercenaria mercenaria* (eggs, S)	TL_m		0.860			64
Mo, Toxaphene	*Mercenaria mercenaria* (egg, S)	TL_m		1.12			2
Mo, Phenol	*Mercenaria mercenaria* (egg, S)	TL_m		52.63			365, 64
Mo, Disulfoton	*Mercenaria mercenaria* (eggs, S)	TL_m		55.28			2
Mo, Dowicide A	*Mercenaria mercenaria* (eggs, S)	TL_m		100.0			64

Chemical	Species	Toxicity test	24 h	48 h	96 h	Chronic exposure	Ref.
Mo, Co-Ral	*Merceneria merceneria* (larvae, S)	TL_m				5.210 (12 d)	64
Mo, Co-Ral	*Merceneria merceneria* (egg, S)	TL_m		9.120			64
Mo, Tetramethyllead	Mussel	LC_{50}				0.27	114
Mo, Tetraethyllead	Mussel	LC_{50}			0.10		114
Mo, Trimethyllead chloride	Mussel	LC_{50}			0.50 (Pb)		114
Mo, Triethyllead chloride	Mussel	LC_{50}			1.1 (Pb)		114
Mo, Ammonium salt of saturated carboxylic acid, a dispersant	*Mytilus edulis*	LD_{50}		500.0			42
Mo, Carbaryl	*Mytilus edulis* (larvae)	EC_{50}			2.3		407
Mo, Nitrilotriacetic acid, Na, H_2O	Bay mussel (S)	TL_{50}				3400.0 (168 h)	37
Mo, Nitrilotriacetic acid,Na,H_2O	Bay mussel (S)	TL_{50}			6100.0		37
Mo, Malathion	Mussel (larvae)	NOEL				0.0001	39
Mo, Isopropylbenzene	*Mytilus edulis* (larvae)	NOE				1.0–50.0	368
Mo, Toxaphene	*Rangia cuneata*	LC_{50}		0.067			515
Mo, Toxaphene	*Rangia cuneata*	LC_{50}	940.0	699.0	460.0	480.0 (72 h)	741
O, Phenylmercuric acetate	*Anthocidaris crassispina* (eggs)	NOAE				0.009 (12 h)	1073
O, Sodium fluoride	*Anthopleura aureodiata* ($^1/_2$S, 19°C, pH 8)	LC_0				100.0 (144 h)	641
O, Nitrilotriacetic acid, Na, H_2O	*Asterias forbesi* (S)	TL_{50}				3000.0 (168 h)	37
O, Nitrilotriacetic acid, Na, H_2O	*Asterias forbesi* (S)	TL_{50}			3000.0		37
O, Mercuric chloride	*Capitella capitella* (S)	LC_{50}			0.014		1075
O, Dow Corning-561	*Cordylophora caspia* (semiS)	NOAE		100.0			1074
O, *n*-Nitrosodimethylamine	*Dugesia dorotocephala*	LC_{50}			1365.0		567
O, Dibutyl phthalate	*Gymnodium breve*	EC_{50}			0.0034–0.2		366
O, Dibutyl phthalate	*Gymnodium breve*	TL_m			0.02–0.6		366
O, *n*-Proponal	*Hydra oligactis* (S, 17°C, pH 8.4)	LC_0		5100.0			1076
O, *n*-Proponal	*Hydra oligactis* (S, 17°C, pH 8.4)	LC_{50}		6800.0			1076
O, Mercuric chloride	*Neanthes arenaceodentata* (adult)	LC_{50}			0.022		1075
O, Mercuric chloride	*Neanthes arenaceodentata* (juv.)	LC_{50}			0.11		1075
O, Fluoranthene	*Neanthes arenaceodentata*	TL_m			0.5		248

O, Fluorene	*Neanthes arenaceodentata*	TL_m		1.0		248
O, Naphthalene	*Neanthes arenaceodentata* (S, 22°C, SW)	TL_m		3.80		248
O, Nitrilotriacetic acid, Na, H_2O	*Nereis virens* (S)	TL_{50}			5500.0 (168 h)	37
O, Mercuric chloride	*Nereis virens* (S, 20°C, pH 8.0)	NOAEL		0.025		680
O, Mercuric chloride	*Nereis virens* (S, 20°C, pH 8.0)	LC_{50}		0.07		680
O, Nitrilotriacetic acid, Na, H_2O	*Nereis virens* (S)	TL_{50}		5500.0		37
O, Carbaryl	*Oedothorax insecticeps*	LD_{50}			0.840	302
O, Hexachlorobenzene	*Oedothorax insecticeps*	LD_{50}			21.0	302
O, Diazinon	*Oedothorax insecticeps*	LD_{50}			2450.0	302
O, Fenitrothion	*Oedothorax insecticeps*	LD_{50}			3200.0	302
O, Swedish EDC tar	*Ophiura texturata*	LC_{50}	23.0			572
O, Rotenone	Planaria	LC_{100}		0.500		557
O, Nitrilotriacetic acid, Na, H_2O	*Pseudodiaptimus coronatus* (S)	TL_{50}			700.0 (72 h)	37
O, Nitrilotriacetic acid, Na, H_2O	*Tisbe furcata* (S)	TL_{50}			270.0 (72 h)	37
O, Nitrilotriacetic acid, Na, H_2O	*Trigriopus japonicus* (S)	TL_{50}			3200.0 (72 h)	37
P, *p*-Chlorophenol	*Colpoda*	LC_{100}			5.0	8
P, 1,3,5-Trioxane	*Colpoda*	LD_0			1.0	8
P, Glutaric acid	*Colpoda*	LC_{100}			16.0	8
P, *o*-Chlorophenol	*Colpoda*	LC_{100}			30.0	8
P, 4,4′-Diaminodiphenylmethane	*Colpoda*	LC_{100}			124.0	8
P, Succinic acid	*Colpoda*	LD_0			125.0	8
P, Diisopropyl ether	*Colpoda*	NOAE			125.0	8
P, Dimethylamine	*Colpoda*	LC_{100}			250.0	8
P, 2-Butanone oxime	*Colpoda*	LC_{100}			2500.0	8
P, Benzaldehyde	*Entosiphon sulcatum*	TT			0.29	10
P, *o*-Nitrophenol	*Entosiphon sulcatum*	TT			0.40	10
P, 2-Nitro-*p*-cresol	*Entosiphon sulcatum*	TT			0.42	10
P, 2,4-Dichlorophenol	*Entosiphon sulcatum*	TT			0.5	10
P, Furfural	*Entosiphon sulcatum*	TT			0.6	10
P, Cyclohexyl amine	*Entosiphon sulcatum*	TT			0.6	10

Chemical	Species	Toxicity test	24 h	48 h	96 h	Chronic exposure	Ref.
P, *m*-Dinitrobenzene	*Entosiphon sulcatum*	TT				0.76	10
P, *p*-Nitrophenol	*Entosiphon sulcatum*	TT				0.83	10
P, *m*-Nitrophenol	*Entosiphon sulcatum*	TT				0.970	10
P, 2,4-Dinitrotoluene	*Entosiphon sulcatum*	TT				0.98	10
P, 6-Nitro-*m*-cresol	*Entosiphon sulcatum*	TT				1.30	10
P, Salicylaldehyde	*Entosiphon sulcatum*	TT				1.40	10
P, 2,4,6-Trinitrotoluene	*Entosiphon sulcatum*	TT				1.6	10
P, Ethylenediamine	*Entosiphon sulcatum*	TT				1.8	10
P, Nitrobenzene	*Entosiphon sulcatum*	TT				1.90	10
P, Pyridine	*Entosiphon sulcatum*	TT				3.50	10
P, Butyraldehyde	*Entosiphon sulcatum*	TT				4.2	10
P, Ethyleneimine	*Entosiphon sulcatum*	TT				4.3	10
P, 4,6-Dinitro-*o*-cresol	*Entosiphon sulcatum*	TT				5.4	10
P, 4-Nitro-*m*-cresol	*Entosiphon sulcatum*	TT				5.80	10
P, 2,3-Dinitrotoluene	*Entosiphon sulcatum*	TT				5.9	10
P, *o*-Diethylbenzene	*Entosiphon sulcatum*	TT				6.9	10
P, *p*-Nitroaniline	*Entosiphon sulcatum*	TT				6.90	10
P, Allyl chloride	*Entosiphon sulcatum*	TT				8.4	10
P, *n*-Butylamine	*Entosiphon sulcatum*	TT				9.0	10
P, 2,4-Dimethylaniline	*Entosiphon sulcatum*	TT				9.8	10
P, Phenyl acetate	*Entosiphon sulcatum*	TT				10.0	10
P, Hydroquinone	*Entosiphon sulcatum*	TT				11.0	10
P, 2,6-Dinitro toluene	*Entosiphon sulcatum*	TT				11.0	10
P, 2,4-Pentanedione	*Entosiphon sulcatum*	TT				11.0	10
P, 2-Ethylhexyl amine	*Entosiphon sulcatum*	TT				12.0	10
P, Diallyl phthalate	*Entosiphon sulcatum*	TT				13.0	10
P, Tributyl phosphate	*Entosiphon sulcatum*	TT				14.0	10
P, *n*-Pentanol	*Entosiphon sulcatum*	TT				17.0	10
P, *o*-Cresol	*Entosiphon sulcatum*	TT				17.0	10

P, Propargyl alcohol	*Entosiphon sulcatum*	TT	17.0	10
P, Diethyl phthalate	*Entosiphon sulcatum*	TT	19.0	10
P, 2,4-Dinitrophenol	*Entosiphon sulcatum*	TT	20.0	10
P, Acrylic acid	*Entosiphon sulcatum*	TT	20.0	10
P, Formaldehyde	*Entosiphon sulcatum*	TT	22.0	10
P, Allylamine	*Entosiphon sulcatum*	TT	23.0	10
P, Aniline	*Entosiphon sulcatum*	TT	24.0	10
P, Benzyl chloride	*Entosiphon sulcatum*	TT	25.0	10
P, *n*-Butyric acid	*Entosiphon sulcatum*	TT	26.0	10
P, Acetone	*Entosiphon sulcatum*	TT	28.0	10
P, Acetic anhydride	*Entosiphon sulcatum*	TT	30.0	10
P, Isooctanol	*Entosiphon sulcatum*	TT	30.0	10
P, Acetic anhydride	*Entosiphon sulcatum*	TT	30.0	10
P, Benzonitrile	*Entosiphon sulcatum*	TT	30.0	10
P, *m*-Cresol	*Entosiphon sulcatum*	TT	31.0	10
P, 1-Heptanol	*Entosiphon sulcatum*	TT	31.0	10
P, Phenol	*Entosiphon sulcatum*	TT	33.0	10
P, Ethyleneglycol acetate	*Entosiphon sulcatum*	TT	34.0	10
P, Epichlorhydrin	*Entosiphon sulcatum*	TT	35.0	10
P, EDTA	*Entosiphon sulcatum*	TT	36.0	10
P, *n*-Propanol	*Entosiphon sulcatum*	TT	38.0	10
P, Lauryl sulfate	*Entosiphon sulcatum*	TT	40.0	10
P, Ethylamine	*Entosiphon sulcatum*	TT	45.0	10
P, *n*-Butanol	*Entosiphon sulcatum*	TT	55.0	10
P, Triethanolamine	*Entosiphon sulcatum*	TT	56.0	10
P, Benzotrichloride	*Entosiphon sulcatum*	TT	56.0	10
P, *o*-Dichlorobenzene	*Entosiphon sulcatum*	TT	64.0	10
P, Ethanol	*Entosiphon sulcatum*	TT	65.0	10
P, Butyl digol	*Entosiphon sulcatum*	TT	73.0	10
P, *n*-Hexanol	*Entosiphon sulcatum*	TT	75.0	10

Chemical	Species	Toxicity test	24 h	48 h	96 h	Chronic exposure	Ref.
P, *o*-Toluidine	*Entosiphon sulcatum*	TT				76.0	10
P, Acetic acid	*Entosiphon sulcatum*	TT				78.0	12
P, Acetic acid	*Entosiphon sulcatum*	TT				78.0	10
P, Chloral	*Entosiphon sulcatum*	TT				79.0	10
P, *o*-Chlorotoluene	*Entosiphon sulcatum*	TT				80.0	10
P, Vinyl acetate	*Entosiphon sulcatum*	TT				81.0	10
P, Butyl cellosolve	*Entosiphon sulcatum*	TT				91.0	10
P, *n*-Propyl acetate	*Entosiphon sulcatum*	TT				97.0	10
P, Acetamide	*Entosiphon sulcatum*	TT				99.0	10
P, 2-Hexanol	*Entosiphon sulcatum*	TT				116.0	10
P, Cyclohexyl acetate	*Entosiphon sulcatum*	TT				120.0	10
P, Ethylbenzene	*Entosiphon sulcatum*	TT				140.0	10
P, Diethanolamine	*Entosiphon sulcatum*	TT				160.0	10
P, 3-Hexanol	*Entosiphon sulcatum*	TT				182.0	10
P, Ethyl acetate	*Entosiphon sulcatum*	TT				202.0	10
P, Benzoic acid	*Entosiphon sulcatum*	TT				218.0	10
P, *n*-Amyl acetate	*Entosiphon sulcatum*	TT				226.0	10
P, Furfuryl alcohol	*Entosiphon sulcatum*	TT				227.0	10
P, Cyclopentanone	*Entosiphon sulcatum*	TT				232.0	10
P, Ethyl butyrate	*Entosiphon sulcatum*	TT				236.0	10
P, Styrene	*Entosiphon sulcatum*	TT				256.0	10
P, Ethyl-*sec*-amyl ketone	*Entosiphon sulcatum*	TT				256.0	10
P, Cyclopentanol	*Entosiphon sulcatum*	TT				290.0	10
P, Isobutanol	*Entosiphon sulcatum*	TT				295.0	10
P, Ethanolamine	*Entosiphon sulcatum*	TT				300.0	10
P, *n*-Butyl acetate	*Entosiphon sulcatum*	TT				321.0	10
P, Chlorobenzene	*Entosiphon sulcatum*	TT				390.0	10
P, Ethyl acetoacetate	*Entosiphon sulcatum*	TT				391.0	10
P, Isobutyl acetate	*Entosiphon sulcatum*	TT				411.0	10

P, Toluene	*Entosiphon sulcatum*	TT				456.0	10
P, Isopropyl acetate	*Entosiphon sulcatum*	TT				460.0	10
P, Citric acid	*Entosiphon sulcatum*	TT				485.0	10
P, Cyclohexanone	*Entosiphon sulcatum*	TT				545.0	10
P, Ethyl propionate	*Entosiphon sulcatum*	TT				560.0	10
P, Benzene	*Entosiphon sulcatum*	TT				700.0	10
P, Nitrilotriacetic acid	*Entosiphon sulcatum*	TT				800.0	10
P, Trichloroacetic acid	*Entosiphon sulcatum*	TT				800.0	10
P, *t*-Butyl acetate	*Entosiphon sulcatum*	TT				970.0	10
P, 1,1,2-Trichloroethane	*Entosiphon sulcatum*	TT				1040.0	10
P, Ethylene dichloride	*Entosiphon sulcatum*	TT				1127.0	10
P, Trichloroethylene	*Entosiphon sulcatum*	TT				1200.0	10
P, *sec*-Butanol	*Entosiphon sulcatum*	TT				1280.0	10
P, Diacetone alcohol	*Entosiphon sulcatum*	TT				1400.0	10
P, Methyl cellosolve	*Entosiphon sulcatum*	TT				1715.0	10
P, Acetonitrile	*Entosiphon sulcatum*	TT				1810.0	10
P, Glycerol	*Entosiphon sulcatum*	TT				3200.0	10
P, Isopropanol	*Entosiphon sulcatum*	TT				4930.0	10
P, Chloroform	*Entosiphon sulcatum*	TT				6560.0	10
P, Triethylene glycol	*Entosiphon sulcatum*	TT				10,000.0	10
P, Ethylene glycol	*Entosiphon sulcatum*	TT				10,000.0	10
P, Diethylene glycol	*Entosiphon sulcatum*	TT				10,745.0	10
P, Allylthiourea	*Entosiphon sulcatum*	TT				13.0	10
P, Mercuric chloride	*Euplotes vannus* (S, 28°C, syn. seaW)	NOAE		0.10			1077
P, Dipropyl phthalate	*Gymnodinium breve*	EC_{50}				0.9–2.4	366
P, Diethyl phthalate	*Gymnodinium breve*	EC_{50}				3.0–6.1	366
P, *o*-Dimethyl phthalate	*Gymnodinium breve*	EC_{50}				54.0–96.0	366
P, Dipropyl phthalate	*Gymnodinium breve*	TL_m			1.3-6.5		366
P, Diethyl phthalate	*Gymnodinium breve*	TL_m	23.5		33.0		366
P, *o*-Dimethyl phthalate	*Gymnodinium breve*	TL_m			125.0–185.0		366

Chemical	Species	Toxicity test	24 h	48 h	96 h	Chronic exposure	Ref.
P, Catechol	*Paramecium caudatum*	LD_0				35.0	8
P, *n*-Butyric acid	*Paramecium caudatum*	LC_{100}				250.0	8
P, Benzyl chloride	*Paramecium caudatum*	LC_{100}				800.0	8
P, Furfural	*Paramecium caudatum*	PertL				1200.0	8
P, Formic acid	*Paramecium caudatum*	LC_{100}				6000.0	8
P, 2,6-Dimethylquinoline	*Tetrahymena pyriformis*	LC_{100}	1.27 mmol/l				291
P, *m*-Toluidine	*Tetrahymena pyriformis*	LC_{100}	1.9 mmol/l				291
P, *p*-Ethylphenol	*Tetrahymena pyriformis*	LC_{100}	2.07 mmol/l				291
P, 3,5-Xylenol	*Tetrahymena pyriformis*	LC_{100}	2.30 mmol/l				291
P, 2,6-Xylenol	*Tetrahymena pyriformis*	LC_{100}	2.660 mmol/l				291
P, *m*-Cresol	*Tetrahymena pyriformis*	LC_{100}	3.5 mmol/l				291
P, *o*-Cresol	*Tetrahymena pyriformis*	LC_{100}	3.7 mmol/l				291
P, *p*-Cresol	*Tetrahymena pyriformis*	LC_{100}	3.7 mmol/l				291
P, *m*-Xylene	*Tetrahymena pyriformis*	LC_{100}	3.770 mmol/l				291
P, *p*-Xylene	*Tetrahymena pyriformis*	LC_{100}	3.770 mmol/l				291
P, Toluene	*Tetrahymena pyriformis*	LC_{100}	5.97 mmol/l				291
P, Quinoline	*Tetrahymena pyriformis*	LC_{100}	6.19 mmol/l				291
P, Isoquinoline	*Tetrahymena pyriformis*	LC_{100}	6.19 mmol/l				291
P, Phenol	*Tetrahymena pyriformis*	LC_{100}	6.37 mmol/l				291
P, Benzene	*Tetrahymena pyriformis*	LC_{50}	12.8 mmol/l				291
P, Hydroquinone	*Tetrahymena pyriformis*	LC_{100}	15.4 mmol/l				291
P, 2,6-Dimethylpyridine	*Tetrahymena pyriformis*	LC_{100}	32.7 mmol/l				291
P, α-Picoline	*Tetrahymena pyriformis*	LC_{100}	64.4 mmol/l				291
P, Pyridine	*Tetrahymena pyriformis*	LC_{100}	113.8 mmol/l				291
P, 4,6-Dinitro-*o*-cresol	*Uronema parduczi*	TT				0.012	36
P, Lead acetate	*Uronema parduczi*	TT				0.07	36
P, 2,4-Dinitrophenol	*Uronema parduczi*	TT				0.22	36
P, Hydrazinium hydroxide	*Uronema parduczi*	TT				0.240	36
P, 4-Nitro-*m*-cresol	*Uronema parduczi*	TT				0.260	36

P, Isopropylbenzene hydroperoxide	*Uronema parduczi*	TT	0.35	36
P, Acrolein	*Uronema parduczi*	TT	0.44	36
P, 2,4-Dinitrotoluene	*Uronema parduczi*	TT	0.55	36
P, Lauryl sulfate	*Uronema parduczi*	TT	0.75	36
P, *m*-Dinitrobenzene	*Uronema parduczi*	TT	0.79	36
P, *p*-Nitrophenol	*Uronema parduczi*	TT	0.89	36
P, 2,3-Dinitrotoluene	*Uronema parduczi*	TT	1.6	36
P, 2,4-Dichlorophenol	*Uronema parduczi*	TT	1.6	36
P, 1-Heptene	*Uronema parduczi*	TT	1.8	36
P, Diphenylmethane	*Uronema parduczi*	TT	2.2	36
P, *o*-Nitrophenol	*Uronema parduczi*	TT	2.90	36
P, *p*-Nitroaniline	*Uronema parduczi*	TT	3.10	36
P, *m*-Nitrophenol	*Uronema parduczi*	TT	3.40	36
P, 6-Nitro-*m*-cresol	*Uronema parduczi*	TT	5.30	36
P, Salicylaldehyde	*Uronema parduczi*	TT	5.50	36
P, 2-Nitro-*p*-cresol	*Uronema parduczi*	TT	5.80	36
P, 2,4-Pentanedione	*Uronema parduczi*	TT	5.90	36
P, Trinitrotoluene	*Uronema parduczi*	TT	5.9	36
P, Formaldehyde	*Uronema parduczi*	TT	6.5	36
P, 2-Ethylhexylamine	*Uronema parduczi*	TT	8.0	36
P, *n*-Butanol	*Uronema parduczi*	TT	8.0	36
P, Furfural	*Uronema parduczi*	TT	11.0	36
P, Acrylic acid	*Uronema parduczi*	TT	11.0	36
P, 2,4-Dimethyl aniline	*Uronema parduczi*	TT	12.0	36
P, Nitrobenzene	*Uronema parduczi*	TT	15.0	36
P, *o*-Diethylbenzene	*Uronema parduczi*	TT	16.0	36
P, 1-Heptanol	*Uronema parduczi*	TT	17.0	36
P, Phenyl acetate	*Uronema parduczi*	TT	17.0	36
P, EDTA	*Uronema parduczi*	TT	17.0	36
P, Tri-*n*-butyl phosphate	*Uronema parduczi*	TT	21.0	36

Chemical	Species	Toxicity test	24 h	48 h	96 h	Chronic exposure	Ref.
P, *n*-Butylacrylate	*Uronema parduczi*	TT				21.0	36
P, *o*-Toluidine	*Uronema parduczi*	TT				21.0	36
P, Hydroquinone	*Uronema parduczi*	TT				21.0	36
P, Diallyl phthalate	*Uronema parduczi*	TT				22.0	36
P, Benzaldehyde	*Uronema parduczi*	TT				22.0	36
P, 2,6-Dinitrotoluene	*Uronema parduczi*	TT				23.0	36
P, Picric acid	*Uronema parduczi*	TT				26.0	36
P, Ethyleneimine	*Uronema parduczi*	TT				27.0	36
P, Benzoic acid	*Uronema parduczi*	TT				31.0	36
P, *o*-Cresol	*Uronema parduczi*	TT				31.0	36
P, Cycloheptene	*Uronema parduczi*	TT				40.0	36
P, *n*-Butyl ether	*Uronema parduczi*	TT				40.0	36
P, Diethyl phthalate	*Uronema parduczi*	TT				48.0	36
P, Benzyl chloride	*Uronema parduczi*	TT				50.0	36
P, Cyclohexene	*Uronema parduczi*	TT				50.0	36
P, Cyclohexane	*Uronema parduczi*	TT				50.0	36
P, Ethylene diamine	*Uronema parduczi*	TT				52.0	36
P, Isooctanol	*Uronema parduczi*	TT				55.0	36
P, Epichlorhydrin	*Uronema parduczi*	TT				57.0	36
P, Acetaldehyde	*Uronema parduczi*	TT				57.0	36
P, *m*-Cresol	*Uronema parduczi*	TT				62.0	36
P, Ethyl *sec*-amyl ketone	*Uronema parduczi*	TT				65.0	36
P, *o*-Dichlorobenzene	*Uronema parduczi*	TT				80.0	36
P, Benzotrichloride	*Uronema parduczi*	TT				80.0	36
P, *o*-Chlorotoluene	*Uronema parduczi*	TT				80.0	36
P, Chloral	*Uronema parduczi*	TT				86.0	36
P, Vinyl acetate	*Uronema parduczi*	TT				91.0	36
P, Aniline	*Uronema parduczi*	TT				91.0	36
P, *n*-Hexanol	*Uronema parduczi*	TT				93.0	36

P, Butyraldehyde	*Uronema parduczi*	TT	98.0	36
P, Ethylbenzene	*Uronema parduczi*	TT	110.0	36
P, Benzonitrile	*Uronema parduczi*	TT	119.0	36
P, *n*-Butyric acid	*Uronema parduczi*	TT	129.0	36
P, Phenol	*Uronema parduczi*	TT	144.0	36
P, *n*-Pentanol	*Uronema parduczi*	TT	144.0	36
P, *o*-Xylene	*Uronema parduczi*	TT	160.0	36
P, Isobutanol	*Uronema parduczi*	TT	169.0	36
P, Pyridine	*Uronema parduczi*	TT	183.0	36
P, Styrene	*Uronema parduczi*	TT	185.0	36
P, Cyclohexyl amine	*Uronema parduczi*	TT	200.0	36
P, Allyl chloride	*Uronema parduczi*	TT	240.0	36
P, 3-Hexanol	*Uronema parduczi*	TT	246.0	36
P, Cyclohexanone	*Uronema parduczi*	TT	280.0	36
P, 2-Hexanol	*Uronema parduczi*	TT	335.0	36
P, Furfuryl alcohol	*Uronema parduczi*	TT	384.0	36
P, Chlorobenzene	*Uronema parduczi*	TT	392.0	36
P, Cyclohexyl acetate	*Uronema parduczi*	TT	400.0	36
P, Butyl digol	*Uronema parduczi*	TT	420.0	36
P, Trichloroacetic acid	*Uronema parduczi*	TT	435.0	36
P, Toluene	*Uronema parduczi*	TT	450.0	36
P, Butyl cellosolve	*Uronema parduczi*	TT	463.0	36
P, Benzene	*Uronema parduczi*	TT	486.0	36
P, *n*-Amyl acetate	*Uronema parduczi*	TT	550.0	36
P, *n*-Propanol	*Uronema parduczi*	TT	568.0	36
P, *n*-Butyl acetate	*Uronema parduczi*	TT	574.0	36
P, Citric acid	*Uronema parduczi*	TT	622.0	36
P, Ethyl propionate	*Uronema parduczi*	TT	665.0	36
P, Isobutyl acetate	*Uronema parduczi*	TT	727.0	36
P, Nitrilotriacetic acid	*Uronema parduczi*	TT	800.0	36

Chemical	Species	Toxicity test	24 h	48 h	96 h	Chronic exposure	Ref.
P, Cyclopentanol	*Uronema parduczi*	TT				800.0	36
P, Propargyl alcohol	*Uronema parduczi*	TT				800.0	36
P, *n*-Propyl acetate	*Uronema parduczi*	TT				843.0	36
P, Tetrahydrofuran	*Uronema parduczi*	TT				858.0	36
P, Ethyl butyrate	*Uronema parduczi*	TT				916.0	36
P, Trichloroethylene	*Uronema parduczi*	TT				960.0	36
P, 1,1,2-Trichloroethane	*Uronema parduczi*	TT				1040.0	36
P, Ethylene dichloride	*Uronema parduczi*	TT				1050.0	36
P, Cyclopentanone	*Uronema parduczi*	TT				1210.0	36
P, Acetic acid	*Uronema parduczi*	TT				1350.0	36
P, Acetic acid	*Uronema parduczi*	TT				1350.0	36
P, *sec*-Butanol	*Uronema parduczi*	TT				1416.0	36
P, Isopropyl acetate	*Uronema parduczi*	TT				1602.0	36
P, Ethyl acetate	*Uronema parduczi*	TT				1620.0	36
P, Acetone	*Uronema parduczi*	TT				1710.0	36
P, Acetone	*Uronema parduczi*	TT				1710.0	36
P, Diethanol amine	*Uronema parduczi*	TT				1720.0	36
P, *n*-Butylamine	*Uronema parduczi*	TT				1752.0	36
P, *t*-Butyl acetate	*Uronema parduczi*	TT				1850.0	36
P, Ethanolamine	*Uronema parduczi*	TT				2945.0	36
P, Allylamine	*Uronema parduczi*	TT				3140.0	36
P, Isopropanol	*Uronema parduczi*	TT				3425.0	36
P, Ethylamine	*Uronema perduczi*	TT				3500.0	36
P, 1,4-Dioxane	*Uronema parduczi*	TT				5620.0	36
P, Acetonitrile	*Uronema parduczi*	TT				5825.0	36
P, Ethylene glycol acetate	*Uronema parduczi*	TT				5910.0	36
P, Ethanol	*Uronema parduzci*	TT				6120.0	36
P, Chloroform	*Uronema parduczi*	TT				6560.0	36
P, Diethylene glycol	*Uronema parduczi*	TT				8000.0	36

P, Glycerol	*Uronema parduczi*	TT	10,000.0	36
P, Methyl cellosolve	*Uronema parduczi*	TT	10,000.0	36
P, Triethylene glycol	*Uronema parduczi*	TT	10,000.0	36
P, Triethanolamine	*Uronema parduczi*	TT	10,000.0	36
P, Acetamide	*Uronema parduczi*	TT	10,000.0	36
P, Ethylene glycol	*Uronema parduczi*	TT	10,000.0	36
P, Catechol	*Vorticella campanula*	LD_0	1.6	8
P, *n*-Butyric acid	*Vorticella campanula*	LC_{100}	10.0	8
P, Benzyl chloride	*Vorticella campanula*	LC_{100}	11.0	8
P, Furfural	*Vorticella campanula*	PertL	200.0	8
P, Formic acid	*Vorticella campanula*	LC_{100}	500.0	8

Abbreviations: A = amphibian; Ae = algae; Ba = bacteria; C = crustacean; F = fish; I = insect; M = mammal; Mo = mollusc; O = other; P = protozoa; artf = artificial; d = day(s); dw = distilled water; emb = embryo; exp = exposure; F = flow-through; h = hour; I = intermittent; IF = intermittant flow-through; juv = juvenile; L = larvae; LW = lake water; m = month; mg = milligram; min = minute; ppm = parts per million; RW = river water; S = static; seaW = seawater; semiS = semi static; SW, softW = soft water; syn = synthetic; wk = week; all concentrations are in mg/l unless otherwise specified.

7 References

1. *BECTA6*, 24, 439, 1980.
2. *JAPCA*, Publication # AP, 64, March 1970.
3. *BECTA6*, 22, 159, 1979.
4. Internatl. Experts Discussion Panel on Lead, Rovinj, Yugoslavia, October 1977.
5. *AECTCV*, 5, 1977, "Effects of Tetramethyllead on Freshwater Green Algae".
6. *PNASA6*, USA, 72(12), 5135, 1975.
7. *Environmental Control in the Organic and Petrochemical Industries*, Noyes Data Corp., 1971.
8. "Les Eaux Residuaires Industrielles", 1970.
9. *ESTHAG*, 14(3), 301, 1980.
10. *WATRAG*, 14, 231, 1980.
11. *SIWAAQ*, 23, 1546, 1951.
12. "Vergleichende Befunde der Schadwirkung wassergefahrdender Stoffe gegen Bakerian (*Pseudomonas putida*) und Blaualgen (*Microcystis aeruginosa*)", *Gwf-Wasser/Abwasser*, 1977(9), 1976.
13. *WATRAG*, 10, 231, 1976.
14. *WATRAG*, 10(6), 537, 1976.
15. *ESTHAG*, 10(2), —, 1976, "Absorption Efficiencies for Source Sampling of Hydrogen Sulphide".
16. "Compilation of Odour and Taste Thresholds: Value Data", Amer. Soc. for Testing Materials (ASTM), #48, 1973.
17. Verschuren, K., Ed., *Handbook of Environmental Data on Organic Chemicals*, Van Nostrand Reinhold Co., New York, U.S., 1983.
18. *Water Quality Criteria*, U.S. Public Health Service (PHS) Control Board Pub., 3A, April 1971.
19. *Water*, 6, 21, 1973.
20. *Environmental Effect of Photoprocessing Chemicals*, Vols. I and II, Natl. Assoc. Photographic Manufacturers, Harrison, NY, U.S., 1974.
21. *SIWAAQ*, 24(11), 1397, 1952.
22. *Environmental Control in the Organic and Petrochemical Industries*, Noyes Data Corp., 1971.
23. Stauffer Chemical Company Technical Information, A-10176R, October 1969.
24. Stauffer Chemical Company, Technical Information, A-10104R-71, November 1970.
25. Stauffer Chemical Company, Technical Information, A-10177, July 1973.
26. U.S. Dept. of Interior, Bureau of Fisheries and Wildlife Circ., 143, 1962.
27. *Industrial Hygiene and Toxicology*, Interscience Publishers, Vol. 2, 1967.
28. *PEMNDP*, Worcester, U.K., 1968.
29. *TXAPA9*, 3, 521, 1961.
30. *Pesticide Dictionary, Farm Chemicals Handbook*, Meister Publishing Co., Ohio, 1976.
31. *Environmental Toxicology of Pecticides*, Academic Press, New York, 1972.
32. U.S. Dept. of Fisheries and Wildlife Circ. 84, *Pesticide–Wildlife Review*, 1959.
33. *TXAPA9*, 18, 944, 1971.
34. "Organochlorine Pesticides in Gamebirds of Eastern Tennessee", *Water, Air & Soil Pollution*, 11, 1979.
35. *TXAPA9*, 30, 255, 1974.
36. *Zeit. Wasser/Abwasser Forsch.*, (1), 26, 1980.
37. National Marine Water Quality Laboratory Progress Report F.W.Q.A. Project 18080, GJ4, 1970.
38. U.S. EPA Final Report, EPA-68/01/0151, 1976.

39. *AECTCV,* 3, 410, 1975/76.
40. *MPNBAZ,* 9, 206, 1978.
41. *WATRAG,* 12, 687, 1978.
42. *RIPMAG,* 36(1), 1972
43. *JWPFA5,* 42(8, part 1), 1544, 1970.
44. *AECTCV,* 1(2), 159, 1973.
45. *The Toxicities of Some Insecticides to Four Species of Malocostacan Crustacea,* Fish Pesticide Res. Lab., Columbia, Mo., Bureau of Sport Fish and Wildlife, 1972.
46. *WATRAG,* 8, 739, 1974.
47. *BECTA6,* 21, 439, 1979.
48. *Bestimmung niedermolekularer Chlorkohlenwasserstoffe in Wassern und Schlammen mittels Gas Chromatographie,* Ruhrverband.
49. Faust, S. and Hunter, J., Eds., *Organic Compounds in Aquatic Environment,* Marcel Dekker, New York, 1971.
50. *JWPFA5,* 38(9), 1419, 1966.
51. *JWPFA5,* 46(1), 1, 1974.
52. *SIWAAQ,* 24, 1397, 1975.
53. *STEVA8,* 4, 185, 1975.
54. *MPNBAZ,* 5(8), 116, 1974.
55. "A Report on the Cooperative Blue Crab Study-South Atlantic States", U.S. Dept. of the Interior, Bureau of Commercial Fisheries, 1970.
56. *CPSCAL,* 18, 224, 1977.
57. "Effects of a seawater-soluble fraction of Cook Inlet crude oil and its major aromatic components on larval stages of the Dungeness crab, *Cancer magister dana*", in *Proceedings of NOAA-EPA Symposium on Fate and Effects of Petroleum Hydrocarbons,* Pergamon Press, Oxford, 1977.
58. *JOOEDU,* EE5, 1043, 1976.
59. "Acute Toxiciteitstoetsen met 1,2-dichloroethaan, fenol, acrylonitrile en alkylbenzenesulfonaat in zeewater", Central Laboratory TNO, Delft, Netherlands, 1976.
60. *CRUSAP,* 16(3), 302, 1969.
61. Verschuren, K., Ed., *Handbook of Environmental Data on Organic Chemicals,* Van Nostrand Reinhold Company, New York, U.S., 1983.
62. "Shell Industrie Chemicalien Gids", Shell Nederland Chemie, Nederland, 1.1.1975.
63. Unpublished data, Minnesota Mining and Manufacturing Company.
64. *FSYBAY* 67(2), 383, 1969.
65. *JTEHD6,* 1, 955, 1976.
66. "Toxicity of Pesticides to Crustacean, *Gammarus lacustris*", Bureau of Sport Fisheries and Wildlife technical paper 25, U.S. Government Printing Office, Washington, D.C., 1969.
67. *TAFSAI,* 95, 165, 1966.
68. *JEENAI,* 60, 1228, 1967.
69. *TAFSAI,* 95(2), 165, 1966.
70. "Recherches sur l'elimination des Hydrocarbures par voie Biologique", *Material und Organismen,* 10(2), 1975.
71. Fish Pesticide Research Lab. Annual Report, Columbia, MO, 1971.
72. *ESTHAG,* 12(9), 1062, 1978.
73. *ESTHAG,* 13(5), 594, 1979.
74. *SIWAAQ,* 25(7), 845, 1953.
75. Unpublished work from Louisiana Petroleum Refiners Waste Control Council, taken from reference 61.
76. *SIWAAQ,* 28, 12, 1475, 1956.
77. *JWPFA5,* 37(9), 1310, 1965.
78. National Water Quality Laboratory, Duluth, MN, 1971.
79. See reference 17.
80. Pflanzenschutzmittel-Ruckstande,Verlag E, Ulmer, Stuttgart, 1955.
81. *Botyu Kagaku* (Scientific Pest Control), 36, 189, 1971.
82. "The Effect of Chlorination on Selected Organic Chemicals", U.S. EPA-WPC Research Series, 12020 EXG, 03/72.
83. *BECTA6,* 21, 849, 1979.

84. "Water Quality Criteria", Resources Agency for California, State Water Quality Control Board, 1963.
85. "Toxicology of Thiodan in Several Fish and Aquatic Invertebrates", Bureau of Sport Fisheries and Wildlife, Investigation in Fish Control 35, U.S. Government Printing Office, Washington, D.C., 1970.
86. *Botyu-Kagaku,* 32, 5, 1967.
87. *BECTA6,* 23, 158, 1979.
88. *BECTA6,* 21, 521, 1979.
89. *TJSCAU,* 9(3), 270, 1957.
90. *ENVPAF,* 18, 51, 1979.
91. National Water Quality Laboratory, Duluth, MN, 1971.
92. *AECTCV,* 5, 501, 1977.
93. *WATRAG,* 13, 137, 1979.
94. *BECTA6,* 18(6), 674, 1977.
95. *Sources, Effects and Sinks of Hydrocarbons in the Aquatic Environments,* 519–539, American Institute of Biological Sciences, Washington, D.C., 1976.
96. *Water Quality Criteria Data Book,* Vol. 5, *Effects of Chemicals on Aquatic Life,* U.S. EPA, PB 234435, 1973.
97. *BECTA6,* 16, 508, 1976.
98. *TAFSAI,* 98(3), 438, 1969.
99. Annual Progress Report 1970, Fish-Pesticide Research Laboratory, Bureau of Sport Fish. Wildl., U.S. Dept. of the Interior, Columbia, MO, U.S., 1976.
100. *BECTA6,* 22, 767, 1979.
101. *AECTCV,* 6, 355, 1977.
102. *HYCJAC,* 10, 45, 1972.
103. *TAFSAI,* 78, 1948, "Toxicity of Ferro- and Ferricyanide Solutions to Fish and Determination of the Cause of Mortality".
104. Unpublished report on "Organophosphate Pesticides: Specific Level of Brain AChE Related to Death in Sheepshead Minnows".
105. See reference 17.
106. *BECTA6,* 20, 275, 1978.
107. *WAPLAC,* 9, 323, 1978.
108. *AECTCV,* 5, 353, 1977.
109. Rao, K. R., Ed., *Pentachlorophenol,* Plenum Press, 147, 1978.
110. *BECTA6,* 17, 399, 1977.
111. *BECTA6,* 20, 233, 1978.
112. "Toxicity and Distribution of Aroclor 1254 in Pink Shrimp (*Penaeus duorrum*)", Gulf Breeze Laboratory, EPA, Gulf Breeze, FL.
113. "Some Investigations into the Effect of 1,2-Dichloroethane on Marine Life", Brixham Laboratory report BL/B/1571, 1974.
114. "The Acute Toxicity and Bioaccumulation of Some Lead Alkyl Compounds in Marine Animals", Internatl. Experts Discussion Mtg. on Lead: Occurrence, Fate, and Pollution in the Marine Environment, October 1977.
115. "Acute Toxicities of Organochlorine and Organophosphorus Insecticides to Estuarine Fishes", Bureau of Sport Fish. Wildl., Technical Paper 46, U.S. Government Printing Office, Washington, D.C., 1970.
116. "Nature and Origins of Pollution in Aquatic Systems by Pesticides", in *Pesticides in Aquatic Environments,* Plenum Press, New York, U.S., 1977.
117. *BECTA6,* 18, 361, 1977.
118. *Industrial Pollution Control Handbook,* 14-21, Table 2, McGraw-Hill, 1971.
119. *BECTA6,* 18, 35, 1977.
120. U.S. Dept. of the Interior, Fish & Wildlife Ser., Washington, D.C., Investigations into Fish Control, #73, 1977.
121. U.S. Dept. of the Interior, Fish & Wildlife Ser., Washington, D.C., Investigation into Fish Control, #76, 1977.
122. *TAFSAI,* 99 (1), 20, 1970.
123. *IECHAD,* 46, 324, 1954.
124. "The Toxicity of Phenol, *o*-Chlorophenol and *o*-Nitrophenol to Bluegill Sunfish", Purdue Univ. Eng. Bull. Ext. Ser., 106, 541-555, Lafayette, IN, 1960.

125. *The Biological Effects of Water Pollution* by C. G.Wilber, published by Charles C. Thomas, Springfield, IL, 1969.
126. *IECHAD,* 46(2), 324, 1954.
127. *TAFSAI,* 99(1), 20, 1970.
128. *JWPFA5,* 73, 396, 1974.
129. *AHACBAU,* 6, 137, 1978.
130. *TAFSAI,* 99(1), 20, 1970.
131. *WEEDAT,* 7, 397, 1959.
132. *OCANB6,* October 1970.
133. *CPSCAL,* 18, 227, 1977.
134. "The Effect of Some Organic Cyanides (Nitriles) on Fish", Purdue Univ. Eng. Bull. Ext. Ser., 106, 120-130, Lafayette, IN, 1960.
135. *BECTA6,* 24, 543, 1980.
136. *TAFSAI,* 105, 442, 1976.
137. *AECTCV,* 7, 325, 1978.
138. *AECTCV,* 4, 18, 1976.
139. *BECTA6,* 19(2), 250, 1978.
140. *AECTCV,* 4, 18, 1976.
141. U.S. Fish & Wildl. Ser. Cir. 226, "The Effect of Pesticides on Fish and Wildlife", 51-64, 1965.
142. *BECTA6,* 13, 377, 1975.
143. *PGWTA2,* 7, 599, 1975.
144. "Acute Toxicity of Selected Toxicants to Six Species of Fish", EPA-600/3-76-008, 1976.
145. *CHINAG,* 21, June, 1975, "The Acute Toxicity of 102 Pesticides and Miscellaneous Substances to Fish".
146. "Effects of Temperature on Aquatic Organism Sensitivity to Selected Chemicals", Bulletin No. 106, Virginia Water Resources Res. Ctr., Blacksburg, VA, 1978.
147. *MOSQAU,* 38, 256, 1978.
148. *TAFSAI,* 106, 386, 1977.
149. *JWPFA5,* 48, 2570, 1976.
150. *HYDRDA,* 3, 201, 1978.
151. "Chemical Forest Fire Retardants: Acute Toxicity to Five Freshwater Fishes and a Scud", Tech. Rept. No. 91, U.S. Dept. of the Interior Fish & Wildl. Ser., Washington, D.C., 1977.
152. "Effects of Propylene Oxide on Selected Species of Fish", Rept. No. AFATL-TR-74-183, Air Force Armament Lab. Eglim Air Force Base, FL, U.S., 1974.
153. *ENVRAL,* 10, 397, 1975.
154. From reference 61(p. 1230).
155. *WATRAG,* 11, 811, 1977.
156. "Toxicity of Diazinon to Brook Trout and Fathead Minnows", EPA-600/3-77-060, U.S. EPA, Duluth, MN, 1977.
157. U.S. Dept. of the Interior, Fish & Wildl. Ser., Washington, D.C., Investigations in Fish Control #77, 1977.
158. *HYDRB8,* 56, 49, 1977.
159. *JWPFA5,* 73, 396, 1974.
160. *JFIBA9,* 9, 441, 1976.
161. *PFCUAY,* 37, 143, 1975.
162. "Toxicity of Endrin to Some Species of Aquatic Vertebrates", *Priodoved Pr. Ustavu Cesk. Akad. Ved Brne* (Czech), 11, 1, 1977.
163. *Noyaku Seisan Gijutsu,* 23, 1, 1971.
164. *IJEBAG,* 12, 334, 1974.
165. *IJEBAG,* 16, 689, 1978.
166. *BECTA6,* 23, 725, 1979.
167. "Toxaphene: Chronic Toxicity to Fathead Minnows and Channel Catfish", EPA-600/3-77-069, U.S. EPA, 1977.
168. *BECTA6,* 18, 267, 1977.
169. *AECTCV,* 7, 159, 1978.
170. From reference 61.

171. *BECTA6,* 23, 711, 1979.
172. *MPNBAZ,* 7(9), 1976.
173. "Effects of Crude Oil and Some of Its Components on Young Coho and Sockeye Salmon", EPA-660/3-73-018, January 1974.
174. *PGWTA2,* 9, 787, 1978.
175. Unpublished data.
176. *JFRBAK,* 32, 411, 1975.
177. *BECTA6,* 20, 167, 1978.
178. *Gegenbaurs Morph. Jahrb.,* (Leipzig), 121, 38, 1975.
179. *Gegenbaurs Morph. Jahrb.,* (Leipzig), 120, 439, 1974.
180. *WATRAG,* 13, 207, 1979.
181. *MPNBAZ,* 10, 166, 1979.
182. *JFBRAK,* Tech. Rept. 472, 1974.
183. *WATRAG,* 7, 929, 1973.
184. *La Tribune Du Cebadeau,* 28(374), 3-11, January 1975.
185. *JFRBAK,* 33, 1671, 1976.
186. "Acute Toxicity of Four Organochlorine Insecticides to Two Species of Surf Perch", Fish-Pesticide Res. Lab. Bureau, Sport Fish & Wildl., U.S. Dept. of the Interior, Columbia, MO.
187. *BECTA6,* 16, 376, 1976.
188. *EVETBX,* 7, 428, 1978.
189. "Chronic Toxicities of Methoxychlor, Malathión and Carbofuran to Sheepshead Minnows (*Cyprinodon variegatus*)", EPA Ecological Res. Ser., EPA-600/3-77-059, 1977.
190. "Chronic Toxicity of Chlordane, Trifluralin, and Penta Chlorophenol to Sheepshead minnows (*Cyprinodon variegatus*)", U.S. EPA Ecological Res. Ser., EPA-600/9-78-010, 121, 1978.
191. Int. Register of Potential Toxic Chemicals (IRPTC)—Databases, Health and Welfare Canada, Ottawa, Canada, 1990.
192. *CPSCAL,* 17, 125, 1976.
193. *BECTA6,* 19(3), 312, 1978.
194. *PFCUAY,* 29(2), 67, 1967.
195. *LIOCAH,* 4, 53, 1959.
196. "Acute Toxicity of Selected Organic Compounds to Fathead Minnows", EPA-600/3-76-097, 1976.
197. *WATRAG,* 100, 685, 1976.
198. *JFRBAK,* 33, 1303, 1976.
199. *TAFSAI,* 104, 567, 1975.
200. *WATRAG,* 11, 31, 1977.
201. "Toxicity of Four Pesticides to Water Fleas and Fathead Minnows", EPA-600/3-76-099, U.S. EPA, Duluth, MN, 1976.
202. *WATRAG,* 10, 165, 1976.
203. *BECTA6,* 21, 576, 1979.
204. "A Rapid Assessment of the Toxicity of Three Chlorinated Cyclodiene Insecticide Intermediates to Fathead Minnows", EPA-600/3-77-099, U.S. EPA, Duluth, MN, 1977.
205. *JFRBAK,* 33, 209, 1976.
206. *ENPBBC,* 5, 361, 1975.
207. Konemann, W. H., "Quantitative Structure-Activity Relationships for Kinetics and Toxicity of Aquatic Pollutants and Their Mixtures", Univ. Utrecht, Netherlands, 1979.
208. *WATRAG,* 9, 211, 1975.
209. *BECTA6,* 2(3), 147, 1967.
210. "Some Effects of Endrin on Estuarine Fishes", *Proc. Southeast Assoc. Game Fish Commissioners,* 19, 271, 1965.
211. *TAFSAI,* 90(3), 264, 1961.
212. "Metabolism of Pentachlorophenol in Fish", in *Pesticide and Xenobiotic Metabolism in Aquatic Organisms,* American Chemical Society Symposium Series 99, 1979.
213. "Uber die Wirkung der Grenzflachenaktivitat auf Fische", *Munch. Beitr. Abwass. Ficherei-und Flussbiol.,* 9(2), 1967.
214. *BECTA6,* 14, 61, 1975.
215. McGauhey, P. H., *Engineering Management of Water Quality,* McGraw-Hill, New York, 1968.

216. Stauffer Chemical Company Technical Information, A–10423, March 1972.
217. "Toxicity of Hexachlorobutadiene in Aquatic Organisms", in *Sublethal Effects of Toxic Chemicals in Aquatic Animals* (Koeman and Strik, Eds.), 167, Elsevier, NY, U.S., 1975.
218. *WATRAG,* 10, 165, 1976.
219. "Detection and Measurement of Stream Pollution", U.S. Bureau of Fish., Bull. No. 22, XLVIII, 365, U.S. Dept. of Commerce, Washington, D.C., USA, 1937.
220. U.S. Dept. of Commerce, Bureau of Fisheries, Bull. 22, XLVII, 1937.
221. *EVETBX,* 5, 1053, 1976.
222. *BECTA6,* 15, 720, 1976.
223. "Gerusch- und Geschmacks-Schwellen-Konzentrationen von Phenolkorper", *Gwf-Wasser/Abwasser,* 119(6), 1978.
224. Dawson, G. W., Jennings, A. L., Drozdowski, D., and Rider, E., "The Acute Toxicity of 47 Industrial Chemicals to Fresh and Saltwater Fishes", *J. Hazardous Materials,* 1, 1975/77.
225. *PEMNDP,* U.K., 1968.
226. *WATRAG,* 11, 889, 1977.
227. Hughes, J. and Davis, J., "Variation in Toxicity to Bluegill Sunfish of Phenoxy Herbicides", *Weeds,* 11(1), 50, 1963.
228. "Comparative Toxicity to Bluegill Sunfish of Granular and Liquid Herbicides", *Proceedings of the 16th Annual Conference of Southeast Game and Fish Commissioners,* 319-323, 1962.
229. "Effect of Selected Herbicides on Bluegill Sunfish", *Proceedings of the 18th Annual Conference of Southeastern Association of Game and Fish Commissioners,* 480-482, 1964.
230. "Toxicity of Various Herbicidal Materials to Fish", R. A. Taft Sanitory Eng. Ctr. Tech. Rep. W603, 96-101, 1960.
231. *PNASA6,* 23, 77, 1960.
232. *TAFSAI,* 88(1), 23, 1959.
233. *Notulae Natur.* (Philadelphia), No. 370, 1-10, 1964.
234. "Biological Problems in Water Pollution", Third Seminar(1961), U.S. Public Health Service, Publication No. 999-WP-25, 247-249, 1961.
235. *TAFSAI,* 91(2), 175, 1962.
236. *WATRAG,* 4, 673, 1971.
237. *TAFSAI,* 96(2), 185, 1967.
238. *PFCUAY,* 24(4), 164, 1962.
239. *Relationship Between Organic Chemical Pollution of Freshwater and Health,* Fresh Water Quality Association, Report #71632, A.D. Little, December 1970.
240. The Institute of Advanced Sanitation Research International, *Hazardous Chemicals Handling and Disposal,* Noyes Data Corp., 1971.
241. "Factors Affecting Pesticide-Induced Toxicity in Estuarine Fish", Bureau of Sprt. Fish & Wildl., Technical Paper No. 45, U.S. Government Printing Office, Washington, D.C., 1970.
242. *BECTA6,* 14, 281, 1975.
243. *SIWAAQ,* 29(6), 695, 1957.
244. *PCBPBS,* 7, 297, 1977.
245. "The Disposition and Biotransformation of Organochlorine Insecticides in Insecticide-Resistant and -Susceptible Mosquito Fish", in *Pesticide and Xenobiotic Metabolism in Aquatic Organism,* American Chemical Society Symposium Ser. 99, 1979.
246. *MOSQAU,* 36, 322, 1976.
247. "Acute Toxicity to and Bioconcentration of Endosulfan by Estuarine Animals", in *Aquatic Toxicology and Hazard Evaluation* (Mayer and Hamelink, Eds.), ASTM, 241, 1977.
248. *MPNBAZ,* 9, 220, 1978.
249. *AQFEDI,* 44, 1973.
250. *BECTA6,* 23, 153, 1979.
251. *IRPTC Databases,* Health and Welfare Canada, Ottawa, Canada, 1990.
252. *TAFSAI,* (3), 264, 1961.
253. *IRPTC Databases,* Health and Welfare Canada, Ottawa, Canada, 1990.
254. *ENPBBC,* 4, 226, 1974.
255. *CRUSAP,* 16, 302, 1969.
256. *BECTA6,* 20, 344, 1978.

257. *IRPTC Databases,* Health and Welfare Canada, Ottawa, Canada, 1990.
258. *NATUAS,* 233, 120, 1971.
259. *DOWCC,* June 1974.
260. *BECTA6,* 5(5), 408, 1970.
261. *TJSCAU,* 3, 391, 1951.
262. *TAFSAI,* 99(3), 489, 1970.
263. *LIOCAH,* 13, 112, 1968.
264. Unpublished data on Toxicity of Solvents on Rainbow Trout Fingerlings, Tiburon Laboratory, NOAA, July, 1974.
265. Pesticide Wildlife Studies, U.S. Dept. of the Interior Fish & Wildl. Ser. Circ. 199, 1963.
266. *BECTA6,* 23, 250, 1979.
267. *WATRAG,* 13, 217, 1978.
268. *CMSHAF,* 7, 215, 1978.
269. *JFRBAK,* 32(8), 1249, 1975
270. *TAPPAP,* 59, 129, 1976.
271. *BECTA6,* 13, 518, 1975.
272. *WATRAG,* 10, 303, 1976.
273. *JFIBA9,* 10, 575, 1977.
274. *JFRBAK,* 30, 1047, 1973.
275. *WATRAG,* 14, 515, 1980.
276. *TAFSAI,* 105, 322, 1976.
277. *WATRAG,* 9, 1163, 1975.
278. *JWPFA5,* 48, 183, 1976.
279. *WATRAG,* 10, 303, 1976.
280. *JJIND8,* 60, 1127, 1978.
281. *JFRBAK,* 32, 2556, 1975.
282. *IRPTC Databases,* Health and Welfare Canada, Ottawa, Canada, 1990.
283. *BECTA6,* 13, 518, 1975.
284. "Residues of Emulsified Xylene in Aquatic Weed Control and Their Impact on Rainbow Trout, *Salmo gairdneri*", Rep. No. REC-ERC-77-11, Engineering & Res. Ctr. Bureau of Reclamation, Denver, CO, U.S., 1977.
285. Dissertation Abstr. 39, 7813246, 1978.
286. *HYDRDA,* 3, 201, 1978.
287. *Trans. Soc. Min. Engr.* 256, 337, 1974.
288. *AECTCV,* 7, 317, 1978.
289. *BECTA6,* 18, 35, 1977.
290. *Industrial Water Pollution Control* (Eckenfelder,W.W.), McGraw-Hill, 1970.
291. *AECTCV,* 7, 457, 1978.
292. *JFIBA9,* 10, 575, 1977.
293. *TAFSAI,* 90(4), 394, 1961.
294. *IJEBA6,* 13, 185, 1975.
295. *IRPTC Databases,* Health and Welfare Canada, Ottawa, Canada, 1990.
296. *UBZAD4,* 40, 824, 1946.
297. *TAFSAI,* 93(4), 357, 1964.
298. *JWPFA5,* 38(8), 1273, 1966.
299. "In Vivo and In Vitro Studies of Mixed Function Oxidase in an Aquatic Insect, *Chironomus Riparius*", in *Pesticide and Xenobiotic Metabolism in Aquatic Organisms,* American Chemical Society, Symp. Ser. 99, 1979.
300. *BECTA6,* 1, 1973, "Biological Concentration of the Photo-isomers of Cyclodiene Insecticides and Their Metabolites".
301. *AECTCV,* 5, 63, 1976.
302. *AECTCV,* 5, 353, 1977.
303. *IRPTC Databases,* Health and Welfare Canada, Ottawa, Canada, 1990.
304. *Federal Register,* 39(125), subpart G, June 1974.
305. *IRPTC Databases,* Health and Welfare Canada, Ottawa, Canada, 1990.
306. U.S. Dept. HEW, (Bioassay of 1,4-Dioxane), TRS No. 80, NIH Pub. No. 78-1330, 1978.

307. *Proceedings of TCDD Pollution,* Commission of the European Communities, September 30, 1977.
308. *TXAPA9,* 18, 398, 1971.
309. *Toxikologisch-Arbeitmedizinische Begrundung von MAK-Verte* (Henschler, D.), Verlag Chemie.
310. *BECTA6,* 20, 819, 1978.
311. "Aldehydes and Acetals", in *Industrial Hygiene and Toxicology* (Patty, F. A., Ed.), Vol. II, John Wiley, 1963.
312. *STEVA8,* 5, 253, 1976.
313. *AIHAAP,* 38(5), —, 1977, "The Toxicity of 1,3-Dichloropropene as Determined by Repeated Exposure of Laboratory Animals".
314. *STEVA8,* 9, 1, 1978.
315. *DOWCC,* "Material Safety Data Sheet", 1978.
316. Summer, W., *Odour Pollution of Air, Causes and Control,* CRC Press, Cleveland, OH, 1971.
317. *CENEAR,* Feb. 12, 1979.
318. "Action des Hydrocarbures Chlorofluroes(HCCF) Utilises dans les Aerosols. Problemes de leur Retention par l'Organisme apres Inhalation", *Cebedeau* (Nov. 1970), Liege, Belgium.
319. *Farmakol i. Toksikol.* 26, 750, 1963; *Chem. Abstr.* 60:1377b, 1964.
320. "The Toxicity of Hexachloroethane in Laboratory Animals", *AIHAJ,* 40(3), 1979.
321. "Fluorocarbons in the Los-Angeles Basin", *J. Air Poll. Control Assoc.,* 24(6), 1974.
322. *The Determination of Toxic Substances in Air* (Strafford, Strouts, and Stublings), Haffer, Cambridge, U.K.
323. *ATENBP,* 7, 551, 1973.
324. *Effects of Contaminants Other than Sulfur Dioxide, on Vegetation and Animals,* (Katz), Canadian Council of Resource Ministers, Montreal, P.Q. Canada, Oct.-Nov., 1966.
325. "L'Hygiene et Securite dans la Grande Industrie Chemique", Inst. Natl. de Recherche et de Securite pour la Prevention des Accidents du Travail et des Maladies Professionnelles, Paris, France, 1971.
326. "Review of the Toxicity of Hexachlorobenzene and Hexachlorobutadiene", Spl. Release, Dow Chemical, MI, U.S., 1971.
327. "Toxicological Study of the Insecticide, Hexachlorobutadiene", Chem. Abstr. 62:13757c, 1965.
328. *AIHAAP,* 40(4), —, 1979, "Respiratory Retention and Acute Toxicity of Furan".
329. "Evaluation of Toxicity of Pesticides Residues in Food", FAO Mtg. Rept. No. PL/1965/10, WHO/Food Add/26, 1965.
330. *AEHLAU,* 15, 739, 1967.
331. *AECTCV,* 5, 97, 1976.
332. *AIHAAP,* 40(2), A31, 1979.
333. "Analysis of Kepone", Carcinogen Assessment Group, U.S. EPA, July 1976.
334. U.S. Dept. HEW, "Bioassay of Dicofol", TRS no. 90, Publication No. (NIH) 78-1340, 1978.
335. *AIHAAP,* 40(3), —, 1979, "A Short-Term Test to Predict Acceptable Levels of Exposure to Air-Borne Sensory Irritants".
336. "Up-to-date Conclusions and Comments on the Long-Term Carcinogenicity Bioassays of Dichloroethane", Tumor Centre and Institute Oncology of Bologna, Italy, Report to the EEC, May 8, 1978.
337. U.S. Testing Company Inc., Report of Test No. 4299, Feb. 28, 1949.
338. *IRPTC Databases,* Health and Welfare Canada, Ottawa, Canada, 1990.
339. *JPETAB,* 38, 161, 1930
340. *BJIMAG,* 27, 1970, "The Subacute Inhalation Toxicity of 109 Industrial Chemicals".
341. *JAFCAU,* 20, 944, 1972.
342. *STEVA8,* 12, 101, 1979.
343. *ENVRAL,* 19, 460, 1979
344. "Occupational Bladder Tumors and Carcinogens", in *Bladder Cancer: A Symposium* (Deichmann and Lampe, Eds.), Aesculapius Publications, Birmingham, AL, 1967.
345. *Analytical Methods for Organic Compounds in Sewage Effluents,* Water Research Centre, Stevenage Laboratory, Stevenage, Herts, U.K., 1975.
346. "Acute and Sub-Acute Inhalation Toxicity of Peroxyacetyl Nitrate and Ozone in Rats", in *VDI-Berichte* No. 270. Ozon und Begleitsubstanzen in Photochemischen Smog, 101–109, 1977.
347. *AIHAAP,* 38(11), 589, 1977, "Results of a Two Year Chronic Toxicity Study With Hexachlorobutadiene in Rats".

348. *AOHYA3,* 3, 226, 1961.
349. Unpublished Data from Ethyl Corporation.
350. *AECTCV,* 5, 1977.
351. "Report on the Toxicity of Hexachlorobutadiene Vapour", Communication Release, Kettering Laboratory, University of Cincinnati, 1948.
352. *Kievsk. Med. Inst.* 158, 1964.
353. *NATUAS,* 189, 449, 1961.
354. *JWPFAF,* 47(1), 57, 1975, "Removal of Wastewater Organics by Reverse Osmosis".
355. *Veszelyes Novenyvedoszerek,* 6 Edn., Mezogard Kiad, Budapest, 1967.
356. *Die Insetizide,* 2 Edn., Huthig Verlag, Heidelberg, Germany, 1968.
357. *Pesticides in the Modern World,* p. 19, Newgate, London, U.K., 1972.
358. Unpublished Report, Lederle Division, American Cyanamid Company.
359. *BECTA6,* 22, 745, 1979.
360. "Suggested Guide for the Use of Insecticides to Control Insects Affecting Crops, Livestock, Households, Stored Products, Forests and Forest Products–1968", U.S. Dept. of Agriculture Handbook 331, 1968.
361. *BECTA6,* 22, 293, 1979.
362. *IRPTC Databases,* Health and Welfare Canada, Ottawa, Canada, 1990.
363. "Bioassay of Malathion", TRS No. 24, U.S. Dept. HEW Publication No. (NIH) 78-824, 1978.
364. *EXPEAM,* 29(5), 622, 1973.
365. *APMBAY,* 10(6), 532, 1962.
366. *BECTA6,* 20, 149, 1978.
367. *WATRAG,* 13, 285, 1979.
368. Rapp. P. v. Reunm. Cons. Int. Explor. Mer. 171, 189, 1977.
369. *JTEHD6,* 1, 485, 1976.
370. *AQCLAL,* 7, 293, 1976.
371. British Weed Control Conf., Brighton, England, 12, 239, 1974.
372. *CMSHAF,* 11, 891, 1982.
373. *EVHPAZ,* 14, 26, 1976.
374. *ABWQC,* 81/117327, B-15, 1980.
375. *ABWQC,* B-27, 1979.
376. *ABWQC,* 81/117863, B-2, 1980.
377. *JFRBAK,* 35, 1366, 1978.
378. *MPNBAZ,* 14, 303, 1983.
379. *AJMFA4,* 31, 795, 1980.
380. *PRNEN,* 70-15, Table 11, 1971.
381. *CRCMY,* 1, 29, 1981.
382. *EVETBX,* 5, 701, 1976.
383. *BECTA6,* 25, 69, 1980.
384. *ENVCO,* VIR-214, 23, 1980.
385. *AQTODG,* 8, 163, 1986.
386. *EPEBD7,* 38, 273, 1985.
387. *AQTODG,* 5, 245, 1984.
388. *JWPFA5,* 46, 63, 1974.
389. *JWPFA5,* 55, 63, 1983.
390. *ESTHAG,* 17, 107, 1983.
391. *ZANZA9,* 48, 325, 1981.
392. *DOWAS,* 1981.
393. *WATRAG,* 12, 687, 1978.
394. *BIBUX,* 145, 340, 1973.
395. *PERAS,* 1, 253, 1980.
396. *NRCNDA,* 14098, 39, 45, 46, 1975.
397. *47EXAM,* 33–7, 1980.
398. *BECTA6,* 21, 439, 1979.
399. *VOLNAB,* 898, 112, 1980.
400. *AHCBAU,* 14, 627, 1986.

401. *APTODG,* 4, 73, 1983.
402. *RPFWDE,* 137, 27, 1980.
403a. *EPAPW,* 77, 1972.
403b. *RPFWDE,* 137, 31, 33, 1980.
404. *RPFWDE,* 160, 1986.
405. *XBTBT,* 66, 1-19, 1972.
406. *BECTA6,* 21, 643, 1979.
407. *TAFSAI,* 96(1), 25, 1967.
408. *ETOCDK,* 5, 831, 1986.
409. *RPFWDE,* 137, 63, 1980.
410. *RPFWDE,* 137, 34, 1980.
411. *RPFWDE,* 137, 66, 1980.
412. *BECTA6,* 33, 325, 1984.
413. *EESADV,* 7, 552, 1983.
414. *SARIA3,* 67, 171, 1982.
415. *TAFSAI,* 96, 25, 1967.
416. *PNSFAN,* 51, 23, 1960.
417. *XFWCAW,* 167, 11, 1963.
418. *CPSCAL,* 18, 224, 1976.
419. *EPAQC,* 176, 1976.
420. *NRCNDA,* 16079, 153, 159, 1978.
421. *EVHPAZ,* 1, 159, 1974.
422. *EPAMK,* 88, 1979.
423. *ABWQC,* B14, 1978.
424. *AQTODG,* 2, 301, 1982.
425. *EPMMM,* 600/3-76, 007, 1976.
426. *BECTA6,* 21, 76, 1979.
427. *ABWQC,* 81/117834, 1980.
428. *AECTCV,* 3, 371, 1975.
429. *EPAMK,* 72, 1978.
430. *JFRBAK,* 26, 1969.
431. *CRAFG,* 1970.
432. *WDFPH,* 210, 1977.
433. *MPNBAZ,* 12, 305, 1981.
434. *EPRDC,* 1977.
435. *MPNBAZ,* 12, 385, 1981.
436. *AECTCV,* 8, 383, 1979.
437. *37BKAP,* 210, 1977.
438. *MPNBAZ,* 3, 190, 1972.
439. *OUBUC,* 56, 1970.
440. *MBIOAJ,* 49, 113, 1978.
441. *XCSPP,* 87-11801, 1135, 1985.
442. *CRRPC3,* E:13, 1983.
443. *MAFFF,* 22, 1971.
444. *EPAWG,* 5, A-399, 1973.
445. *BECTA6,* 25, 802, 1980.
446. *RUOMAY,* 43, 79, 1976.
447. *ENVPAF,* 8, 1, 1975.
448. *AJMFA4,* 31, 75, 1980.
449. *EPAMM,* 1976.
450. CRSBAW, 173, 1105, 1979.
451. *ETOCDK,* 5, 303, 1986.
452. *NETSR,* R-84/59-10455, 1984.
453. *CRUSAP,* 16, 302, 1969.
454. *CRRPC3,* 4, 22, 1974.
455. *CAFGAX,* 63, 204, 1977.

456. *CMSHAF,* 8, 53, 1979.
457. *BECTA6,* 25, 921, 1980.
458. *ABWQC,* 11757, B-29, 1980.
459. *FRTDDJ,* 70, 1982.
460. *NRCNDA,* 18789, 112, 123, 1982.
461. *EPAFP,* 2, A-307, 1981.
462. *MPNBAZ,* 5, 116, 1984.
463. *RREVAH,* 62, 159, 1976.
464. *TNORE,* MD-NE76/1, 1976.
465. *SAX,* 2, 33, 1982.
466. *BECTA6,* 22, 796, 1979.
467. "Effects of Dursban in Shiner Perch", in *Effects of Pesticides on Estuarine Organisms,* U.S. Public Health Service Research Report, 63-76, U.S. Government Printing Office, Washington, D.C., 1969.
468. *XENOBH,* 15, 1103, 1985.
469. *NRCNDA,* 18075, 178, 1978.
470. *JWPFA5,* 42, 1544, 1969.
471. *ENVRAL,* 15, 357, 1978.
472. *ACPTF,* PB-257-80, 1976.
473. *HYDRA8,* 59, 141, 1978.
474. *BECTA6,* 14, 214, 1976.
475. *RPFWDE,* 37, 10, 11, 15, 1980.
476. *TAFSAI,* 103, 562, 1974.
477. *AECTCV,* 13, 595, 1984.
478. *AECTCV,* 9, 53, 1980.
479. *ABWQC,* 81/117764, B-22, 1980.
480. *49WEAZ,* 3, 537, 1983.
481. *NRCNDA,* 18979, 112, 1982.
482. *RPFWDE,* 137, 11, 1980.
483. *RPFWDE,* 137, 15, 1980.
484. *RPFWDE,* 137, 50, 1980.
485. *RPFWDE,* 137, 58, 1980.
486. *AECTCV,* 12, 661, 1983.
487. *JFRBAK,* 31, 1556, 1974.
488. *BECTA6,* 19, 465, 1970.
489. *JFRBAK,* 32, 411, 1975.
490. *TJEWP,* 1-47, 1982.
491. *BECTA6,* 27, 309, 1981.
492. *PSSCBG,* 12, 417, 1981.
493. *DGMTAO,* 27, 77, 1983.
494. *ZWABAQ,* 15, 1, 1982.
495. *CMSHAF,* 10, 891, 1982.
496. *RREVAH,* 77, 189, 1981.
497. *JEENAI,* 56, 1043, 1965.
498. *APTOA6,* 37, 94, 1975.
499. *BECTA6,* 33, 598, 1984.
500. *SCIEAS,* 154, 289, 1966.
501. *ZWABAQ,* 10, 161, 1977.
502. *BECTA6,* 24, 684, 1980.
503. *CMSHAF,* 9, 753, 1980.
504. *DOEAAH,* 20, 6, 1985.
505. *EINUDQ,* 72, 47, 1983.
506. *TAFSAI,* 95, 165, 1966.
507. *MEPCE,* 1981.
508. *NEPHBW,* 15, 29, 1976.
509. *EHCRDN,* 65, 1987.
510. *CCHAT,* 16, 582, 1969.

511. *RPFWDE,* 137, 98, 1980.
512. *PRIAM,* 66-10936, 1965.
513. *RPFWDE,* 137, 49, 50, 1980.
514. *AHBPAX,* 23, 183, 1981.
515. *ORNLT,* 81/13240, 5-75, 1979.
516. *ORGAD2,* 186, 1981.
517. *TPFSDO,* 1, 1983.
518. *EPAPW,* 85, 1972.
519. *BECTA6,* 35, 546, 1985.
520. *NAPFA,* 279, 1981.
521. *RPFWDE,* 137, 53, 58, 1980.
522. *ASTTA8,* 854, 87, 1985.
523. *MONSD,* 1985.
524. *IFCRAG,* 35, 1, 1970.
525. *ASTTA8,* STP802, 30, 1983.
526. *HYDRB8,* 59, 125, 1978.
527. *RPFWDE,* 137, 47, 1980.
528. *RREVAH,* 76, 173, 1980.
529. *ABWQC,* 81/117657, B-6, 1980.
530. *GPOPI,* 146, 1971.
531. *CMSHAF,* 12, 1121, 1983.
532. *RPFWDE,* 137, 68, 1980.
533. *ASTTA8,* STP737, 1981.
534. *EPAPW,* 76, 1972.
535. *CHAVF,* PB-255439, 20, 1976.
536. *EPAFP,* A-95, 1981.
537. *RPFWDE,* 137, 24, 1980.
538. *ABWQC,* 81-11741, B-1, 1980.
539. *ABWQC,* 81-117673, B-1, 1980.
540. *ABWQC,* 81-117871, B-1, 1980.
541. *AQTODG,* 7, 145, 1985.
542. *CHAVF,* PB-255439, 1-26, 1976.
543. *JFRBAK,* 29, 1691, 1972.
544. *DTVCA,* 81-111098, 1-17, 1980.
545. *NRCNDA,* 15015, 267, 1977.
546. *EPEBD7,* 24, 21, 1981.
547. *OTSEA,* 1978.
548. *HYDRB8,* 59, 135, 1978.
549. *AECTCV,* 12, 661, 679, 1983.
550. *ABWQC,* 81/117855, B-5, 1980.
551. *AQTODG,* 5, 143, 1984.
552. *AQTODG,* 37, 94, 1975.
553. *EPAQC,* 134, 1976.
554. *HYBUD9,* 9, 102, 1975.
555. *CHIPS,* 7, 1985.
556. *WJTFA,* 73, 1982.
557. *JEPTDQ,* 1, 315, 1978.
558. *JEENAI,* 60, 1228, 1967.
559. *PANYAZ,* 24, 583, 1977.
560. *37KXA7,* 3, 599, 1980.
561. *CINFS,* 1984.
562. *AECTCV,* 8, 269, 1979.
563. *CBINA8,* 9, 245, 1974.
564. *GLCTD,* 1984.
565. *ABWQC,* 81/117285, B-5, 1980.
566. *ZANCA8,* 192, 94, 1963.

567. *JTEHD6,* 5, 985, 1979.
568. *MBIOAJ,* 48, 215, 1978.
569. *MBIOAJ,* 66, 179, 1982.
570. *PRLBA4,* 189, 305, 1975.
571. *MPNBAZ,* 8, 138, 1977.
572. *WATRAG,* 6, 1181, 1972.
573. *WATRAG,* 9, 607, 1975.
574. *AECTCV,* 7, 129, 1978.
575. *EVETBX,* 23, 1251, 1982.
576. *RREVAH,* 71, 159, 1979.
577. *NRCNDA,* 22494, 126, 1985.
578. *IVLBDQ,* B677, 1982.
579. *BECTA6,* 7, 182, 1972.
580. *EVHPAZ,* 1, 162, 1972.
581. *EESADV,* 12, 233, 1987.
582. *BROAW,* 146, 1978.
583. *NRCNDA,* 16079, 158, 1978.
584. *NRCNDA,* 14094, 65, 1974.
585. *XBTPBT,* 25, 1969.
586. *TPFSDO,* 25, 1969.
587. *AECTCV,* 7, 139, 1978.
588. *RREVAH,* 57, 98, 1975.
589. *ZANZA9,* 48, 325, 1961.
590. *JANCA2,* 55, 895, 1972.
591. *AGROB2,* 19, 197, 1979.
592. *FWPCA,* 62, 1968.
593. *GPOPI,* 12, 1971.
594. *NRCNDA,* 14094, 64, 1974.
595. *TAFSAI,* 113, 74, 1984.
596. *RPFWDE,* 22, 1980.
597. *RPFWDE,* 137, 35, 1980.
598. *EPAPW,* 68, 1972.
599. *RPFWDE,* 137, 62, 1980.
600. *RPFWDE,* 137, 36, 1980.
601. *XPTPBT,* 25, 1-18, 1969.
602. *RPFWDE,* 137, 38, 1980.
603. *EPAPW,* 62, 1972.
604. *RPFWDE,* 137, 18, 1980.
605. *EPAPW,* 70, 1972.
606. *RPFWDE,* 137, 69, 1980.
607. *RPFWDE,* 137, 75, 1980.
608. *AECTCV,* 10, 321, 1981.
609. *RPFWDE,* 137, 12, 15, 1980.
610. *TPFSDO,* 1, 1983.
611. *CPPCBB,* 52C, 75, 1975.
612. *MOSQAU,* 36, 294, 1976.
613. *BECTA6,* 23, 91, 1979.
614. *BECTA6,* 19, 250, 465, 1978.
615. *RREVAH,* 57, 90, 1975.
616. *MPNBAZ,* 14, 213, 1983.
617. *CMSHAF,* 11/12, 849, 1979.
618. *PRMBP,* 185, 1979.
619. *NASPM,* 79, 1975.
620. *JFRBAK,* 31, P1556, P1949, 1974.
621. *CJZOAG,* 54, 1231, 1976.
622. *MPNBAZ,* 3, 105, 1972.

623. *BECTA6,* 21, 338, 1979.
624. *BECTA6,* 16, 508, 1976.
625. *BECTA6,* 10, 305, 1973.
626. *HBEDC,* 906, 1173, 1983.
627. *TAFSAI,* 98, 438, 1969.
628. *EHCRDN,* 36, 37, 1984.
629. *XNCEL,* 1, 74, 1960.
630. *XNCEL,* 1, 82, 1960.
631. *XNCEL,* 1, 96, 1960.
632. *XNCEL,* 1, 115, 1960.
633. *XNCEL,* 1, 117, 1960.
634. *XNCEL,* 1, 64, 1960.
635. *XNCEL,* 1, 65, 1960.
636. *XNCEL,* 1, 97, 1960.
637. *NRCNDA,* 14104, 71, 1975.
638. *CRUSAP,* 481, 1985.
639. *TAFSAI,* 107, 493, 1978.
640. *AZOFAO,* 16, 209, 1979.
641. *EPEBD7,* 23, 299, 1980.
642. *AQTODG,* 7, 25, 1985.
643. *ASTME,* 109, 1977.
644. *ABWQC,* 81, 117327, B-23, 1980.
645. *ABWQC,* 81/117608, B-5, 1980.
646. *ABWQC,* B-20, 1978.
647. *ACTBP,* 1981.
648. *ABWQC,* B-16, 1978.
649. *ABWQC,* 81-117608, B-4, 1980.
650. *ABWQC,* 81-117392, B-10, 1980.
651. *ABWQC,* 81-117541, B-2, B-10, 1980.
652. *EPAME,* 29, 1981.
653. *EPARL,* 1980.
654. *EPAIT,* 1981.
655. *ESTHAG,* 4, 301, 1980.
656. *NJRPH,* 21, 1981.
657. *MBIOAJ,* 27, 75, 1974.
658. *JAFCAU,* 31, 104, 1983.
659. *HYBOD9,* 93, 179, 1982.
660. *ABWQC,* 81-117509, 1980.
661. *ABWQC,* 81-117434, B-7, 1980.
662. *ABWQC,* 81-117749, B-2, B-3, 1980.
663. *ABWQC,* 81-117830, B-2, 1980.
664. *ABWQC,* 81-117855, B-10, 1980.
665. *EESADV,* 14, 260, 1987.
666. *CMSHAF,* 8, 843, 1979.
667. *CMSHAF,* 8, 846, 1979.
668. *CMSHAF,* 13, 613, 1984.
669. *CHIPS,* 21, 1985.
670. *MPNBAZ,* 9, 238, 1978.
671. *MPNBAZ,* 13, 125, 1982.
672. *EESADV,* 81-117327, 127, 1986.
673. *IVLBDQ,* B676,1982.
674. *EVHPAZ,* 1, 159, 1972.
675. *JPRCD2,* B14, 579, 1979.
676. *EPAWG,* 139, 1976.
677. *TAFSAI,* 107, 825, 1978.
678. *RPFWDE,* 137, 46, 1980.

679. *CUSCAM,* 50, 334, 1981.
680. *AECTCV,* 6, 315, 1977.
681. *AECTCV,* 7, 23, 1978.
682. *USDI3,* 10, 1970.
683. *XBRPAI,* 106, 10, 1970.
684. *RAORK,* 193, 1978.
685. *BECTA6,* 9, 129, 1973.
686. *WALPAC,* 25, 33, 1985.
687. *BECTA6,* 15, 297, 1976.
688. *WALPAC,* 9, 323, 1978.
689. *ORNLT,* 81-13240, 5-73, 1979.
690. *TAFSAI,* 99, 696, 1970.
691. *PPMOV,* 285, 1974.
692. *EPAPQ,* 135, 1976.
693. *JTEHD6,* 1, 955, 1976.
694. *AECTCV,* 5, 353, 1977.
695. *ECMSC6,* 6, 365, 1978.
696. *WATRAG,* 13, 137, 1979.
697. *BECTA6,* 17, 399, 1977.
698. *RAORK,* 141, 147, 1978.
699. *RAORK,* 181, 1978.
700. *45ZOAI,* 37, 1981.
701. *TATHE,* 28, 1975.
702. *PLEPO,* 39, 1974.
703. *NRCNDA,* 17583, 85, 1980.
704. *JHYDA7,* 51, 259, 359, 1981.
705. *AJMFA4,* 29, 1, 1978.
706. *GPOPI,* 118, 1971.
707. *AJMFA4,* 33, 459, 1982.
708. *MAFFF,* 22, 1971.
709. *PSEBAA,* 154, 151, 1977.
710. *NRCNDA,* 17589, 94, 95, 1981.
711. *EPAWP,* 101, 1972.
712. *NAWTA6,* 36, 171, 1971.
713. *MBIOAJ,* 25, 191, 1971.
714. *EPAQC,* 130, 132, 1976.
715. *MBIOAJ,* 37, 75, 1976.
716. *BECTA6,* 5, 171, 1970.
717. *EHCRDN,* 36, 1984.
718. *AQCLAL,* 7, 293, 1976.
719. *TJSCAO,* 9, 270, 1957.
720. *XFWCAW,* 20, 1970.
721. *CCRDR,* 21349, 1966.
722a. *EPAQC,* 262, 1976.
722b. *CECPP,* 36, 1979.
723. *EPAWG,* 3, B-107, 1971.
724. *EPAWG,* 3, B-130, 1971.
725. *EPAPI,* 81-216815, 15, 1980.
726. *EPAME,* 28, 1981.
727. *AECTCV,* 6, 355, 1977.
728. *JTEHD6,* 1, 485, 1976.
729. *ORNLT,* 81-13240, 5-69, 1979.
730. *ORNLT,* 81-13240, 5-66, 1979.
731. *CMSCAU,* 22, 193, 1979.
732. *CHECDY,* 1, 131, 1982.
733. *PCGFAM,* 28, 179, 1975.

734. *PGWTAZ,* 7, 579, 1975.
735. *EPAMK,* 1–78, 74, 1979.
736. *BECTA6,* 6, 89, 1971.
737. *JEENAI,* 66, 70, 1973.
738. *BECTA6,* 8, 334, 1972.
739. *TAFSAI,* 92, 428, 1963.
740. *LOAGA2,* 23, 8–9, 1980.
741. *BECTA6,* 14, 281, 1975.
742. *LOAGA2,* 16, 14, 1973.
743. *PFCUAY,* 42, 169, 1980.
744. *WSWOAC,* 12, 276, 1965.
745. *EPARL,* 600/3–79, 104, 1979.
746. *CRTXB2,* 14, 159, 1984.
747. *WALPAC,* 20, 69, 1983.
748. *ECOCDK,* 4, 343, 1985.
749. *38RGAW,* 95, 1977.
750. *MBIOAJ,* 53, 281, 1979.
751. *ABWQC,* 81-117780, B-19, 1980.
752. *MERSDW,* 21, 109, 1987.
753. *ORNLT,* 81-13240, 5-60, 1979.
754. "The Toxicity of 120 Substances to Marine Organisms", Shellfish information leaflet, Fisheries Experimental Station, Conway, N. Wales, Ministry of Agriculture, Fisheries and Food, September 1970.
755. *FWPCA,* 234, 1968.
756. *JAPEAI,* 3rd Suppl., 36, 1966.
757. *ABWQC,* 81-117632, B-4, 1980.
758. *RPFWDE,* 137, 28, 1980.
759. *FSYBAY,* 75, 633, 1977.
760. *EPAMK,* 89, 1979.
761. *49HEAS,* 531, 1982.
762. *PRMBP,* 37, 1977.
763. *VCCIB,* 521-2, 1979.
764. *NRCNDA,* 17589, 85, 1981.
765. *NRCNDA,* 22494, 113, 1985.
766. *DOWSF,* 7, 1981.
767. *ESTUDO,* 5, 158, 1982.
768. *CMSHAF,* 8, 849, 1979.
769. *APICR,* 55, 1981.
770. *APICR,* 58, 1981.
771. *GPOPI,* 119, 1971.
772. *BECTA6,* 18, 361, 1977.
773. *APTUAO,* 56, 307, 1979.
774. *APICR,* 59, 1981.
775. *CUSCAM,* 36, 397, 1967.
776. *EXPEAM,* 37, 1327, 1981.
777. *CECDS,* 2, 459, 1982.
778. *EPARC,* 90, 1975.
779. *TSCMA9,* 7–8, 329, 1985.
780. *BECTA6,* 12, 475, 1974.
781. *ZWABAQ,* 15, 49, 1982.
782. *AECTCV,* 14, 1, 1985.
783. *ORNLT,* 81-13245, 6, 1979.
784. *NRCNDA,* 15105, 26, 1977.
785. *NRCNDA,* 14094, 66, 1974.
786. *PSTAB,* 7, 316, 1974.
787. *BECTA6,* 17, 720, 1977.

788. *EPAHP,* 146, 1976.
789. *EPAMO,* 52, 1975.
790. *ABWQC,* 81-117749, B-7, 1980.
791. *NRCNDA,* 15015, 264, 265, 266, 1977.
792. *NRCNDA,* 14098, 48, 1975.
793. *BECTA6,* 21, 409, 1979.
794. *ABWQC,* 81-117749, B-17, 1980.
795. *WATRAG,* 13, 623, 1979.
796. *TAFSAI,* 52, 219, 1982.
797. *ABWQC,* 81-117590, B-1-4, 1980.
798. *WATRAG,* 13, 627, 1979.
799. *BSPEAM,* 57, 124, 1964.
800. *BECTA6,* 15, 726, 1976.
801. *ORNLT,* 81-13240, 5, 1979.
802. *NRCNDA,* 136, 1975.
803. *NRCNDA,* 13684, 97, 99, 100, 1974.
804. *HBAGC,* 1983.
805. *ABWQC,* 81-117632, B-3, 1980.
806. *BROAW,* 175, 194, 1978.
807. *BECTA6,* 27, 775, 1981.
808. *TAFSAI,* 91, 175, 1962.
809. *31ZOAD,* 256, 258, 1983.
810. *EPATF,* 1979.
811. *HBAGC,* 33, A014, 1983.
812. *BECTA6,* 13, 194, 1975.
813. *31ZOAD,* 7, 243, 1983.
814. *JWPFA5,* 38, 1419, 1966.
815. *31ZOAD,* 382, 1983.
816. *NRCNDA,* 18572, 121, 1982.
817. *ABWQC,* 81-117392, B-9, 1980.
818. *BECTA6,* 27, 775, 1981.
819. *BMGHAC,* 130, 80, 1978.
820. *BECTA6,* 33, 127, 1984.
821. *NRCNDA,* 15105, 273, 1977.
822. *GPOPI,* 108, 1971.
823. *NRCNDA,* 16454, 115, 1979.
824. *PJZOAN,* 7, 135, 1975.
825. *CPECDM,* 5, 100, 1980.
826. *CRCTP,* 1, 1986.
827. *AHCBAU,* 6, 137, 1978.
828. *AQTODG,* 6, B-5, 1985.
829. *APTOA6,* 50, 398, 1982.
830. *BECTA6,* 24, 527, 1981.
831. *WATRAG,* 9, 601, 1975.
832. *NRCNDA,* 17587, 62, 1981.
833. *NOAAR,* 3, 1, 1980.
834. *AECTCV,* 11, 335, 1982.
835. *ABWQC,* 81-117798, B-37, 1980.
836. *ABWQC,* 81-117576, B-7, 1980.
837. *ABWQC,* 81-117848, B-2, 1980.
838. *ABWQC,* 81-117707, B-12, 1979.
839. *ABWQC,* 81-117632, B-5, 1980.
840. *CPSCAL,* 18, 227, 1976.
841. *EPACT,* 33, 1978.
842. *BECTA6,* 27, 596, 1981.
843. *XPARDG,* 82-136333, 1982.

844. *EHCRDEN,* 43, 34, 1984
845. *JTEHD6,* 8, 225, 1981.
846. *EPACT,* 23, 1978.
847. *EPAFP,* 81, 9, 1985.
848. *EPAIP,* 1978.
849. *RREVAH,* 94, 49, 1985.
850. *BECTA6,* 21, 202, 1979.
851. *BECTA6,* 27, 196, 1981.
852. *BECTA6,* 26, 295, 446, 1981.
853. *ABWQC,* 81-117780, B-9, 1980.
854. *ABWQC,* B-15, 1978.
855. *ABWQC,* 81-117392, B-10, 1980.
856. *JACCC,* 7, 27, 1986.
857. *WJTFA,* 81, 1982.
858. *WJTFA,* 1982.
859. *CJFSDX,* Abstr. 7, 7Q10524, 1977.
860. *RYKHAK,* 4, 28, 1975.
861. *ABWQC,* 81-117576, B-28, 1980.
862. *EPAPI,* 80-216815, 15, 1980.
863. *ASCHAN,* 35, 5, 1962.
864. *XPBRCA,* 52, 1974.
865. *SCPBR,* 241801, 40, 1975.
866. *TSDTAZ,* 12, 316, 1975.
867. *SAX,* 2, 84, 1982.
868. *AHBPAX,* 23, 183, 1981.
869. *ZANZA9,* 48, 87, 1961.
870. *CASIJ,* 161, 1973.
871. *GPOPI,* 147, 1971.
872. *SAX,* 3, 50, 1983.
873. *NRCNDA,* 15105, 277, 1977.
874. *NRCNDA,* 15105, 269, 1977.
875. *JACCC,* 2, 8, 1983.
876. *JFRBAK,* 35, 1060, 1978.
877. *49HEAS,* 311, 1982.
878. *BECTA6,* 28, 298, 1982.
879. *OEEAN,* 398, 1979.
880. *EPAMK,* 90, 91, 1979.
881. *KTPMN,* 1977.
882. *CECDS,* 5-27, 1978.
883. *GPOPI,* 113, 1971.
884. *CEUSI,* 5, 1980.
885. *JTSCDR,* 7, 193, 1982.
886. *AECTCV,* 5, 415, 1977.
887. *ABWQC,* 81/117327, B-20, 1980.
888. *BJASAS,* 22, 21, 1977.
889. *ABWQC,* 81-117897, B-26, 1980.
890. *APICR,* 54, 1981.
891. *TXAPA9,* 43, 1, 1978.
892. *SARIA3,* 67, 299, 1982.
893. *GESAM,* 6, 60, 1977.
894. *WATRAG,* 10, 303, 1976.
895. *SARIA3,* 70, 11, 1985.
896. *EVETBX,* 7, 428, 1978.
897. *PCCMAN,* 45, 56, 1977.
898. *LUDPA,* UM70-16553, 1970.
899. Vom Wasser, XXXVII, Band, Verlag Chemie, 1970.

900. *KHFKDF,* 9, 146, 1981.
901. *EVETBX,* 5, 1053, 1976.
902. *JWPFA5,* 52, 1717, 1980.
903. *NRCNDA,* 16740, 138, 1979.
904. *PRSQAG,* 89, 25, 1980.
905. *XADRGH,* AD/A003637, 1974.
906. *JFRBAK,* 35, 997, 1978.
907. *ENDKAC,* 77, 173, 1981.
908. *TAFSAI,* 99, 20, 1970.
909. *SCPEAT,* 269, 1, 1977.
910. *JWPFA5,* 46, 1575, 1974.
911. *JAPEAI,* 3(suppl.), 36, 1966.
912. *SCPCF,* 258700, 98, 100, 1976.
913. *GPOPI,* 199, 1971.
914. *HERBH,* 1979.
915. *AECTCV,* 4, 8, 18, 1976.
916. *JWMAA9,* 39, 807, 1975.
917. *RPFWDE,* 160, 325, 1986.
918. *PSSTAB,* 12, 269, 1979.
919. *URTDS,* 7, 297, 1981.
920. *EPAFP,* 1, A162, 1981.
921. *XBRPAI,* 137, 81, 1980.
922. *Handbuch der Frichwasser- und Abwasserbilogie,* Oldenbourgh-Munchen, 1962.
923. *WATRAG,* 11, 927, 1977.
924. *ABWQC,* 81-117985, B-8, 1980.
925. *ACTBP,* 1981.
926. *ABWQC,* B-28, 1979.
927. *ABWQC,* 81-117764, 1980.
928. *JJIND8,* 55, 129, 1975.
929. *TXCYAC,* 19, 209, 1981.
930. *HBDEC,* 303, 1983.
931. *NRCNDA,* 14098, 395, 1975.
932. *AHCBAU,* 14, 169, 1983.
933. *JFHYD8,* 15FASC.3, 249, 1984.
934. *NRCNDA,* 16075, 181, 184, 1978.
935. *URTDS,* 1981.
936a. *AQTODG,* 7, 145, 1985.
936b. *HBAGC,* 85, 1971.
937. *PSSTAB,* 13, 526, 1980.
938. *XFWCAW,* 199, 5, 1964.
939. "Acute Toxicity of Chlordane to Fish and Invertebrates", EPA-600/3-77-099, U.S. EPA, Duluth, MN, 1977.
940. "Chronic Toxicity of Atrazine to Selected Aquatic Invertebrates and Fishes", EPA 600/3-76-047, 1976.
941. *BECTA6,* 15, 756, 1981.
942. *ICENI,* 275078, 437, 444, 1976.
943. *EPAPW,* 82, 1973.
944. *GPOPI,* 102, 1971.
945. *SCPBR,* 241801, 40, 1975.
946. *OJSCA9,* 66, 508, 1966.
947. *XSWRBX,* 1971.
948a. *RPFWDE,* 137, 79, 1980.
948b. *AECTCV,* 7, 325, 1978.
948c. *Toxicoligie Fibel,* George Thieme, Verlag, Stuttgart, 1971.
949. *AECTCV,* 6, 385, 1977.
950. *BECTA6,* 27, 588, 1981.
951. *TAFSAI,* 109, 304, 1980.

952. *RPFWDE,* 137, 86, 1980.
953. *MLEBC,* 82-247583, 51, 1982.
954. *JHMAD9,* 1, 303, 1977.
955. *TAFSAI,* 87, 79, 1957.
956. *JWPFA5,* 50, 1615, 1978.
957. *CEHDDV,* 80EHD62, 1-37, 1980.
958. *BECTA6,* 31, 309, 1983.
959. *31ZOAD,* 55, 1975.
960. *HBAGC,* A019, 1983.
961. *RPFWDE,* 137, 1980.
962. *TPSDO,* 51, 1982.
963. *EHCRDN,* 76, 68, 1985.
964. *ETOCDK,* 5, 273, 1986.
965. *ABWQC,* 81/117871, 132, 1980.
966. *RREVAH,* 63, 67, 1976.
967. *HERBH,* 48, 1979.
968. *PFCUAY,* 165, 1962.
969. *CMSHAF,* 13, 813, 1984.
970. *ZWABAQ,* 11, 161, 1978.
971. *ZWABAQ,* 11, 813, 1978.
972. *TAKHAA,* 39, 28, 1980.
973. *ASTME,* 145, 1985.
974. *ABWQC,* 81-117392, B-18, B-22, 1980.
975. *JFIDDI,* 5, 13, 1982.
976. *WHOEV,* 112, 1982.
977. *FAATDF,* 3, 353, 1983.
978. *JTEHD6,* 8, 687, 1981.
979. *JAFCAU,* 60, 128, 1974.
980. *TAFSAI,* 114, 861, 1985.
981. *NRCNDA,* 16081, 39, 1977.
982. *EPARC,* EPA-600/3-75, 015, 1975.
983. *XFWCAW,* 167, 224, 1963.
984. *CHIPS,* 12, 1982.
985. *JEBIAM,* 21, 1, 1935.
986. *WATRAG,* 10, 869, 1976.
987. *ASTTA8,* 921, 277, 1986.
988. *AQTODG,* 12, 73, 1985.
989. *DPIRDV,* 3, 5, 1983.
990. *ETOCDK,* 3, 143, 1984.
991. *EHDTP,* 82-EHD-73, 109, 1982.
992. *JRGSAW,* 1, 499, 1973.
993. *XPARD6,* 82-119561, 1981,
994. *JJIND8,* 60, 1127, 1978.
995. *GPOPI,* 41, 1971.
996. *BECTA6,* 23, 250, 1979.
997. *FSYBAY,* 73, 207, 1975.
998. *PWPSA8,* 21, 475, 1978.
999. *COPAAR,* 2, 326, 1975.
1000. *EPAFP,* 2, 9, 1981.
1001. *AQTODG,* 6, 113, 1983.
1002. *CECAE,* 2, VI-14, 1982.
1003. *BECTA6,* 22, 508, 1979.
1004. *CECDS,* 2, X-8, 1979.
1005. *NRCNDA,* 19246, 68, 1982.
1006. *JWPFA5,* 51, 1623, 1978.
1007. *ORNLK,* 14, 1977.

1008. *CJFSDX,* 39, 1426, 1982.
1009. *AECTCV,* 17, 313, 1988.
1010. *PFCUAY,* 46, 1, 1984.
1011. *DABSAQ,* 39, 78-13246, 538, 1978.
1012. *31ZOAD,* 89, 1983.
1013. *JPFCD2,* 17, 51, 1982.
1014. *BECTA6,* 30, 614, 1983.
1015. *WATRAG,* 13, 631, 1979.
1016. *PFCUAY,* 40, 30, 1978.
1017. *HBDEC,* 1173, 1983.
1018. *BECTA6,* 21, 521, 1979.
1019. *TAFSAI,* 112, 205, 1983.
1020. *MONSI,* 1985.
1021. *BECTA6,* 28, 7, 1982.
1022. *BECTA6,* 17, 66, 1977.
1023. *APCRAW,* 8, 227, 1968.
1024. *TAFSAI,* 95(1), 1, 1966.
1025. *BECTA6,* 6, 113, 1971.
1026. *CAHWS,* 1985.
1027. *JFIBA9,* 27(suppl. A), 197, 1985.
1028. *MERSDW,* 5, 295, 1981.
1029. *WATRAG,* 10, 303, 1976.
1030. *CRUSAP,* 16, 302, 1969.
1031. *JWPFA5,* 50, 1604, 1978.
1032. *BUBCE,* 28, 11, 1955.
1033. *ABWQC,* 81-117632, B-10, B-12, 1980.
1034. *AECTCV,* 11, P699, P1991, 1982.
1035. *JEVQAA,* 13, 494, 495, 1984.
1036. See reference 61.
1037. *JFRBAK,* 35, 997, 1978.
1038. Unpublished data, National Water Quality Laboratory, Duluth, MN, U.S., 1971.
1039. *JEVQAA,* 13, 3, 1984.
1040. *JAFCAW,* 21, 576, 1979.
1041. *INCPC,* 1987.
1042. *HBEDC,* 726, 1983.
1043. *CJFSDX,* 41, 141, 1984.
1044. *EPEBD7,* 27, 179, 1982.
1045. *PGATP,* 81-231144, 1981.
1046. *EPEBD7,* 35(Ser. A), 362, 1984.
1047. *RPFWDE,* 160, 831, 1986.
1048. *BECTA6,* 20, 344, 1978.
1049. *ASTAA8,* 802, 90, 1983.
1050. *ATSOC,* 600/3-76-097, 1976.
1051. *CJFSDX,* 40, 743, 1983.
1052. *BECTA6,* 35, 225, 1985.
1053. *CARDEM,* 13, 31, 1980.
1054. *ICIBR,* BL/B205H, B206I, 1980.
1055. *IPCLB2,* 11, 29, 1969.
1056. *CHINAG,* 12, 523, 1975.
1057. *NRCNDA,* 22846, 152, 1984.
1058. *BECTA6,* 37, 258, 1986.
1059. *EPAFP,* A-250, 1981.
1060. *JFIBA9,* 22, 215, 1983.
1061. *TAFSAI,* 107, 361, 1978.
1062. *JWPFA5,* 53, 1044, 1981.
1063. *ORGAD2,* 18, 1979.

1064. *NFGJAX,* 26, 37, 1977.
1065. *NFGJAX,* 11(2), 106, 1964.
1066. *BECTA6,* 19, 465, 1978.
1067. *SIWAAP,* 24, 1397, 1952.
1068. *IJMNBF,* 12, 64, 1983.
1069. *JWPFA5,* 50, 1604, 1978.
1070. *ABWQC,* 81-117475, B-54, 1979.
1071. *JFRBAK,* 25, 1631, 1968.
1072. *TXAPA9,* 42, 85, 1977.
1073. *PSMBAG,* 18, 379, 1971.
1074. *DOWAE,* 41, 1980.
1075. *WATRAG,* 10, 299, 1977.
1076. *AQTODG,* 4, 73, 1983.
1077. *RUOMAY,* 37-38, 125, 1975.
1078. *BECTA6,* 22, 767, 1979.
1079. *ETOCDK,* 11, 187, 1992.
1080. *ETOCDK,* 9, 623, 1990.
1081. *ETOCDK,* 10, 1229, 1991.
1082. *ETOCDK,* 10, 1351, 1991.
1083. *ETOCDK,* 9, 1045, 1990.
1084. *ETOCDK,* 12, 1261, 1993.
1085. *ETOCDK,* 12, 925, 1993.
1086. *ETOCDK,* 11, 513, 1992.
1087. *ETOCDK,* 8, 623, 1989.
1088. *ETOCDK,* 8, 545, 1989.

8 Key To Alphanumeric (Coded) References

31ZOAD	*Pesticide Manual*, Worcheshire, England, British Crop Council, 1968.
37BKAP	"Fate of Effluent Hydrocarbons on Marine Ecosystems"
37KXA7	*Water Chlorination: Environmental Impact and Health Effects, Proceedings on the Conferences on the Environment*, Ann Arbor, MI, Ann Arbor Science Pub., 1975.
38RGAW	"Pollutants Effects on Marine Organisms", Proceedings of Workshop, May 16–19, 1976, College Station, Texas.
45ZOAI	"Biological Monitoring of Marine Pollutants", Proceedings
47EXAM	Proceedings of the Third International Conference, "Bioindicators Deteriorisations Regionis".
49HEAS	"Physiological Mechanisms of Marine Pollutant Toxicity", Proceedings of the Symposium, p. 452.
49WEAZ	"Pesticide Chemistry: Human Welfare and the Environment", Proceedings of the International Conference of Pesticide Chemistry, Pergamon Press, Ltd., 1983.
ABWQC	Ambient Water Quality Criteria document.
ACPTF	"Acute and Chronic Parathion Toxicity to Fish and Invertebrates"
ACTBP	Acephate, Aldicarb, Carbophenothion, DEP, EPN, Ethoprop, Methyl Parathion, and Phorate: Their Acute and Chronic Toxicity, Bioconcentration Potential and Persistence as Related to Marine Environments, U.S. EPA.
AECTCV	Archives on Environmental Contamination and Toxicology, Springer-Verlag, New York, V. 1 (1973).
AEHLAU	Archives of Environmental Health, Heldreff Publications, Washington, D.C., V. 1 (1960).
AHBPAX	*Acta Hydrobiologica*
AHCBAU	*Acta Hydrochimica et Hydrobiologica*
AIHAAP	*American Industrial Hygiene Association Journal*, American Industrial Hygienists Association, Akron, OH, V. 19 (1958).
AJMFA4	*Australian Journal of Marine and Freshwater Research*
AOHYA3	*Annals of Occupational Hygiene*, Pergamon Press, Oxford, U.K., V. 1 (1921).
APCRAW	*Advances in Pest Control Research*, New York, V. 1–8 (1957–1968), discontinued.
APICR	American Petroleum Institute — Chromium
APJUA8	*Acta Pharmaceutica Jugoslavica*, V. 1 (1951).
APMBAY	*Applied Microbiology*, American Society of Microbiology, Washington, D.C., V. 1-30 (1953–1975).
APTOA6	*Acta Pharmacologica et Toxicologica* (Munksgaard, Copenhagen K, Denmark), V. 1 (1945).

AQCLAL	*Aqua Culture*
AQFEDI	*Aqua Fennica*, Water Association, Helsinki, Finland, V. 1 (1971).
AQTODG	*Aquatic Toxicology*
ASTME	*Aquatic Toxicology and Hazard Evaluation*, ASTM Publication.
ASTTA8	*Aquatic Toxicology and Hazard Assessment*, ASTM Special Publication, American Society of Testing Materials, Philadelphia, No. 1 (1911).
ATENBP	*Atmospheric Environment*, Pergamon Press, Oxford, U.K., V. 1 (1967).
AZOFAO	*Annales Zoologici Fennici*
BECTA6	*Bulletin of Environmental Contamination and Toxicology*, Springer-Verlag, NY, U.S., V. 1 (1966).
BIBUBX	*Biological Bulletin*, Marine Biological Laboratories, Woods Hole, MA, U.S., V. 1 (1898).
BJASAS	*Bulletin, New Jersey Academy of Sciences*
BMGHAC	*Bamidgeh*
BJIMAG	*British Journal of Industrial Medicine*, British Medical Association, London, U.K., V. 1 (1944).
BROAW	*Ecology of Pesticides*
BSPEAM	*Bulletin de la Societé de Pathologie Exotique*
BUBCE	*Bulletin de Belge Condument Eaux*
CAFGAX	*California Fish and Game*
CAHWS	Chemical, Physical, and Biological Properties of Compounds Present at Hazardous Waste Sites.
CASIJ	Pyrethrum, The Natural Insecticide
CBINA8	*Chemico-Biological Interactions*, Elsevier Publishers, Amsterdam, The Netherlands, V. 1 (1969).
CCHAT	*Handbook of Analytical Toxicology*
CCRDR	*Chem-Agro Report*
CECAE	Study of Noxious Effects of Dangerous Substances Recorded in List II (EEC)
CECDS	Noxious Effects of Dangerous Substances in the Aquatic Environment—Final Report (EEC)
CECPP	CEC, Evaluation de l'impact des PCB et PCT sur l'Environment Aquatique (Rapport Final)
CEHCDV	Environmental Health Directorate (Canada) Report
CENEAR	*Chemical Engineering News*
CEUSI	Impact of Organosilicon Compounds on the Aquatic Environment: Statement Adopted by the Environmental Committee of the Centre European des Silicones.
CHAVF	Chronic Toxicity of Atrazine to Selected Aquatic Invertebrates and Fish
CHECDY	*Chemistry and Ecology.*
CHINAG	*Chemistry and Industry,* Society of Chemical Industry, London, England, V. 1–2, 1923–1943; V. 1 (1944).
CHIPS	*TSCA Chemical Assessment Series: Chemical Hazard Information Profiles*, U.S. EPA, Office of Toxic Substances, Washington, D.C.
CINFS	Chemical Information Fact Sheet
CJFSDX	*Canadian Journal of Fisheries and Aquatic Sciences*, Ottawa, Canada
CJZOAG	*Canadian Journal of Zoology*, Ottawa, Canada
CLTNO	Acute Toxiciteitstoetsen mit 1,2-Dichloroethaan, fenol, acrylonitrl et alkylbenzenesulfonaat in zeewater
CMSHAF	*Chemosphere*, Pergamon Press, Oxford, England, V. 1 (1971).
COPAAR	*Copeia.*

CPECDM	*Comparative Physiology and Ecology*
CPSCAL	*Chesapeake Science*
CRAFA	Effects of Some Pesticides on Larvae of the Market Crab, *Cancer magister*, and the Red Crab, *Cancer productus*, and a Bioassay of Industrial Wastes with Crab Larvae.
CRCMY	Mycotoxins and *N*-Nitroso Compounds: Environmental Risks
CRCTP	*Toxicology of Pesticides in Animals*
CRRPC3	Cooperative Research Report—International Council for the Exploration of the Sea
CRSBAW	*Comptes Rendus des Sciences de la Societé de Biologie de et Ses Filales*, Paris, France, V. 1 (1849).
CRTXB2	*CRC Critical Reviews in Toxicology*, CRC Press, FL, V. 1 (1971).
CRUSAP	*Crustaceana* (Leiden)
CUSCAM	*Current Science*, Raman Research Institute, Bangalore, India, V. 1 (1932).
DABSAQ	*Dissertation Abstracts, B: The Sciences and Engineering*, University Microfilms, Ann Arbor, MI.
DGMTAO	*Deutsche Gewasserkundliche mitteilungen*
DOEAAH	*Down to Earth, A Review of Agricultural Chemical Progress*, Dow Chemical USA, Midland, MI, V. 1 (1945).
DOWAE	Study of the Fate and Impact of Polydimethylsiloxane Introduced into the Aquatic Environment, Dow Chemical, U.S.
DOWCC	*Dow Chemical Company Reports*, Dow Chemical USA, Health and Environmental Research,Toxicology and Research Laboratory, Midland, MI.
DOWAS	Essai de Toxicité à court-terme des silicones vis-à-vis de artemia-salina.
DOWSF	Dow-Corning Bulletin 22-069B-01, Information About Silicone Fluids.
DPIRDU	*Dangerous Properties of Industrial Materials Report*, Van Nostrand Reinhold Co., New York, V. 1 (1981).
DTVCA	Toxicity of 1,1-Dichloroethylene (Vinylidene Chloride) to Aquatic Organisms—Final Report.
ECMSC6	*Estuarine and Coastal Marine Science*
EESADV	*Ecotoxicology and Environmental Safety*, Academic Press, New York, V. 1 (1977).
EHCRDN	*Environmental Health Criteria*
EHDTP	Review of Environmental and Health Aspects of Triallyl/Alkyl Phosphates
EINUDQ	Eau, l'industrie, les nuisances
ENDKAC	*Endokrinologie*, Johann Ambrosius Barth Verlag, Leipzig, Germany, V. 1–80 (1928–1982).
ENPBBC	*Environmental Physiology and Biochemistry*, Copenhagen, Denmark, V. 2–5 (1972–1975), discontinued.
ENVCO	Hazard Information Review on Phenylenediamine
ENVPAF	*Environmental Pollution*, Applied Science Publishers Ltd., Essex, London, U.K., V. 1 (1970).
ENVRAL	*Environmental Research*, Academic Press, New York, V. 1 (1967).
EPACT	Chronic Toxicity of Chlordane, Trifluarilin and Pentachlorophenol to Sheepshead Minnows, U.S. EPA.
EPAFP	Environmental Fate and Impact of Major Forest Use Pesticides, U.S. EPA.
EPAGD	Intermedia Priority Pollutants—Guidance Documents, U.S. EPA.
EPAHP	Investigation of Selected Potential Environmental Contaminants, U.S. EPA.
EPAMI	Health Effects Assessment for Methylisobutyl Ketone, U.S. EPA.

EPAIP	In-Depth Studies on Health and Environmental Impacts of Selected Water Pollutants, U.S. EPA.
EPAIT	Initial Toxicological Assessment of Ambush, Bolera, Bux, Dursban, Fentrifanil, Larvin, and Pydrin: Static Acute Toxicity Tests with Selected Estuarine Algae, Invertebrates, and Fish.
EPAME	Acephate, Aldicarb, Carbophenothion, DEP, EPN, Ethoprop, Methyl Parathion, and Phorate: Their Acute and Chronic Toxicity, Bioconcentration Potential and Persistence as Related to Marine Environments, U.S. EPA.
EPAMK	Reviews of the Environmental Health Effects of Pollutants: Mirex and Kepone, U.S. EPA.
EPAMM	Effects of Mirex, Methoxychlor, and Malathion on the Development of Crabs
EPAMO	Initial Scientific and Microeconomic Review of Monuron, U.S. EPA.
EPAPI	Pesticides in the Illinois Waters of Lake Michigan, U.S. EPA.
EPAPW	Effects of Pesticides in Water—A Report to the States, U.S. EPA.
EPAQC	Quality Criteria for Water, U.S. EPA.
EPARL	Ecological Research Series, Environmental Research Laboratory, U.S. EPA.
EPATF	Toxicity of Organic Chemicals to Embryo-Larval Stages of Fish, U.S. EPA.
EPATM	*Treatability Manual*, U.S. EPA.
EPAWG	Water Quality Criteria, U.S. EPA.
EPAWP	Pollution Potential in Pesticide Manufacturing, Pesticide Study Series No. 5.
EPEBD7	*Environmental Pollution Series A: Ecological and Biological*, Applied Science Publishers, England.
EPRDC	Biological Effects of Pesticides on the Dungeness Crab, Gulf Breeze, FL.
ESTHAG	Environmental Science and Technology, American Chemical Society, Washington, D.C., V. 1 (1967).
ESTUDO	Estuaries.
ETOCDK	*Environmental Toxicology and Chemistry*, V. 1, Pergamon Press, (1981).
EVETBX	*Enviromental Entomology*, V. 1, Entomological Society of America, College Park, MD, (1972).
EVHPAZ	*Environmental Health Perspectives*, Dept. of Health and Human Services Publications, U.S. Government Printing Office, Washington, D.C., U.S., V. 1 (1972).
EXPEAM	*Experientia*, Birkhauser Verlag, Basel, Switzerland, V. 1 (1945).
FAATDF	*Fundamental and Applied Toxicology*, Society of Toxicology, Akron, OH, V. 1 (1981).
FSYBAY	*Fishery Bulletin*
FWPCA	Report of the National Technical Administrative Committee to Secretary of the U.S. Department of the Interior
GESAM	Gesamp Reports and Studies
GLCTD	Toxicity Data Sheets
GPOPI	Ecological Effects of Pesticides on Non-Target Species
HBAGC	*Agrochemicals Handbook*
HERBH	*Herbicide Handbook of the Weed Science Society of America*
HYCJAC	*Hyacinth Control Journal*, Aquatic Plant Management Society, Vicksburg, MS, (*changed to Journal of Aquatic Plant Management*).
HYBUD9	*Hydrobiological Bulletin*
HYDRDA	*Hydrometallurgy*, Elsevier Science Publishers B.V., Amsterdam, Netherlands, V. 1 (1975).
ICAPDG	*Indian Journal of Comparative Animal Physiology*
ICENI	Investigations of Selected Potential Environmental Contaminants—Nitroaromatics
ICIBR	Imperial Chemical Industries—I
IFCRAG	*Investigations in Fish Control*

IJEBA6	*Indian Journal of Experimental Biology*
IJMNBF	*Indian Journal of Marine Sciences*
IPCLBZ	*International Pest Control*, McDonald Publications, Middlesex, England, V. 5 (1962).
IVLBDQ	*Institutet fuer Vatten Uch Luftvardsforskning*
JACCC	Joint Assessment of Commodity Chemicals, European Chemical Industry Ecology and Toxicology Centre (ECIET), Brussels, Belgium.
JAFCAU	*Journal of Agricultural and Food Chemistry*, American Chemical Society Publications, Washington, D.C., V. 1 (1953).
JANCA2	*Journal of the Association of Official Analytical Chemists*, AOAC Publication, Washington, D.C.,V. 49 (1966).
JAPCA	*Journal of Air Pollution Control Association*, Pittsburgh, PA, V. 1–39 (1958–1989); replaced by *Journal of Air and Waste Management.*
JAPEAI	*Journal of Applied Ecology*
JEBIAM	*Journal of Experimental Biology*, Portland Press, Colchester, U.K., V. 7 (1930).
JEENAI	*Journal of Economic Entomology*, Entomological Society of America, College Park, MD, V. 1 (1908).
JEPTDQ	*Journal of Environmental Pathology and Toxicology*, Park Forest South, IL, V. 1–5 (1977–1981).
JEVQAA	*Journal of Environmental Quality*
JFHYD8	*Journal Francais d'Hydrologie (French Journal of Hydrology).*
JFIBA9	*Journal of Fish Biology*, Academic Press, London, England, V. 1 (1969).
JFIDDI	*Journal of Fish Diseases*
JFRBAK	*Journal of the Fisheries Research Board of Canada*
JHMDA9	*Journal of Hazardous Materials*, Elsevier Science, Amsterdam, The Netherlands.
JHYDA7	*Journal of Hydrology*
JJIND8	*Journal of the National Cancer Institute*, U.S. Government Printing Office, Washington, D.C., V. 61 (1978).
JOOEDU	*Journal of Environmental Engineering*, American Society of Civil Engineers, New York.
JPETAB	*Journal of Pharmacolgy and Experimental Therapeutics*, Williams and Wilkins Company, Baltimore, MD, V. 1 (1909).
JPFCD2	*Journal of Environmental Science and Health, Part B: Pesticides, Food Contaminants and Agricultural Wastes*, V. B11 (1976).
JRGSAW	*Journal of the Research of the U.S. Geological Survey*
JTEHD6	*Journal of Toxicology and Environmental Health*, Hemisphere Publishing, Washington, D.C., V. 1 (1975/76).
JTSCDR	*Journal of Toxicological Sciences*, Higashi Nippon Gakuen University, Sapporo, Japan, V. 1 (1976).
JWPFA5	*Journal of the Water Pollution Control Federation*, London, England.
JWMAA9	*Journal of Wildlife Management*, Wildlife Society, MD, V. 1 (1937).
KHFKDK	*K'o Hseuh Fa Chan Yueh K'an (Progress in Science)*, Taipei, Taiwan, V. 1 (1973).
KTPMN	*Khimiya I Technologiya Pestitsidov (Chemistry and Technology of Pesticides)*
LIOCAH	*Limnology and Oceanography*
LUDPA	*Mechanism of Resistance to Parathion in Mosquito Fish*
MAFFF	USSR Ministry of Agriculture, Fisheries and Food Shellfish, Information Centre, leaflet.
MEPCE	*Les Produits Chimiques dans l' Environment*
MERSDW	*Marine Environmental Research*
MLEBC	*Environmental Implications of Changes in the Brominated Chemicals*

MONSD	Material Safety Data Sheet on Saniticizer, Monsanto Chemical Co.
MONSI	Monsanto Company Toxicity Information, Bloomington, DE.
MOSQAU	*Mosquito News*
MPNBAZ	*Marine Pollution Bulletin*
NAPFA	Formaldehyde and Other Aldehydes: Committee on Aldehydes; Board on Toxicology and Environmental Hazards.
NASPM	*Petroleum in Marine Environment*
NATUAS	*Nature,* Macmillan Journals Ltd., Basinstoke, U.K., V. 1 (1869).
NAWTA6	*Transactions of the North American Wildlife and Natural Resources Conference*
NEPHBW	*Neuropharmacology*, Pergamon Press, Oxford, England, V. 1 (1970).
NETSR	Netherlands Organization for Applied Scientific Research--Report
NFGJAX	*New York Fish and Game Journal*
NJRPH	Response of Marine Animals to Petroleum and Specific Petroleum Hydrocarbons
NRCNDA	National Research Council of Canada Publication, Ottawa, Canada.
OCANB6	*Oceanology International,* Industrial Research Park, Beverly Shores, Indiana, U.S., V 1-5 (1966–1970); Continued as *Oceanology International/Offshore Technology*, V. 6 (1971).
OJSCA9	*Ohio Journal of Science*
ORGAD2	*Organika*
ORNLK	Kepone: I. A Literature Summary; II. An Abstracted Literature Collection.
ORNLT	Reviews of Environmental Effects of Pollutants: Toxaphene
OTSEA	Evaluation of FR-651-A in the Aquatic Environment.
OUBUC	Effects of the Insecticide "Sevin" on the Dungeness Crab, *Cancer magister* Dana.
PCBPBS	*Pesticide Biochemistry and Physiology*, Academic Press, New York, V. 1 (1971).
PCCMAN	Proceedings of the Annual Conference of the California Mosquito Control Association.
PCGFAM	Proceedings of the Annual Conference of U.S. Southeastern Association of Game and Fish Commissions.
PEMNDP	*Pesticide Manual*, British Crop Protection Council, Thornton Heath, Worcester, U.K., V. 1 (1968).
PERAS	Brine Shrimp *Artemia*
PFCUAY	*Progressive Fish Culturist*
PGATP	Acute Toxicity of Phenol and Substituted Phenols to the Fathead Minnows
PGWTA2	*Progress in Water Technology*, Pergamon Press, Oxford, U.K..
PJZOAN	*Pakistan Journal of Zoology*
PLEPO	Effects of Pollutants on Marine Organisms
PNSFAN	*Proceedings of the National Shellfisheries Association*
PNASA6	*Proceedings of the National Academy of Sciences of the U.S.A.*, Academy Printing and Publishing Office, Washington, D.C., U.S., V. 1 (1915).
PPMOV	Pollution and Physiology of Marine Organisms
PRIAM	Accumulation and Metabolism of DDT, Parathion, and Endrin by Aquatic Foodchain Organisms
PRLBA4	*Proceedings of the Royal Society of London, Series B, Biological Sciences*, The Royal Society, London, England, V. 76 (1905).
PRMBP	*Physiological Responses of Marine Biota to Pollutants*
PRNEN	*Pesticide Registration Notice*
PRSQAG	*Proceedings of the Royal Society of Queensland*
PSEBAA	*Proceedings of the Society of Experimental Biology and Medicine,* Academic Press, New York, V. 1 (1903–1904).
PSMBAG	Publications of the Seto Marine Biological Laboratory.

PSSCBG	*Pesticide Science*, Blackwell Scientific Publications, Oxford, England, V. 1 (1970).
PSSTAB	*Pesticide Abstracts*
PWPSA8	*Proceedings of the Western Pharmacolgy Society,* Western Pharmacolgy Society, California, V. 1 (1958).
RAORK	Pentachlorophenol: Chemistry, Pharmacology and Environmental Toxicology
RIPMAG	*Revue des Travaux Institut des Peches Maritimes*, Montrouge, France; renamed *Aquatic Living Resources*, V. 1 (1988).
RPFWDE	U.S. Fish and Wildlife Service Resource publication
RREVAH	*Residue Reviews*, Springer-Verlag, New York, V. 1 (1962).
RYKHAK	Rynoe Khozyaistov: Effect of "Herban" on Freshwater Fish
SARIA3	*Sassia (Nordic Journal of Marine Biology)*
SCIEAS	*Science*, American Association for the Advancement of Science, Washington, D.C., V. 1 (1895).
SCPBR	Initial Scientific and Microeconomic Review of Bromacil
SCPCF	Initial Scientific and Microeconomic Review of Carbofuran
SCPEAT	Science et Peche
SIWAAQ	*Sewage and Industrial Wastes*
STEVA8	*Science of the Total Environment,* Elsevier Science Publishers, Amsterdam, Netherlands, V. 1 (1978).
TAFSAI	*Transactions of the American Fisheries Society*, American Fisheries Society, Bethesda, MD, U.S., V. 1 (1870).
TAKHAA	*Takeda Kenkyusho Ho*, Journal of the Takeda Research Laboratories, Osaka, Japan, V. 29 (1970).
TAPPAP	*Tappi*, One Dunwoody Park, Atlanta, Georgia, USA, V. 32 (1949).
TATHE	Toxicity and Physiological Effects of Oil and Petroleum Hydrocarbons on Estuarine Grass Shrimp (*Palaemonotes pugio*)
TCEBAA	*La Tribune Cebadeau*
TJSCAU	*Texas Journal of Science*, Texas Academy of Science, College Station, TX, V. 1 (1949).
TJWEP	*Evaluation of Water Pollutants*
TMENAE	*Transactions of the Society of Mining Engineers*, Society of Mining Engineers, Littleton, CO.
TNORE	Central Laboratory TNO Report
TPFSDO	U.S. Fish and Wildlife Services technical papers
TSCMA9	*Techniques et Sciences Municipales (Techniques and Municipal Science)*
TXAPA9	*Toxicology and Applied Pharmacology Journal*
TXCYAC	*Toxicology*, Elsevier/North-Holland Publishers Ltd., New York, V. 1 (1973).
UBZHD4	*Ukrainian Biochemistry Journal*
URTDS	Unpublished report of toxicology data sheets
USDI3	*Effects of Pesticides on Aquatic Animals in the Estuarine and Marine Environment*
VCCIB	Dicamba (Banvel) Herbicide, Technical Information Bulletin.
WALPAC	*Water, Air and Soil Pollution*, Kluwer Academic Publishers, Dordrecht, Netherlands, V. 1 (1971).
WATRAG	*Water Research*, Pergamon Press, Oxford, England, V. 1 (1967).
WEEDAT	*Weeds*, Weed Society of America, Champaign, IL, V. 1 (1968); title changed to *Weed Science.*
WDFPH	Fate and Effects of Petroleum Hydrocarbons in Marine Organisms and Ecosystems
WHOEV	Waste Discharge into the Marine Environment: Principles and Guidelines for the Mediterranean Action Plan

WJTFA	Toxicology and Fate of Selected Industrial Chemicals in the Aquatic Ecosystems (Final Report)
WSWOAC	*Water and Sewage Works*
XADRCH	U.S. Technical Information Service Report
XBRPAI	U.S. Bureau of Sport Fisheries and Wildlife, Resource Publication.
XBTPBT	U.S. Bureau of Sport Fisheries and Wildlife, Technical Paper.
XCSPP	Captan, Special Review Position document 2/3.
XENOBH	*Xenobiotica*, The Fate of Foreign Compounds in Biological Systems, Taylor and Francis Ltd., London, England, V. 1 (1971).
XFWCAW	U.S. Fish and Wildlife Circular
XNCEL	U.S. Naval Civil Engineering Laboratory, Interim Report.
XPARD6	U.S. EPA, Office of Research and Development, Report.
XPBRCA	U.S. National Technical Information Service (PB # Series)
XSWRBX	U.S. Bureau of Sport Fisheries and Wildlife, Research Report.
ZANCA8	*Zeitschrift fur Analitische Chemie*
ZANZA9	*Zeitschrift fur Angewandte Chemie*
ZWABAQ	*Zeitschrift fur Wasser und abwasser Forschung*

9 Glossary Of Terms Used

Absorption Transport of a chemical to the inside of an abiotic or biotic system.

Acaricide A chemical used for the control of mites in plants.

Acclimation A physiological adaptation of an organism to selected experimental conditions, including any adverse stimulus.

Actinomycetes A group of bacteria common to soil. These are filamentous and reproduce by terminal spores.

Acute effects A health effect manifested quickly, usually of short-term duration.

Acute toxicity The harmful effects of a chemical that are demonstrated within a short period (hours to days) of exposure; relevant to lethal effects.

Adsorption The concentration of a chemical on the surface of an abiotic or biotic system.

Algicide A chemical used for the control of algae.

Benthic Used to describe aquatic organism that grow or live in or close to the substrate material.

Bioaccumulation Storage of a chemical within an organism at a concentration higher than detected in the environment. This process is not necessarily harmful.

Bioassay This term is used for toxicity testing with different organisms; it is usually reserved for formalized testing of toxic potency of chemicals.

Bioconcentration Accumulation of a chemical directly from the water, to a higher concentration in an aquatic organism. The bioconcentration results from simultaneous processes of uptake and depuration with the establishment of a dynamic equilibrium.

Chlorinated Presence of one or more chlorine atoms in a chemical compound.

Chronic Prolonged. Can refer to the effect or the duration of exposure. In aquatic toxicology, is sometimes used to mean a full life-cycle test.

Chronic effect A prolonged health effect that may involve irreversible change or damage.

Dose-response Dose (concentration received inside the animal body) is plotted against the response of the test animal to derive a dose-response curve.

EC_{50} Median effective concentration; the concentration of a chemical that produces some effect in one-half of a test population. The effect could be lethal or nonlethal. Effect and exposure time must be specified.

Ecosystem An interacting system of living organisms in a circumscribed region of similar characteristics and the nonliving substrate, nutrients, energy, and other environmental components.

Effluent A liquid or gaseous discharge of waste material into the environment.

Half-life The length of time required for the quantity or activity of a chemical to be reduced by one-half of its original concentration or activity.

Herbicide An agent that kills plant life.

IC_{50} An estimate of the sample concentration causing a 50% reduction in the growth of the algal population compared to a control.

Insecticide An agent that kills insects.

LC_{50} Median lethal concentration; the concentration lethal to one-half of a test population. Duration of exposure must be specified.

LD_{50} Median lethal dose; the dose delivered inside the body that is lethal to one-half of a test population.

Mutagen An agent that possesses the ability to cause an alteration of the inherited genetic material.

NOAE No observed adverse effect.

NOEL No observed effect level. The concentration level below which the chemical does not cause significant effect(s). Similar to NOAEL (no-observed-adverse-effect level).

pH The negative logarithm of the hydrogen ion concentration.

Risk Expected frequency of undesirable effects resulting from exposure to a chemical.

Safety factor A numerical value applied to NOEL (or NOAEL) to arrive at an acceptable daily intake (ADI) value. This value compensates for inadequacies in the estimate of NOEL (or NOAEL).

Sorption Uptake of a chemical by abiotic or biotic systems by both absorption and adsorption processes.

Sublethal A concentration level that would not cause death. An effect that is not directly lethal.

Synergism Accentuation of the effects of one chemical by another one; this explains the increased toxicity of chemical mixture when compared to the calculated individual toxicities.

Teratogen An agent that possesses the ability to cause alteration in the developing cells, tissues, or organs at the embryonic stage of development.

Threshold The point on a dose-response curve above which effects are observed and below which adverse effects are observable.

TT Toxicity threshold; commonly used to measure the level at which the inhibition of algal cell multiplication is initiated.

Xenobiotic A synthetic chemical or substance found in biological systems but of foreign origin.